Exploring Creation

with

Biology
2nd Edition

by Dr. Jay L. Wile and Marilyn F. Durnell

Exploring Creation With Biology, 2ⁿᵈ Edition

Published by
Apologia Educational Ministries, Inc.
1106 Meridian Plaza, Suite 220
Anderson, IN 46016
www.apologia.com

Manufactured in the United States of America
Third Printing 2007

ISBN: 1-932012-54-0

Printed by CJK, Cincinnati, OH

*Cover photos: zebra (Dawn Strunc), cell Illustration (© Imagineering / Custom Medical Stock Photo),
leaf in background (© Creatas, Inc.), cells in background (© Brand X Pictures)*

Cover design by Kim Williams

Need Help?

Apologia Educational Ministries, Inc. Curriculum Support

If you have any questions while using Apologia curriculum,
feel free to contact us in any of the following ways:

By Mail: Dr. Jay L. Wile
Apologia Educational Ministries, Inc.
1106 Meridian Plaza, Suite 220
Anderson, IN 46016

By E-MAIL: help@apologia.com

On The Web: http://www.apologia.com

By FAX: (765) 608 - 3290

By Phone: (765) 608 - 3280

*Illustrations from the MasterClips collection
and the Microsoft Clip Art Gallery*

Are you ready to be impressed? If not, you'd better get ready. Why? Because in this course, you are going to get a broad overview of God's creation. As you begin to learn its secrets, you will become more and more impressed with its majesty and complexity. The sheer grandeur of it all should leave you in awe of God's mighty power. If you learn nothing else in this course, learn to appreciate the wonder of God's creation!

If you are like most students, this will be the first truly rigorous science course that you have ever taken. Thus, you might find it difficult to adjust to the time and patience required by a course like this. If you find yourself getting frustrated or discouraged, remember that whether you decide to go to college, go straight into the workforce, or get married and become a full time parent, there will be many tasks more rigorous than the experience of studying biology. Thus, you need to stick with it, because life is full of challenges!

Pedagogy of the Text

This text contains 16 modules. Each module should take you about two weeks to complete, as long as you devote 45 minutes to an hour of every school day to studying biology. The one exception to this rule is Module #6, which will take you three weeks to complete. At this pace, you will complete the course in 33 weeks. Since most people have school years which are longer than 33 weeks, there is some built-in "flex time." You should not rush through a module just to make sure that you complete it in two weeks. Set that as a goal, but be flexible. Some of the modules might come harder to you than others. On those modules, take more time on the subject matter.

To help you guide your study, there are two sets of student exercises that you should complete:

> ➤ The "On Your Own" questions should be answered as you read the text. The act of answering these questions will cement in your mind the concepts you are trying to learn. Answers to these questions appear at the end of the module. Once you have answered an "On Your Own" question, turn to the end of the module and check your work. If you did not get the correct answer, study the answer to learn why.

> ➤ You should answer the questions in the study guide at the end of the module after you have completed the module. This will allow you to review the important concepts from the module and help you prepare for the test.

Your teacher/parent has the answers to the study guides.

Any information that you must memorize is centered in the text and put in boldface type. In addition, all definitions presented in the text must be memorized. Words that appear in boldface type (centered or not) in the text are important terms that you should know. Some of the information in the course is presented in the form of tables or figures. Whether or not you will be given such information on the test depends on the information itself. If the study guide tells you that you can use a particular figure or table, you will also be able to use that figure or table on the test. If a study guide question requires that you know the information in a table or figure and the study guide does not tell you that you can use it, you will not be able to use it on the test.

Learning Aids

Extra material is available to aid you in your studies. For example, Apologia Educational Ministries, Inc. has produced a multimedia companion CD that accompanies this course. It contains videos of many organisms (including microscopic ones) that you have probably not seen before. These videos will help you better understand the organisms that they show. In addition, it contains animated white board solutions of example problems such as the Punnett square problems you must solve in Module #8. There is also an audio explanation that is different from the explanation given in this book. Thus, if you are having trouble understanding how we worked a certain example problem, you might find more explanation on the multimedia CD. The following graphic in the book:

indicates that there is a video or animation on the CD that relates to what you are reading.

Finally, the CD contains audio pronunciations of the technical words used in this book. Even though the book gives pronunciation guides for most of the technical words used, nothing beats actually hearing someone say the word! As you read through the book, you will see words that have pronunciation guides in parentheses. If you would like to hear one of those words pronounced for you, you will find it on the multimedia companion CD.

In addition to the multimedia companion CD, there is a special website for this course that you can visit. The website contains links to web-based materials related to the course. These links are arranged by module, so if you are having trouble with a particular subject in the course, you can go to the website and look at the links for that module. Most likely, you will find help there. In addition, there are answers to many of the frequently-asked questions regarding the material. For example, many people ask us for examples of how to properly record experiments in your laboratory notebook. Those examples can be found at the website. Finally, if you are enjoying a particular module in the course and would like to learn more about it, there are links which will lead you to advanced material related to that module.

To visit the website, go to the following address:

http://www.apologia.com/bookextras

When you get to the address, you will be asked for a password. Type the following into the password box:

Myfathersworld

Be sure that you do not put spaces between any of the letters and that the first letter is capitalized. When you click on the button labeled "Submit," you will be sent to the course website. You must use Internet Explorer 5.1 or higher to view this website.

There are also several items at the end of the book that you will find useful in your studies. There is a glossary that defines many of the terms used in the course and an index that will tell you where topics can be found in the course. In addition, there are three appendices in the course. Appendix A compiles some of the tables and figures that are found throughout the reading. Appendix

B contains a summary of each module in the course. These summaries are presented as fill-in-the-blank, matching, and identification exercises. If you are having trouble studying for the tests, these summaries might help you. Your parent/teacher has the answers to the exercises in Appendix B. Appendix C contains a complete list of all of the supplies you need to perform the experiments in this course.

Experiments

The experiments in this course are designed to be done as you are reading the text. I recommend that you keep a notebook of these experiments. This notebook serves two purposes. First, as you write about the experiment in the notebook, you will be forced to think through all of the concepts that were explored in the experiment. This will help you cement them into your mind. Second, certain colleges might actually ask for some evidence that you did, indeed, have a laboratory component to your biology course. The notebook will not only provide such evidence but will also show the college administrator the quality of your biology instruction. I recommend that you perform the experiments in the following way:

➤ When you get to an experiment, read through it in its entirety. This will allow you to gain a quick understanding of what you are to do.

➤ Once you have read the experiment, start a new page in your laboratory notebook. The first page should be used to write down all of the data taken during the experiment. What do I mean by "data"? Any observations or measurements you make during the experiment are considered data. Thus, if you see an organism during an experiment, you need to either describe it or draw it. If you measure the length of something during the experiment, that is part of the experiment's data and should be written down. In addition, any data analysis that you are asked to do as a part of the experiment should be done on this page.

➤ When you have finished the experiment and any necessary analysis, write a brief report in your notebook, right after the page where the data and calculations were written. The report should be a brief discussion of what was done and what was learned. You should not write a step-by-step procedure. Instead, write a brief summary that will allow someone who has never read the text to understand what you did and what you learned.

PLEASE OBSERVE COMMON SENSE SAFETY PRECAUTIONS! The experiments in this course are no more dangerous than most normal, household activity. Remember, however, that the vast majority of accidents do happen in the home. Chemicals used in the experiments should never be ingested; hot beakers and flames should be regarded with care; and all experiments should be performed while wearing eye protection such as safety glasses or goggles.

Laboratory Equipment

Exploring Creation With Biology contains laboratory exercises for you to perform. The laboratories come in three types: microscope labs, dissection labs, and household labs. The household labs use only household equipment, and you should definitely do them. The microscope labs, however, require expensive equipment, while the dissection labs require an additional kit. As a result, we do not *require* you to perform those labs. They will be beneficial to you, but they are not absolutely necessary. Thus, you should not feel pressured into purchasing the microscope or dissection equipment. Do so only if you can afford it!

If you have the financial means, we recommend that you order from Nature's Workshop. We have worked very hard with them to provide you with a quality set at a very low price. To order, simply call or write Nature's Workshop:

> Nature's Workshop Plus
> 1-888-393-5663 (toll free)
> P.O. Box 220
> Pittsboro, IN 46167-0220
> http://www.naturesworkshopplus.com

You can also order lab equipment through Apologia Educational Ministries, Inc. at 888-524-4724 or at http://www.apologia.com.

The microscope set contains a microscope and a microscope slide set. If you wish to order the entire set, simply ask for:

#40030: Slide Set With Microscope $260.00

If you have a quality microscope, you can order just the microscope slide set. To order this, simply ask for:

#40026:   Microscope Slide Set $80.00

Please note that in order for your microscope to work for this course, it must have at least three separate magnifications and a maximum magnification of at least 400x. It must also have a "fine focus," which is separate from the "coarse focus."

If you want to do the dissection experiments, there is a separate dissection kit, which comes complete with specimens. You can order this kit through Nature's Workshop as well. Just ask for

#40031 Dissection Kit With Specimens $36.00

Once again, you can order the individual items in this kit if you already have some dissection equipment.

If you have some biology lab equipment and need to know whether or not it will work for this course, please visit Nature's Workshop Plus on the web at http://www.naturesworkshopplus.com. If you click on "Apologia" on the left-hand side of the page and then click on "Next" until you find the "Apologia Labware" page, you can find detailed lists of everything in the various lab sets for this course. If you have some of the items in these lists, you can order only the items you need.

All prices are subject to change!!!!!!!

Exploring Creation With Biology
Table of Contents

MODULE #1: Biology: The Study of Life

Introduction

In this course, you're going to take your first detailed look at the science of biology. Biology, the study of life itself, is a vast subject, with many subdisciplines that concentrate on specific aspects of biology. Microbiology, for example, concentrates on those biological processes and structures that are too small for us to see with our eyes. Biochemistry studies the chemical processes that make life possible, and population biology deals with the dynamics of many life forms interacting in a community. Since biology is such a vast field of inquiry, most biologists end up specializing in one of these subdisciplines. Nevertheless, before you can begin to specialize, you need a broad overview of the science itself. That's what this course is designed to give you.

What Is Life?

If biology is the study of life, we need to determine what life is. Now to some extent, we all have an idea of what life is. If we were to ask you whether or not a rock is alive, you would easily answer "No!" On the other hand, if we were to ask you whether or not a blade of grass is alive, you would quickly answer "Yes!" Most likely, you can intuitively distinguish between living things and nonliving things.

Even though this is the case, scientists must be a little more deliberate in determining what it means to be alive. Thus, scientists have developed several criteria for life. Not all scientists agree on all of these criteria, but in general, most biology courses will list at least some of the following criteria for life:

1. **All life forms contain deoxyribonucleic (dee ahk' see rye boh noo klay' ik) acid, which is called DNA.**

2. **All life forms have a method by which they extract energy from the surroundings and convert it into energy that sustains them.**

3. **All life forms can sense changes in their surroundings and respond to those changes.**

4. **All life forms reproduce.**

If something meets all of these criteria, we can scientifically say that it is alive. If it fails to meet even one of the criteria, we say that it is not alive. Now if you're not sure exactly what each of these criteria means, don't worry. We will discuss each of them in the next few sections of this module.

DNA and Life

Our first criterion states that all life contains DNA. Now we're sure you've at least heard about DNA. It is probably, however, still a big mystery to you at this point. Why is DNA so special when it comes to life? Basically, DNA provides the information necessary to take a bunch of lifeless chemicals and turn them into an ordered, living system. Suppose, for example, we were to analyze an organism and determine every chemical that made up the organism. Suppose further that we went into a laboratory and made all of those chemicals and threw them into a big pot. Would we have made something that is alive? Of course not. We would not even have made something that resembles the organism we studied. Why not?

In order to make life, we must take the chemicals that make it up and organize them in a way that will promote the other life functions mentioned in our list of criteria for life. In other words, just the chemicals themselves cannot extract and convert energy (criterion #2), sense and respond to changes (criterion #3), and reproduce (criterion #4). In order to perform those functions, the chemicals must be organized so that they work together in just the right way. Think about it this way: suppose you go to a store and buy a bicycle. The box says, "Some assembly required." When you get it home, you unpack the box and pile all of the parts on the floor. At that point, do you have a bicycle? Of course not. In order to make the bicycle, you have to assemble the pieces in just the right way, according to the instructions. When you get done with the assembly, all of the parts will be in just the right place, and they will work together with the other parts to make a functional bike.

In the same way, DNA is the set of instructions that takes the chemicals which make up life and arranges them in just the right way so as to produce a living system. Without this instruction set, the chemicals that make up a life form would be nothing more than a pile of goo. However, directed by the information in DNA, these molecules can work together in just the right way to make a living organism. Now of course, the exact way in which DNA does this is a little complicated. Nevertheless, in an upcoming module, we will spend some time studying DNA and how it works in detail.

<u>Energy Conversion and Life</u>

In order to live, organisms need energy. This is why our second criterion states that all life forms must be able to absorb energy from the surroundings and convert it into a form of energy that will sustain their life functions. The production and use of this energy is called **metabolism** (muh tab' uh liz uhm).

<u>Metabolism</u> – The sum total of all processes in an organism which convert energy and matter from outside sources and use that energy and matter to sustain the organism's life functions

Metabolism can be split into two categories: **anabolism** (uh nab' uh lizm) and **catabolism** (kuh tab' uh lizm).

<u>Anabolism</u> – The sum total of all processes in an organism which use energy and simple chemical building blocks to produce large chemicals and structures necessary for life

<u>Catabolism</u> – The sum total of all processes in an organism which break down chemicals to produce energy and simple chemical building blocks

Although these definitions might seem hard to understand, think about them this way: when you eat food, your body has to break it down into simple chemicals in order to use it. Once it is broken down, your body will either burn those simple chemicals to produce energy or use them to make larger chemicals. The entire process of breaking the chemicals down and then burning them to produce energy is part of your body's catabolism. Once your body has that energy, it will use some of it to take simple chemicals and build large, complex chemicals that are necessary for your body to work correctly. The process of making those complex chemicals from simple chemicals is part of your body's anabolism. As we progress throughout the course, we will discuss specific examples of anabolism and catabolism, and that will help you better understand the distinction between them. One way to remember these two definitions is to notice that "catabolism" has the same prefix as "catastrophe," so they both involve things being broken down.

Obviously, then, the energy that an organism gets from its surroundings is important. Where does it come from? Ultimately, almost all of the energy on this planet comes from the sun, which bathes the earth with its light. When you take chemistry, you'll learn a lot more about light. For right now, however, all you need to know is that light is a form of energy and that it is the main energy source for all living organisms on our planet. Green plants (and some other things you will learn about later) take this energy and, by a process called **photosynthesis** (foh' toh sin thuh' sis), convert that energy into food for themselves.

Photosynthesis – The process by which green plants and some other organisms use the energy of
                 sunlight and simple chemicals to produce their own food

We'll be looking at photosynthesis in great detail in a later module. Thus, if the definition is a little confusing to you, don't worry about it. What you need to know at this point is that photosynthesis allows plants and certain other organisms to convert the energy of sunlight into food. Photosynthesis is a part of anabolism, because the organism takes simple chemicals and converts them into food, which is composed of larger chemicals.

If plants and other photosynthetic organisms absorb their energy from the sun, where do other life forms get their energy? Well, that depends. Some organisms eat plants. By eating plants, these organisms take in the energy that plants have stored up in their food reserves. Thus, these organisms are indirectly absorbing energy from the sun. They are taking the energy from plants in the form of food, but that food ultimately came from sunlight. Organisms that eat only plants are called **herbivores** (ur' bih vorz).

Herbivores – Organisms that eat only plants

So you see that even though herbivores don't get their energy directly from sunlight, without sunlight there would be no plants, and therefore there would be no herbivores.

If an organism does not eat plants, it eats organisms other than plants. These organisms are called **carnivores** (kar' nih vorz).

Carnivores – Organisms that eat only organisms other than plants

Even though carnivores eat other organisms, their energy ultimately comes from the sun. After all, the organisms that carnivores eat have either eaten plants or have eaten other organisms that have eaten plants. The plants, of course, get their energy from the sun. In the end, then, carnivores also indirectly get their energy from the sun.

Finally, there are organisms that eat both plants and other organisms. We call these **omnivores** (ahm nih' vors).

Omnivores – Organisms that eat both plants and other organisms

Ultimately, of course, these organisms also get their energy from the sun.

Think about what we just did in the past few paragraphs. We took a large number of the organisms that live on this earth and placed them into one of three groups: herbivores, carnivores, or omnivores. This kind of exercise is called **classification**. When we classify organisms, we are taking a great deal of data and trying to organize it into a fairly simple system. In other words, classification is a lot like filing papers. When you file papers, you place them in folders according to their similarities. In this case, we have taken many of the organisms on earth and put them into one of three folders based on what they eat. This is one of the most important contributions biologists have made in understanding God's creation. Biologists have taken an enormous amount of data and have arranged it into many different classification systems. These classification systems allow us to see the similarities and relationships that exist between organisms in creation. Figure 1.1 illustrates the classification system you have just learned.

FIGURE 1.1
Herbivores, Carnivores, and Omnivores

Giraffe Photo by Dawn Strunc
Tiger Photo © Comstock, Inc.
Woman eating Photo © Tan Kian Khoon

Giraffes eat only plants; they are herbivores.

Tigers eat only meat. This makes them carnivores.

Humans eat both plants and meats; we are omnivores.

In biology, there are hundreds and hundreds of different ways that we can classify organisms, depending on what kind of data we are trying to organize. For example, the classification system we just talked about groups organisms according to what they eat. Thus, organisms that eat similar things are grouped together. In this way, we learn something about how energy is distributed from the sun to all of the creatures on earth.

This is not the only way we can classify organisms to learn how energy is distributed from the sun to all of the creatures on earth. We could, alternatively, classify organisms according to these groups: **producers**, **consumers**, and **decomposers**.

Producers – Organisms that produce their own food

Consumers – Organisms that eat living producers and/or other consumers for food

<u>Decomposers</u> – Organisms that break down the dead remains of other organisms

In this system, plants are producers because they make their own food from chemicals and the sun's light. Omnivores, herbivores, and carnivores are all consumers, because they eat producers and other consumers. Certain bacteria and fungi (the plural of "fungus"), organisms we'll learn about in detail later, take the remains of dead organisms and break them down into simple chemicals. Thus, these creatures are decomposers. Once the decomposers have done their job, the chemicals that remain are once again used by plants to start the process all over again. This classification scheme, illustrated in Figure 1.2, gives us a nice view of how energy comes to earth from the sun and is distributed to all creatures in God's creation.

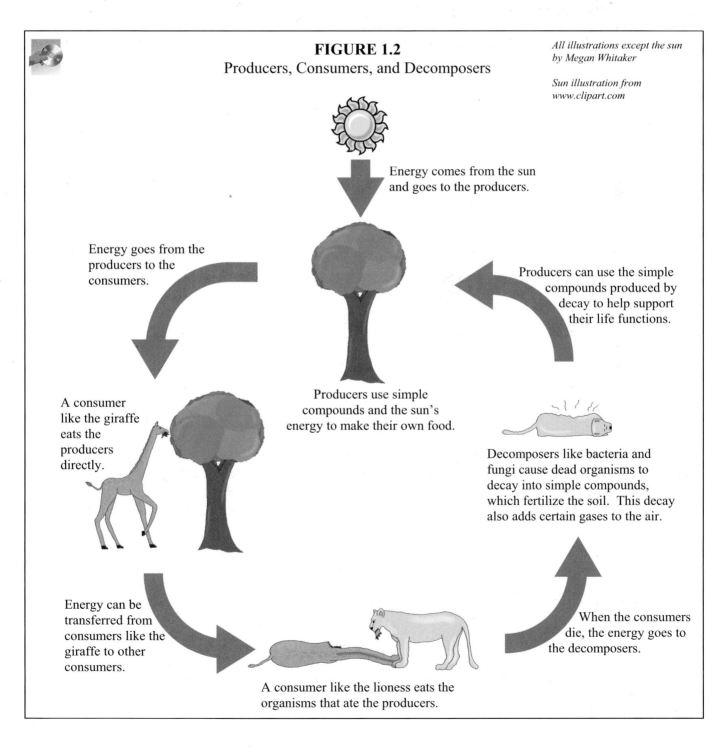

FIGURE 1.2
Producers, Consumers, and Decomposers

All illustrations except the sun by Megan Whitaker

Sun illustration from www.clipart.com

Energy comes from the sun and goes to the producers.

Producers can use the simple compounds produced by decay to help support their life functions.

Energy goes from the producers to the consumers.

A consumer like the giraffe eats the producers directly.

Producers use simple compounds and the sun's energy to make their own food.

Decomposers like bacteria and fungi cause dead organisms to decay into simple compounds, which fertilize the soil. This decay also adds certain gases to the air.

Energy can be transferred from consumers like the giraffe to other consumers.

When the consumers die, the energy goes to the decomposers.

A consumer like the lioness eats the organisms that ate the producers.

There are, of course, differences between this classification system (producers, consumers, and decomposers) and the one you learned previously (omnivores, herbivores, and carnivores). The first difference you should notice between this classification scheme and the one you just studied is that, using this system, we include plants, bacteria, and fungi in the classification. In the previous classification system, we could only classify organisms that ate plants or ate other organisms. There was no grouping in which to put the plants, the bacteria, or the fungi. Does this mean that the second classification system is better than the first? Not really. Each one tells us different information. For example, if we need to look at the differences that exist among animals, then the first classification scheme is best. Some animals are herbivores (cows, for example), some animals are carnivores (lions, for example), and some animals are omnivores (gorillas, for example). In the second classification system, all animals are consumers. So the second classification system doesn't tell us much about the differences that exist among animals. If, however, we want to study how energy flows from the sun to every creature in creation, the second classification system gives more information about this process.

As a point of terminology, producers are often called **autotrophs** (aw' toh trohfs), the Greek roots of which literally mean "self-feeder." Consumers and decomposers, on the other hand, are often called **heterotrophs** (het' er uh trohfs), which literally means "other-feeder."

<u>Autotrophs</u> – Organisms that are able to make their own food

<u>Heterotrophs</u> – Organisms that depend on other organisms for their food

In a little while, these two terms will become very important, so you need to know them.

Before you go on to the next section, answer the "On Your Own" questions below. These questions will be scattered throughout the modules in this course. They allow you to reflect on the things you have just read about, cementing the concepts in your mind.

ON YOUR OWN

1.1 Classify the following organisms as herbivores, carnivores, or omnivores:

    a. tigers   b. cows   c. humans   d. sheep

1.2 Classify the following organisms as producers, consumers, or decomposers:

    a. rose bushes   b. yeast (a fungus)   c. lions   d. humans

<u>Sensing and Responding to Change</u>

Our third criterion for life is that it senses and responds to changes in its surroundings. It is important to realize that in order to meet this criterion, an organism's ability to sense changes is just as important as its ability to respond. After all, even a rock can respond to changes in its environment. If a boulder, for example, is perched on the very edge of a cliff, even a slight change in the wind patterns around the boulder might be enough for it to fall off of the cliff. In this case, the boulder is responding to the changes in its surroundings. The reason a boulder doesn't meet this criterion for life is that the boulder cannot sense the change.

Living organisms are all equipped with some method of receiving information about their surroundings. Typically, they accomplish this feat with receptors.

Receptors – Special structures that allow living organisms to sense the conditions of their internal or external environment

Your skin, for example, is full of receptors. Some allow you to distinguish between hard and soft substances when you touch them. Other receptors react to hot and cold temperatures. If you have your hand under a stream of water coming from a water faucet, for example, your receptors react to the temperature of the water. The receptors send information to your brain, and you can then react to the temperature. If the water is too hot or too cold, you can remove your hand from the stream to avoid the discomfort.

A living organism's ability to sense and respond to changes in its surrounding environment is a critical part of survival, because God's creation is always changing. Weather changes, seasons change, landscape changes, and the community of organisms in a given region changes. As a result, living organisms must be able to sense these changes and adapt, or they would not be able to survive.

All Life Forms Reproduce

Our final criterion for life says that all living organisms reproduce. Although the necessity of reproduction for the perpetuation of life is rather obvious, it is truly amazing how many different ways God has designed the organisms on earth to accomplish this feat. Some organisms, for example, can split themselves apart under the right circumstances. The two parts can then grow into wholly separate organisms. This is an example of **asexual reproduction**.

Asexual reproduction – Reproduction accomplished by a single organism

Other organisms, however, require a male and female in order to reproduce. This method of reproduction (which occurs in most of the life forms with which you are familiar) is called **sexual reproduction**.

Sexual reproduction – Reproduction that requires two organisms

As we go along in this course, we will be studying both of these methods a bit more closely, because there is a great deal of variety among the different means of sexual and asexual reproduction.

Reproduction always involves the concept of **inheritance**. Although this word has several different meanings, in biology the definition is quite specific.

Inheritance – The process by which physical and biological characteristics are transmitted from the parent (or parents) to the offspring

In asexual reproduction, the characteristics and traits inherited by the offspring are, under normal circumstances, identical to the parent. Thus, the offspring is essentially a "copy" of the parent. In sexual reproduction, under normal circumstances, the offspring's traits and characteristics are, in fact, some mixture of each parent's traits and characteristics. Of course, the parents' traits and characteristics are a mixture of each of their parents' traits and characteristics, and their parents' traits and characteristics are a mixture of each of their parents' traits and characteristics, and so on. In the end, then, the inheritance process in sexual reproduction is quite complicated, and leads to offspring that often can be noticeably different from both parents.

Notice that in describing inheritance for both modes of reproduction, we used the phrase "under normal circumstances." This is because every now and again, offspring can possess traits that are incredibly different from their parents. These incredibly different traits are the result of **mutations**.

Mutation – An abrupt and marked change in the DNA of an organism compared to that of its parents

The study of mutations is quite interesting, and we will focus on it later on in the course.

Before we leave this discussion of reproduction, it is important to note that some living organisms cannot actually produce viable offspring. When a horse and a donkey mate, for example, they can produce an offspring called a mule. Adult mules, however, cannot produce offspring of their own. Nevertheless, mules *do not* fail to meet the reproduction criterion for life. Even though they cannot produce *offspring*, their cells (we will discuss cells more thoroughly in a while) reproduce quite frequently so that the mules can grow, repair wounds, etc. Thus, they satisfy the reproduction criterion on the cellular level.

ON YOUR OWN

1.3 A biologist studies an organism and then two of its offspring. They are all identical in every possible way. Do these organisms reproduce sexually or asexually?

Life's Secret Ingredient

Well, now that we have a good idea of whether or not something is alive, another question should come to mind. What gives life the characteristics that we learned in the previous sections? As we said before, if we chemically analyzed an organism, gathered together all of the chemicals contained in it, and threw them in a pot, we would not have a living organism. Those chemicals would be useless without the information stored in the organism's DNA. However, even if we were able to isolate a full set of the organism's DNA and were to throw it into the pot as well, we would still not have a living organism.

You see, life is more than a collection of chemicals and information. There is something more. Scientists have tried to understand what that "something more" is, but to no avail. The secret ingredient that separates life from nonlife is still a mystery to modern science. Of course, to believers, that secret ingredient is rather easy to identify. It is the creative power of God. In Genesis 1:20-27, the Bible tells us that God created all creatures, and then He created man in His own image. Think about it this way. Suppose you had a bunch of engine and metal parts and you also had instructions that led you through all of the steps necessary to take those parts and make a working motorcycle. Could you just throw the parts and the instructions into a pile and make a motorcycle? Of course not.

Even if you had all of the necessary parts as well as all of the instructions, you would still need to exercise some of your own creative power to follow those instructions and make the motorcycle.

If we were talking about a living organism instead of a motorcycle, we could say that chemicals are the "parts" that make up the organism and DNA is the instruction set that contains the information necessary to assemble the parts properly. Nevertheless, if you just threw the chemicals and the DNA into a big pot, you would not make a life. Some creative power must be exercised in order to take lifeless chemicals and use the information in DNA to make a living organism. Of course, only God has such creative power, and that is why all life comes from Him.

So you see, science will never be able to uncover the "secret ingredient" that makes life possible. At some point in the future, scientists might be able to catalog every chemical that makes up a living organism. Scientists might even decode the information stored in DNA and determine all of the instructions necessary to form those chemicals into a living organism. Even after those incredible feats, however, science would be no closer to creating life. Without the creative power of God, lifeless chemicals will never become a living organism.

This little discussion brings us to probably the most important thing that you will ever learn in your academic career: **science has its limitations**. We say that this is probably the most important thing that you will ever learn because we know a great many people whose lives have been ruined because they put too much faith in science. They think that because of all the wonderful advances we have made in recent years, science has no limitations. As a result, they live their lives looking to science as the ultimate answer to every question. This leads them down a path of spiritual destruction. Had they only placed their faith in God, who has no limitations, they would have lived fulfilling lives and spent eternity with the ultimate Life-Giver! Read the next section carefully, so that you will understand the limitations of science.

The Scientific Method

Real science must conform to a system known as the scientific method. This system provides a framework in which scientists can analyze situations, explain certain phenomena, and answer certain questions. The scientific method starts with observation. Observation allows the scientist to collect data. Once enough data have been collected, the scientist forms a **hypothesis** that attempts to explain some facet of the data or attempts to answer a question that the scientist is trying to answer.

Hypothesis – An educated guess that attempts to explain an observation or answer a question

Once he forms a hypothesis, the scientist (typically with help from other scientists) then collects much more data in an effort to test the hypothesis. These data are often collected by performing experiments or by making even more observations. If the data are found to be inconsistent with the hypothesis, the hypothesis might be discarded, or it might just be modified a bit until it is consistent with all data that have been collected. If a large amount of data is collected and the hypothesis is consistent with all of the data, then the hypothesis becomes a **theory**.

Theory – A hypothesis that has been tested with a significant amount of data

Since a theory has been tested by a large amount of data, it is much more reliable than a hypothesis. As more and more data relevant to the theory are collected, the theory can be tested over and over again. If several generations of collected data are all consistent with the theory, it eventually attains the status of a **scientific law**.

Scientific law – A theory that has been tested by and is consistent with generations of data

An example of the scientific method in action can be found in the work of Ignaz Semmelweis, a Hungarian doctor who lived in the early-to-mid-1800s. He was appointed to a ward in Vienna's most modern hospital, the Allegemeine Krankenhaus. He noticed that in his ward, patients were dying at a rate that far exceeded that of the other wards, even the wards with much sicker patients. Semmelweis observed the situation for several weeks, trying to figure out what was different about his ward as compared to all others in the hospital. He finally determined that the only noticeable difference was that his ward was the first one that the doctors and medical students visited after they performed autopsies on the dead.

Based on his observations, Semmelweis hypothesized that the doctors were carrying something deadly from the corpses upon which the autopsies were being performed to the patients in his ward. In other words, Dr. Semmelweis exercised the first step in the scientific method. He made some observations and then formed a hypothesis to explain those observations.

Semmelweis then developed a way to test his hypothesis. He instituted a rule that all doctors had to wash their hands after they finished their autopsies and before they entered his ward. Believe it or not, up to that point in history, doctors never thought to wash their hands before examining or even operating on a patient! Dr. Semmelweis hoped that by washing their hands, doctors would remove whatever was being carried from the corpses to the patients in his ward. He eventually required doctors to wash their hands after examining *each patient* so that doctors would not carry something bad from a sick patient to a healthy patient.

Although the doctors did not like the new rules, they grudgingly obeyed them, and the death rate in Dr. Semmelweis's ward decreased significantly! This, of course, was good evidence that his hypothesis was correct. You would think that the doctors would be overjoyed. They were not. In fact, they got so tired of having to wash their hands before entering Dr. Semmelweis's ward that they worked together to get him fired. His successor, anxious to win the approval of the doctors, rescinded Semmelweis's policy, and the death rate in the ward shot back up again.

Semmelweis spent the rest of his life doing more and more experiments to confirm his hypothesis that something unseen but nevertheless deadly can be carried from a dead or sick person to a healthy person. Although Semmelweis's work was not appreciated until after his death, his hypothesis was eventually confirmed by enough experiments that it became a scientific theory. As time went on, more and more data were gathered in support of the theory. With the aid of the microscope, scientists were able to characterize the deadly bacteria and germs that can be transmitted from person to person, and the theory became a scientific law. Nowadays, doctors do all that they can to completely sterilize their hands, clothes, and instruments before performing any medical procedure.

Before we leave this story, it might be interesting to note that the Old Testament contains meticulous instructions concerning how a priest is to cleanse himself after touching a dead body. These rituals, some of which are laid out in Numbers 19, are quite effective in removing germs from the skin and clothing. As Dr. S. I. McMillen, a medical doctor and international lecturer, says, "In 1960, the Department [of Health in New York State] issued a book describing a method of washing the hands, and the procedures closely approximate the Scriptural method given in Numbers 19" (S. I. McMillen, *None of These Diseases,* [Old Tappam, NJ: Fleming H. Revell Company, 1963], 18). This, of course, should not surprise you. After all, God knows all about germs and bacteria. Thus, it only makes sense that He would lay down instructions as to how His people can protect themselves from germs and bacteria. If only doctors had the sense to follow those rules in the past centuries. Countless lives would have been saved!

So you see, the scientific method (summarized in Figure 1.3) provides a methodical, logical way to examine a situation or answer a question. If a theory survives the scientific method and becomes a law, it can be considered reasonably trustworthy. Even a scientific theory which has not been tested enough to be a law is still pretty reliable, because it is backed up by a lot of scientific data.

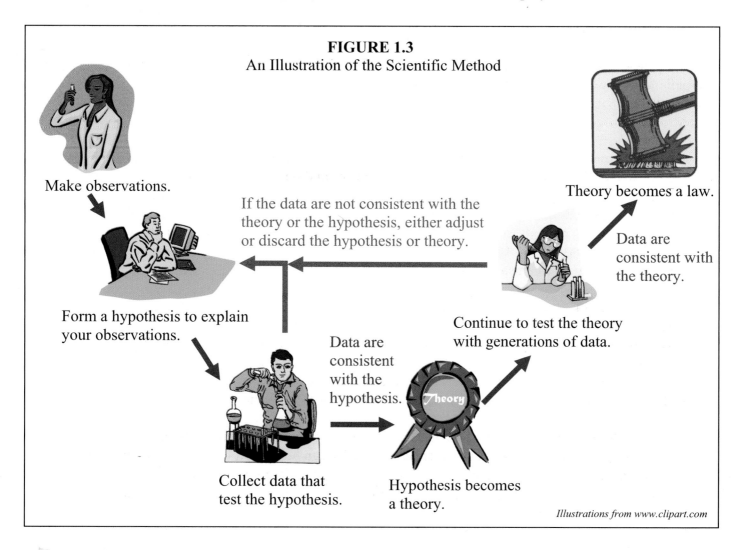

FIGURE 1.3
An Illustration of the Scientific Method

Make observations.

Form a hypothesis to explain your observations.

If the data are not consistent with the theory or the hypothesis, either adjust or discard the hypothesis or theory.

Collect data that test the hypothesis.

Data are consistent with the hypothesis.

Hypothesis becomes a theory.

Continue to test the theory with generations of data.

Theory becomes a law.

Data are consistent with the theory.

Illustrations from www.clipart.com

ON YOUR OWN

1.4 When trying to convince you of something, people will often insert "Science has proven..." at the beginning of a statement. Can science actually prove something? Why or why not?

1.5 A scientist makes a few observations and develops an explanation for the observations that she has made. At this point, is the explanation a hypothesis, theory, or scientific fact?

Limitations of the Scientific Method

At the end of the previous section, we said that if a theory survives the scientific method and becomes a scientific law, it is "reasonably trustworthy." Why did we say "reasonably?" Aren't all scientific laws completely trustworthy? If a hypothesis survives scientific scrutiny and becomes a theory, and the theory goes through more significant scientific scrutiny and becomes a law, isn't it 100% reliable? No, it is not. You see, in order to test hypotheses and theories, scientists must gather data. In order to gather data, they must perform experiments and make observations. Since these experiments and observations are designed and performed by imperfect humans, the data collected might, in fact, be flawed. As a result, even though there might be an enormous amount of data supporting a scientific law, if the data are flawed, the law is most likely wrong! In addition, it is simply impossible, even after centuries of experimentation, to test all implications of a scientific law completely. Thus, even though years and years of experimentation exist in support of a scientific law, some clever person somewhere might devise an experiment that produces data which contradict the law. Thus, scientific laws can be demonstrated false when the experiments that support them are shown to be flawed or when someone finds a new kind of experiment that contradicts the law. Both of these situations occur frequently in the pursuit of science, and they are best studied by example.

Scientific laws are constantly being overthrown due to the fact that it is impossible to test them completely. For example, prior to 1938, it was considered scientific law that the coelacanth (see' luh kanth), a type of fish, was extinct. After all, many fossils of the fish had been uncovered, but no live specimen had ever been found, even after much searching. Since almost 100 years of searching for this fish never turned up a live specimen, the hypothesis that it was extinct was eventually accepted as a theory and then as a scientific law. All scientists agreed: the coelacanth was extinct. Imagine their surprise when, in 1938, a live coelacanth was found in the net of a fishing boat off the coast of South Africa! We now know that the coelacanth is relatively plentiful in the western Indian Ocean. In this case, then, a scientific law was overthrown due to the fact that it was impossible to test it completely. One would think that since 100 years of careful searching for the coelacanth had never turned up a live specimen, the law stating that it was extinct should be rather reliable. However, no one had looked carefully enough in the Indian Ocean off the coast of Africa, and therefore a scientific law turned out to be quite wrong!

Scientific laws are also overthrown because the experiments that support them are flawed. For example, in about 350 B.C., the famous Greek philosopher Aristotle observed that if a person left meat out in the open and allowed it to decay, maggots would appear on the meat within a few days. From that observation, he formed the hypothesis that living maggots were formed from nonliving meat. We call this idea **spontaneous generation**, and Aristotle postulated that this is how many life forms originate. He made many other observations that seemed to support his hypothesis. For example, he showed that eels have a similar smell and feel as the slimy ooze at the bottom of rivers. He considered this evidence that eels spontaneously formed from the ooze.

As time went on, many more experiments were performed that seemed to support the hypothesis of spontaneous generation. As a result, the hypothesis was quickly accepted as a theory. Of course, the experimentation did not stop there. As late as the mid-1600s, a biologist named Jean Baptist van Helmont performed an experiment in which he placed a sweaty shirt and some grains of wheat in a closed wooden box. Every time he performed the experiment, he found at least one mouse gnawing out of the box within 21 days. Think about it. A hypothesis that was formed around 350 B.C. was quickly accepted as a theory due to the fact that all experiments performed seemed to support it. Experiments continued for a total of 1,900 years, and they all seemed to support the theory! As a result of this overwhelming amount of data in support of the theory of spontaneous generation, it became accepted as a scientific law.

About that same time, however, Francesco Redi, an Italian physician, questioned the law of spontaneous generation. Despite the fact that this law was universally accepted by the scientists of his day, and despite the fact that his fellow scientists laughed at him for not believing in the law, Redi challenged it. He argued that Helmont could not tell whether the mice that supposedly formed from a sweaty shirt and wheat grains had gnawed into the box or out of the box. He said that in order to really test this law, you would have to completely isolate the materials from the surroundings. That way, any life forms that appeared would have definitely come from the materials and not from the surroundings. He performed experiments in which he put several different types of meat in sealed jars and left them to decay. No maggots appeared on the meat. He claimed that this showed that maggots appear on meat not because they are formed by the meat, but instead because they get on the meat.

Of course, the scientists of his day said that by sealing the jars, Redi was cutting off the air supply, which would stop the maggots from forming. Thus, Redi redesigned his experiment. Instead of sealing the jars, he covered them with a fine netting. The netting was fine enough to keep insects out but allow air in. Still, no maggots formed on the meat, even long after it was decayed. What these experiments showed was that the previous experiments which purportedly demonstrated that maggots could form from decaying meat were simply flawed. If one were to adequately isolate the meat from the surroundings, maggots would never form.

These experiments sent shock waves throughout the scientific community. A scientific law, one which had been supported by nearly 1,900 years of experiments, was wrong! Of course, many scientists were simply unwilling to accept this. Yes, they agreed, perhaps maggots did not come from decaying meat, but surely there were some types of organisms that could spontaneously generate from nonliving things.

In the 1670s, some scientists thought that Anton van Leeuwenhoek had found such organisms. He had fashioned his own microscope and had used it to study water. As a result, he discovered the world of microorganisms.

Microorganisms – Living creatures that are too small to see with the naked eye

In the next module, we will begin studying this fascinating world in more depth. For right now, you just need to know that because these creatures cannot be seen without the aid of a microscope, scientists prior to 1670 had no idea that they existed.

Leeuwenhoek and many others showed that microorganisms did, indeed, seem to generate spontaneously. For example, in the mid-1700s, John Needham did experiments very similar to Redi's. Needham made a liquid broth of nutrient-rich material such as chicken broth. Such broths were called "infusions," and Needham showed that if you boiled an infusion for several minutes, you could kill all microorganisms in it. If you then put a cork in the flask that held the infusion, microorganisms would appear in the infusion within a few days. Needham concluded that since he had put a cork in the flask, the infusion was isolated from the surroundings. These experiments were hailed as support for the beleaguered law of spontaneous generation.

Lazzaro Spallanzani, a contemporary of Needham, did not like Needham's experiments. He thought that either Needham did not boil the infusion long enough to completely kill off the microorganisms or that Needham's corks allowed air to leak into the flask, bringing microorganisms in with it. Spallanzani repeated Needham's experiments, but he boiled the infusions for a long time and sealed the flasks by actually melting their openings shut. That made a truly airtight seal. In these experiments, no microorganisms formed. Of course, those who still held to the law of spontaneous generation argued that once again, without air, nothing could live. Thus, by completely sealing the flask before the infusion was boiled, Spallanzani cut off the process of spontaneous generation.

In 1859, however, the great scientist Louis Pasteur finally demonstrated that even microorganisms cannot spontaneously generate. In his experiments, illustrated in Figure 1.4, Pasteur stored the infusion in a flask that had a curved neck. The curved neck allowed air to reach the infusion, but because microorganisms are heavier than air, any microorganisms present would be trapped at the bottom of the curve. When Pasteur repeated Needham's experiments in the curved flask, no microorganisms appeared. In a final blow, Pasteur even showed that if you tipped the flask once to allow any microorganisms that might be trapped to fall into the infusion, microorganisms would appear in the infusion. Thus, Pasteur showed that even microorganisms cannot spontaneously generate.

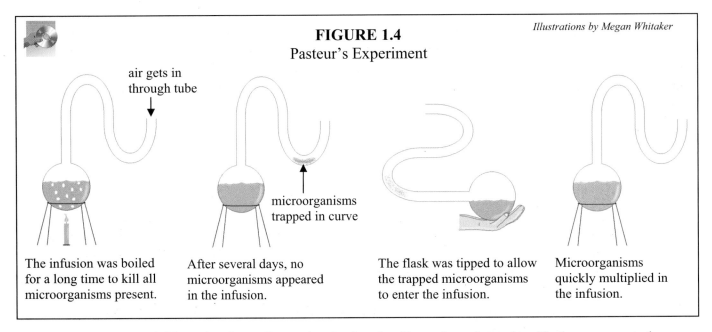

FIGURE 1.4
Pasteur's Experiment

Illustrations by Megan Whitaker

air gets in through tube

microorganisms trapped in curve

The infusion was boiled for a long time to kill all microorganisms present.

After several days, no microorganisms appeared in the infusion.

The flask was tipped to allow the trapped microorganisms to enter the infusion.

Microorganisms quickly multiplied in the infusion.

The point of this rather long discussion is simple. Even though a scientific law seems to be supported by hundreds of years of experiments, it might still be wrong because those experiments might be flawed. All of the experiments that were used to support the law of spontaneous generation were flawed. The scientists who conducted the experiments did not adequately isolate them from the surroundings. Thus, the life forms that the scientists thought were being formed from non-living substances were, in fact, simply finding their way into the experiment.

These two discussions, then, show the limits of science and the scientific method. First, even scientific laws are not 100% reliable. Most likely, some of the things that you learn in this book will someday be proven to be wrong. That is the nature of science. Because it is impossible to fully test a scientific law and because laws are tested by experiments that might be flawed, scientific laws are not necessarily true. They represent the best conclusions that science has to offer, but they are nevertheless not completely reliable. Of course, if you are working with something that is a theory, it is even less reliable. Thus, putting too much faith in scientific laws or theories will end up getting you in trouble, because some of the laws and many of the theories that we treasure in science today will eventually be shown to be wrong.

Well, if scientific laws are not 100% reliable, what is? The only thing in the universe that is 100% reliable is the Word of God. The Bible contains truths that will never be shown to be wrong, because those truths come directly from the Creator of the universe. So much misery and woe have come to this earth because people put their faith in something that is not reliable, like science. In the end, they are spiritually deprived because what they believe is, to one extent or another, wrong (Romans 1:21-25). Those who put their faith in the Bible, however, are not disappointed, because it is never wrong.

If science isn't 100% reliable, why study it? The answer to that question is quite simple. There are many interesting facts and much useful information not contained in the Bible. It is worthwhile to find out about these things. Even though we will probably make many, many mistakes along the way, finding out about these interesting and useful things will help us live better lives. Because of the advances made in science, wonderful technology like vaccines, the television, and the computer exist. Thus, there is nothing wrong with science. In fact, it is even a means by which we can celebrate the awesomeness of God. When we learn how well the world and its organisms are designed, we can better appreciate the gift that God has given to us in His creation. The problem occurs when certain people who are enamored with science end up putting too much faith in it. As a pursuit of flawed human beings, science will always be flawed. Because the Bible was inspired by One who is perfect, the Bible is perfect. As long as we keep this simple fact in mind, our study of science will be very rewarding!

Spontaneous Generation: The Faithful Still Cling to It!

After that long story, it might surprise you to learn that there are many scientists who still believe in spontaneous generation. Now of course, there is no way that they can argue with the conclusions of Pasteur's experiments, so they do not believe that microorganisms can spring from non-living substances. Nevertheless, they still do believe that life can spring from nonlife! These scientists believe in a theory known as **abiogenesis** (āye' bye ōh jĕn' uh sis).

Abiogenesis – The idea that long ago, very simple life forms spontaneously appeared through chemical reactions

Some scientists say that since all life is made up of chemicals, it is possible that long ago on the earth, there was no life; there were just chemicals. These chemicals began reacting and, through the reaction of these chemicals, a "simple" life form suddenly appeared.

As we go through this course, you'll see how such an idea is simply inconsistent with everything that we know about life. At this time, however, we want to make a simple point regarding abiogenesis. Back when scientists believed in spontaneous generation, they had experiments which allegedly backed up their claim. Even before Pasteur's authoritative refutation of spontaneous generation, these experiments were shown to be flawed. Rather than giving up on their law, however, those who fervently believed in spontaneous generation just said, "Well, okay, these experiments are wrong. However, look at these other experiments. Although we now know that life forms which we see with our own eyes cannot spontaneously generate, microorganisms can."

Do you see what the proponents of spontaneous generation did? Because they wanted so badly to believe in their theory, they simply pushed it into an area in which they did not have much knowledge. The whole world of microorganisms was new to scientists back then. As a result, there was a lot of ignorance regarding how microorganisms lived and reproduced. Because of the ignorance surrounding microorganisms, it was relatively easy to say that spontaneous generation occurred in that world. After about 200 years of study, however, scientists began to understand microorganisms a little better, and that paved the way for Louis Pasteur's famous experiments.

Well, nowadays, scientists have pushed the theory of spontaneous generation back to another area that we are rather ignorant about. They say that although Pasteur's experiments show that microorganisms can't arise from nonliving substances, some (unknown) simple life form might have been able to spontaneously generate from some (unknown) mixture of chemicals at some (unknown) point way back in earth's history. Well, since we have very little knowledge about things that happened way back in earth's history, and since we have only partial knowledge about the chemicals that make up life, and since we have no knowledge of any kind of simple life form that could spring from nonliving chemicals, the proponents of abiogenesis are pretty safe. The fact that we are ignorant in these areas keeps us from showing the error in their theory.

Of course, there are a few experiments that lend some support to the theory of abiogenesis. A discussion of these experiments is beyond the scope of this module, but for right now we will just say that they are not nearly as convincing as the ones that van Helmont and Needham performed. In fact, they do not even produce anything close to a living organism, as van Helmont's and Needham's experiments seemed to. They just produce some of the simplest chemicals that are found in living organisms. Nevertheless, those who cling to the idea of spontaneous generation casually disregard the flaws that can be easily pointed out in these experiments and trumpet their results as data that support their theory. However, if you look at the track record of spontaneous generation throughout the course of human history, it is safe to conclude that at some point, the version of spontaneous generation known as abiogenesis will also be shown to be quite wrong.

Biological Classification

Now that we've spent considerable time on the limitations of science, it's time to turn our attention to some of the strengths of science. Classification is probably one of the greatest accomplishments of science. In the study of biology, we uncover many, many facts. For example, there are many, many organisms on the earth, and they have many, many properties and characteristics. Some of their characteristics they have in common with other organisms, and some of their characteristics are unique. All of these facts make up a huge volume of data that, by itself, would be hard to understand and virtually impossible to use. Much like we have split this book into modules and have further split the modules into sections, "On Your Own" questions, study guides, and tests, we need to take all of the data in biology and split them up into an organized system.

Now there are many different classification systems in biology. You have already seen that all organisms can be split into three groups: producers, consumers, and decomposers. You have also seen that we can split most consumers into herbivores, carnivores, and omnivores. Those classification systems were rather simple. They took many, many different organisms and lumped them into only a few groups. Now we need to get more detailed. We need to learn a classification system that takes all organisms and splits them into several groups. The number of groups that we split the organisms into must be large enough so that we are not grouping incredibly different organisms into the same group. At the same time, however, there cannot be too many groups, because the classification system must make the data easier to understand than they were originally. With too many groups, the classification system becomes almost as complex as the data themselves.

The classification system that we will use most frequently is multi-level. It starts by splitting up all organisms into five different groups known as **kingdoms**. The organisms within each kingdom can then be further divided into different groups called **phyla** (fye' luh), the singular of which is **phylum** (fye' lum). Each phylum can be further divided into **classes**, which can be further divided into **orders**. Within an order, organisms can be divided into **families**, which can be further divided into **genera** (jon' ur uh), the singular of which is **genus** (jee' nus), which can finally be broken down into **species**. This multi-level (often called "hierarchical") classification scheme is summarized in the figure below. The specific classification of a bald eagle is given in parentheses so that you can see how the classification scheme is used for a particular organism.

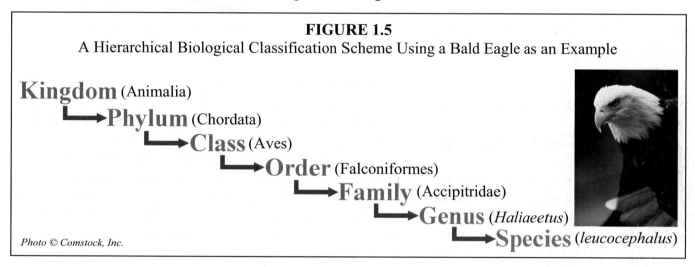

FIGURE 1.5
A Hierarchical Biological Classification Scheme Using a Bald Eagle as an Example

Kingdom (Animalia)
→ Phylum (Chordata)
→ Class (Aves)
→ Order (Falconiformes)
→ Family (Accipitridae)
→ Genus (*Haliaeetus*)
→ Species (*leucocephalus*)

Photo © Comstock, Inc.

To make sure that you can remember the names and orders of this classification system, you can use the following mnemonic:

King Philip Cried Out, "For Goodness Sake!"

Since the first letter of each word in this sentence can stand for a group in our classification system, you can use it to remember the order in which we place these groups. It is important to note that the classification of organisms is so complicated that we often split these groups into subgroups. Thus, do not be confused if you run across a term like "subphylum." A subphylum is simply used to split organisms in a phylum into smaller groups before they are split into classes. There are also subclasses, suborders, and subfamilies. Another issue to remember is that some classification schemes use "division" instead of "phylum" for certain kingdoms. Although we will not do that, you need to be aware that others might.

Now that we know the groups and their respective orders, it's time to see how we use this system to classify organisms in nature. As we mentioned before, we generally split all of the organisms in nature into five separate kingdoms. The names of these kingdoms are **Monera** (muh nihr' uh), **Protista** (pro tee' stuh), **Fungi** (fun' jye), **Plantae**, and **Animalia**. The proper names of all our classification groups are Latin, and when we use those names, we capitalize them to emphasize that these are proper classification names.

How do we know what organisms go into what kingdom? Well, we group organisms together based on similar characteristics. Since the first step in classification deals with placing organisms in kingdoms, the common characteristics that organisms in the same kingdom share are pretty basic. You will learn about that in the next section.

ON YOUR OWN

1.6 Suppose you chose two organisms at random out of a list of the members of kingdom Plantae, then you chose two organisms at random out of a list of the members of family Pinaceae. In which case would you expect the two organisms to be the most similar?

1.7 You compare several organisms from different orders within a given class. You then compare organisms from different classes. In which case would you expect the differences to be greatest?

Characteristics Used to Separate Organisms into Kingdoms

The first and most basic distinction that we make between organisms is based on the number and type of cells that the organism has. Now you have probably learned a few things about cells from your earlier studies in science. You probably learned that all living creatures are made up of at least one cell, and that cells are the basic building blocks of life. We will be making a detailed study of cells throughout the next few modules, so for right now, we don't want to spend a lot of time on them. The only thing that we want to concentrate on right now is the fact that cells come in two basic types: **prokaryotic** (pro' kehr ee aht' ik) and **eukaryotic** (yoo' kehr ee aht' ik).

Prokaryotic cell – A cell that has no distinct, membrane-bounded organelles

Eukaryotic cell – A cell with distinct, membrane-bounded organelles

Now of course, these definitions mean nothing unless you know what **organelles** (or guh nelz') are and what "membrane-bounded" means.

In order to live, a cell must perform certain functions. As two of our criteria for life say, living things must have an energy conversion mechanism as well as reproductive capacity. In order to carry out these functions, cells must complete many different tasks. In eukaryotic cells, the individual tasks needed to complete the functions of life are carried out by distinct structures within the cell. These structures are called organelles. In order to stay distinct, they must be surrounded by something that separates them from the rest of the cell. We call this a membrane. Thus, a "distinct, membrane-bounded organelle" is simply a structure within a cell that performs a specific task. Prokaryotic cells do not contain these internal structures. Nevertheless, they still can perform all of the necessary functions of life. You might wonder how that is possible. Well, you'll learn about these fascinating organisms in the next module. For right now, just familiarize yourself with the distinction between prokaryotic and eukaryotic cells with the figure below.

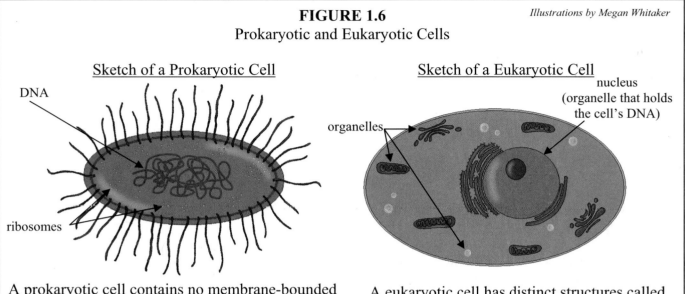

FIGURE 1.6

Illustrations by Megan Whitaker

Prokaryotic and Eukaryotic Cells

Sketch of a Prokaryotic Cell

Sketch of a Eukaryotic Cell

DNA

ribosomes

organelles

nucleus
(organelle that holds
the cell's DNA)

A prokaryotic cell contains no membrane-bounded organelles. The main features of a prokaryotic cell are the DNA visible throughout the cell and the ribosomes, which have no membranes.

A eukaryotic cell has distinct structures called organelles. They have their own tasks to perform in order to maintain the life of the cell. The nucleus, for example, holds the cell's DNA.

Now that we know the distinction between these two basic cell types, we can finally discuss how to split organisms into the five different kingdoms. Kingdom Monera contains all organisms that are composed of either one prokaryotic cell or a simple association of prokaryotic cells. What do we mean when we say "a simple association" of cells? Well, if cells work together in order to complete the tasks necessary for life, they can do so in one of two ways. They can either be highly specialized, each taking on a specific set of tasks needed for the organism to survive, or they can simply work together as a group, each performing essentially the same tasks, but doing so as a group. The cells in a person, for example, work together in the first way. The cells that make up your eyes specialize in the detection of light and the transmission of light-induced information to your brain, while red blood cells specialize in transporting oxygen to other cells. These cells perform different functions, each of which is necessary for the support of life. Blue-green algae (also known as cyanobacteria), however, simply group themselves together in chains. The cells in the chain are usually bound together by mucus, but they each do essentially the same tasks. They simply find strength and survivability in numbers. This is an example of a "simple association" of cells. Blue-green algae and bacteria are both members of kingdom Monera.

The next kingdom is called Protista. It contains those organisms that are composed of only one eukaryotic cell or a simple association of eukaryotic cells. Amoebae, paramecia, and algae are members of kingdom Protista. Kingdoms Monera and Protista together contain most of the microorganisms that exist on earth. Even though we are rarely aware of them, the members of these two kingdoms make up the *vast majority* of life on this earth.

Moving out of the microscopic world (for the most part) and into the macroscopic world (the world we can see with the naked eye), we come to kingdom Fungi. This kingdom is mostly made up of decomposers. If you remember our previous discussion, decomposers are those organisms that feed off of dead organisms, decomposing them into their constituent chemicals so that they can be used again by producers. Members of kingdom Fungi have eukaryotic cells. In addition, most fungi are multicellular, but there are a few single-celled fungi. Mushrooms and bread molds are examples of the organisms in kingdom Fungi.

The next kingdom, Plantae, is mostly composed of autotrophs (organisms that produce their own food). The members of kingdom Plantae are multicelled organisms with eukaryotic cells. Even though we say that members of kingdom Plantae are autotrophs, there are a few exceptions. Some parasitic organisms are considered members of kingdom Plantae. As you have probably already guessed, members of kingdom Plantae are often called "plants." Thus, trees, grass, flowering bushes, etc., are all members of kingdom Plantae.

The last kingdom, Animalia, contains multicellular organisms with eukaryotic cells. Members of kingdom Animalia are separated from kingdom Plantae by the fact that they are heterotrophs (dependent on other organisms for food) but are not decomposers (decomposers are in kingdom Fungi). Of course, members of kingdom Animalia are called "animals." Grasshoppers, birds, cats, fishes, and snakes are all members of kingdom Animalia.

ON YOUR OWN

1.8 An organism is made up of one eukaryotic cell. To what kingdom does it belong?

1.9 An organism is multicellular and an autotroph. To what kingdom does it belong?

1.10 An organism is multicellular with eukaryotic cells. It is also a decomposer. To what kingdom does it belong?

The Definition of Species

After reading the last section, you should have noticed a few things about classifying organisms. It's not very easy or clear-cut. To separate organisms into five separate groups, we have already run into exceptions. Kingdom Plantae, for example, is supposed to contain autotrophs. There are, however, some parasites that belong to that kingdom as well. In addition, we use the word "mostly" quite a lot, because although the majority of the members in a kingdom have a certain characteristic, there will be some members that do not. Thus, classification of organisms into kingdoms gets a little complicated.

As you might expect, classifying organisms in phyla, classes, orders, families, genera, and species becomes even more difficult. After all, as you move down the hierarchy in our classification scheme, you are getting more and more specific. While kingdoms have many, many members, those members are split into phyla. Thus, each phylum has fewer members than does the kingdom of which it is a part. In the same way, classes have fewer members than the phylum that they are in, orders have even fewer members, families have even fewer, and genera have still fewer. By the time you get to species, you have a very small group of organisms.

Since classification gets more and more difficult as you go down the hierarchy, splitting organisms into species becomes incredibly hard. If you thought that our definitions for what organisms go into each of the five kingdoms were bad, it is so hard to classify at the species level that biologists can't even agree on a definition for what the classification "species" really means! There is a lot of work going on right now in the field of biology trying to figure out a good way to define this difficult classification. For our purposes, however, we must have a definition, so we will go with the most commonly accepted one:

Species – A unit of one or more populations of individuals that can reproduce under normal conditions, produce fertile offspring, and are reproductively isolated from other such units

Although this definition is not perfect, it is the one that we will use for now. What does it mean? Basically, if organisms can reproduce and their offspring can also reproduce (that's what "fertile" means), these organisms belong to the same species. Any other organism with which this species cannot reproduce is said to be "reproductively isolated" from this species and therefore must belong to a different species.

Notice that in the previous section, we gave you the characteristics by which you can separate all organisms on earth into the five kingdoms of our classification system. Then, in this section, we skipped over all of the other classification groups except for species. For that classification group we gave a definition. Why did we leave out the other classification groups? Well, we didn't want to overwhelm you with information. There are (depending on whose classification system you use) nearly 100 different phyla in creation. Members of each phyla have their own characteristics, and we would have to go through each phylum individually to give you a good feel for how to classify organisms into these groups. Of course, since each phylum is split into several classes, there are even more of those. Thus, to go through and give you a view of each kingdom, phylum, class, order, family, and genus would be an incredibly long discussion! When we get to species, however, the classification is so specific that we can actually come up with a weak definition for it. That's why we skipped from kingdom all the way to species.

Biological Keys

If a discussion of all groups in our classification system is prohibitively long, how will we ever be able to classify organisms? In order to classify organisms, biologists often refer to biological keys. These keys help you to classify organisms without having to memorize the characteristics of all groups within the classification scheme. A simple biological key is given below. It also is given in Appendix A at the back of the book.

FIGURE 1.7
A Simple Biological Key

1. Microscopic...2
    Macroscopic (visible with the naked eye)................................3
2. Eukaryotic cell.. *kingdom Protista*
    Prokaryotic cell... *kingdom Monera*
3. Autotrophic...*kingdom Plantae*...................**4**
    Heterotrophic..**5**

4.   Leaves with parallel veins ....*phylum Anthophyta*.................... *class Monocotyledoneae*
     Leaves with netted veins ....*phylum Anthophyta*.................. *class Dicotyledoneae*
5.   Decomposer..*kingdom Fungi*
     Consumer...*kingdom Animalia*.................. **6**
6.   No backbone... **7**
     Backbone...*phylum Chordata*...................**22**
7.   Organism can be externally divided into equal halves (like a pie), but it
     has no distinguishable right and left sides...**8**
     Organism either can be divided into right and left sides that are
     mirror images or cannot be divided into two equal halves........................ **9**
8.   Soft, transparent body with tentacles .... *phylum Cnidaria*
     Firm body with internal support; covered with scales or spiny plates;
     tiny, hollow tube feet used for movement....................................*phylum Echinodermata*
9.   External plates that support and protect.......*phylum Arthropoda* **14**
     External shell or soft, shell-less body..**10**
10. External shell..*phylum Mollusca*...................**11**
     No external shell...**12**
11. Coiled shell...*class Gastropoda*
     Shell made of two similar parts...*class Bivalvia*
12. Wormlike body without tentacled receptors on head.............................. *phylum Annelida*
     Non-wormlike body or tentacled receptors on head....*phylum Mollusca*.....**13**
13. Wormlike body with tentacled receptors on head...*class Gastropoda*
     Non-wormlike body with 8 or more tentacles used for grasping.................*class Cephalopoda*
14. More than 3 pairs of legs...**15**
     3 pairs of walking legs.................................*class Insecta*....................**16**
15. 4 pairs of walking legs, body in two divisions...*class Arachnida*
     More than 4 pairs of walking legs..*class Malacostraca*
16   Wings..**17**
     No wings...**21**
17. All wings transparent...**18**
     Nontransparent wings...**19**
18. Capable of stinging from back of body...*order Hymenoptera*
     Cannot sting (may be able to bite)..*order Diptera*
19. Large, sometimes colorful wings...*order Lepidoptera*
     Thick, hard, leathery wings...**20**
20. Pair of hard wings covering a pair of folded, transparent wings.................*order Coleoptera*
     Pair of leathery wings covering a pair of transparent wings.......................*order Orthoptera*
21. Piercing, sucking mouthparts for obtaining blood.......................................*order Siphonaptera*
     Mouthparts for chewing...*order Hymenoptera*
22. Jaws or beak..**23**
     No jaw or beak..*class Agnatha*
23. Skin covered with scales...**24**
     No scales on skin..**26**
24. Fins and gills...**25**
     No fins; breathes with lungs...*class Reptilia*
25. Mouth on lower part of body...*class Chondrichthyes*
     Mouth on front part of body...*class Osteichthyes*

26. No scales, no hair, no feathers; skin is slimy......*class Amphibia*................ **27**
    Feathers or hair..**28**
27. Tail..*order Caudata*
    No tail..*order Anura*
28. Feathers on body... *class Aves*
    Hair on body..*class Mammalia*................ **29**
29. Hooves...**30**
    No hooves..**31**
30. Odd number of toes...*order Perissodactyla*
    Even number of toes..*order Artiodactyla*
31. Carnivore..**32**
    Herbivore..**33**
32. Teeth..*order Carnivora*
    No teeth, eats insects..*order Insectivora*
33. Enlarged front teeth for gnawing.. **34**
    No enlarged front teeth for gnawing.. **35**
34. Legs for crawling...*order Rodentia*
    Hind legs for jumping...*order Lagomorpha*
35. Enlarged trunk, used for breathing and grasping......................*order Proboscidea*
    Tendency to stand erect on two hind limbs.................................*order Primates*

Now don't get overwhelmed by this key. It is actually quite simple to use once you are led through it. You see, a biological key is just a series of questions that you can answer by looking at the major features of the organism you are studying. Based on the answer to a question, you are led to other questions until you eventually run out of questions. At that point, you have classified the organism as well as the biological key allows. For example, consider the elephants shown below.

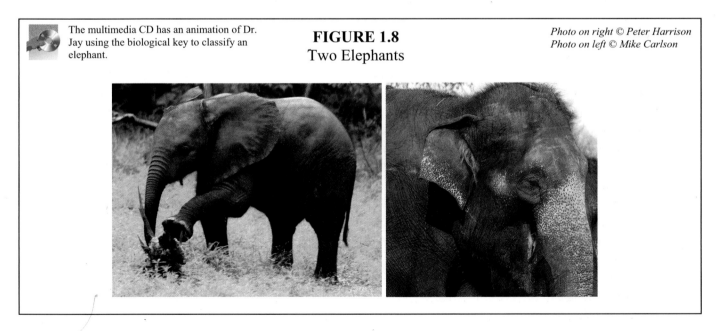

The multimedia CD has an animation of Dr. Jay using the biological key to classify an elephant.

FIGURE 1.8
Two Elephants

Photo on right © Peter Harrison
Photo on left © Mike Carlson

To classify any organism (including an elephant), you would just start at the top of the key. When you answer the question, you proceed to the number that follows that characteristic. You continue to do this until you reach a classification that is not followed by a number.

So, we start at the top of the key. Key 1 asks about size. Since we don't need to magnify an elephant in order to see it, the elephant is macroscopic. This means that we move to key 3, because a "3" follows the term macroscopic. In key 3, we are asked whether or not the elephant is autotrophic (uses photosynthesis to make food) or heterotrophic (eats other organisms). Clearly, the elephant is heterotrophic; it eats plants in order to live. This means we move to key 5, where we need to determine whether it is a decomposer or a consumer. Since the elephant eats plants, it is a consumer. That tells us that our first classification is kingdom Animalia.

Now of course, this should be no surprise. An elephant is an animal. The key also tells us to move on to key 6 for a more detailed classification. Here, we determine whether or not it has a backbone. Now from the picture, you might not be able to tell, but all you have to do is think. Have you seen pictures or movies of people riding on elephants' backs or elephants carrying heavy loads on their backs? They must have a backbone to do that, so we learn that the elephant is in phylum Chordata (kor dah' tuh), and we move on to key 22.

Key 22 asks if the animal has a jaw or beak. Since the elephant's mouth opens and closes up and down, it has a jaw. Thus, we move to key 23, which asks if there are scales on the skin. There are not, so we move to key 26. This key asks about hair or feathers. The picture on the right shows hair on the head. Thus, we move to key 28, which distinguishes between hair and feathers. Based on that distinction, we learn that the elephant is in class Mammalia, and we move to key 29.

In key 29, we must decide whether or not the elephant has hooves. Looking at the picture on the left, the feet have skin all the way to the bottom, so there are no hooves. This means we go to key 31, which asks whether the elephant is a herbivore or carnivore. Although not readily apparent from the picture, you should probably already know that elephants eat plants, making them herbivores. That means we move to key 33, which asks about teeth. There are certainly no enlarged teeth apparent in the picture on the right, so we move to key 35. In this key, we are asked whether there is an enlarged trunk. Yes, there is. Thus, we know that the elephant is in order Proboscidea (pro' boh sid' ee uh). This is as detailed a classification as we can make with this key. As far as this key is concerned, then, the elephant classification is:

Kingdom: Animalia   **Phylum:** Chordata   **Class:** Mammalia   **Order:** Proboscidea

Note that in the case of the elephant, we went all the way to the end of the key. This will rarely be the case. You continue on in the key until you run out of numbers. At that point, you have as detailed a classification as is possible with that key. Take your own turn at classification by performing Experiment 1.1.

EXPERIMENT 1.1
Using a Biological Key

Supplies:

♦ Photographs on the next page
♦ Biological key in Figure 1.7

Object: Identify fifteen living things by using the biological key in the text. Keys vary in their style and content. This key is applicable to all five kingdoms, made especially for use in this course. A good library exercise would be to check other keys and how they are used.

Procedure:

The chart below gives you an example of how to identify the elephant that was described for you in the text. Reread the section on how to identify the elephant and note how the chart has been completed for the elephant. Once you understand how the chart is filled in, identify each of the pictures below by working through the key. As you work through the key, make a chart in your laboratory notebook like the one given below:

Number	Specimen	Specimen Classification		Numbers Used from the Key
Example	Elephant	K. Animalia C. Mammalia P. Chordata O. Proboscidea		1, 3, 5, 6, 22, 23, 26, 28, 29, 31, 33, 35
1.	Butterfly	K. C. P. O.		

Continue the chart so that you have an entry for each specimen. Please note that you may not be able to answer every question in the biological key based on the picture alone. You might have to do a little research to classify some of the specimens. Also, because of the nature of the key, you will not have a kingdom, phylum, class, and order for every specimen. For some specimens, listing the kingdom may be the best that you can do. Once you have completed the chart in your laboratory notebook, check the answers that are provided after the answers to the "On Your Own" questions.

Specimens for the lab:

1. Butterfly

2. Chipmunk

3. Grapevine

4. Swan

5. Spider

6. Tiger

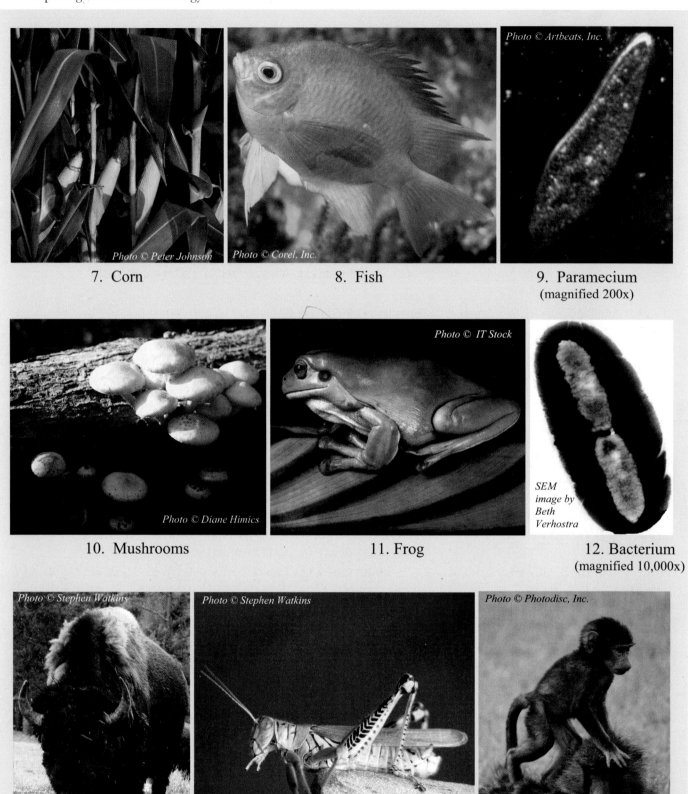

7. Corn

8. Fish

9. Paramecium
(magnified 200x)

10. Mushrooms

11. Frog

12. Bacterium
(magnified 10,000x)

13. Bison

14. Grasshopper

15. Baboon

Naming Organisms Based on Classification

Of course, with a more complicated key, you could continue your classification of an organism right down to species. Why bother? Well, as we said before, classification is a way of ordering the diverse data in biology into some reasonably understandable system. This is such an important practice that an entire field of biology is devoted to it. We call this field **taxonomy** (taks ahn' uh mee).

Taxonomy – The science of classifying organisms

Taxonomy is a very important part of biology because, in order to give a scientific name to an organism, we must know both its species and its genus. In biology, we name things with **binomial** (bye no' mee ul) **nomenclature** (no' mun klay chur).

Binomial nomenclature – Naming an organism with its genus and species name

People, for example, are called *Homo sapiens*. *Homo* is the genus to which humans belong, and *sapiens* is the species. Notice that in binomial nomenclature, we italicize the genus and species name. This is to emphasize that we are using binomial nomenclature. In fact, whenever we use a genus or species name alone, we still italicize it, just to emphasize that it is a part of binomial nomenclature.

So, in order to properly name an organism, we need to know its genus and species. For example, if you were classifying oak trees, you would find that all oak trees are in genus *Quercus*. A red oak is given the species name *rubra*, while a white oak is given the species name *alba*. Notice that while we have capitalized all classification names up to this point, we do not capitalize the species name. This is a convention that makes binomial nomenclature a bit clearer. Thus, the scientific name of the red oak is *Quercus rubra*, whereas the scientific name of the white oak is *Quercus alba*. As a point of notation, once we have introduced a genus name, we are allowed to abbreviate it in discussions that follow. Thus, we could say that the red oak is *Q. rubra* and the white oak is *Q. alba*.

Now why bother to do this? Why not just call a white oak a white oak and a red oak a red oak? Wouldn't that be easier? Well, yes and no. You see, English is constantly changing. What we mean by "oak" today may not mean the same thing in 100 years. That's because a spoken language continues to change. Latin, however, is a dead language. It will never change. Since a lot of binomial nomenclature is based on Latin, the binomial names of organisms do not change. Thus, *Q. rubra* will mean the same thing 100 years from now that it means today. Also, by using the genus name in the name of the organism, we have a start at being able to figure out other organisms that are similar to it. Any other organism that belongs to genus *Quercus* will be very similar to the red or white oaks. In addition, if we find out what family the genus *Quercus* comes from, we can find other organisms that are also similar to the white and red oaks. That's why we use this complicated naming system.

Alternate Forms of Taxonomy

Before we leave this discussion of taxonomy, it is important for you to know that the classification that you see in this course will *not necessarily* be the classification system that you see in another biology course. That's because biologists have different ideas about classification. As a result, we might classify an organism in a particular kingdom, for example, but another biology book might classify that same organism in a different kingdom. It might be hard for you to believe that there can be arguments about which kingdom a particular organism belongs in, but there are! In Module #4, for example, you will learn about the slime molds. Some biology books place these odd creatures in kingdom Protista, while others place them in kingdom Fungi. You will learn why this disagreement occurs when you study slime molds.

Not only is there disagreement about which kingdoms, phyla, classes, etc., to place organisms in, there is also disagreement on *which classification system to use*. The classification system that we have just taught you is based on one first developed in the 1700s by a devout Christian, **Carrolus** (kair' uh lus) **Linnaeus** (lih nay' us). It is typically called the **five-kingdom system**, because it uses five kingdoms. It is the one used by the majority of biology courses on the high school and college level. However, it is important to realize that there are other classification systems that are used by some biologists and some biology courses.

For example, some biologists propose that five kingdoms really are not enough. They suggest that kingdom Monera, for example, contains organisms that are just too different from one another to justify putting them in the same kingdom. As a result, they propose splitting Monera into two kingdoms. In addition, some propose splitting kingdom Protista into two kingdoms as well, because some unicellular (single-celled) eukaryotic organisms do not have certain organelles that are present in most other unicellular eukaryotic organisms. These biologists therefore think that the lack of certain "standard" organelles is reason enough to put these organisms into a completely separate kingdom. As biologists start splitting one or more of the standard five kingdoms into several smaller kingdoms, the number of kingdoms, of course, goes up. Some biologists recommend using an eight-kingdom system. Some biologists propose classification systems that have more than *twenty* kingdoms. However, since the five-kingdom system is the most widely used system, we will stick with it.

Not only is the number of kingdoms in creation a point of disagreement among biologists, some biologists propose that we should scrap the five-kingdom classification system altogether and move to what is called the **three-domain system**. Since this system has been gaining some popularity in the field of biology, we should discuss it to some extent, even though we will not use it in this course. The three-domain system classifies all living things into one of three large domains: **Archaea** (ar kay' uh), **Bacteria**, and **Eukarya** (yoo' kair ee' uh).

The Eukarya domain contains all organisms with eukaryotic cells. From our five-kingdom classification system, then, the Eukarya domain would contain all members of kingdoms Protista, Fungi, Plantae, and Animalia. The organisms that our five-kingdom system puts in kingdom Monera would go into either the Archaea domain or the Bacteria domain, depending on certain characteristics. Those prokaryotic organisms that live in very extreme environments such as boiling hot springs or incredibly salty lakes belong in domain Archaea, while those prokaryotic organisms that live in more "normal" environments would belong in domain Bacteria.

Once you have decided the domain in which an organism should be placed, you then assign it a kingdom, phylum, class, etc. Most users of the three-domain system have only one kingdom in Archaea and only one kingdom in Bacteria, but they have many kingdoms in Eukarya. Some users of the three-domain system have kingdoms Protista, Fungi, Plantae, and Animalia in the Eukarya domain, but most split domain Eukarya into many, many different kingdoms. Let's summarize the three-domain system with a figure so that you can understand it a bit better.

Top two photos © Dennis Kunkel / Phototake   *Bottom four photos © Kathleen J. Wile*

FIGURE 1.9
The Three-Domain System

Archaea
Prokaryotic Organisms That Live in Extreme Environments

This domain holds some members of kingdom Monera. The magnified image to the left, for example, is of organisms from genus *Halobacterium*. These organisms are prokaryotic and live in very salty environments.

Bacteria
The Other Prokaryotic Organisms in Creation

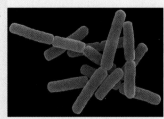

This domain holds the other members of kingdom Monera. The magnified image to the left, for example, is of *Bacillus anthracis*, which is the bacterium that causes anthrax.

Eukarya
All Organisms with Eukaryotic Cells

Members of kingdom Protista, like diatoms (magnified)   Members of kingdom Fungi, like shelf fungi   Members of kingdom Plantae, like rose bushes   Members of kingdom Animalia, like meerkats

Now you might ask yourself what the reasoning behind the three-domain system is. After all, it is similar to the five-kingdom system in that it still uses kingdom, phylum, class, etc. However, it simply adds a grouping called "domain," and depending on those who use it, the system might have several more than five kingdoms. Well, the main rationale behind the three-domain system is that those who use it believe in the hypothesis of **evolution**, which we will discuss in detail in an upcoming module. In this hypothesis, all life on earth descended from one (or a few) "simple" life form (or forms) that lived on earth billions of years ago and was (or were) formed through abiogenesis. As a result, all organisms are "related" to one another in some way, and the three-domain system tries to separate organisms based on those relationships. The Archaea are supposed to be most closely-related to the original life form or forms that were the result of abiogenesis, while the Bacteria are more distantly related, and the Eukarya are even more distantly related.

As you will learn when we study the hypothesis of evolution in depth, there is precious little evidence for such an idea and quite a bit of evidence against it. As a result, it does not make sense to us to base a classification system on such a tenuous hypothesis. Instead, it makes more sense to base our classification system on the observable similarities among organisms. This is the essence of what Carrolus Linnaeus developed in the 1700s, and it has served biology well since that time.

Since we have touched on a classification system that has been inspired by the hypothesis of evolution, we should at least mention a classification system that has been proposed by those who believe that the earth and the life on it were specially created out of nothing by God. This classification system, usually called **baraminology** (bear' uh min ol' uh jee), attempts to determine the kinds of creatures that God specifically created on earth. Indeed, the word "baraminology" comes from two Hebrew words used in Genesis: *bara*, which means "create," and *min*, which means "kind." Thus, baraminology is the study of created kinds.

Those who work with baraminology think that God created specific kinds of creatures and that He created them with the ability to adapt to their changing environment. As time went on, then, these created kinds did change within strict limits that we will discuss later on in the course. This led to a greater diversity of life on the planet than what existed right after creation. As a result, baraminologists think that all organisms we see on the planet today came from one of the many kinds of creatures that God created during the creation period discussed in the first chapter of Genesis. Baraminologists, then, try to define groupings called "baramins." Any organisms that exist within a baramin came from the same originally-created organism. For example, some baraminologists place domesticated dogs, wild dogs, and wolves into the same baramin because they believe that God created a basic kind of creature called a "dog," and the various forms of dogs and wolves that we see today are simply the result of that basic kind of creature adapting to a changing environment.

Although we think that there is a lot of evidence in favor of this new classification scheme, we still do not think that it should be used in this course. It is still relatively new and not fully developed. We doubt that it will be fully developed for many, many years to come. As a result, we think that the five-kingdom system still provides the best overall means by which to classify the organisms of God's creation, and we will limit ourselves to that system. Nevertheless, we will mention the other systems (the three-domain system and baraminology) from time to time, so it is important that you understand the basics of each.

The Microscope

We'll be revisiting classification in nearly every module, so don't worry. It won't go away. However, this brief introduction allows us to get started exploring creation. In the next two modules, we will be taking an in-depth look at kingdoms Monera and Protista. Since these kingdoms are composed of microorganisms, the labs we do in those two modules are heavily microscope-oriented. If you don't have a microscope, however, don't be concerned. We will have drawings or pictures of everything that you need to know, so a microscope isn't essential for taking this course. It does, however, help to make things clearer and more interesting. So for those who do have one, you need to perform Experiment 1.2. If you don't have a microscope, please read through the experiment so that you get a basic idea of what it covers.

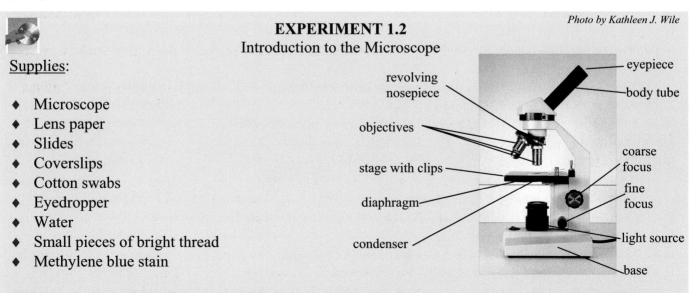

EXPERIMENT 1.2
Introduction to the Microscope

Photo by Kathleen J. Wile

Supplies:

♦ Microscope
♦ Lens paper
♦ Slides
♦ Coverslips
♦ Cotton swabs
♦ Eyedropper
♦ Water
♦ Small pieces of bright thread
♦ Methylene blue stain

revolving nosepiece
objectives
stage with clips
diaphragm
condenser

eyepiece
body tube
coarse focus
fine focus
light source
base

Object: To learn the various parts of the microscope and to learn to use the microscope properly

<u>Procedure</u>:

A. Place the microscope on your table with the arm of the microscope nearest you. With the aid of the illustration, locate all the parts of the microscope and become familiar with them.

1. The **eyepiece (called the ocular)** is what you look through. It usually contains a 10x lens.
2. The **body tube** starts at the eyepiece and runs to the part that holds the revolving nosepiece.
3. The **revolving nosepiece** is the disc that holds the lenses (which are called objectives).
4. The **coarse focus** is controlled by two large knobs on each side of the microscope. It allows for quick focus, but it does not make the image as sharp as it could be.
5. The **fine focus** knobs are used to produce sharp focus. They are usually smaller and lower than the coarse focus knobs, but in some scopes they are mounted on top of the coarse focus knobs.
6. The **arm** supports the body and stage and is attached to the base.
7. The **base** is the heavy structure at the bottom that supports the microscope and makes it steady.
8. The **stage with clips** is a platform just below the objectives and above the light source. The clips are used to hold the slide in place.
9. The **objectives** are found on the revolving nosepiece. They are metal tubes that contain lenses of varying powers, usually 4x, 10x, and 40x. Some microscopes have a 100x objective as well.
10. The **diaphragm** regulates the amount of light that passes through the specimen. It is located between the stage and the light source. It might be a disc that has several holes (a disc diaphragm), or it might be a single hole whose diameter can be varied (an iris diaphragm).
11. The **condenser** is also located between the light source and stage. It is a lens system that bends and concentrates the light coming through the specimen.
12. The **light source** is on the base and provides necessary light for the examination of specimens.

Magnification is an important feature of any microscope. In your laboratory notebook, write down the magnifications of the objectives on your microscope. You calculate the total magnification of the scope by taking the power of the ocular (usually 10x) and multiplying it by the power of each objective. Thus, if your ocular is 10x and your objectives are 4x, 10x, and 40x, your three magnifications are 40x, 100x, and 400x. Label your three magnifications as low, medium, and high.

B. Now that you are familiar with the parts of the microscope, you are ready to use it to view thread.

1. Rotate the low-power objective so that it is in line with the eyepiece. Listen for a click to make sure it is in place.
2. Turn your light on. If you have a mirror instead of a light, look through the eyepiece and adjust the mirror until you see bright light.
3. Using the coarse focus, raise the stage (or lower the body tube) until it can move no more. (*Never* force the knobs.)
4. Place a drop of water on a clean slide and add several short pieces of brightly-colored thread.
5. Add a coverslip (a thin piece of plastic or glass that will cover the water and press it against the slide). This works best if you hold the coverslip close to the drops of water and then drop it gently. If air bubbles form, tap the coverslip gently with the lead of your pencil.
6. Put the slide on the stage and clip it down, making sure the coverslip is over the hole in the stage.
7. Looking in the eyepiece, gently move the stage down (or body tube up) with the coarse focus. If you do not see anything after a couple of revolutions, move your slide a little to make sure the threads are in the center of the hole in the stage. This indicates that the threads are in the field of view.

8. Once you have the image in focus using the coarse focus, "fine tune" it with the fine focus.
9. Place the threads in the very center of the field of view by moving the slide as you look at it through the microscope. Make sure that the threads are at the center of the field, or you will lose them when you change to a higher magnification.
10. Turn the nosepiece so that the medium-power objective is in place. Until you are very familiar with any microscope, do not turn the nosepiece without checking to make sure it will not hit the slide. Always move the nosepiece slowly, making sure that it does not touch the slide in any way. A lens can easily be damaged if it hits or breaks a slide.
11. Once the medium-power objective is in place, you should use only the fine focus to make the image sharp. Once again, move the slide so that the thread is at the center of the field.
12. Again, watching to make sure you don't hit the slide, turn the nosepiece so that the high magnification objective is in place. You should use only the fine focus to refocus.
13. (Optional) If you like, repeat steps 1-12 using a strand of your own hair rather than thread.

If we wanted to look at the threads at high magnification, why didn't we just start with the high-power objective? Had we tried to bring the threads into focus under high magnification without first looking at them under low and then medium magnification, we almost certainly would have never found the threads. When you look at the slide at high magnification, you are looking at a very, very tiny portion of the slide, and it is unlikely that what you are looking for will be there. As a result, you should always start your microscope investigation with the lowest magnification and then work your way up, centering the specimen in the field of view each time before you increase magnification.

C. Now it is time to get your first look at cells! (The course website discussed in the "Student Notes" section of this book has some magnified images of cheek cells. They may be of some help to you.)

1. Collect some cheek cells by rubbing a cotton swab back and forth on the walls of your cheek inside your mouth. Use only one side of the swab.
2. Remove the swab carefully without getting a lot of saliva on it.
3. Rub the wet side of the swab on the slide. You should see a smear where you rubbed the slide.
4. If you were to look at the cells under the microscope right now, it would be hard to find them, because they are almost transparent. To help make them easier to see, you will add a dye to them. This dye is called a **stain**, and it will help contrast the cells against the light, making them much easier to see. Place a drop of methylene blue stain on the area where you placed the cells. (This stain will not come out of most fabric, so use it with care.)
5. Add the coverslip carefully.
6. Place the slide on the microscope and begin the procedure outlined in section B, looking at the cells under low, then medium, and then high magnifications. At low magnification, the cells will look like dots. Once you find some dots, center them and increase the magnification. At high magnification, you should see a dark blob (the nucleus) and a ring outlining the cell (the plasma membrane). Note the irregular shape of the cells. Draw what you see at each magnification.
7. Rinse the slides that you used in water and wipe them dry with a paper towel. Wipe the lenses of the scope with lens paper, and put everything away. Clean up any mess you made.

    Believe it or not, we are at the end of the first module. Now you need to take a look at the study guide. On a separate sheet of paper, write out all of the definitions listed in the study guide, and answer all of the questions. After you have completed the study guide, check your work with the solutions. When you are confident that you understand the material covered in the study guide, you are ready to take the test.

ANSWERS TO THE "ON YOUR OWN" PROBLEMS

1.1   a. <u>Carnivores</u> - Tigers eat only meat; thus, they are carnivores.
       b. <u>Herbivores</u> - Cows eat grass. This makes them herbivores.
       c. <u>Omnivores</u> - Humans eat plants and meat. This makes us omnivores.
       d. <u>Herbivores</u> - Sheep graze on grasses. This makes them herbivores.

1.2   a. <u>Producers</u> – Rose bushes have green stems and leaves to produce food via photosynthesis.
       b. <u>Decomposers</u> - Almost all fungi are decomposers.
       c. <u>Consumers</u> - Lions depend on other organisms for food.
       d. <u>Consumers</u> - Humans depend on other organisms for food.

1.3 These organisms reproduce <u>asexually</u>. If they reproduced sexually, the offsprings' traits would be a blend of both parents' traits. Since these offspring are identical to the organism that produced them, this must be asexual reproduction.

1.4 <u>Science cannot prove anything. The best science can say is that all known data support a given statement.</u> However, since all data come from experiments which might be flawed, there is no way that science can prove anything. If the experiments that produced the data which support a particular statement are flawed, the statement might be quite wrong.

1.5 It is a <u>hypothesis</u>. The explanation will have to be tested with a significant amount of data before it can even be considered a theory.

1.6 In a hierarchical classification scheme like ours, the further you go down the classification groups, the more similar the organisms within the groups become. This is because each group is made by splitting the previous group into smaller groups. Thus, since kingdoms are split into several phyla, we expect the organisms within the phyla to be more similar than those in the entire kingdom. Since family is several steps down from kingdom, <u>the organisms in the same family should be much more similar</u>.

1.7 Since going down the hierarchical scheme tells us that the organisms are getting more similar, going up the hierarchical should enhance the differences. Since class is one step higher than order, <u>the organisms from different classes should have more differences</u>.

1.8 <u>Protista</u> - This kingdom has the single-celled eukaryotes.

1.9 <u>Plantae</u> - Almost all autotrophs belong in this kingdom.

1.10 <u>Fungi</u> – Most decomposers are in this kingdom.

ANSWERS TO EXPERIMENT 1.1

Number	Specimen	Specimen Classification	Numbers from the Key
1.	Butterfly	K. Animalia C. Insecta P. Arthropoda O. Lepidoptera	1, 3, 5, 6, 7, 9, 14, 16, 17, 19
2.	Chipmunk	K. Animalia C. Mammalia P. Chordata O. Rodentia	1, 3, 5, 6, 22, 23, 26, 28, 29, 31, 33, 34
3.	Grapevine	K. Plantae C. Dicotyledonae P. Anthophyta O.	1, 3, 4
4.	Swan	K. Animalia C. Aves P. Chordata O.	1, 3, 5, 6, 22, 23, 26, 28
5.	Spider	K. Animalia C. Arachnida P. Arthropoda O.	1, 3, 5, 6, 7, 9, 14, 15
6.	Tiger	K. Animalia C. Mammalia P. Chordata O. Carnivora	1, 3, 5, 6, 22, 23, 26, 28, 29, 31, 32
7.	Corn	K. Plantae C. Monocotyledonae P. Anthophyta O.	1, 3, 4
8.	Fish	K. Animalia C. Osteichthyes P. Chordata O.	1, 3, 5, 6, 22, 23, 24, 25
9.	Paramecium	K. Protista C. P. O.	1, 2
10.	Mushroom	K. Fungi C. P. O.	1, 3, 5
11.	Frog	K. Animalia C. Amphibia P. Chordata O. Anura	1, 3, 5, 6, 22, 23, 26, 27
12.	Bacterium	K. Monera C. P. O.	1, 2
13.	Bison	K. Animalia C. Mammalia P. Chordata O. Artiodactyla	1, 3, 5, 6, 22, 23, 26, 28, 29, 30
14.	Grasshopper	K. Animalia C. Insecta P. Arthropoda O. Orthoptera	1, 3, 5, 6, 7, 9, 14, 16, 17, 19, 20
15.	Baboon	K. Animalia C. Mammalia P. Chordata O. Primates	1, 3, 5, 6, 22, 23, 26, 28, 29, 31, 33, 35

STUDY GUIDE FOR MODULE #1

1. On a separate sheet of paper, write down the definitions for the following terms. You will be expected to have them memorized for the test!

a. Metabolism
b. Anabolism
c. Catabolism
d. Photosynthesis
e. Herbivores
f. Carnivores
g. Omnivores
h. Producers
i. Consumers
j. Decomposers
k. Autotrophs
l. Heterotrophs
m. Receptors
n. Asexual reproduction

o. Sexual reproduction
p. Inheritance
q. Mutation
r. Hypothesis
s. Theory
t. Scientific law
u. Microorganisms
v. Abiogenesis
w. Prokaryotic cell
x. Eukaryotic cell
y. Species
z. Taxonomy
aa. Binomial nomenclature

2. What are the four criteria for life?

3. An organism is classified as a carnivore. Is it a heterotroph or an autotroph? Is it a producer, consumer, or decomposer?

4. An organism has receptors on tentacles that come out of its head. If those tentacles were cut off in an accident, what life function would be most hampered?

5. A parent and two offspring are studied. Although there are many similarities between the parent and the offspring, there are also some differences. Do these organisms reproduce sexually or asexually?

6. What is wrong with the following statement?

"Science has proven that energy must always be conserved."

7. Briefly explain the scientific method.

8. Why does the story of spontaneous generation illustrate the limitations of science?

9. Where does the wise person place his or her faith: science or the Bible?

10. Why is the theory of abiogenesis just another example of the idea of spontaneous generation?

11. Name the classification groups in our hierarchical classification scheme in order.

12. An organism is a multicellular consumer made of eukaryotic cells. To what kingdom does it belong?

13. If we were using the three-domain system of classification, in which domain would the organism in question #12 belong?

14. An organism is a single-celled consumer made of prokaryotic cells. To what kingdom does it belong?

15. If we were using the three-domain system of classification, could you determine the domain of the organism in question #14? If so, give the domain. If not, give the possible domains in which it could be placed.

16. Use the biological key in the appendix to classify the organisms pictured below:

a.

Owl
Photo © Comstock, Inc.

b.

Fly
Photo © Jason Ng

Note: Since the study guide specifically tells you that you can use the biological key to classify the creatures shown above, you know that if such a question is asked on the test, you will be able to use the biological key on the test as well.   This is how you can use the study guide to determine what you must memorize and what you will be able to reference during the test. Had we asked you to classify these creatures without telling you to use the biological key, you would have known that you would be required to memorize the biological key for the test.

MODULE #2: Kingdom Monera

Introduction

With this module, you will begin to explore the incredible microscopic world that exists all around us. This world, unknown to human beings until the 1670s when Anton van Leeuwenhoek crafted a crude microscope, is home to an incredibly large number of microorganisms. Amazingly enough, the combined weight of all microscopic organisms far exceeds the combined weight of all other living organisms on earth! Bringing this fact a little closer to home, the number of organisms from kingdom Monera that live in your body exceeds the number of cells that make up your body! Thus, even though microorganisms are small, they are an important part of life on earth.

In this module, we will concentrate on the organisms that make up kingdom **Monera**, which some biologists have renamed **Prokaryota** (pro kehr ee aht' uh) . Since, as we learned in Module #1, members of kingdom Monera are all composed of prokaryotic cells, this name does make sense. Nevertheless, in this course, we will still refer to the kingdom of prokaryotic cells by its traditional name, Monera.

Organisms in kingdom Monera are interesting on many different levels. First of all, these organisms are not well understood by biologists. There are several facets of their structure and function that we simply do not understand. Second, some of these organisms can survive in habitats that are deadly to other organisms. For example, some members of kingdom Monera live on dust particles floating 6 kilometers (20,000 ft.) above the surface of the earth. Others thrive and multiply in temperatures that are too extreme for any other organism, and some survive in the presence of radiation that would kill most other organisms. In fact, microbiologists have found certain organisms from kingdom Monera living in the water core of a nuclear reactor! Clearly, these interesting organisms are worth studying.

Bacteria

The general name **bacteria** (singular is bacterium) can essentially be applied to all of the organisms in kingdom Monera. When you hear this term, you probably think about the disease and suffering caused by bacteria. Although bacteria are responsible for many illnesses that plague humanity (and other organisms), there are also many forms of bacteria that are beneficial to people. For example, there are bacteria in your gut that synthesize B vitamins and vitamin K, which your body uses to stay healthy. To make the distinction between harmful and beneficial bacteria, we use the term **pathogen** (path' uh jen) to indicate those bacteria that are harmful.

Pathogen – An organism that causes disease

Thus, a "pathogenic bacterium" is a bacterium that causes disease. It is important to realize, however, that there are many, many forms of bacteria which are not only non-pathogenic, but are, in fact, quite beneficial.

Bacteria help us make cheese. When starter bacteria are added to milk, they feed on the milk and produce a chemical called lactic acid. This begins to separate the milk solids (called "curds") from the liquids in the milk (called "whey"). The whey is then drained off, and the curds are used to make the cheese. Depending on the starter bacteria used, the time and temperature over which they act, and how you process the curds, you get different kinds of cheese. Bacteria from species *Lactococcus* (lak' tuh koh kus) *lactis* (lak' tis), for example, can produce cheeses like Colby or cheddar, depending on the time and temperature over which they act. If you change the starter species to *Lactobacillus* (lak' toh buh sil' us) *helveticus* (hel' veh tih kus), you can make cheeses like mozzarella and provolone. Bacteria not only help us make food, but they also help make the earth hospitable to life. Cyanobacteria, for example, use photosynthesis to make their own food, and that gives us oxygen to breathe. In addition, many bacteria are decomposers, performing the necessary job of recycling chemicals from dead organisms so that living organisms can use them.

Since bacteria belong to kingdom Monera, they are prokaryotic. Their cells have no distinct, membrane-bounded organelles. Nevertheless, the cell of a bacterium does have a few recognizable components. A sketch of a "typical" bacterium is given in the figure below.

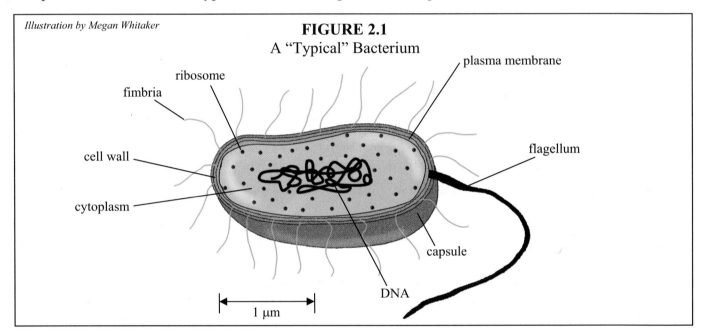

Illustration by Megan Whitaker

FIGURE 2.1
A "Typical" Bacterium

The word "typical" is in quotes because bacteria are so diverse that no sketch could be made which represents all bacteria. Thus, as you read through this discussion, you will find a lot of exceptions to this "typical" bacterium. Nevertheless, it is worthwhile to try to develop some general characteristics for bacteria, so we will discuss this figure in depth.

Some bacteria have a **capsule** that surrounds the cell wall. This capsule is composed of an organized layer of sticky sugars that help the bacteria adhere to surfaces. For example, the bacteria that cause tooth decay are able to cling to teeth using their sticky capsules. The capsule usually does more than just that, however. In general, the capsule is the bacterium's protective layer which tends to deter infection-fighting agents. When a pathogenic bacterium enters an organism, that organism has an infection-fighting mechanism which activates. The capsule of the pathogenic bacterium, however, can offer a significant amount of protection from the agents of this infection-fighting mechanism.

Most bacteria also have a **cell wall**. As you will learn later, the absence or presence of a cell wall as well as the composition of the cell wall (if it exists) are used to classify bacteria. The cell wall holds the contents of the bacterium together, regulates the amount of water that a bacterium can absorb, and holds the cell into one of three basic shapes, as illustrated in Figure 2.2 on the next page.

FIGURE 2.2
Bacterial Shapes

SEM images © Dennis Kunkel/Phototake

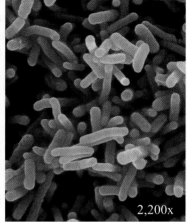
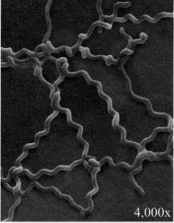

2,300x   2,200x   4,000x

Spherical bacteria are known as **coccus** (kahk' us), the plural of which is cocci.

Rod-shaped bacteria are known as **bacillus** (ba sil' us), the plural of which is bacilli.

Helical bacteria are known as **spirillum** (spuh ril' um), the plural of which is spirilla.

These images are made with a scanning electron microscope, which is much more powerful than the microscope that you are using in this course. Also, since an electron microscope uses electron beams instead of light, the images that it produces do not have any color. The colors you see in the images above have been added after the image was acquired and do not reflect reality. They are simply there for contrast. As a result, images like these are sometimes called **false-color images**.

Underneath the bacterium's cell wall is the **plasma membrane**. As we will learn in Module #6, this membrane is made up of certain chemicals called phospholipids and other chemicals called proteins. These chemicals regulate what the bacterium takes in from the outside world. This is truly a remarkable piece of chemical engineering! The plasma membrane protects the interior of the cell by sensing the nature of the chemicals in the surroundings and then determining whether or not they can enter the cell. In some cases, if the cell needs the chemicals, the plasma membrane will actually *force* the chemicals into the cell, whether they "want" to come in or not!

Inside the plasma membrane, we find a semifluid substance known as **cytoplasm** (sy' tuh plaz um). Cytoplasm exists throughout the interior of the cell, supporting the **DNA** and the **ribosomes** (rye' buh sohms). As we already learned, DNA holds all of the information required to make this mass of chemicals a living entity. Ribosomes, on the other hand, are chemical factories. They make special chemicals known as proteins, which we will take an in-depth look at when we reach Module #5. Positioned throughout the cytoplasm are thousands and thousands of different types of chemicals which aid these parts of the cell in their tasks.

Outside of the cell wall and capsule, many bacteria (but not all) have fibrous bristles called **fimbriae** (fim' bree ay) (singular is fimbria). Although these might look like legs or paddles, they are not used to move the bacterium about. Instead, fimbriae are typically used for grasping. Either they grasp surfaces to help the bacteria adhere to them (as an aid to the sticky capsule), or they grasp other bacteria as a part of reproduction. If they are used in reproduction, they are called **sex pili**.

Locomotion (moving the bacterium from one place to another) is accomplished with the **flagellum** (fluh jel' um), the plural of which is flagella. Not all bacteria have flagella, and some have more than one. If a bacterium does not possess a flagellum, it is not capable of locomotion. The motion of substances in its surroundings might push it back and forth, but it is unable to move towards a goal. Now when you look at this flagellum, it looks a lot like a tail. Thus, you might think that the bacterium moves by swishing the flagellum back and forth. That is not correct, however. The flagellum is much more efficient than that.

The flagellum is truly a remarkable feat of engineering. It is composed of three parts: the **filament**, the **hook**, and the **basal body**. The filament (which is all that you see in Figure 2.1) attaches to the hook, a small tube shaped like an "L." The hook slides onto a rod that sticks through the cell wall. This rod, and all of the structures that attach it to the cell wall, make up the basal body. The rod and hook fit together so smoothly that the hook can actually spin around the rod in circles. When the bacterium wants to move, a series of amazingly-complex chemical reactions make the hook spin. Since the filament is attached to the hook, it begins to spin as well. In other words, the bacterium has its own "outboard motor" that it uses to get from one point to another.

Although a discussion of the details behind the bacterium's "outboard motor" is beyond the scope of this course, I want to give you an idea of just how complex it is. First, we do not really know *how* it works. Several models have been proposed to explain the workings of the bacterial flagellum, but none of them is completely consistent with the facts that we know. Second, what we do know is that the bacterial flagellum's system is incredibly complex. Consider the words of one noted biochemist: "…as biochemists have begun to examine apparently simply structures like…flagella, they have discovered staggering complexity, with dozens if not hundreds of precisely tailored parts" (Michael J. Behe, *Darwin's Black Box*, [New York, NY: Simon and Shuster, 1996], 73). If you would like to learn more about the marvelous design of the bacterial flagellum, you will find some articles about it on the course website which we mentioned in the "Student Notes" section at the beginning of this book.

Now before we go any further, we want you to go back and take a look at the length marked off in Figure 2.1. Basically, the figure indicates that 1 μm is about half the length of a bacterium. Of course, this little bit of information does you no good if you don't know what a "μm" is! The symbol "μm" stands for "micrometer," a metric length measurement. If you aren't familiar with the metric system, you will get a thorough introduction to it when you reach chemistry. For right now, you just need to know that a meter is about 3 feet in length, and a micrometer is one millionth of that size. Thus, bacteria are *really* small. In fact, to give you an idea of exactly how small bacteria are, 1,000 of them would fit in the dot of this "i." Now *that's* small!

Let all of this information sink in for just a moment. In the last couple of pages, we have discussed some pretty complicated stuff. Bacteria have capsules that protect the cell and, along with the help of fimbriae, allow it to attach to surfaces. They have a cell wall that keeps the shape of the bacterium. Inside the cell wall there is a plasma membrane that regulates the passage of substances in and out of the cytoplasm. They have DNA which stores all of the information needed for the cell to be alive, and they have the ability to copy this DNA in a mere 20 minutes! They also have ribosomes which can produce complicated chemicals called proteins. Finally, they have an incredibly advanced system of locomotion which, in fact, strongly resembles the design of an outboard motor. All of these things are contained in a cell that is barely 2 μm long!

Now here's the really amazing part: *Bacteria represent the "simplest" form of life on earth!* Think about it. This tiny, complex organism is as simple as life gets. Are you beginning to see why the whole idea of spontaneous generation is so absurd? Whether we are talking about mice forming from sweaty shirts and grain or a "simple" life form being generated by random chemical reactions, life is just too complicated to form itself spontaneously. Even the simplest life form has obviously been designed by an incredibly intelligent Designer. Think about it this way. It took human beings nearly 3,000 years of scientific inquiry to develop the technology that could make a motorized propeller that moves a boat. The simplest life form on the planet has a much more efficiently-designed propeller as a part of its "standard equipment." Clearly, life is the result of design. To believers, of course, this is rather obvious. Since God is all-powerful and all-knowing, it only makes sense that He could design and create some incredibly complicated life forms. What is really amazing is that scientists can study these life forms all of their lives and not reach such an obvious conclusion!

ON YOUR OWN

2.1 Two different species of bacteria attempt to infect an organism. One bacterium succeeds, while the other is destroyed by the organism's infection-fighting mechanisms. What is most likely the major difference between these two bacteria?

2.2 A bacterium is poisoned by a substance that is allowed into the interior of the cell. What bacterial component did not do its job?

2.3 If a bacterium cannot move, what structure is it missing?

The Eating Habits of Bacteria

As we already mentioned, there is really no such thing as a "typical" bacterium. From a cellular point of view, kingdom Monera is probably the most diverse kingdom in creation. This is best illustrated by the diversity of eating habits among bacteria. Remember from Module #1 that all life must have a means of converting energy from its surroundings into energy that will sustain the processes necessary for life. Some organisms do this by producing their own food (we call them producers); some do it by eating other organisms (we call them consumers); and some do it by decomposing dead organisms (we call them decomposers). It turns out that there are producers, consumers, and decomposers in kingdom Monera.

Most bacteria are decomposers. Since decomposers rely on other (dead) organisms for food, we can also say that they are heterotrophs. These bacteria are referred to as **saprophytes** (sap' roh fytes).

Saprophyte – An organism that feeds on dead matter

Saprophytic bacteria are an integral part of nearly every aspect of nature. Without the aid of these microscopic decomposers, many of the materials on earth that are necessary for life would be contained in the bodies of only a few generations of dead organisms. Since bacteria decompose dead organisms, however, these materials are recycled back into creation so living organisms can use them.

Not all heterotrophic bacteria are saprophytic, however. Some bacteria (the ones that cause disease, for example) are **parasites** (pehr' uh sytes).

Parasite – An organism that feeds on a living host

Many parasitic bacteria lack the ability to digest nutrients, so they need to absorb nutrients that have already been digested. In addition, they often lack the ability to manufacture the complex chemicals necessary for life. As a result, they must also absorb those chemicals from their host.

Even though most bacteria are heterotrophic, there are many forms of autotrophic bacteria as well. In general, there are two different means by which autotrophic bacteria manufacture their own food: **photosynthesis** or **chemosynthesis**. In Module #1 you learned that photosynthesis uses the energy from sunlight along with certain chemicals to make food for the autotrophic organism. In green plants and most of the other photosynthetic organisms, the byproduct of photosynthesis is oxygen. The photosynthesis that occurs in some bacteria, however, does not have oxygen as a byproduct. This is because the chemicals that those bacteria use in photosynthesis are quite different from the chemicals used in the photosynthesis that is carried out by plants. Nevertheless, the byproducts of photosynthesis from these bacteria are useful to certain organisms. You'll learn more about the details of photosynthesis in Module #5.

Chemosynthetic bacteria use a different process for the manufacture of their food. The main difference between that process and photosynthesis is the source of energy. In chemosynthesis, rather than using energy that comes from sunlight, the bacteria promote chemical reactions which release energy. The bacteria then use that energy, along with other chemicals, to manufacture their food. Usually, the chemical reactions that provide energy to chemosynthetic bacteria also convert chemicals that living organisms can't use into chemicals that certain living organisms can use. Thus, even though there are only a few types of chemosynthetic bacteria, they perform an essential function for many living organisms.

Of course, once a bacterium (or any living organism) has food, there must be some process by which it can then convert that food into energy to be used to support life. In human beings, for example, we break down our food into small molecules that our bodies can absorb. That's called **digestion**. However, digestion does not give us energy. It simply breaks food down into smaller molecules. In order to get energy, we must then combine these molecules with oxygen in a long, complicated chemical process that you will learn about later. In fact, that's why we breathe. We breathe so that we get the oxygen we need in order perform this complicated chemical process.

Since we use oxygen in order to get energy from our food, humans are called **aerobic** (ehr oh' bik) **organisms**.

Aerobic organism – An organism that requires oxygen

Most of the living things with which you are familiar are aerobic organisms. This is why the earth must have a plentiful supply of oxygen in order to support life. There can be food aplenty, but without oxygen, the aerobic organisms would not be able to effectively convert it into useable energy.

There are certain types of bacteria, however, that do not require oxygen in order to convert their food into energy. These bacteria are **anaerobic** (an uh roh' bik) **organisms**.

<u>Anaerobic organism</u> – An organism that does not require oxygen

Typically, they live in areas that are barren of oxygen such as deep underground or in the muck at the bottom of a swamp. These bacteria are essential to life, because they either decompose dead organisms or convert useless chemicals into chemicals that can be used by other life forms. So even though anaerobic organisms are probably unfamiliar to you, they are another essential part of the system that the Creator designed to support life.

Before we leave this section, we want you to stop for a moment and think about what you have just learned. Bacteria have many different ways of getting their food, but they all, in some way, promote life on earth. Without the saprophytic bacteria, the chemicals necessary for life would not be recycled from dead organisms to live ones. As a result, in just a few generations, it would be impossible for life to exist on the planet. Photosynthetic (autotrophic) bacteria, on the other hand, produce certain useful chemicals that no other process on earth can produce. In addition, chemosynthetic (autotrophic) bacteria are adept at converting chemicals that are useless to living organisms into substances which are essential for life. Even pathogenic bacteria have a role in supporting life. Often, a group of organisms gets too large and could destroy the balance of nature. Typically, disease caused by pathogenic bacteria will kill off a large number of the organisms, making sure that the balance of nature remains intact.

Now if you think about it, these kinds of processes need to occur everywhere on earth. For example, when organisms die in a swamp, they might sink to the bottom of the swamp and slowly be engulfed by the muck at the bottom. If there were no decomposing bacteria in that muck, the chemicals in that dead organism would never be recycled back into creation. Because certain bacteria can be anaerobic, even though there is no oxygen down there, there are bacteria that will decompose the dead organism. Thus, the diverse food-gathering processes present in bacteria and the fact that some bacteria do not require oxygen assure that there are bacteria everywhere doing their jobs.

Isn't that amazing? Bacteria are needed everywhere to do many different jobs in order to make sure that life can continue on earth. As a result, bacteria have different means by which they get their food, and some don't even need oxygen in order to live. This ensures that they can be everywhere they are needed. Can you think of a more well-designed system for the support of life? Suppose a team of brilliant scientists spent their careers trying to devise a system that recycles the materials necessary for life, manufactures the chemicals essential for life, and assists other life forms in digestion. Do you think that they could come up with a plan this elegant and this well designed? Of course not!

Indeed, a team of scientists tried to design and build a self-contained system for supporting life. It was called "Biosphere 2," and it was supposed to be a microcosm of life on earth. It contained a variety of animals and plants and was designed to be completely self-supporting. The scientists thought they had it all worked out. They spent 7 years and $200 million designing and building this airtight, enclosed facility that spans 3.15 acres in Arizona. Despite the best that technology and science had to offer, Biosphere 2 could not support life for even two years! After about one year and four months, oxygen levels could not be maintained. They had to start pumping oxygen in from the outside. Many of the animal species that had been put in Biosphere 2 became extinct, while the population of others boomed. In the end, Biosphere 2 was a failure. At least part of the reason behind this failure is that the scientists who designed Biosphere 2 did not take into account the incredibly valuable role that bacteria and other microscopic organisms play in creation. Since these microorganisms were not present in the right amounts, Biosphere 2 could not maintain itself.

Of course, none of this should be surprising to anyone. The planet earth is an intricate web of hundreds of millions of processes that work together to support life. The best talent and technology that humans have could not possibly mimic what the earth does naturally. Why? The answer is very simple. An omniscient, omnipotent God designed the earth. He could foresee all of life's needs, even the tiny bacteria needed to support it. Limited human beings, on the other hand, just do not have the ability to design and create what God has designed and created, even on a very small scale. Biosphere 2 was a failure, and it stands in stark contrast to the grandeur and elegance of God's creation!

FIGURE 2.3
Biosphere 2 in Arizona

Photo from www.clipart.com

Although it was a failure, Biosphere 2 is now a hands-on, interactive science museum. The course website mentioned in the "Student Notes" section of the book has a link to the Biosphere 2 website.

ON YOUR OWN

2.4 Can saprophytic bacteria be autotrophic?

2.5 Can an aerobic bacterium be chemosynthetic?

<u>Asexual Reproduction in Bacteria</u>

As you should recall from Module #1, another criterion for life is reproduction. For an organism to be considered alive, it must be able to reproduce so that life can continue. We also learned in Module #1 that there are two modes of reproduction: sexual and asexual. In order to produce offspring, however, bacteria can only reproduce asexually. Although there are several ways in which this asexual reproduction can occur, we will concentrate on the most popular form, which is often called **binary fission**. Figure 2.4 on the next page illustrates this form of asexual reproduction.

FIGURE 2.4

Illustration by Megan Whitaker

Asexual Reproduction in Bacteria

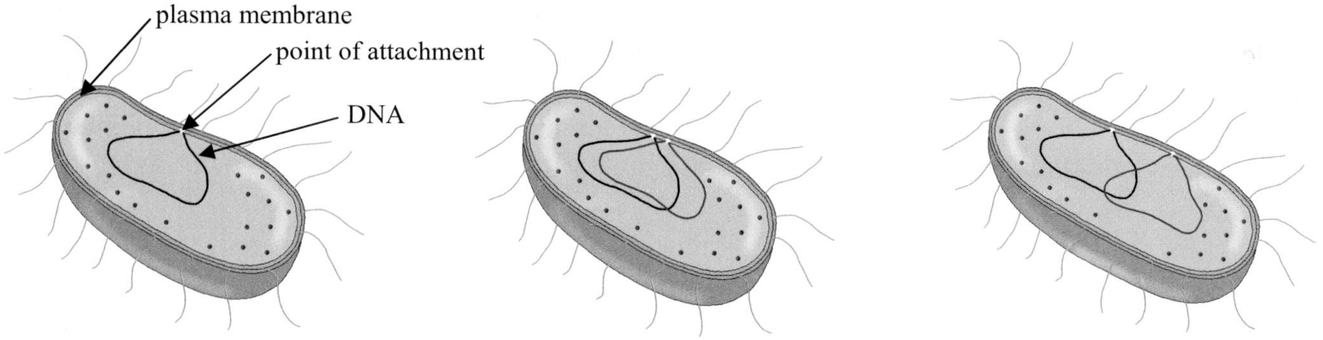

plasma membrane
point of attachment
DNA

a. When a bacterium reproduces, the DNA loop is attached to a point on the plasma membrane.

b. The DNA is then copied, and the copy is attached to the plasma membrane close to the place where the original is attached.

c. After the DNA copy is made, the cell wall elongates, separating the two strands of DNA.

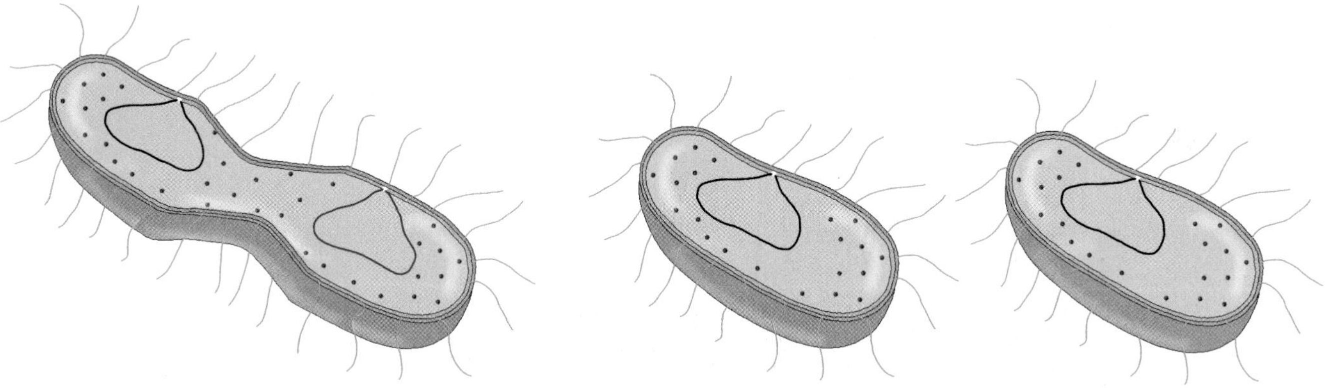

d. Once the two strands of DNA are separated, new cell wall material and plasma membrane begin to grow, closing the two strands off from each other.

e. When the cell wall and plasma membrane growth is complete, two bacteria (a parent and an offspring) exist where there was once just one. As long as no mutation took place, the two bacteria are identical.

In the figure, we have removed some of the features of a typical bacterium so that we can concentrate on those parts that are essential to reproduction. As you can see from the figure, reproduction begins when the DNA forms a loop attached to the plasma membrane. The DNA then copies itself. It should make sense to you that in order to reproduce, the bacterium must create new DNA. After all, DNA is required for life. Since reproduction forms a new living organism, new DNA must be formed.

Once the DNA is copied, the copy attaches to the plasma membrane close to the original. The cell wall and plasma membrane then elongate, separating the copy from the original. Once they are separated, the cell wall and plasma membrane begin to grow. They grow in between the two strands of DNA, eventually closing them off from one another. At that point, there are two bacteria. As we learned in Module #1, asexual reproduction usually results in offspring that are identical to the parents. Now you should see why. The DNA in the offspring (sometimes called the "daughter") is the same as the DNA in the parent. Since DNA holds all of the information that makes up an organism, the two bacteria must be identical. If, however, a mistake was made while the DNA was being copied, there will be a marked difference between the offspring and the parent. This is what we called a "mutation" in Module #1. Thus, mutations in asexual reproduction result when the DNA of the parent is not copied correctly.

One of the interesting aspects of bacterial reproduction is the speed at which it occurs. Under ideal conditions, a bacterium can divide in about 20 minutes. Of course, once it divides, the new bacterium *and the old one* can divide again in about 20 minutes. If this process continued indefinitely, one bacterium could multiply into more than a billion bacteria in about 10 hours. If this continued for a whole week, the bacteria formed from that first, single bacterium would have a combined weight that is larger than the entire planet! A situation like this can never occur, however, because of several factors.

Suppose, for example, a saprophytic bacterium found its way to a dead cow. The bacterium would feed on the cow and begin to reproduce rapidly. In just a day or so, there would be billions and billions of them. They would begin to spread out across the remains of the cow, so that they could all have access to the nutrients that they need. At some point, however, bacteria would cover all of the remains of the cow. What then? Well, some bacteria would be cut off from their food supply, and they would die, decreasing the population. Some bacteria would actually try to eat other bacteria, which would further decrease the population. Now, of course, while all this was going on, bacteria would still be reproducing, so new bacteria would be forming and other bacteria would be dying. For a while, the bacteria population would reach a point where the number of new bacteria that form equaled the number of bacteria that die. We call that the **steady state** of the population.

Steady state – A state in which members of a population die as quickly as new members are born

The population of bacteria would reach a steady state for a while, but then something would happen. The cow remains would start to dwindle due to all of the bacteria feeding on them. At some point, the remains of the cow would not support the population, and more bacteria would begin to starve to death. In addition, the incidents of bacteria eating other bacteria would increase. Thus, the population of bacteria would dwindle along with the remains of the cow. Eventually, when all of the cow remains were gone, all of the bacteria would be dead. This situation is illustrated in Figure 2.5.

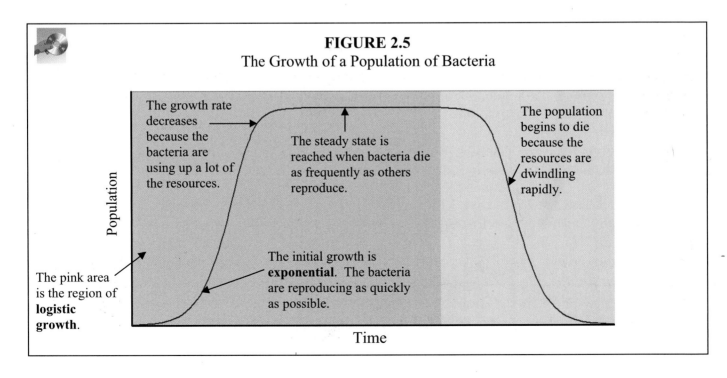

FIGURE 2.5
The Growth of a Population of Bacteria

The growth rate decreases because the bacteria are using up a lot of the resources.

The steady state is reached when bacteria die as frequently as others reproduce.

The population begins to die because the resources are dwindling rapidly.

The initial growth is **exponential**. The bacteria are reproducing as quickly as possible.

The pink area is the region of **logistic growth**.

Population

Time

So, even though bacteria can populate an area quickly, a limitation of resources will eventually slow the growth of the population until, when all resources run out, the population dies.

Before we leave this discussion, we want to point out a few things about the graph in Figure 2.5. During the initial stage of population growth, the bacteria are reproducing unchecked. There are plenty of resources for the population, so the population grows as quickly as the bacteria can reproduce. This is called **exponential growth**.

Exponential growth – Population growth that is unhindered because of the abundance of resources for an ever-increasing population

However, as the population begins to reach the limits of the environment's resources, it can no longer grow unchecked. The population growth is controlled by the limited resources of the environment. This is referred to as **logistic growth**.

Logistic growth – Population growth that is controlled by limited resources

If you look at Figure 2.5 again, you will see that there is a region that is highlighted in pink. That is the region over which logistic growth occurs. When you graph a population that experiences logistic growth, you get the S-shaped curve (often called a **sigmoidal** [sig moy' duhl] **curve**) that is shown in the pink section of the graph.

Under certain circumstances, a population might never experience a decline. If the environment continually renews its resources, the population might stay at its steady state indefinitely. In our example, suppose a small amount of dead cow was added periodically to replace the amount of dead cow that had been consumed by the bacteria. If that were to happen, the bacteria would never run out of resources, so the population could continue at its steady state. This can occur in nature. Consider, for example, a population of herbivores. They would eat the plants in their environment, but as long as enough new plants grew to replace the old plants, the population would not experience a decline in food resources, and it might continue at its steady state for quite some time.

ON YOUR OWN

2.6 A population of bacteria grown from a single "starter" bacterium is rather fragile. When conditions are changed, the population dies quickly. Based on what you have just learned, develop a hypothesis for why this is the case.

2.7 A population of bacteria reaches a steady state and then, after several days, the population actually increases dramatically. What could cause such an event?

Genetic Recombination in Bacteria

When bacteria reproduce asexually, the offspring is a genetic copy of the parent. However, it is often beneficial for bacteria to exchange genetic information in order to increase the genetic diversity of the population. Bacteria do this by **genetic recombination**, which can occur in one of three ways: **conjugation**, **transformation**, or **transduction**. Let's start by discussing conjugation.

<u>Conjugation</u> – A temporary union of two organisms for the purpose of DNA transfer

Conjugation is different from sexual reproduction. When two bacteria conjugate, no offspring is formed. Instead genetic information is transferred from a donor bacterium to a recipient.

How does this strange process occur? Well, in a population of bacteria, there may be a few individuals with traits that are different from the traits of the others. If these traits are desirable, it would benefit the population if they were passed on to its other members. In order to do this, conjugation can occur. The desirable traits that are going to be passed on are usually contained in a small, circular strand of DNA called a **plasmid**. The plasmid is an extra component, separate from the rest of the bacterium's DNA.

<u>Plasmid</u> – A small, circular section of extra DNA that confers one or more traits to a bacterium and can be reproduced separately from the main bacterial genetic code

If the traits conferred to the bacterium are beneficial, that bacterium (usually referred to as the "donor") will conjugate with a bacterium that does not possess these traits (usually referred to as the "recipient"). The mechanism of this form of reproduction is illustrated in the figure below.

FIGURE 2.6
Conjugation in Bacteria

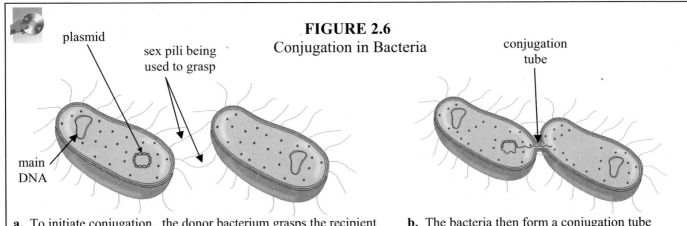

a. To initiate conjugation, the donor bacterium grasps the recipient with its pili. **NOTE:** The plasmid's size is exaggerated for the illustration. It is significantly smaller than what is shown here.

b. The bacteria then form a conjugation tube between them, and the donor sends one strand of the plasmid to the recipient.

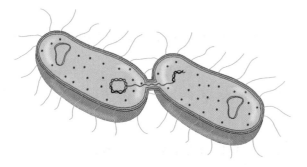

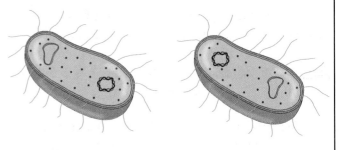

c. The plasmid strand travels through the conjugation tube, and both bacteria use the one plasmid strand to make a second plasmid strand. As you will learn later on, DNA is composed of two strands, but if an organism has one strand, it can produce the other strand using the first as a template.

d. The conjugation tube then collapses. Now the recipient has the same plasmid that the donor had.

Illustration by Megan Whitaker

The first step in conjugation among bacteria involves the donor grasping the recipient with its sex pili. Remember, "sex pili" is the term we use for fimbriae that are used in reproduction. Research indicates that the donor bacterium is the one that initiates conjugation. However, we are presently unsure of how it "knows" which bacterium to conjugate with. Once the donor grasps the recipient, a **conjugation tube** is formed between the bacteria. At that point, the donor bacterium transfers one of the plasmid's strands to the recipient.

As you will learn in Module #5, DNA is made up of two strands that are joined in the shape of a double helix. Thus, the plasmid that the donor has is composed of two strands. It turns out that due to the chemical nature of DNA, if an organism has one strand, it can construct the other strand using the first as a template. Thus, as the strand is being transferred, the donor bacterium begins making a second strand for the plasmid to replace the strand that it is donating. As the recipient bacterium gets the new strand, it also constructs a second strand so that it will have a complete plasmid as well. When the plasmid strand has been transferred, the conjugation tube collapses, and each bacterium finishes constructing the plasmid strand that it was making. In the end, then, there are two bacteria, each with the plasmid.

Although this process might look something like sexual reproduction, it not like the sexual reproduction found among the rest of the organisms in creation. First, no offspring are formed. Second, consider the genetics involved. In most cases of sexual reproduction, each parent contributes DNA, and the offspring's DNA is a blend of parents' DNA. Thus, each parent contributes to the new DNA being formed. In bacterial conjugation, this is not the case. No DNA travels from the recipient to the donor. Thus, the traits conferred by the donor's plasmid are given to the recipient, and the recipient gives nothing back. In conjugation among bacteria, then, only one organism donates DNA.

ON YOUR OWN

2.8 A population of bacteria are living in a lake. Due to volcanic activity nearby, the lake's temperature begins to increase. In the population, there are some bacteria that are resistant to low temperatures (call them type A) and another type that are resistant to high temperature (call them type B). Which type will be the donor and which the recipient as the population begins to conjugate?

Transformation and Transduction

If conjugation wasn't strange enough for you, consider another means by which bacteria can diversify their genetic code: **transformation**.

Transformation – The transfer of a DNA segment from a nonfunctional donor cell to that of a functional recipient cell

If a bacterium dies, its cell wall falls apart and the components of the cell (including the DNA) flow into the surroundings. By yet another process that we do not fully understand, the dead bacterium's DNA seems to break into small pieces, and a living bacterium might "sense" that one or more of those pieces contains a trait or traits that it could use. The living bacterium then absorbs what it needs, incorporating the new genetic information as a plasmid. Of course, once a bacterium has such a plasmid, it might then engage in conjugation to pass the plasmid on to other bacteria.

The last method of genetic recombination among bacteria is transduction, which occurs with help from an outside source.

Transduction – The process in which infection by a virus results in DNA being transferred from one
    bacterium to another

You will learn more about viruses in a later module, so we don't want to spend much time on this now. Just realize that a virus can pick up DNA from a bacterium during infection and that DNA can be inserted into another bacterium in a subsequent infection.

Endospore Formation

Suppose you have a sample of water that is filled with bacteria. You boil the water to try to kill the bacteria. Will that get rid of the population? Not necessarily. Do you remember our discussion of spontaneous generation from Module #1? John Needham did experiments that seemed to provide evidence for spontaneous generation, but one problem that Pasteur noted was that Needham did not boil his infusions long enough. Although boiling water will kill the bacteria, they can actually survive the high temperature for some period of time by producing an **endospore**.

Endospore – The DNA and other essential parts of a bacterium coated with several hard layers

The endospore is formed *inside* the bacterium's plasma membrane. Once an endospore forms, the bacterium itself might die, but as long as the endospore survives, the bacterium can reform.

An endospore can withstand extreme situations for a lot longer than the bacterium itself. For example, endospores can withstand boiling or freezing for a while. In addition, they can withstand up to a few days of extreme dryness, pressure, or exposure to toxic chemicals. If the endospore can survive through the extreme situation, when favorable conditions return, the hard layers surrounding the DNA will deteriorate, and the cell will burst from the endospore, ready to grow and reproduce again. Once again, our understanding of the process by which bacteria know when and how to form endospores is still limited. Much exciting research is being done in these areas.

If you were following the news in late 2001, you probably heard about anthrax spores that were sent to political leaders through the mail. *Bacillus anthracis* is a bacterium that is often called "anthrax," because that's the disease it causes. The spores that were sent through the mail were endospores that this bacterium had formed in order to survive harsh conditions (most likely a lack of water). The spores could survive a trip through the mail, and if they could get into a person's body, they could form the actual bacterium again, causing anthrax.

Bacterial Colonies

Up to this point, our discussion has focused on bacteria as individuals. However, many bacteria exist in colonies. Remember in Module #1 when we discussed a "simple association" of single-celled creatures? A colony is an example of such a simple association. In a bacterial colony, the individual bacteria group together, but they all still exist as individuals. If the colony gets broken apart, the individual bacteria can still live and function. Thus, they do not *have* to be a part of the colony to survive. Nevertheless, since there is strength in numbers, the bacteria's ability to survive is usually enhanced when they form a colony.

As you might expect, bacteria form colonies in a variety of different ways. In our discussion of a "typical" bacterium, we learned that bacteria take on one of three basic shapes: coccus (spherical), bacillus (rod-shaped), or spirillum (helical). The most common bacterial colonies are made up of either cocci or bacilli bacteria. Figure 2.7 shows the most common forms of bacterial colonies that exist.

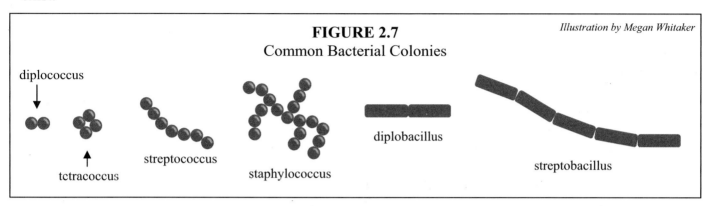

FIGURE 2.7
Common Bacterial Colonies

Illustration by Megan Whitaker

diplococcus

tetracoccus

streptococcus

staphylococcus

diplobacillus

streptobacillus

Notice that the shape of the bacteria is contained in the name. If the bacteria are spherical, the name ends in "coccus." If they are rod-shaped, the name ends in "bacillus." This is often the case. Bacteria names often contain the bacterium shape so that you automatically know something about the organism when you see its name. Although we do not require you to memorize the names in Figure 2.7, we do expect you to be able to determine the shape of a bacterium when it is a part of the name.

As we mentioned before, a bacterial colony is really just a simple association of individual bacteria. Each bacterium in the colony could survive on its own if separated from the rest of the colony. Despite this fact, however, some colonies do possess strikingly collective behavior. Some colonies, for example, will secrete a capsule-like substance that surrounds and protects the entire colony. This substance gives the bacterial colony a slimy feel and holds the colony together. Bacterial colonies encapsulated like this often look like mats floating on the top of water. Stagnant ponds are a common place to find such colonies. Other colonies have bacteria that actually work together to capture and eat prey. A *Myxococcus xanthus* colony, for example, moves as a unit in search of prey, typically other bacterial colonies. When they find their prey, they engulf it and, as a group, secrete a substance that digests it. The entire colony then feeds on the nutrients.

Another interesting aspect about the bacteria that make up colonies is the fact that their DNA seems to tell them what colony they should form. In other words, if two spherical bacteria end up close to one another, they will not necessarily form a diplococcus colony. Their DNA determines whether they "want" to exist as an independent bacterium or as a member of a colony. Furthermore, if the bacterium has the DNA of a colony-dweller, the DNA determines what kind of colony in which it will live. Thus, if a streptococcus bacterium is separated from its colony, it won't join up with just any other colony. It will continue to live independently until it can find another streptococcus colony. If it never finds one, it will live out its life as an independent bacterium.

ON YOUR OWN

2.9 A bacterial colony is called staphylobacillus. What shape do the bacteria in the colony possess: spherical, rod-shaped, or helical?

Before we move on to the next section, we need to begin an experiment that will last through the next module. Most of the first part of the experiment deals with collecting specimens that we will later view under a microscope. If you do not have a microscope, you should still do the parts of the experiment that involve visiting a pond, studying the ecosystem, and making drawings in your lab notebook. You just don't need to collect the samples.

EXPERIMENT 2.1
Pond Life, Part A

<u>Supplies:</u>

♦ Four jars with lids (You do not want a lot of light to get into the jars. Thus, jars made of darkened glass or plastic work really well. If you can't find that kind of jar, cover your jars with paper or foil to keep the light out.)
♦ A small amount of chopped hay (Dried grass will work as a substitute.)
♦ Uncooked white rice (Brown rice will not work as well.)
♦ Egg yolk (This should come from a boiled egg so that the yolk is cooked.)
♦ A small amount of rich soil
♦ A long-handled ladle (A good one can be made by attaching a kitchen ladle to a broom handle with duct tape.)
♦ A pond or small body of water (A still creek will do in a pinch, but it will not be ideal.)
♦ Something to rest your lab notebook on while you draw in it
♦ Colored pencils

Object: To study the ecosystem of a pond and collect specimens for the next two experiments

<u>Procedure:</u>

1. After locating a proper body of water, plan a field trip of one to two hours.
2. Before leaving, prepare your four jars as follows:

   • Label one jar "hay" and place a tablespoon of hay in it.
   • Label one jar "rice" and place a teaspoon of rice in it.
   • Label the third jar "egg yolk" and place ¼ teaspoon of egg yolk in it.
   • Label the last jar "soil" and place 2 teaspoons of soil in it.

3. At the pond, use your ladle to collect pond water. Take your samples near the bottom of the pond. Fill each of the four jars half full of the water. Put the lids back on the jars when you are done.
4. Set the jars aside and walk slowly around the pond, sitting occasionally to observe everything that is around you. Make sure you not only look, but also listen carefully.
5. As you note those things around you, draw each of them in your notebook. Don't forget to look under rocks and other hiding places.
6. Take the jars home and place them in a reasonably warm (never under 60 °F) area with subdued light. The items that you placed in the jars are food for the microorganisms that are present in the pond water. As time goes on, the microorganisms will grow and reproduce, increasing the population. When we take a few microorganisms and try to increase their population this way, biologists say that we are **culturing** the microorganisms. Thus, these jars are often called **cultures**, because microorganisms are being cultured inside them.

7. If you have placed the jars in an area where bad smells will not bother anyone, you can poke some holes in the lids of the jars. That will allow for air flow. Although not absolutely critical (there is already air in the jars and the water, and microorganisms do not use a lot of it), it can increase the population of microorganisms in your cultures, which will make for better results. It will get pretty smelly, though, so you should probably not do this if you are keeping the cultures in the house!
8. The cultures are to be used in the next two experiments. They will reach their peak of growth usually between days 3 and 5, but even after 2 weeks some organisms will still be active.

Classification in Kingdom Monera

Now that we have discussed aspects that are common to many of the organisms in kingdom Monera, it is time to start talking about the differences between individual members. To do this, we will discuss how to classify bacteria. One of the first things you must learn is that since kingdom Monera is so incredibly diverse, the classification of its members is a rather difficult task. As a result, we will not delve as deeply as we might into the classification groups that make up kingdom Monera. In fact, our discussion of classification will not go much deeper than splitting bacteria into classes. Thus, we will not really talk about the different orders, families, genera, and species classifications that exist in kingdom Monera.

To learn how we split bacteria into different phyla, you need to learn a little history. In the late 1800s, a Danish physician named Hans Christian Gram was studying bacteria. In order to make them show up better under a microscope, he developed several different types of stains. In Experiment 1.2, you should have stained the sample you took from your cheek in order to make your cheek cells show up better under the microscope. This was the kind of staining that Gram was doing. He noticed that after a multistep staining procedure, later called the **Gram stain**, certain bacteria looked blue when viewed under the microscope, whereas others looked red, as shown in the figure below.

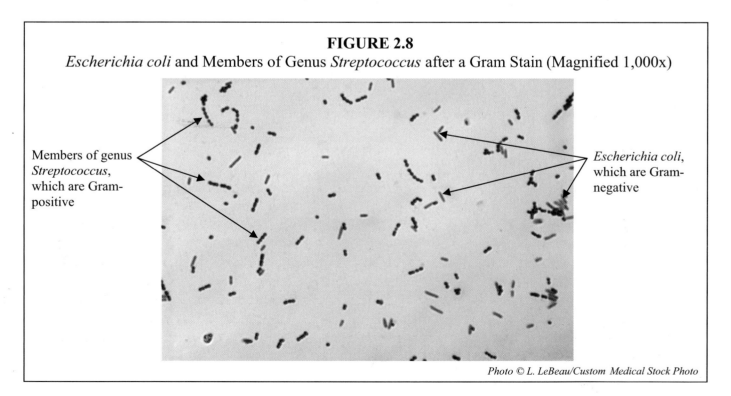

FIGURE 2.8
Escherichia coli and Members of Genus *Streptococcus* after a Gram Stain (Magnified 1,000x)

Members of genus *Streptococcus*, which are Gram-positive

Escherichia coli, which are Gram-negative

Photo © L. LeBeau/Custom Medical Stock Photo

As you can see in the photo, the difference between the bacteria is quite noticeable. The *Escherichia coli* (a species you will learn about in a moment) are clearly pinkish-red, while the members of genus *Streptococcus* (you should recognize them as one of the standard colony types shown in Figure 2.7) are clearly blue.

This is now used as a means of classifying bacteria. **Gram-negative** bacteria are the ones that look red after the Gram stain, while **Gram-positive** bacteria are the ones that look blue following the Gram stain. Now please realize that at first, this classification was completely arbitrary. Scientists had no idea what was causing some bacteria to be red and others to be blue. They just assumed that there was some fundamental aspect of the bacteria that caused them to react with the Gram stain in a certain way. Thus, they assumed that this would be a way of classifying bacteria. Well, it turns out that they were right. The difference in reaction to Gram stain is caused by differences in the cell walls of these bacteria. Although those differences are a bit too detailed to go into in this course, we can say that bacteria which are Gram-positive have cell walls which retain the Gram stain so that it stays in the cell. Gram-negative bacteria, however, have cell walls that do not retain the Gram stain. As a result, the stain leaves those cells, allowing a red stain added at the end of the procedure to takes its place.

The first way we separate the organisms in kingdom Monera, then, is by their cell walls. If a bacterium has a cell wall that doesn't retain the Gram stain (making it Gram-negative), it belongs to phylum **Gracilicutes** (gruh' sil uh kyoo' teez). If a bacterium has a cell wall that retains the Gram stain (making it Gram-positive), the bacterium belongs to phylum **Firmicutes** (fir' muh kyoo' teez). As time has gone on, we have determined two other ways to classify bacteria. Some bacteria have no cell walls at all. We put these bacteria into phylum **Tenericutes** (ten' uh ruh kyoo' teez). Finally, some bacteria possess cell walls, but the compounds which form these walls are rather different from the compounds that form the cell walls of Gram-positive and Gram-negative bacteria. Thus, we give them a separate phylum, called **Mendosicutes** (men' doh suh kyoo' teez).

You need to be a bit familiar with these phyla, so please complete the following "On Your Own" problems to make sure you understand each phylum and its characteristics.

ON YOUR OWN

2.10 A bacterium has no cell wall. To what phylum does it belong?

2.11 A bacterium is classified as Gram-positive. To what phylum does it belong?

2.12 A bacterium appears red after a Gram stain. To what phylum does it belong?

2.13 A bacterium has a cell wall that is different from both Gram-positive and Gram-negative bacteria. To what phylum does it belong?

Classes in Kingdom Monera

Remember our mnemonic: King Philip cried out, "For goodness sake!" Now that we have the phyla in Kingdom Monera, it is time to separate them into individual classes. Phylum Gracilicutes (Gram-negative bacteria) has three classes. We separate members of this phylum based on how they obtain their food. The first, class **Scotobacteria** (skoh' toh bak tehr'ee uh), is composed of the non-photosynthetic bacteria. The bacterium that causes Lyme disease (*Borrelia burgdorferi*) is a member of this class. In fact, many pathogenic bacteria can be found here. The second class, **Anoxyphotobacteria** (an ox' ee foh' toh bak tehr'ee uh), is composed of photosynthetic bacteria that do not produce oxygen. Typically, these bacteria live in the sediments of lakes or rivers. The last class, **Oxyphotobacteria** (ox' ee foh' toh bak tehr'ee uh), is made up of photosynthetic bacteria that produce oxygen. The most common example of this class is the blue-green algae (known as cyanobacteria) that you see floating on the top of stagnant ponds.

In phylum Firmicutes, there are only two classes. We do not separate the bacteria into these classes based on the way they obtain their food; rather, we do so based on their shape. Class **Firmibacteria** (fir' muh bak tehr'ee uh) contains cocci and bacilli bacteria, while class **Thallobacteria** (thal' oh bak tehr'ee uh) is made up of any other shape. One of the bacteria that cause pneumonia (*Streptococcus pneumoniae*) is a member of class Firmibacteria, while many of the members of class Thallobacteria are responsible for dental and gum disease.

You'll be happy to know that the last two phyla (Tenericutes and Mendosicutes) have only one class each. This is because there are few bacteria that fall into these phyla. Phylum Tenericutes has the class **Mollicutes** (mol' uh kyoo' teez) which contains another kind of pneumonia-causing bacteria, and phylum Mendosicutes contains the class **Archaebacteria** (ar kee' uh bak tehr'ee uh), which holds all of the bacteria with exotic cell walls. These exotic cell walls allow members of class Archaebacteria to live in rather odd locations, such as deep-ocean hydrothermal vents or incredibly brackish seas like the Great Salt Lake. Many places that are uninhabitable to other organisms are populated with members of this class.

Now working through this classification scheme might have been a little tedious, but it is important for you to get an overview of the different types of bacteria that exist. The classification scheme, as we have discussed it, is summarized in Table 2.1.

TABLE 2.1
Partial Classification of Kingdom Monera
(from *Bergy's Manual of Systematic Bacteriology,* 9th Edition)

Phylum Gracilicutes: Gram-negative bacteria
   Class Scotobacteria: Non-photosynthetic bacteria
   Class Anoxyphotobacteria: Non-oxygen-producing photosynthetic bacteria
   Class Oxyphotobacteria: Oxygen-producing photosynthetic bacteria

Phylum Firmicutes: Gram-positive bacteria
   Class Firmibacteria: Bacilli or cocci
   Class Thallobacteria: All others that are not bacilli or cocci

Phylum Tenericutes: Bacteria lacking a cell wall
   Class Mollicutes: The only class in this phylum

Phylum Mendosicutes: Bacteria with exotic cell walls
   Class: Archaebacteria: The only class in this phylum

Please spend a few minutes looking at Table 2.1 so that you can learn this classification system. On the test, you will be expected to classify bacteria given only their characteristics. Thus, you will have to commit this information to memory. To help you do this, the "On Your Own" question below asks you to use Table 2.1 as a guide in constructing your own biological key for the classification of bacteria. Making a biological key can be difficult when you have not done it before. If you have trouble answering the question, read the beginning of the answer so that you can get an idea of where to start.

☞ ON YOUR OWN

2.14 Construct a biological key that separates bacteria into their different classes. You can assume that the only organisms the key will be used to analyze are bacteria. (HINT: The first question should determine whether or not the bacterium has a cell wall.)

A Few Words on Other Classification Systems

As we mentioned in the first module of this course, there is controversy surrounding the classification system that biologists should use. We have settled on the five-kingdom classification system, so we have discussed bacteria in the context of that classification system. However, there are other views of classification, and much of the controversy between those views centers on members of kingdom Monera.

As we mentioned in Module #1, there is a three-domain approach to biological classification that splits kingdom Monera into two domains: Archaea and Bacteria. Thus, if you happen to run across the three-domain approach to biological classification, you will not find kingdom Monera. Instead, you will find that most of the members in domain Archaea can be found in what we have called phylum Mendosicutes, class Archaebacteria. Most of the rest of what we call kingdom Monera is placed in domain Bacteria.

Since there is so much controversy surrounding the classification of prokaryotic organisms, it shouldn't surprise you that some have wanted to compromise between the three-domain approach and the five-kingdom system. Thus, some biologists have proposed a **six-kingdom system**. This system retains kingdoms Protista, Fungi, Plantae, and Animalia, and it splits kingdom Monera into two separate kingdoms: **Archaebacteria** and **Eubacteria**. You can probably guess how this classification scheme works. Most of the organisms that we have placed in phylum Mendosicutes, class Archaebacteria are put into kingdom Archaebacteria, while most of the rest are placed in kingdom Eubacteria. Do you see how this is a compromise between the three-domain system and the five-kingdom system? It does what the three-domain system does in terms of splitting up the members of kingdom Monera, but it does so in the "style" of the five-kingdom system by just adding a sixth kingdom. Now remember that we will stick with the five-kingdom system, but you need to be aware that there are other views out there.

Specific Bacteria

Now that we've gone through all of the generalities, it is time to look at some specific types of bacteria. It is important for you to realize that what follows is just a sampling of the many different kinds of bacteria that exist. Entire books have been written devoted solely to discussing the various species of bacteria and what they do. We, of course, will not go into that kind of detail. Rather, we want to discuss these bacteria because either we think that they are important representatives of kingdom Monera or because we think that you might already be somewhat familiar with them.

Class Archaebacteria

As we mentioned before, the members of this class contain exotic substances in their cell walls that allow them to live in conditions that would be uninhabitable for most other organisms. For example, a species of bacteria from this class was found living in the water core of a nuclear reactor. Anaerobic chemosynthetic bacteria also belong to this class. Certain species of these bacteria live in the muck at the bottom of marshes and swamps. Because they are anaerobic, they do not require oxygen to convert their food into energy. Instead, they use hydrogen and carbon dioxide. One of the byproducts of this process is methane, which is the principal component of what is called "marsh gas" or "swamp gas." Because methane is flammable (and even explosive at times), swamp gas can ignite and even explode under the right conditions.

Blue-Green Algae (al' jee)

These organisms live in freshwater lakes and ponds and are most prevalent in stagnant waters. They form the blue-green mats that float on the surface of the water. Later on, you will learn that algae are, in fact, colonies of eukaryotic cells. However, since these organisms looked like algae, they were at first incorrectly named as such. Today, we recognize them as colonies of Gram-negative prokaryotic cells and thus classify them in kingdom Monera. The new (better) name for them is **cyanobacteria** (sye' an oh bak tehr'ee uh). This properly identifies them as bacteria. Nevertheless, some biologists still refer to them as "blue-green algae," so you need to be aware of both names.

Cyanobacteria are photosynthetic and are one of the few types of bacteria that require light in order to survive. Some bacteria cannot live under intense light, and most like their habitat to be dim or completely dark. Since cyanobacteria require light energy for metabolism, however, they represent a notable exception. These bacteria are also colonial, living in long, thin strands of cells. The strands encapsulate themselves, giving cyanobacteria a slimy feel.

Clostridium (claw strid' ee um) *botulinum* (bot' yool in um)

This bacterium, which belongs to class Firmibacteria in phylum Firmicutes, can be a source of food poisoning. The endospores that it forms are quite heat resistant, so when food has not been properly cooked, this bacterium can survive. It produces a toxic chemical that causes the disease known as **botulism**. Mild cases of botulism involve severe nausea, diarrhea, and high fever. However, severe cases of botulism can cause respiratory failure and death.

Salmonella (sal muh nell' uh) *typhimurium* (tye' fim ur ee um) **and *S. enteriditis*** (en' ter uh dye' tus)

These two species belong to the same genus (remember, if we write the genus name once, we can abbreviate it afterward). We place them in phylum Gracilicutes and class Scotobacteria. You have probably heard them both referred to by their genus name, *Salmonella*. These bacteria are common contaminants of eggs and poultry. People can be infected by one of these two bacteria when the poultry or eggs that they have eaten have not been cooked thoroughly. Since these bacteria are present in nearly every fowl, about the only way to make sure you get rid of them is by cooking the meat or eggs thoroughly enough to kill the bacteria and their endospores. Once again, while mild cases of *Salmonella* poisoning cause just nausea and diarrhea, severe cases can cause death.

Escherichia (esh' ur ee' kee uh) *coli* (koh' lye)

This species also belongs in phylum Gracilicutes and class Scotobacteria. It is a very common bacterium, living in your gut. In fact, that's how it gets its species name, as "coli" refers to the "colon," which is a part of your gut. One of the byproducts of *E. coli*'s metabolism is Vitamin K, a substance your body needs. It also secretes a chemical that helps your body digest fat. In addition, its activities in your gut actually keep food-borne pathogenic bacteria from colonizing there. Thus, the presence of this bacterium in your gut keeps out other, pathogenic bacteria!

Although the *E. coli* bacterium found in your gut is nonpathogenic, there are pathogenic forms of *E. coli*. If you are infected with pathogenic *E. coli*, it can give you severe diarrhea. Infants infected with such *E. coli* can actually die of dehydration due to the diarrhea. In some developing countries, this is the leading cause of infant death!

Now wait a minute. How can there be both pathogenic and nonpathogenic *E. coli*? They are all the same species. Well, when bacteria develop in different habitats, they can develop different traits. We will discuss this more in a later module. For right now, you just have to realize that the traits one organism has might be different from the traits of another organism *from the same species,* if these organisms developed in different habitats. As a result, not all *E. coli* are the same. When two organisms from the same species have markedly different traits, we say that they are from two different **strains**.

Strains – Organisms from the same species that have markedly different traits

We will learn in a later module why strains develop. For right now, just realize that there can be many strains of a bacterial species. The differences between these strains can be so extreme that one strain is beneficial while another strain is pathogenic.

Conditions for Bacterial Growth

Even though bacteria are hardy creatures and can survive in a lot of habitats, there are still some conditions that must be present in order for populations of bacteria to grow. These are:

1. *Moisture*: Almost all bacteria require some moisture to grow. It is important to note, however, that the capsules which some bacteria form will protect them during periods of dryness. In addition, if a bacterium can form an endospore, it can withstand even longer dry spells.

2. *Moderate temperature*: Most bacteria prefer moderate temperatures, 27 °C - 38 °C (80 °F - 100 °F). This makes summer the most ideal time for bacterial growth. Once again, it is worth noting that many bacteria (especially those from class Archaebacteria) can survive in quite extreme temperatures.

3. *Nutrition*: Bacteria must have nutrition. Since most are heterotrophic, removing the nutrition will destroy them. In the case of photosynthetic bacteria, you must remove light in order to kill them. Chemosynthetic bacteria can be killed by removing the chemicals they convert to energy.

4. *Darkness*: Many bacteria grow best in darkness or limited light. There are, of course, exceptions to this rule, photosynthetic bacteria being the obvious one.

5. *The proper amount of oxygen*: Aerobic bacteria require oxygen to survive. On the other hand, some anaerobic bacteria are killed by the presence of oxygen. In addition, some anaerobic bacteria actually prefer the presence of oxygen even though they don't absolutely need it in order to survive. For a given species, then, the presence or absence of oxygen will be critical for survival.

Once again, since there is great diversity among bacteria, these conditions are only the most important conditions for bacterial growth. In general, we find that bacteria are very specialized in terms of their habitat. One bacterium might be able to withstand extreme temperatures, but it might not be able to survive in even the slightest light. Thus, unlike many species, bacteria often can only live in a certain habitat. Take them out of that habitat, and they might die.

Preventing Bacterial Infections

Since some bacteria are pathogenic, it is worthwhile to spend a few minutes discussing how to prevent bacterial infections. A common means of becoming infected by bacteria is through the food we eat. Since bacteria are everywhere (their endospores are even in the dust particles that fly through the air), all food is exposed to bacteria. One way we can eliminate the bacteria is to expose the food to extreme heat or radiation. This will kill the bacteria (and their endospores) and then, if we seal the food in a container before exposing it to fresh air, no bacteria will be present. This is what happens in canned foods. Thus, canned foods can sit on a shelf for a long time, because all bacteria that were in the food have been destroyed, and the can does not let any air in, so new bacteria can't get to the food.

Another way to prevent bacterial infection in food is to dehydrate it. When a food is dehydrated, almost all of the water is removed. Since moisture is a condition for bacterial growth, as long as the dehydrated food stays dry, no bacteria can grow. Many campers who stay in the wilderness for a long time take dehydrated food with them. These foods are lightweight and bacteria-free, making them ideal for long trips.

If you can't dehydrate or can something, you can freeze it. Prolonged freezing will kill many of the bacteria that exist in the food. Of course, many foods do not taste good after freezing, so it is not the solution for all kinds of food. If, instead, you put food in the refrigerator, the rate of bacterial growth will slow, but it will not stop. Thus, food placed in the refrigerator will keep for a longer time than food left out on the counter or in the cupboard. However, since the cool temperatures just slow down the growth of bacterial populations, refrigerated food will eventually become contaminated with a significant number of bacteria.

One very bacteria-prone food is milk. From the time milk leaves the cow, it is laden with bacteria. If something is not done, the bacteria count will rise to a dangerous level in just a few days, even if it is stored in a refrigerator. If you heat milk to a temperature sufficient for killing the bacteria, however, the milk curdles. This used to be a real problem. Louis Pasteur (the scientist who debunked the law of spontaneous generation), however, developed a process called **pasteurization**, which he originally used to keep wine from souring. This process is now applied to milk, hence the term "pasteurized milk."

In the pasteurization process, milk is heated either to a moderate temperature of 63 °C (145 °F) for thirty minutes or 72 °C (161 °F) for 15 seconds. This temperature is not sufficient to kill all of the bacteria; however, after the heating is finished, the milk is quickly cooled to refrigerator temperature. This rapid change in temperature is sufficient for killing about 95% of all bacteria in the milk. Thus, with most of the bacteria gone, it takes many more days for the milk to sour. Still, exposure to air will allow bacteria back into the milk, so it should be covered and refrigerated at all times.

<u>Take a Look at the Microscopic World</u>

Even though we have come to the end of the second module, we are not done with the microscopic world. In the next module, we will look into the tiny world of kingdom Protista. To give you an introduction into this world, as well as to let you see some live bacteria like those that we have been studying, perform the following experiment. If you do not have a microscope, you should still spend some time looking over the pictures shown in the experiment instructions. This will give you some idea of what kinds of microorganisms exist.

EXPERIMENT 2.2
Pond Life, Part B

<u>Supplies:</u>

♦ Microscope
♦ Slides
♦ Coverslips for the slides
♦ 4 jars of water collected at the pond (Remember, they are called **cultures.**)
♦ 4 eyedroppers (one for each jar)
♦ A small amount of cotton (from a cotton ball, for example)

Object: You have studied the things that you were able to see by just using your eyes and ears while you were at the pond. That was your *macroscopic* observation of the pond. In this experiment, you will begin making *microscopic* observations of the pond.

<u>Procedure:</u>

1. Set up the microscope as described in Experiment 1.2.
2. Prepare a slide by adding a drop or two of your first culture (it doesn't matter which one) to the slide. When you open each jar, be prepared for a *mighty* stench! It is best to take your sample from the bottom of the jar, close to the food source in the culture. In fact, it is best to get a small bit of the food in the drop that you put on the slide, as that will give you something to guide your focus when you begin scanning the slide.
3. If you did not get a bit of food in the one or two drops that are on the slide, add a fine strand of cotton to the water on the slide.
4. Carefully place a coverslip over the water.
5. Place the slide on the microscope stage, and arrange it so that the bit of food (or cotton) is at the center of the hole on the stage.

6. Set the objectives so that you are using low power, and use the coarse focus to look at the bit of food (or cotton) that is in the water. You might have to move the slide around to find it.

7. Once you have focused in on the bit of food or cotton, *do not adjust the focus any more.* By focusing on the food or cotton, you have ensured that you are focusing at the right depth. In other words, your focus is in the sample of water. That is important, because some students end up focusing above or below the sample of water, and that makes it very hard to see what is in the sample.

8. Begin scanning the slide *on low power*, moving the slide slowly by creating a pattern as you move the slide. A good pattern is shown below:

First, scan up and down the slide:

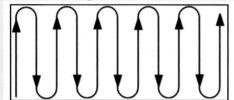

Next, scan left and right:

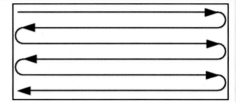

9. As you scan, you should see organisms. At low power, they will look like dots that are moving or wiggling. Please note that you must spend your time scanning the slide at *low* power, because that way you are viewing a large portion of the slide. At higher magnification, you are looking at a much smaller portion of the slide, and that will make it hard to find organisms.

10. Once you find organisms at low power, center them and then bring in the higher magnifications to get a closer look at them.

11. If the organisms are moving too fast for you to center easily, add a few fine strands of cotton. Add the cotton by lifting the coverslip, dropping the cotton on the slide, and replacing the coverslip. You might need to add another drop of the culture before replacing the coverslip. The cotton strands will form barriers that will keep the organisms from moving as much.

12. As you locate an organism, examine it on medium and high powers. Use the same technique that you learned in Module #1. Sketch what you see in your laboratory notebook, and try to identify each organism using the pictures below and on the next page. In this experiment, look especially for the presence of bacteria and algae. As shown in the photos below, bacteria will be the smallest things that you see. When you find something that is moving, try to determine how it moves.

13. Here are some photographs that will help you identify what you might see. If you cannot identify something right now, don't worry about it. You will learn about more microorganisms in the next module, and what you learn then will help you identify what you see now. The sketches that you make as you observe the organisms will help you remember what they look like later on when you are trying to identify them.

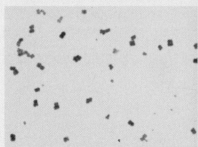

Coccus bacteria (x400)

Photo © M. Peres/Custom Medical Stock Photo

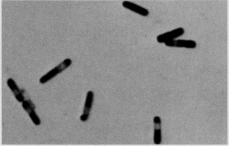

Bacillus bacteria (x400)

Photo © L.LeBeau/Custom Medical Stock Photo

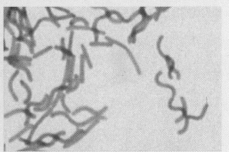

Spirillum bacteria (x400)

Photo © M. Peres/Custom Medical Stock Photo

Algae (x100)

Photo © Educational Images/Custom
Medical Stock Photo

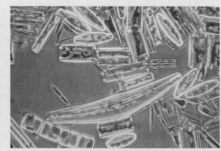

Diatoms (x400)

Photo © Educational Images/Custom
Medical Stock Photo

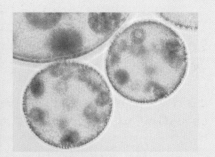

Volvox (x100)

Photo © J. L. Carson/Custom Medical
Stock Photo

Euglena (x400)

Photo © J. L. Carson /Custom Medical
Stock Photo

Amoeba (x400)

Photo © J. L. Carson /Custom Medical Stock
Photo

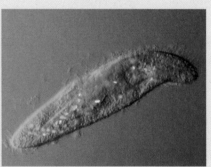

Paramecium (x100)

Photo © J. L. Carson /Custom Medical
Stock Photo

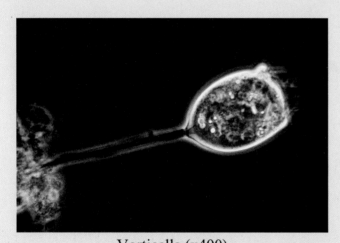

Vorticella (x400)

Photo © Educational Images/Custom Medical Stock Photo

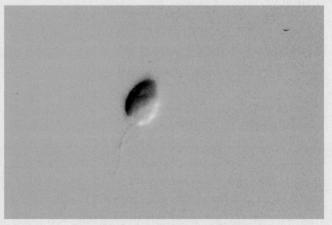

Chlamydomonas (x400)

Photo © J. L. Carson /Custom Medical Stock Photo

14. Examine each of the four cultures. Be sure to note which cultures produced which organisms.
15. Make sure all slides, coverslips, and eyedroppers are washed carefully with soap and water, dried, and stored back in their containers.
16. Close and save the cultures for the next lab.
17. Put the microscope away, and clean up any mess that you made.

ANSWERS TO THE "ON YOUR OWN" PROBLEMS

2.1 <u>The bacterium that succeeded most likely has a capsule while the other does not</u>. In bacteria, the capsule helps to protect the organism from infection-fighting mechanisms.

2.2 <u>The plasma membrane</u> did not do its job. Since it regulates what moves in and out of the cell, it should not have allowed the toxin to enter.

2.3 <u>It is missing a flagellum</u>. The fimbriae are not for movement; thus, if a bacterium has no flagellum, it cannot move.

2.4 <u>No</u>. Saprophytic means that it feeds on dead matter. Autotrophic organisms make their own food.

2.5 <u>Yes</u>. Aerobic and anaerobic deal with how the organism converts its food into useful energy. Chemosynthetic deals with how the organism gets the food to begin with. Thus, the bacterium can make the food chemosynthetically and then convert it to useable energy aerobically.

2.6 <u>Since asexual reproduction allows no variation in the DNA, an entire population of bacteria started from a single bacterium have all the same traits. If the environment changes, an organism might need new traits to survive. Since the whole population has essentially the same traits, there is no way to get the needed new traits, and the population dies</u>. The variability that exists in sexual reproduction usually makes a population much more resistant to changes in the habitat.

2.7 <u>More resources (most likely food) were added to the habitat</u>. The only way you can get population growth *after* the steady state would be due to an influx of new resources. In our cow example, maybe another cow died and fell on top of the first cow's body!

2.8 <u>Type B will be the donors and Type A will be the recipients</u>. Since the bacteria will need to survive in high temperatures, they need resistance to high temperatures. That's what will be donated.

2.9 <u>They are rod-shaped</u>, since "bacillus" means rod-shaped.

2.10 <u>Tenericutes</u> is the phylum for bacteria without a cell wall.

2.11 <u>Firmicutes</u> contains all Gram-positive bacteria.

2.12 <u>Gracilicutes</u> contains all Gram-negative bacteria. Since the bacterium appears red after the Gram stain, it is Gram-negative.

2.13 <u>Mendosicutes</u> contains all bacteria with exotic cell walls.

2.14 To make a biological key, we ask a series of questions that separate bacteria according to their traits. Since Table 2.1 summarizes those traits, we should use it as a guide. Let's start with whether or not the bacterium has a cell wall. If it has no cell wall, we know the phylum and the class. If it does have a cell wall, however, we will have to ask other questions. Thus, key 1 looks like this:

1. Cell wall **2**
   No cell wall ....*phylum Tenericutes*..........*class Mollicutes*

If we have a cell wall, it is either Gram-positive, Gram-negative, or neither. If it's either Gram-positive or Gram-negative, we will need to ask more questions. If it is neither, we are done.

2. Gram-positive or Gram-negative...**3**
   Neither ....*phylum Mendosicutes*........ *class Archaebacteria*

Now we deal with Gram-negative or Gram-positive. We'll have to ask more questions either way, so both alternatives will have to lead to more keys. We have no idea how many more, so we'll lead to key 4 on one and leave the other with question marks that we will go back and replace later. I'll send Gram-positive to key 4 because it has fewer classes; thus, it will get done quicker.

3. Gram-positive...............................*phylum Firmicutes*............**4**
    Gram-negative...........................*phylum Gracilicutes*..........??

Key 4 needs to make the distinction between the two classes in Firmicutes, so we ask about shape.

4. Bacillus or coccus .... *class Firmibacteria*
    Neither .... *class Thallobacteria*

Now we are ready to deal with Gram negatives. Thus, we know that the question marks on key 3 should be replaced with a "5." In key 5, we need to make the distinction between three classes. Two have photosynthetic bacteria in them, so we should ask whether or not the bacterium is photosynthetic:

5. Non-photosynthetic... *class Scotobacteria*
    Photosynthetic.. **6**

Now we just determine whether or not the bacteria produce oxygen, and we are done.

6. Produces oxygen..*class Oxyphotobacteria*
    Does not produce oxygen................................... *class Anoxyphotobacteria*

The final key, then, is as follows:

1. Cell wall **2**
    No cell wall ....*phylum Tenericutes*....... *class Mollicutes*
2. Gram-positive or Gram-negative...**3**
    Neither ....*phylum Mendosicutes*.......*class Archaebacteria*
3. Gram-Positive...............................*phylum Firmicutes*............**4**
    Gram-Negative...........................*phylum Gracilicutes*.........**5**
4. Bacillus or Coccus... *class Firmibacteria*
    Neither .... *class Thallobacteria*
5. Non-photosynthetic... *class Scotobacteria*
    Photosynthetic.. **6**
6. Produces oxygen..*class Oxyphotobacteria*
    Does not produce oxygen................................... *class Anoxyphotobacteria*

Now your key does not have to be in the same order, but it should contain all of these questions. It might have a few more steps, or the order of the questions might be different. That's fine.

STUDY GUIDE FOR MODULE #2

1. On a separate sheet of paper, write down the definitions for the following terms:

a. Pathogen
b. Saprophyte
c. Parasite
d. Aerobic organism
e. Anaerobic organism
f. Steady state
g. Exponential growth
h. Logistic growth
i. Conjugation
j. Plasmid
k. Transformation
l. Transduction
m. Endospore
n. Strains

2. Label all of the indicated structures on the bacterium below:

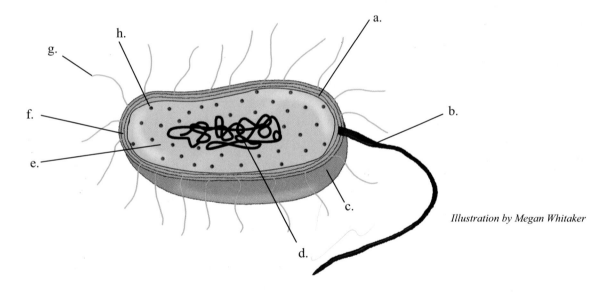

Illustration by Megan Whitaker

3. Describe the functions of each of the components you labeled in problem #2.

4. What is the most popular means by which bacteria obtain food?

5. If a bacterium is parasitic, is it heterotrophic or autotrophic?

6. List the basic steps in asexual reproduction among bacteria.

7. A sample of food is dehydrated, which kills all of the bacteria. However, in a few days, bacteria-free water is added to the food in a bacteria-free environment. Nevertheless, microscopic investigations indicate that bacteria are in the food. How did the bacteria get there?

8. What kind of growth does a population of bacteria experience when just a few of them are placed in an environment rich with resources?

9. A population of bacteria experiences logistic growth. What can you say about the resources of their environment?

10. Even though genetic recombination among bacteria does not result in offspring, it can significantly affect the growth of the population. Why?

11. What are the technical names of the three common bacterial shapes?

12. A bacterium is heterotrophic and Gram-negative. To what phylum and class does it belong?

13. A spirillum bacterium is Gram-positive. To what phylum and class does it belong?

14. A bacterium has no cell wall. To what phylum and class does it belong?

15. A bacterium lives in boiling-hot water. To what phylum and class does it most likely belong?

16. In the six-kingdom classification system, what replaces kingdom Monera?

17. What conditions are ideal for most bacteria to grow and reproduce?

18. What methods exist for reducing the chance of bacterial contamination of food?

**"I want to cross a rabbit with a bacterium.
The result should be able to multiply and divide."**

Cartoon by Speartoons

MODULE #3: Kingdom Protista

Introduction

In the previous module, we looked at the diverse kingdom Monera. In this module, we will move on to an equally fascinating kingdom: **Protista** (proh tee' stuh). Both of these kingdoms are made up of microorganisms, but they differ rather dramatically. The main difference is that organisms in kingdom Monera are prokaryotic, whereas those in Protista are eukaryotic. Now as you learned in Module #1, eukaryotic cells have distinct, membrane-bounded organelles, while prokaryotic cells do not. This fact adds a level of complexity to the organisms in kingdom Protista that doesn't exist in kingdom Monera.

Why do we say this? Well, both prokaryotes and eukaryotes must perform certain basic functions (those discussed in Module #1) in order to live. In prokaryotic cells, these functions are performed by the cell as a whole. In eukaryotic cells, however, each organelle performs its own set of tasks. Only when each organelle cooperates with the other organelles will the cell be able to perform all of the functions necessary for life. Thus, in eukaryotic cells, the organelles must be able to work together. This adds a level of complexity to the system.

This added level of complexity makes these organisms even more fascinating to study, but it also causes a bit of a problem for us. You see, when studying kingdom Monera, we could discuss the "typical" bacterium. Of course, there were many exceptions, but most bacteria resemble that which was labeled as "typical." Because of the increased complexity in kingdom Protista, however, there is just no way to describe what is typical for these organisms. Thus, in order to study this fascinating kingdom, you should start with another microscope experiment. We will then give you a short introduction to classification within Protista, which will be followed by a detailed discussion of the organisms in each major phylum.

If you do not have a microscope, read through the experiment to get an idea of what it covers. You can also get a glimpse of this fascinating world using the multimedia companion CD that accompanies this course. The multimedia companion CD has several video clips of live microscopic organisms. These video clips (found mostly in the sections that cover Modules #2 and #3) show live microscopic organisms, and a narrator tells you a bit about what you are seeing as you watch each video. Even if you perform the experiment, the videos on the CD might help you understand even more about the microscopic world.

EXPERIMENT 3.1
Pond Life, Part C

Supplies:

♦ Microscope
♦ Slides
♦ Coverslips
♦ The 4 culture jars used in Experiment 2.2
♦ 4 eyedroppers (one for each jar)
♦ A small amount of cotton (from a cotton ball, for example)

Object: To further the study the cultures set up in Experiment 2.2. You will hopefully see different organisms this time.

Procedure:

1. Set up your microscope and make slides of each culture as you did in Experiment 2.2. Prepare the slides one at a time, as you are ready to observe them.
2. Once again, try taking your sample from the very bottom of the jar, where the food is. Do not disturb the water in the process.
3. Make sure you scan the slide systematically, as you did in Experiment 2.2, to observe all organisms in the sample.
4. Examine each organism (on all three powers) as you find it. Remember, you can use cotton to hinder an organism's movement, as you did in Experiment 2.2. When you find an organism different from the ones you saw in Experiment 2.2, examine it with all three magnifications and draw it. Note the culture from which the organism came and how the organism moves.
5. Try to identify the organism by looking at the pictures in Experiment 2.2 and in this module.
6. Unless you wish to study the cultures further on your own, take your cultures outside and dump them. Discard the containers. Clean all slides, coverslips, and eyedroppers carefully. Put the microscope away.

Classification in Kingdom Protista

Kingdom Protista is divided into two main groups: **protozoa** (proh' tuh zoh' uh) and **algae** (al' jee). These groups are *not* phyla. Instead, each group contains several phyla; thus, they are often called **subkingdom Protozoa** and **subkingdom Algae**. Most protozoa exist as individual, single-celled creatures, while most algae form colonies. Whereas most protozoa are heterotrophic, most algae are autotrophic. Finally, most protozoa have a means of locomotion, while most algae simply float on or near the top of a body of water.

Within the two subkingdoms, there is an amazing amount of diversity. There are four major phyla in subkingdom Protozoa: phylum **Mastigophora** (mas tih gah' for uh), phylum **Sarcodina** (sar kuh die' nuh), phylum **Ciliophora** (sil ee ah' for uh), and phylum **Sporozoa** (spor' uh zoh' uh). These phyla are distinguished from one another based on their organisms' method of locomotion. Subkingdom Algae, on the other hand, contains five major phyla: phylum **Chlorophyta** (klor' uh fye' tuh), phylum **Chrysophyta** (cry' so fye' tah), phylum **Pyrrophyta** (pie' roh fye' tuh), phylum **Phaeophyta** (fay' uh fye' tuh), and phylum **Rhodophyta** (roh' duh fye' tuh). Organisms are separated into these phyla based on habitat, organization, and type of cell wall.

Now you need not worry about memorizing the names of the phyla that we have just discussed. As time goes on, you will get more familiar with them, but we will not require you to remember them. When answering questions on an exercise or a test, we will provide you with a table that will list the subkingdoms and phyla in kingdom Protista. You should, instead, concentrate on trying to remember what types of organisms go into each phylum. That's what you will be required to know for the test. In order to help you get an idea of the organisms that exist in kingdom Protista, study Figure 3.1. Don't worry that some of the terms are unfamiliar; we'll get to them in a while. For now, just try to get an overview of what the organisms in kingdom Protista look like and how they are separated into phyla.

FIGURE 3.1
Representatives of Kingdom Protista

Subkingdom Protozoa

Phylum: Sarcodina
Locomotion: Pseudopods
Example: genus *Amoeba*
(uh mee' buh)

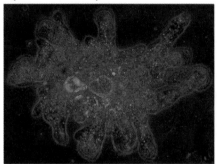

© A. Rakosy/Custom
Medical Stock Photo

Phylum: Mastigophora
Locomotion: Flagellum
Example: genus *Euglena*
(yoo glee' nuh)

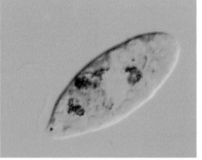

© JL Carson/Custom
Medical Stock Photo

Phylum: Sporozoa
Locomotion: None
Example: genus *Plasmodium*
(plas moh' dee uhm)

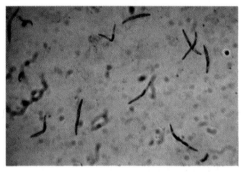

© NMSB/Custom
Medical Stock Photo

Phylum: Ciliophora
Locomotion: Cilia
Example: genus *Paramecium* (pehr uh mee' see um)

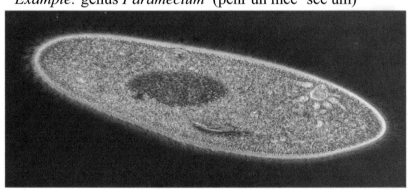

© Roland Birke/Phototake

Subkingdom Algae

Phylum: Chlorophyta
Habitat: Fresh water
Organization: Single cells
Cell Wall: Cellulose
Example: genus *Cosmarium*
(kos mair' ee uhm)

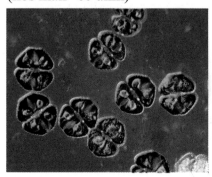

© A. Rakosy/Custom Medical
Stock Photo

Phylum: Chrysophyta
Habitat: Marine and fresh water
Organization: Single cells,
some colonies
Cell Wall: Silicon dioxide
Example: Diatoms (several genera)

Photo by Kathleen J. Wile

Phylum: Pyrrophyta
Habitat: Marine
Organization: Single cells
Cell Wall: Cellulose or atypical
Example: genus *Peridinium*
(pehr' uh din' ee uhm)

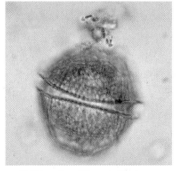

© Educational Images/Custom
Medical Stock Photo

Phylum: Phaeophyta
Habitat: Cold marine waters
Organization: Multiple cells
Cell Wall: Cellulose and alginic acid
Example: genus *Macrocystis* (mah' kroh sis' tus)

© Cybersea Images

Phylum: Rhodophyta
Habitat: Warm marine waters
Organization: Multiple cells
Cell Wall: Cellulose
Example: genus *Corallina* (kor' uh lee' nuh)

© Bob Ford / Nature Portfolio

As we discuss these phyla in a little more detail, you will notice something about the way that we refer to organisms within this kingdom. Sometimes, we refer to organisms using just the name of a classification group to which they belong, typically their genus. For example, when we talk about phylum Sarcodina, we will study the *Amoeba proteus* (proh' tee us). It turns out that most of the species within genus *Amoeba* are so similar that when we see them under the microscope, they look almost identical. Thus, we often refer to each members of this genus as simply "amoeba." When we do this, we do not capitalize or italicize the word because we are not talking about the classification groups used in binomial nomenclature; we are instead using a generic term to refer to all members within the classification group.

We often use a generic term with other classification groups as well. When we were discussing the two subkingdoms Protozoa and Algae, we began referring to their members with the generic term "protozoa" and "algae." This happens quite a bit when we study kingdom Protista, so you will just have to get used to it.

ON YOUR OWN

3.1 Construct a biological key for separation of organisms in kingdom Protista into phyla. You can assume that any organism for which you use the key is already known to be in kingdom Protista. You can also assume that if an organism is completely autotrophic it belongs in subkingdom Algae, but if it is at all heterotrophic it belongs in subkingdom Protozoa. With those assumptions, you should be able to use the characteristics in Figure 3.1 to construct the questions for the key.

Subkingdom Protozoa

In general, members of subkingdom Protozoa have a mechanism that allows them to move around. This is one aspect that sets them apart from the algae, which in general have no locomotion. Another quality that separates protozoa from algae is the fact that most protozoa are heterotrophic, while most algae are autotrophic. There are, of course, exceptions to both of these rules, which adds to the confusion when it comes to classifying organisms within kingdom Protista. For example, you will soon learn that the genus *Volvox* belongs to phylum Mastigophora, which means that we consider creatures in this genus to be protozoa. Genus *Volvox* is placed in phylum Mastigophora because its members have flagella and can use those flagella to move around. Nevertheless, members of this genus use photosynthesis to produce their own food. Thus, some might consider them algae, because they are autotrophic. So, even though we might lay down some general rules for classification, please realize that there will be exceptions to them, like there are to most rules in biology!

Phylum Sarcodina

The most striking feature of the organisms in this phylum is that they have no standard body shape. They are enclosed in a flexible plasma membrane that allows them to change shape at will. When resting, they are usually somewhat spherical. However, when they wish to move, they form extensions of their bodies called **pseudopods** (soo' doh podz) or "false feet."

Pseudopod – A temporary, foot-like extension of a cell, used for locomotion or engulfing food

You will learn more about pseudopods in a moment. First, however, let's take a look at a typical member of phylum Sarcodina, *Amoeba proteus*, the common amoeba (plural: amoebae).

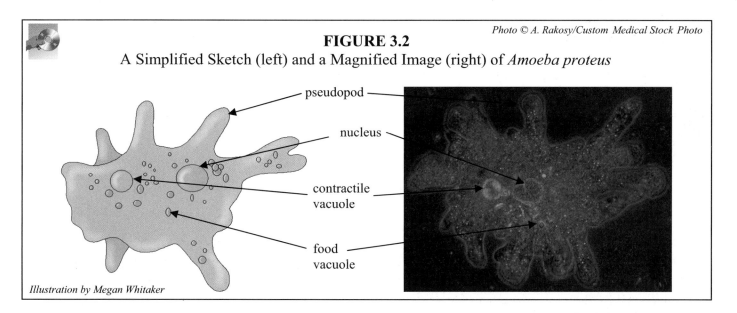

FIGURE 3.2
A Simplified Sketch (left) and a Magnified Image (right) of *Amoeba proteus*

Photo © A. Rakosy/Custom Medical Stock Photo

pseudopod

nucleus

contractile vacuole

food vacuole

Illustration by Megan Whitaker

As you can see from the figure, this organism is considered eukaryotic because it has a distinct, membrane-bounded **nucleus** (new' clee us).

Nucleus – The region of a eukaryotic cell that contains the cell's main DNA

Note that the nucleus contains the cell's *main* DNA. It turns out that DNA can be found in another organelle called the **mitochondrion** (my' tuh kahn' dree uhn), but we don't want to dwell on that at the moment. For right now, just understand that most of a cell's DNA exists in the nucleus, and as a result, reproduction is controlled by the nucleus.

In the case of *Amoeba proteus* (as well as other members of phylum Sarcodina), the nucleus not only controls reproduction, but it also controls the organism's metabolism. The other organelle typically visible in the amoeba is the **vacuole** (vak' yoo ohl).

Vacuole – A membrane-bounded "sac" within a cell

The amoeba has two types of vacuoles. The **food vacuoles** hold and store food while it is being digested, and the **contractile vacuoles** regulate the amount of water in the cell. If the cell absorbs too much water (we will discuss how this can happen in Module #5), it could potentially become over-pressurized and end up exploding! To release the pressure, the contractile vacuoles collect excess water in the cell and release it into the amoeba's surrounding environment.

The cytoplasm (that jelly-like substance inside the cell) of the amoeba is divided into two parts. Near the plasma membrane, we find the **ectoplasm** (ek' toh plas uhm), which is thin and watery.

Ectoplasm – The thin, watery cytoplasm near the plasma membrane of some cells

In the interior of the cell, the cytoplasm becomes much more dense. In this region, it is called the **endoplasm** (en' doh plas uhm).

Endoplasm – The dense cytoplasm found in the interior of many cells

The amoeba has a very interesting method of locomotion. If you have a microscope, you may have had the opportunity to observe this fascinating movement. While at rest, the amoeba looks like a blob. When it needs to move, the amoeba forces endoplasm to a region near the plasma membrane. This dense liquid deforms the membrane, causing a bulge, called a pseudopod, to form. If the amoeba wants to move, all of the endoplasm can be forced into the pseudopod, carrying the cell contents with it. Thus, the amoeba moves by forming a pseudopod and pushing itself into that pseudopod. This can be done over and over, resulting in a laborious form of locomotion.

The amoeba uses its pseudopods for other tasks as well. Consider the figure below.

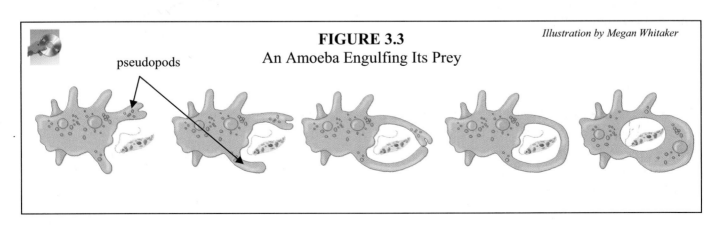

FIGURE 3.3   *Illustration by Megan Whitaker*
An Amoeba Engulfing Its Prey

pseudopods

When an amoeba comes into contact with a food source (mostly other, smaller microorganisms), it can use its pseudopods to surround and then engulf its prey, as shown in Figure 3.3. Once engulfed, the food is held in a food vacuole where digestion can begin. Once the food is digested by chemicals that are released into the vacuole, substances that can be used by the organism are sent into the endoplasm, while substances that cannot be used are ejected from the cell.

Although amoebae inhabit watery environments with little current such as ponds, lakes, and slow-moving rivers, they do not swim. Their method of locomotion is simply too laborious for such activity. As a result, amoebae typically adhere to surfaces in the water. When you touch a rock that is in a pond, the slimy coating that you feel is often the result of amoebae that are stuck to the rock. In fact, when an amoeba is floating in water, it will try to adhere to almost any surface with which it comes into contact.

When an amoeba reaches its maximum size (typically in about three days), it can reproduce asexually. As far as we know, amoebae do not sexually reproduce. Amoebae can form **cysts** (much like the endospores which bacteria form) in an attempt to survive life-threatening conditions such as dryness or a lack of food.

Other Sarcodines

Although *Amoeba proteus* is a typical sarcodine, there are many other organisms in this phylum. Genus *Entamoeba* (ent' uh mee' buh), for example, contains many organisms that live inside human beings. *Entamoeba gingivalis* (jin' jih val us) is a harmless sarcodine that lives in your mouth, while *E. coli* is a sarcodine that lives in your gut (remember, "coli" refers to the colon). Notice that this *E. coli* is not the same as the *E. coli* that we discussed in the previous module. The bacterium that we discussed in the previous module is *Escherichia coli*, while the sarcodine we are discussing now is *Entamoeba coli*. They both inhabit the human gut, but they are quite different organisms. One is prokaryotic (*Escherichia coli*), and the other is eukaryotic (*Entamoeba coli*). One is a bacterium (*Escherichia coli*), and the other is a protozoan (*Entamoeba coli*). This potential point of confusion illustrates that although you *can* abbreviate the genus name in binomial nomenclature, it is not always wise to do so because many different genera begin with the same letter.

One other organism in this same genus, *Entamoeba histolytica* (his' toh lih tih' cuh), is a gut-dwelling protozoan. This particular species is pathogenic and causes severe dysentery. It is often spread through contaminated water. Countries that do not carefully sterilize their water supply often have lots of *E. histolytica* in their water. This is where the phrase "don't drink the water" comes from. If you drink water that is contaminated with *E. histolytica*, you can experience diarrhea, nausea, exhaustion, and dizziness, which are all symptoms of dysentery. In some cases, *E. histolytica* can spread to the liver and brain, causing death!

ON YOUR OWN

3.2 Suppose you were observing an amoeba under the microscope and it suddenly exploded. What organelle was probably not working properly in the amoeba?

3.3 A biologist is studying an amoeba and sees a vacuole with several small solid objects in it. Most likely, what kind of vacuole is it?

Phylum Mastigophora

As you can see from Figure 3.1, one of the characteristic features of organisms in phylum Mastigophora is a flagellum, which is used for locomotion. Because of this characteristic feature, members of this phylum are often called **flagellates** (flah' gel ayts).

Flagellate – A protozoan that propels itself with a flagellum

Although most flagellates use their flagellum to swim freely in the water, some use it to attach themselves to a solid object. Interestingly enough, this phylum contains all sizes of organisms. Some flagellates are among the smallest protozoa in kingdom Protista, while others form elaborate colonies that are large enough to see with the unaided eye. Members of this phylum can be found in a variety of habitats, including salt water (which is referred to as a **marine** environment), fresh water, and moist soil.

Genus *Euglena* provides an interesting example of organisms from phylum Mastigophora. Members of this genus have the ability to produce their own food via photosynthesis. Despite this fact, we do not necessarily consider them autotrophic, because they also ingest and decompose the remains of dead organisms. Thus, euglenas are both autotrophic and saprophytic! Even when there is plenty of light for photosynthesis, euglenas still tend to absorb some food from their surroundings. When their surroundings become dim, their photosynthesis shuts down, and they become completely saprophytic. In fact, if their environment stays dim for several days or more, the euglenas' photosynthetic capability degenerates. After that happens, they can never produce their own food again, regardless of how bright their surroundings might later become.

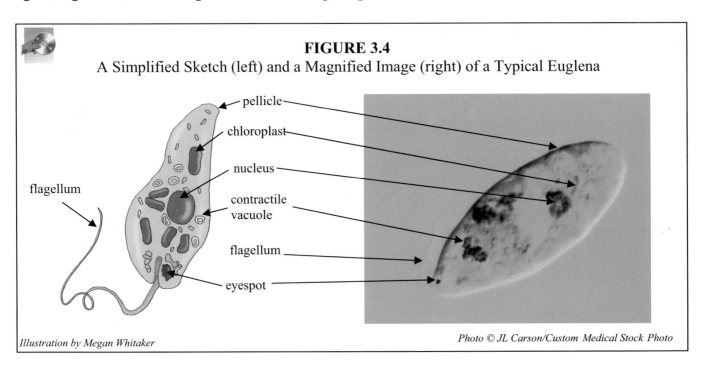

FIGURE 3.4
A Simplified Sketch (left) and a Magnified Image (right) of a Typical Euglena

pellicle
chloroplast
nucleus
contractile vacuole
flagellum
eyespot
flagellum

Illustration by Megan Whitaker   *Photo © JL Carson/Custom Medical Stock Photo*

You should notice that there are many prominent organelles in this creature. Like sarcodines, euglenas have contractile vacuoles that help expel excess water in order to reduce the pressure inside the cell. They also have a nucleus that holds the main DNA and controls both reproduction and metabolism.

Unlike sarcodines, euglenas have a firm but flexible shape-sustaining **pellicle** (pel' ik uhl).

Pellicle – A firm, flexible coating outside the plasma membrane

The pellicle usually keeps a euglena in its spindle-like shape. When it wants to move, however, it can use a worm-like method of locomotion in addition to its flagellum. While whirling its flagellum, a euglena can also draw its cytoplasm into the central region of the cell. This deforms the euglena, making it look like a "+" sign. The euglena then re-extends itself forward. This mode of locomotion is used in addition to whirling the flagellum, as illustrated below.

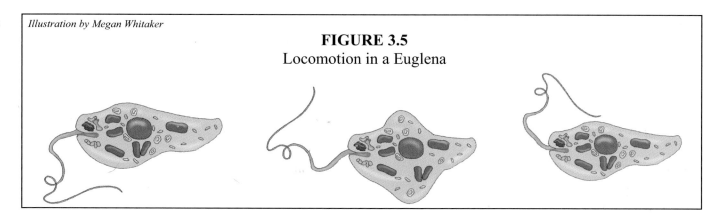

Illustration by Megan Whitaker

FIGURE 3.5
Locomotion in a Euglena

In order to create food by photosynthesis, euglenas have **chloroplasts** (klor' oh plasts), which contain a green pigment called **chlorophyll** (klor' oh fill). You will learn more about this pigment in an upcoming module. For right now, just understand that chlorophyll is a green pigment necessary for photosynthesis.

Chloroplast – An organelle containing chlorophyll for photosynthesis

Chlorophyll – A pigment necessary for photosynthesis

Euglenas also have a red, light-sensitive region known as the **eyespot.**

Eyespot – A light-sensitive region in certain protozoa

Since the eyespot is sensitive to light, euglenas use it to move towards regions of bright light, increasing their photosynthetic output. Understand that although we call this the "eyespot," the euglenas do not really see with it. The eyespot simply allows them to determine where there is a lot of light and where there is little light. The biological structures and processes necessary to actually see are far beyond the reach of a single-celled organism!

Euglenas reproduce only asexually. Under ideal conditions, a single euglena can reproduce once each day.

Other Mastigophorites

Other rather interesting members of phylum Mastigophora come from the genus *Volvox*. Also photosynthetic, members of this genus are colonial and possess two flagella. They join together and create an elaborate latticework out of strands of cytoplasm. The entire colony moves by rolling the latticework. The standard mode of reproduction among *Volvox* is asexual. When this occurs, a daughter colony is formed which often stays within the original parent colony. Under certain conditions, however, individual members of a *Volvox* colony can reproduce sexually. A magnified image of several *Volvox* colonies is shown below.

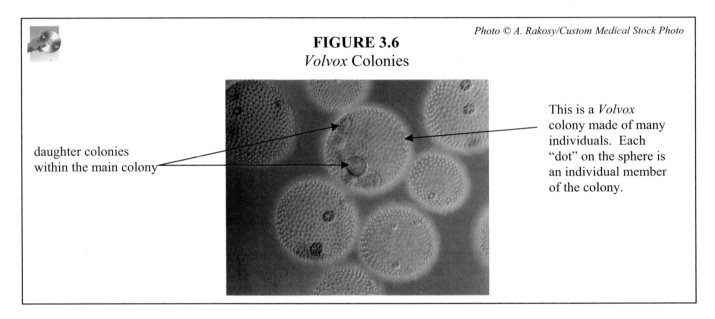

Photo © A. Rakosy/Custom Medical Stock Photo

FIGURE 3.6
Volvox Colonies

daughter colonies
within the main colony

This is a *Volvox* colony made of many individuals. Each "dot" on the sphere is an individual member of the colony.

As is typical with most phyla in this subkingdom, some mastigophorites are harmful to human beings. A small fly known as the **tsetse fly** carries mastigophorites that belong to the genus *Trypanosoma* (try' pan oh soh' muh). These protozoa are passed to humans when the tsetse fly bites a person's skin or when a person eats meat from an animal that has been bitten. The *Trypanosoma* then travel into the bloodstream and cause **African sleeping sickness**, a disease which often ends in death.

Another interesting mastigophorite comes from genus *Trichonympha* (trik' oh nim' fuh). These protozoa live in the gut of a termite and actually enable the termite to survive. You see, termites eat wood. However, they are unable to digest a substance called cellulose, which is a major component in all types of wood. The *Trichonympha* that live in the termite's gut feed on this cellulose. If it were not for these protozoa, the termites would not be able to feed on wood, and they would die of starvation. The *Trichonympha*, however, keep this from happening and, in the process, feed themselves.

The relationship between the termite and the *Trichonympha* is an example of what biologists call **symbiosis** (sim by oh' sis).

Symbiosis – A close relationship between two or more species where at least one benefits

Symbiosis is very common in God's creation. In fact, it is so common that there are many different kinds of symbiosis. The basic kinds of symbiosis are listed below.

Mutualism – A relationship between two or more organisms of different species where all benefit from the association

<u>Commensalism</u> – A relationship between two organisms of different species where one benefits and the other is neither harmed nor benefited

<u>Parasitism</u> – A relationship between two organisms of different species where one benefits and the other is harmed

Mutualism, **commensalism**, and **parasitism**, then, are all specific types of symbiosis. The relationship between the termite and the *Trichonympha*, for example, is symbiosis, but specifically, it is mutualism, because both species benefit. The termite benefits because it can eat wood and not worry about the cellulose building up in its body, and the *Trichonympha* gets food.

If only one species benefits in a symbiotic relationship but the other species is not hurt, we call that commensalism. Notice the barnacles (the white bumps) on the humpback whale pictured below.

Photo © Gary Bell/oceanwideimages.com

FIGURE 3.7
Barnacles on a Whale

Barnacles are living creatures, and some live on the skin of whales. They benefit by living on the body of a whale because they are protected (to some degree) from predators since they move with the whale. The whale does not benefit from the barnacles, but it is also not harmed. As a result, this symbiotic relationship is an example of commensalism.

Parasitism is a symbiotic relationship between two species in which one benefits and the other is harmed. Remember the sarcodine *Entamoeba histolytica*? It lives in the gut of a person, feeding on what is there. The *E. histolytica* clearly benefits, but the human host is harmed. As discussed earlier, *E. histolytica* causes dysentery, which can even lead to death. Thus, we call *E. histolytica* a **parasite**.

Symbiosis is yet another witness to God's handiwork and His plans. Mutualism, for example, is a testimony to how God's design solves problems. You see, termites must eat wood in order to provide balance in nature. However, they cannot digest cellulose, a principle component of wood. The Creator, fully aware of this problem, simply created another organism that could digest cellulose and designed that organism to live in the termite's gut. That way, wood and cellulose could both be digested together. Truly, God's handiwork in creation is astounding! Commensalism demonstrates that God has created the world so rich with resources that some species can help other species without needing to benefit from the relationship. Even parasitism is a natural testimony of Adam and Eve's sin. The Bible tells us that creation "groans" as a result of what happened in the Garden of Eden (Romans 8:22). The sickness and disease caused by parasites is a part of that groaning. Had Adam and Eve not sinned in the Garden, sickness and disease would not exist.

ON YOUR OWN

3.4 A euglena is in dim light. There is a bright spot of light not too far away, but the euglena does not move towards it. Instead, it seems to wander aimlessly. What organelle is not functioning properly in the euglena?

3.5 According to most biologists, an organism must always perform photosynthesis or chemosynthesis to be considered autotrophic. Why do these biologists say that the euglena is not autotrophic?

3.6 In a later section of this module, you will learn about members of genus *Plasmodium*, which can live in people. These microorganisms get food and shelter from the people they inhabit, but they cause malaria, which can be deadly. Is this an example of symbiosis? If so, what kind?

Phylum Ciliophora

Phylum Ciliophora contains protozoa that use **cilia** (sil' ee uh) to move. As a result, these organisms are called **ciliates**.

Cilia – Hairlike projections that extend from the plasma membrane and are used for locomotion

The cilia beat rhythmically to move the organism from place to place. Ciliates can be quite large (for a protozoan), with some species attaining lengths of three millimeters (about one-ninth of an inch). Ciliates vary in shape, ranging from cone-shaped to bell-shaped to foot-shaped. Members of this phylum mostly live in fresh water, preferring stagnant lakes and ponds. Members of the genus *Paramecium* offer a good view of the typical ciliate.

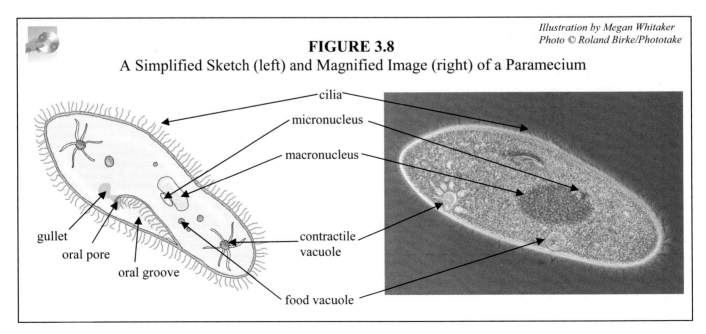

Illustration by Megan Whitaker
Photo © Roland Birke/Phototake

FIGURE 3.8
A Simplified Sketch (left) and Magnified Image (right) of a Paramecium

cilia
micronucleus
macronucleus
gullet
oral pore
oral groove
contractile vacuole
food vacuole

The most obvious feature of the paramecium is its **macronucleus**. This large nucleus controls the paramecium's metabolism. The paramecium and most other ciliates require a large amount of energy to live, mostly because of the rapid movement of their cilia. As a result, the metabolic control center must be large. In fact, the metabolic demands on the macronucleus are so large that there is a smaller **micronucleus** that controls reproduction. That way, the macronucleus can spend all of its resources controlling metabolism.

Like flagellates, paramecia (the plural of paramecium) and other ciliates have a pellicle that holds the cell's shape. Like sarcodines and flagellates, ciliates have contractile vacuoles that control water pressure within the cell. Like sarcodines, paramecia have food vacuoles to hold food while it is being digested. The paramecium, however, has a unique method of feeding. Paramecia have an **oral groove** that is lined with cilia. These cilia sweep food (bacteria, algae, and other ciliates) into the **gullet** through the **oral pore**. When it is full of food, the gullet then pinches off, becoming a food vacuole. The food vacuole then travels throughout the cell, floating in the cytoplasm. That way, the food is spread throughout the cell as it is digested. Any undigested portions of the food are expelled through an **anal pore**.

In addition to the asexual reproduction that is typical among protozoa, paramecia can also engage in a form of **conjugation**. In this process, two paramecia attach to one another at the openings of their oral grooves. Then, they exchange DNA with each other. Thus, unlike the conjugation that occurs among bacteria (which we studied in the previous module), this conjugation results in a mutual exchange of DNA. After the conjugation is complete, the two paramecia separate from each other and then immediately reproduce asexually, each making a new paramecium. Since there was exchange of DNA in the conjugation, the offspring produced in this way will not be exact duplicates of their parents prior to conjugation. Instead, their characteristics will be a blend of the characteristics of the two original paramecia.

Other Members of Phylum Ciliophora

While looking at pond water under the microscope, you might have seen another important ciliate, one from the genus *Stentor*. This ciliate can grow as large as one-tenth of an inch. Common in fresh-water habitats, members of this genus are shaped like a trumpet. They have cilia which surround their gullet. As the cilia beat back and forth, a micro-current in the water is created. This current sweeps food into the gullet, sometimes at the rate of a hundred protozoa each minute! Stentor eat all kinds of protozoa (including paramecia) and bacteria.

 The multimedia CD has microscopic video of an organism from genus Stentor.

Although most ciliates are not pathogenic, one particular species, *Balantidium* (bal' an tid' ee um) *coli* is a parasite of many species, including pigs, rats, and guinea pigs. These ciliates can form cysts that survive in the fecal matter of their host. If an individual eats food or drinks water contaminated with such infected fecal material, the individual can become infected as well. In humans, infection by *B. coli* can result in severe dysentery. *B. coli* infection in humans is not common in the U.S., but it is common in tropical regions of the world where malnutrition occurs and where pigs and humans live in close proximity to one another.

ON YOUR OWN

3.7 A paramecium cannot conjugate. What organelle is not functioning properly?

3.8 A biologist studies a group of bacteria (all one species) and a group of paramecia (all one species). She notices that while the bacteria all seem to be almost exact duplicates of each other, there is a great deal of variation among the paramecia. Why?

Phylum Sporozoa

This phylum contains the protozoa that have no real means of locomotion. There are more than 30,000 species in this phylum, and they are all parasitic. The main characteristic of the sporozoa, however, is the fact that they form **spores** at some point in their life.

Spore – A reproductive cell with a hard, protective coating

Much like a cyst, a spore can survive for quite a while, even in unfavorable conditions. Sporozoa typically form their spores as a result of a unique form of asexual reproduction, illustrated in the figure below.

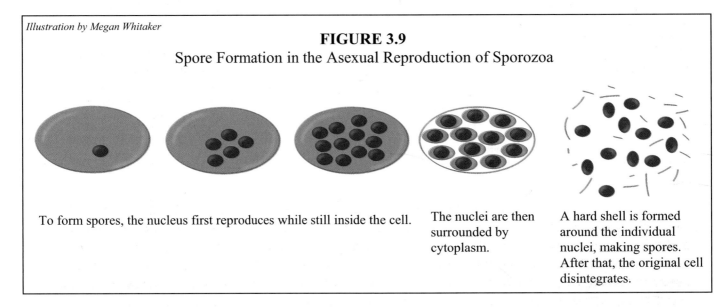

Illustration by Megan Whitaker

FIGURE 3.9
Spore Formation in the Asexual Reproduction of Sporozoa

To form spores, the nucleus first reproduces while still inside the cell.

The nuclei are then surrounded by cytoplasm.

A hard shell is formed around the individual nuclei, making spores. After that, the original cell disintegrates.

In this form of reproduction, the nucleus divides while still inside the cell. This may happen several times, resulting in several different nuclei in one cell. Cytoplasm concentrates around each nucleus, and hard shells form around the cytoplasm. This makes several spores. The cell then disintegrates, allowing the spores to be released so that each can grow into a new sporozoan. Spore formation is not the only type of reproduction in which sporozoa engage, however. The best way to illustrate this is by example.

Genus *Plasmodium* (plaz' moh dee um) is home to very deadly sporozoa. These parasites cause **malaria**, a disease that has claimed countless lives over the years. Spread by the mosquito, members of genus *Plasmodium* have a very interesting and complex life cycle that is typical of the sporozoa. This life cycle is illustrated in Figure 3.10 on the next page.

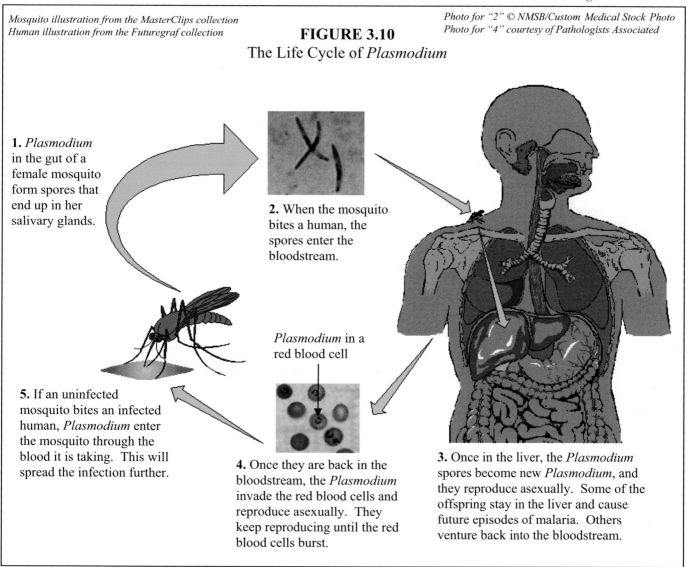

FIGURE 3.10
The Life Cycle of *Plasmodium*

1. *Plasmodium* in the gut of a female mosquito form spores that end up in her salivary glands.

2. When the mosquito bites a human, the spores enter the bloodstream.

Plasmodium in a red blood cell

5. If an uninfected mosquito bites an infected human, *Plasmodium* enter the mosquito through the blood it is taking. This will spread the infection further.

4. Once they are back in the bloodstream, the *Plasmodium* invade the red blood cells and reproduce asexually. They keep reproducing until the red blood cells burst.

3. Once in the liver, the *Plasmodium* spores become new *Plasmodium*, and they reproduce asexually. Some of the offspring stay in the liver and cause future episodes of malaria. Others venture back into the bloodstream.

In the gut of a female mosquito, *Plasmodium* reproduce and form spores which end up in the mosquito's salivary glands. When the mosquito bites a human, it injects saliva into the bite wound. This saliva keeps the blood from clotting, making it easy for the mosquito to ingest blood. The saliva also contains spores of *Plasmodium*. These spores lose their hard shells to become new *Plasmodium*. They then reproduce in the human's liver. After that, they enter the bloodstream, where they invade the red blood cells (cells that carry oxygen throughout the body). The *Plasmodium* reproduce again inside the red blood cells, causing them to burst. As the infection progresses, the symptoms (fever, chills, and shaking) become more and more severe until death occurs. *Plasmodium* spread because once in the bloodstream, they will infect any other mosquito that bites an infected human. Thus, infected humans infect mosquitoes, which in turn infect more humans!

Plasmodium must form spores while they are in the mosquito, because the salivary glands do not provide a habitat that is suitable for their survival. However, once inside the infected human, conditions are ideal for the *Plasmodium*, so future reproduction does not require the formation of spores. When a female mosquito bites an infected human and becomes infected herself, the *Plasmodium* must once again form spores in order to survive until they can be injected into another human. This is why we say that *Plasmodium* sometimes engage in reproduction that forms spores and sometimes engage in reproduction that does not. This type of life cycle is often called **alternation of generations**, because different generations alternate between spore formation and other modes of reproduction. You will learn more about this kind of life cycle in a later module.

Another typical sporozoan is *Toxoplasma* (tox' oh plaz muh) *gondii* (gon' dee). These organisms live in the intestines of mammals (such as cats) and reproduce there sexually. Although they sometimes can cause serious illness, they usually do not harm the cats. However, they often exit the animal with the feces (solid waste). While in the feces, *T. gondii* form spores so that they can survive. These spores are then spread through houseflies, cockroaches, insects, and direct contact with the cat feces. Once they infect a human, they reproduce asexually, causing the disease **toxoplasmosis** (tox' oh plaz moh' sis). This disease causes severe birth defects in pregnant women, which is why doctors tell pregnant women that they should never empty litter boxes or otherwise clean up after cats.

ON YOUR OWN

3.9 What is the difference between cysts (such as those formed by amoebae and ciliates) and spores?

3.10 One way that people fight the spread of malaria is to significantly reduce the population of mosquitoes in their vicinity. Why does this work?

Before we go on to discuss subkingdom Algae, perform the following experiment to review subkingdom Protozoa. If you do not have a microscope, please read through the experiment to get an idea of what it covers.

EXPERIMENT 3.2
Subkingdom Protozoa

Supplies:

♦ Microscope
♦ Prepared slide: amoeba
♦ Prepared slide: paramecium
♦ Prepared slide: euglena
♦ Prepared slide: volvox

Object: To become familiar with subkingdom Protozoa. You will examine the structures inside the organisms, which will help you understand how these organisms live, breathe, feed, move about, and rid themselves of waste.

Procedure:

As you look at each of the prepared slides, try to find several individuals and try to notice differences between them. Observe them at all three magnifications, in order to get a really good feel for what they look like. Once you have gotten a good feel for the general nature of the genus that you are observing, find an individual whose organelles are pronounced and easily visible. Concentrate on that individual when you make your drawing.

A. **Slide 1: Amoeba, Phylum Sarcodina**

1. Observe the prepared slide of amoebae. This slide has been stained to make the organelles easier to view. Scan the slide until you find an amoeba whose organelles are easy to see. Observe it on high magnification.
2. Draw what you see.
3. Try to find the structures listed below. You might not be able to find them all. Just do the best you can. Label the structures that you find in your drawing.
   a. pseudopod
   b. food vacuole
   c. contractile vacuole
   d. ectoplasm
   e. nucleus
   f. endoplasm

B. **Slide 2: Paramecium, Phylum Ciliophora**

1. Observe the stained slide of paramecia. Find a paramecium whose organelles are easy to see. Observe it on high magnification.
2. Sketch the paramecium, labeling the following structures, if you can find them.
   a. macronucleus
   b. micronucleus
   c. food vacuole
   d. contractile vacuole
   e. oral groove
   f. gullet
   g. mouth pore
   h. anal pore
   i. cilia

C. **Slide 3: Euglena, Phylum Mastigophora**

1. Observe the stained slide of euglena. Find a euglena that has organelles which are easy to see. Observe it on high magnification.
2. Sketch the euglena, labeling the following structures if you can find them.
   a. eyespot
   b. contractile vacuole
   c. pellicle
   d. chloroplast
   e. flagellum
   f. nucleus

D. **Slide 4: Volvox, Phylum Mastigophora.**

1. Observe the stained slide of volvox, sketching what you see.
2. The lattice of the colony is formed by individuals. Look for the individuals.

Put away the slides, and clean the microscope lenses with lens paper before you put it away.

Subkingdom Algae

Often called the "grass of the water," algae are organisms that can produce their own food by means of photosynthesis. If you've ever had a fish tank or a swimming pool, you've probably had a bit of experience with algae. If algae are allowed to grow unchecked in a body of water, the water becomes cloudy and takes on a greenish tint. In swimming pools, we get rid of algae by killing it with a chemical, usually chlorine. In fish tanks, we usually do not get rid of *all* algae, but instead impede the growth of the algae population with fish or other creatures that eat the algae.

Of course, natural bodies of water are the best places to find algae. Whether we are talking about marine (salt water) environments or freshwater environments, algae abound. In fact, the "fishy smell" and "slimy feel" of a body of water is usually due not to the fish under the water, but the algae floating on it. In any natural body of water, there are tiny floating organisms called **plankton** (plank' ton).

Plankton – Tiny organisms that float in the water

Biologists separate plankton into two groups: **zooplankton** (zoh' uh plangk' tun) and **phytoplankton** (fye' toh plangk' tun).

Zooplankton – Tiny floating organisms that are either small animals or protozoa

Phytoplankton – Tiny floating photosynthetic organisms, primarily algae

Based on these two definitions, then, we could say that up to this point, we have mostly been discussing zooplankton. Now we are ready to discuss phytoplankton.

As the definition states, phytoplankton are photosynthetic organisms, using the energy of sunlight to make their own food. As we mentioned in Module #1, oxygen is often the byproduct of photosynthesis. This process is, in fact, responsible for replenishing the oxygen that we need to survive. After all, virtually all of the organisms on earth need to breathe oxygen in order to survive. With all of these organisms breathing oxygen, why doesn't the earth run out of oxygen? Well, photosynthetic organisms are constantly making more. We will learn a lot more about photosynthesis in Module #5, so for now, you just need to understand that photosynthetic organisms continually replenish the earth's supply of oxygen.

Now when biologists mention photosynthesis, most people think of green plants, which are classified in kingdom Plantae. As we mentioned in Module #1, green plants do indeed use photosynthesis to create their own food. What you might not know, however, is that the *majority* of photosynthesis on earth is not done by green plants; instead, *it is done by phytoplankton*. This, of course, means that most of the oxygen on earth has been produced not by green plants such as trees and grasses, but instead by phytoplankton such as algae. In fact, biologists estimate that nearly three-fourths of all oxygen on earth is produced by phytoplankton. This is a rather important point. Contrary to popular belief, we could probably have enough oxygen to survive even if all green plants in creation were destroyed, because most of the earth's oxygen is produced by phytoplankton. Conversely, earth would quickly become uninhabitable if the waters of earth became too polluted for phytoplankton!

Algae are important for more reasons than just the oxygen they produce. Algae are also a major food source for many aquatic (water-living) organisms. In addition, humans have found quite a number of uses for algae. Japanese people cultivate it as a food crop. In New England and Hawaii, certain types of marine algae are eaten as vegetables. Other algae are used for food additives. Carrageenan (kehr' uh jee' nun), for example, is a substance extracted from a species of algae commonly called Irish moss. It is used to thicken such things as ice cream, sauces, and dressings. In addition, algae can be used as a natural kind of fertilizer. Finally, many useful products such as potash, iodine, nitrogen, vitamins, and minerals can be made from processed algae.

Many algae exist as individual cells, but most form simple colonies that are held together with slime. A few species of algae form highly complex colonies. A colony of algae is often referred to as a **thallus** (plural is thalli), but that term actually has a much broader definition.

Thallus – The body of a plant-like organism that is not divided into leaves, roots, or stems

We often refer to algae colonies as thalli because they function like a big plant, but there are no real roots, leaves, or stems. If you have trouble picturing this, don't worry. We will show you some pictures in a later section.

Algae, whether in colonies or existing as individuals, have both asexual and sexual reproduction at their disposal. This is one reason that they are so abundant in aquatic environments. In fact, when conditions are ideal, algae will reproduce so rapidly that they essentially take over their habitat, making the water appear the same color as the algae themselves. When this happens, it is referred to as an **algal bloom**.

Now that you've had a general introduction to algae, it is time to examine each individual phylum so that we can learn more about this interesting subkingdom of Protista.

Phylum Chlorophyta

The more than 7,000 species of phylum Chlorophyta are mostly found in fresh water, although some marine species do exist. The most visible feature of these algae is that they contain the pigment chlorophyll, which is green. As a result, they are often referred to as **green algae**. Just like the members of genus *Euglena*, members of this phylum store their chlorophyll in organelles called chloroplasts. Despite their name, most members of this phylum also have yellowish pigments called carotenoids (kuh rot' en oydz), making them appear yellowish green.

The other distinguishing feature of this phylum is that its members have cell walls made of **cellulose** (sel' yoo lohs). We will learn more about cellulose in Module #5. For right now, you just need to know that cellulose is a substance composed of certain types of sugar.

Cellulose – A substance (made of sugars) that is common in the cell walls of many organisms

The cells that make up plants usually have walls made of cellulose. Since green algae have cell walls similar to plants and chlorophyll like most plants, some biologists actually consider them plants and place them into kingdom Plantae. Most biological classification schemes tend to place them in kingdom Protista, however, because they are microscopic and tend to exist as individual cells or simple colonies of cells.

The figure below shows magnified images of algae from three different genera in this phylum.

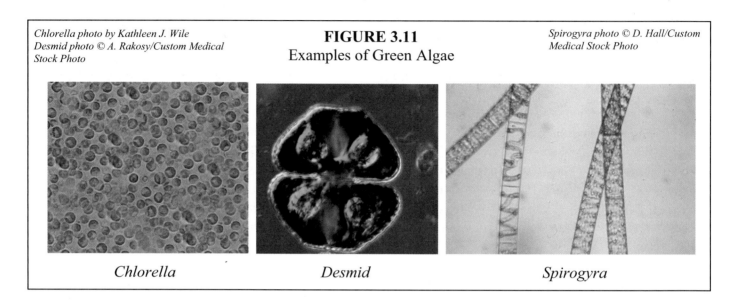

Chlorella photo by Kathleen J. Wile
Desmid photo © A. Rakosy/Custom Medical Stock Photo

FIGURE 3.11
Examples of Green Algae

Spirogyra photo © D. Hall/Custom Medical Stock Photo

Chlorella   *Desmid*   *Spirogyra*

The algae to the far left in the figure are from genus *Chlorella*. These algae exist as individual cells. Although they might clump together, the group does not cover itself with a common slime covering, nor do the individual cells work together in any way. Many of the species in this genus actually live inside other organisms, another example of mutualistic symbiosis. These algae use photosynthesis to produce food for both themselves and the organism in which they live. In return, they are protected from predators.

A species from genus *Desmid* is shown in the middle of the figure. Members of this genus sometimes form simple colonies, but they mostly exist as individual cells. Most species of *Desmid* have interesting shapes, usually being comprised of two halves that are pinched in the middle. These halves are usually mirror images of each other.

One of the more interesting genera of green algae is called *Spirogyra*, shown at the far right of Figure 3.11. Members of this genus get their name and unusual appearance from their spiral chloroplasts. They form colonies of slender, chainlike threads of cells. These colonies, called **filaments**, can reach up to 2 feet long.

ON YOUR OWN

3.11 If an organism is in phylum Chlorophyta, it must have a chloroplast. Why?

3.12 Of the three genera of green algae discussed above, which would you consider the most complex?

Phylum Chrysophyta

Although the green algae are responsible for a large portion of the photosynthesis that occurs on the planet, the greatest producers of oxygen in creation are in phylum Chrysophyta. This phylum contains more than 1,000 different species which are collectively referred to as **diatoms**. Diatoms are a unique type of algae, mostly because their cell walls are composed of **silicon dioxide**, which is the principal component of glass. This makes their cell wall very hard, providing excellent protection. The cell wall is so hard that it remains long after the diatom dies. When the cell wall remains of many dead diatoms clump together, they form a crumbly, abrasive substance called **diatomaceous earth** or diatomite.

Huge deposits of diatomaceous earth exist in most regions of the world. Many creation scientists think that these deposits were laid down in the worldwide Flood described in the Bible (Genesis chapters 6-9). This makes sense because the catastrophic nature of the flood would be responsible for killing huge numbers of organisms, including diatoms. The currents caused by the flood would then tend to sweep the remains of the diatoms together and lay them in one place. Interestingly enough, scientists who do not believe in the Bible have a difficult time explaining these huge deposits of diatomaceous earth.

Diatomaceous earth is actually quite useful for people. Large amounts of it are used by industry as a means of filtering liquids. It is also used as an abrasive. For example, most toothpastes contain an abrasive that helps clean and polish your teeth. Diatomaceous earth is often the abrasive of choice. Finally, crawling insects can be killed by laying down a thin coating of diatomaceous earth. As the insects crawl over the hard, jagged cell walls of these dead diatoms, it cuts their abdomens, killing them. The figure below contains magnified images of some diatoms.

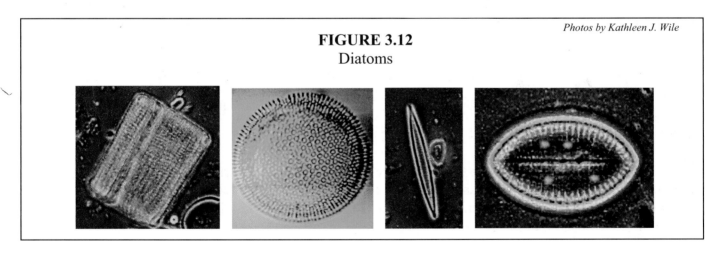

Photos by Kathleen J. Wile

FIGURE 3.12
Diatoms

If you look at the diatom on the far right side of the figure, you will see that there are small, light ovals inside the darker part of the organism. Although some people mistake these for flaws in the picture, they are, in fact another interesting feature of diatoms. When diatoms have excess food, they do not store it in food vacuoles as many other members of kingdom Protista do. Instead, they convert it into oil. The drops that you see in the figure are oil drops, which are stored food for the diatom.

Although diatoms make up a large part of phylum Chrysophyta, there are other organisms that belong to this phylum. Genus *Dynobryon* (dye noh' bree uhn), for example, contains algae that form colonies. These colonies typically contain a few special cells called **holdfasts** which are designed to hold on to objects in the water such as rocks.

<u>Holdfast</u> – A special structure used by an organism to anchor itself

These holdfasts form long strands that attach to a surface in the water, acting like an anchor. The colony is then not at the mercy of the currents. When a colony uses holdfasts, it is usually called a **sessile colony**.

<u>Sessile colony</u> – A colony that uses holdfasts to anchor itself to an object

If you were unfortunate enough to drink water that contained *Dynobryon*, it would taste fishy and feel quite slimy on the tongue.

ON YOUR OWN

3.13 You might say that diatoms are the most important form of algae in creation. Why?

3.14 Suppose you could analyze a water sample for the presence of any chemical. If you were given two samples, one that contained members of phylum *Chlorophyta* and another that contained members of phylum *Chrysophyta*, how could you determine which was which?

<u>Phylum Pyrrophyta</u>

Phylum Pyrrophyta contains a group of single-celled creatures that are often referred to as the **dinoflagellates**. They get their name because most species have two flagella, one of which is in a groove that encircles the cell. The figure below shows a drawing of a dinoflagellate.

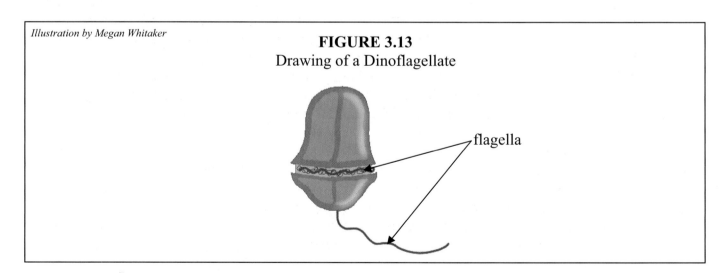

Illustration by Megan Whitaker

FIGURE 3.13
Drawing of a Dinoflagellate

flagella

Some of these organisms are heterotrophic, some are photosynthetic, and they mostly inhabit marine waters. Like the green algae, their cell walls are composed of cellulose. The dinoflagellates that are photosynthetic are an important source of food for many aquatic organisms, including other forms of plankton.

The most important thing to remember about the dinoflagellates, however, is that certain species (*Gymnodinium brevis*, for example) frequently bloom in nutrient-rich waters. Because the species are reddish-brown in color, their bloom tends to turn the water red in their immediate vicinity. As a result, these blooms are often called **red tides**.

Red tides are very deadly to other marine creatures. Hundreds of thousands of fish can be killed in a single bloom of *Gymnodinium brevis*, because these organisms emit a toxin into the water. Under normal conditions, there are few enough dinoflagellates that the toxin never reaches levels that are dangerous to marine creatures. During a dinoflagellate bloom, however, the toxin reaches deadly proportions.

Interestingly enough, although fish and most other marine life find red tides deadly, clams, oysters, and mollusks are immune. Unfortunately, the toxin emitted by the dinoflagellate does build up in their bodies. As a result, heterotrophs that eat clams, oysters, or mollusks which were exposed to a red tide can become poisoned. Some human deaths can be attributed to eating clams, oysters, or mollusks that have been in a red tide. This is why seafood restaurants refuse to serve such dishes when a red tide occurs in the area from which they get their supplies.

ON YOUR OWN

3.15 In the book of Exodus (chapter 7), God caused several plagues to befall Egypt. In the first plague, all of the rivers turned to blood, the fish died, and the Egyptians could not drink from the rivers. Some have said that algae offer a natural explanation for this miracle. What algae are they referring to and why do they think this? Why is this not a good explanation?

Phylum Phaeophyta

Up to this point, the protozoa and algae we have studied have been single-celled creatures that exist either as individuals or as members of simple colonies. In fact, one of the classification rules you learned in Module #1 says that kingdom Protista includes eukaryotic organisms made up of a single cell or a simple association of single-celled organisms. Well, hopefully by now you have seen that every classification rule has its exceptions. We can hardly discuss anything about a kingdom or phylum without using phrases like "usually" or "most of the organisms." Well, the last two phyla we will discuss are exceptions to the general description of the members of kingdom Protista.

Phylum Phaeophyta is made up of about 1,500 species of multicellular organisms that inhabit the cold ocean waters. Now let's make sure that you understand the difference between a colony of single-celled organisms and a multicellular organism. In a colony of single-celled organisms, each organism can exist on its own. Although cells tend to group together for mutual protection and other such benefits, if a cell is split away from its colony, it can survive. In addition, although the cells tend to live together, they mostly perform life functions independently. They may secrete mucus that covers the entire colony, and they may move together as a group, but they perform the majority of life functions without help from the other members of the colony. In a multicellular creature, the individual cells are designed to specialize in individual tasks. The cells work together, each performing the tasks for which it is designed. In the end, the organism survives because the cells work together. Since the cells are specialists in only one or a few of the tasks necessary for life, a single cell that is separated from a multicellular creature usually cannot exist on its own.

As shown in Figure 3.14, members of phylum Phaeophyta, also called the **brown algae**, look a lot like plants. In fact, some biologists classify them as such. Most biologists, however, still think that they have more in common with algae than with plants, so we classify them within kingdom Protista, subkingdom Algae.

If you enjoy ice cream, pudding, salad dressing, or jelly beans, you can thank the members of this phylum. You see, one of the unique characteristics of the organisms in this phylum is that their cell walls contain **alginic acid**, commonly called **algin**. This substance is extracted from brown algae and used to make a thickening agent in the foods described above. In addition, algin is used in cough syrup, toothpaste, cosmetics, paper, and floor polish.

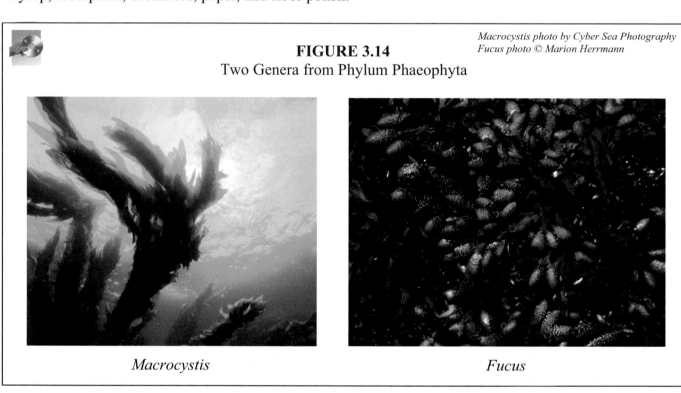

Macrocystis photo by Cyber Sea Photography
Fucus photo © Marion Herrmann

FIGURE 3.14
Two Genera from Phylum Phaeophyta

Macrocystis

Fucus

The most commonly known genus within phylum Phaeophyta is genus *Macrocystis* (ma' kroh sis' tus). Species within this genus are called **kelp** or seaweed, although those terms also seem to be used for many organisms within phylum Phaeophyta. Kelp and most members of phylum Phaeophyta form holdfasts that allow them to anchor themselves to rocks which sit at the bottom of the ocean. Some kelp can grow as long as 100 meters, growing as fast as 2 feet per day under ideal conditions. Kelp is harvested in many parts of the world for food. When many of these organisms grow near one another, large kelp forests result, providing a rich environment for many marine creatures.

Species in genus *Fucus* are often called **rockweed**. These algae are thick and have a leathery feel. They live in the shallow waters along the shoreline, and are generally one to two feet long. On this alga (singular of algae), you can clearly see the **air bladders**, which fill with air to allow the organism to float on top of the water.

ON YOUR OWN

3.16 A biologist has a sample of what looks to be a marine plant. He thinks, however, that it might be an unknown species of brown algae. To test this, he takes part of the "plant," dries it, crushes it into a powder, and mixes it with a solution. The solution thickens. Is this evidence that the organism is a plant or an alga? Why?

Phylum Rhodophyta

The last major phylum in subkingdom Algae is Rhodophyta. Members of this phylum are often called **red algae** because of their striking red color. People often get these algae confused with the dinoflagellates, because they know that some dinoflagellates cause red tides. The only thing that the red algae have in common with those dinoflagellates, however, is their color. Like the brown algae, members of phylum Rhodophyta are multicellular. Unlike the brown algae, however, they tend to live in warm marine waters rather than cold. Figure 3.15 shows two members of phylum Rhodophyta.

FIGURE 3.15

Photos © Bob Ford/Nature Portfolio

Two Genera from Phylum Rhodophyta

Genus *Corallina*   Genus *Hildenbrandia*

The specific species of *Corallina* shown in the figure is *C. officinalis*, which is often called "coral weed." It gets that name because it looks and even feels a lot like coral, a group of organisms common to warm waters. You will learn about coral later in the course. Members of genus *Hildenbrandia* (the red splotches on the rock in the figure) are thin algae that grow in clumps on underwater surfaces.

To finish your study of subkingdom Algae, perform the following experiment. If you do not have a microscope, please read through the experiment to get an idea of what it covers.

EXPERIMENT 3.3
Subkingdom Algae

Supplies:

♦ Microscope
♦ Prepared slide: *Spirogyra*
♦ Prepared slide: Diatoms

Object: To observe subkingdom Algae by observing two of the five phyla in this group

Procedure:

As you did in Experiment 3.2, get a good general feel for each slide by looking at several individuals on all three settings. Then, when you find an individual with pronounced organelles, concentrate on it.

A. **Phylum Chlorophyta: The green algae. Example - genus *Spirogyra***

1. Observe the prepared slide of spirogyra.
2. To get a feel for the three-dimensional nature of this organism, find a single filament and observe it on medium and then high magnification. At each magnification, gently adjust the fine focus. You should see parts of the filament blur out and other parts become more defined as you focus through the organism.
3. Sketch part of one filament, at least 4 cells.

B. **Phylum Chrysophyta: Example- Diatoms**.

1. Observe the prepared slide of diatoms. Note the incredible variety.
2. Sketch 3 or 4 different diatoms. In each case, look for oil spots, which are stored food.

NOTE: You do not have prepared slides of phylum Phaeophyta, phylum Rhodophyta, or phylum Pyrrophyta: The first two contain the macroscopic algae, and the last one contains the dinoflagellates.

Summing Up Kingdom Protista

As you can see, kingdom Protista is quite diverse. The organisms within it range from single-celled individuals (like the amoeba), to colonies of single-celled individuals (like the *Volvox*), to multicellular creatures (like the red and brown algae). Some are heterotrophic (like the paramecium), and some are autotrophic (like the green algae). Some are pathogens (like the *Plasmodium*), some are absolutely necessary for life (like the diatoms), and some produce substances which are incredibly useful to human beings (like kelp). Some of the organisms resemble animals (like the paramecium), and some resemble plants (like kelp).

Although there is a lot of diversity in this kingdom, there are a couple of things that all of these organisms have in common. First of all, they are all composed of eukaryotic cells. Secondly, they are all designed by God. As you look at the intricate features that exist in these organisms, God's incredible handiwork is readily apparent. Only He could produce the amazing diversity and complexity that exist in this kingdom. Remember, however, that the creatures in kingdoms Monera and Protista are, in fact, the ***simplest*** life forms that exist on the planet! As we study more of God's creation, you will become more and more appreciative of the awesomeness of His power!

ANSWERS TO THE "ON YOUR OWN" PROBLEMS

3.1 Biological Key for separating members of kingdom Protista into phyla:

1. Organism is completely autotrophic..*subkingdom Algae*.....…............. **2**
    Organism is heterotrophic ....*subkingdom Protozoa*...…..........**6**

2. Single-celled...……... **3**
    Multicellular **5**

3. Cell wall made of silicon dioxide ....*phylum Chrysophyta*
    Cell wall made of cellulose or is atypical **4**

4. Marine habitat ....*phylum Pyrrophyta*
    Fresh water habitat ....*phylum Chlorophyta*

5.* Cell wall made of cellulose and alginic acid ....*phylum Phaeophyta*
    Cell wall made just of cellulose ....*phylum Rhodophyta*

6. Possesses a means of locomotion **7**
    Possesses no means of locomotion ....*phylum Sporozoa*

7. Uses pseudopods for locomotion ....*phylum Sarcodina*
    Does not use pseudopods for locomotion **8**

8. Uses cilia for locomotion ....*phylum Ciliophora*
    Uses flagella for locomotion ....*phylum Mastigophora*

* This key could ask whether habitat is warm water or cold water. That would separate the two phyla as well.

Your key does not have to be in the same order, but it should contain these same kinds of questions. It might have a few more steps or a few less, depending on how you set up the key.

3.2 Since the contractile vacuoles control pressure in the cell by collecting and removing excess water, those organelles must not have been working in that poor amoeba.

3.3 The biologist must be looking at a food vacuole. A contractile vacuole should contain water. The solid objects in this vacuole are most likely partly-digested food.

3.4 Since it is wandering around, its flagellum is working fine. Since it can't find the light, however, its eyespot must not be working. NOTE: Even if the photosynthetic mechanism of the euglena is destroyed, it will continue to seek light as long as the eyespot is working.

3.5 Euglenas can obtain food either autotrophically (by photosynthesis) or heterotrophically, depending on environmental conditions. Thus, it does not *always* perform photosynthesis.

3.6 <u>This is symbiosis</u>, because the organism from genus *Plasmodium* benefits. It is also <u>parasitism,</u> since the person that the microorganism inhabits is harmed.

3.7 Since conjugation occurs via the <u>oral groove</u>, that organelle must not be functioning properly in the paramecium.

3.8 <u>Paramecia can engage in a form of conjugation that allows DNA to be mixed between organisms.</u> As a result, the offspring formed will not be an exact duplicate of the parents, because the parents have mutually exchanged DNA. <u>When bacteria conjugate, DNA transfer is one-way and the recipient ends up looking like the donor.</u>

3.9 <u>Spores are formed as a natural part of an organism's life</u>. An organism that produces spores will always produce spores as a part of its life cycle. <u>Cysts, on the other hand, are formed only when life-threatening conditions occur</u>. If no life-threatening conditions occur over the course of an organism's lifetime, it will never form a cyst.

3.10 <u>Mosquitoes carry the Plasmodium and spread it by biting humans. Control the mosquito population, and the spread of the disease is controlled as well.</u>

3.11 Chloroplasts hold the chlorophyll that is a part of photosynthesis. <u>Since all members of phylum Chlorophyta have chlorophyll, they must also have chloroplasts to hold it.</u>

3.12 <u>Spirogyra are the most complex because they exist as colonies. Thus, to a very limited degree, the cells work together.</u> The other two exist as individual cells. Any time you get a group of individuals to work together (even to a very small degree), you are adding complexity to the picture.

3.13 <u>Diatoms are responsible for the majority of photosynthesis on earth</u>. Without this photosynthesis, the earth's oxygen supply would quickly dwindle, killing off everything. Thus, diatoms are pretty important!

3.14 <u>You would look for the presence of cellulose or silicon dioxide</u>. The sample that contained members of phylum Chlorophyta would have cellulose, while the sample with members of phylum Chrysophyta would contain silicon dioxide. You might also look for oil, as diatoms store their excess food as oil. You cannot look for chlorophyll, as both diatoms and members of phylum Chlorophyta have chlorophyll, since they both perform photosynthesis.

3.15 <u>They refer to the *Gymnodinium brevis*, because it causes red tide</u>. Since the water in a red tide turns red, it might appear to be blood. Also, red tides are toxic to humans, so people cannot drink the water in a red tide. <u>This is not a good explanation, however, because the Bible does not say that the waters looked like blood; it says they *turned into* blood. Also, the water in containers also turned into blood. It would be hard to explain how a red tide could occur in a container of water that was once clear</u>. It is very dangerous to look for naturalistic explanations for clearly supernatural events!

3.16 <u>This is evidence that the organism is an alga. Since the water thickens, the plant most likely contains alginic acid, a substance that is</u> in the cell walls of brown algae.

STUDY GUIDE FOR MODULE #3

1. Give definitions for the following terms:

a. Pseudopod
b. Nucleus
c. Vacuole
d. Ectoplasm
e. Endoplasm
f. Flagellate
g. Pellicle
h. Chloroplast
i. Chlorophyll
j. Eyespot
k. Symbiosis
l. Mutualism

m. Commensalism
n. Parasitism
o. Cilia
p. Spore
q. Plankton
r. Zooplankton
s. Phytoplankton
t. Thallus
u. Cellulose
v. Holdfast
w. Sessile colony

2. Study the images of the organisms found in Figure 3.1. You will be expected to be able to place each of these organisms into the correct subkingdom and phylum by just looking at its picture. On the test, you will have a list of the subkingdoms and phyla; you will simply have to match them to the picture.

3. Which of the following genera contain organisms with chloroplasts?

Amoeba, Euglena, Paramecium, Spirogyra

4. What is the function of a contractile vacuole? What is the difference between this and a food vacuole?

5. What is the difference between endoplasm and ectoplasm?

6. The amoeba and euglena each have different means of locomotion. How are they different? How are they similar?

7. Name at least three pathogenic organisms from kingdom Protista.

8. For each of the phyla below, list the means of locomotion employed by the organisms in that phyla:

Sarcodina, Mastigophora, Ciliophora

9. What are the main features that separate organisms into phylum Sporozoa?

10. A tapeworm is a parasite that feeds on the nutrients which the host eats, depriving the host of that nutrition. *Trichonympha* is a mastigophorite that lives in the gut of a termite, helping break down chemicals that the termite cannot break down on its own. Name the specific kind of symbiosis in each of these situations.

11. Why do the ciliates have two nuclei (plural of nucleus)? What is the purpose of each?

12. What is the difference between the conjugation that occurs between paramecia and the conjugation that occurs between bacteria?

13. Two microorganism groups are studied. In the first group, the organisms form hard shells around themselves when exposed to life-threatening conditions. If not exposed to those conditions, however, these organisms never form hard shells. The second group form hard shells around themselves as a natural part of their life cycle. Which group would be classified as coming from phylum Sporozoa?

14. What is unique about the way a euglena obtains food?

15. Which phylum (see list in question #2) contains the organisms responsible for most of the photosynthesis that occurs on earth? What generic term is used to refer to these organisms?

16. Give the main function of each of the organelles listed on the left below. Also, choose from the list on the right at least one phylum that has organisms which possess the organelle.

Organelle	Phylum
Food vacuole	Sarcodina
Contractile vacuole	Mastigophora
Flagellum	Ciliophora
Pellicle	Sporozoa
Chloroplast	Chlorophyta
Eyespot	Chrysophyta
Cilia	Pyrrophyta
Nucleus	Phaeophyta
Oral groove	Rhodophyta

17. What are large deposits of diatom remains called? List two uses of these deposits.

18. What is a red tide?

19. What two phyla principally contain macroscopic algae?

20. What substance produced by members of phylum Phaeophyta is useful for thickening ice cream, pudding, salad dressing, and jelly beans?

MODULE #4: Kingdom Fungi

Introduction

Have you ever gone mushroom hunting? We mean other than in a store. Where did you search for the mushrooms? Most likely, you looked for them in decaying mats of leaves, piles of dead tree limbs, or other places where the remains of dead plants and animals are found. Why? Well, mushrooms are a part of kingdom Fungi, and most of the organisms in this kingdom are saprophytic. In other words, they are the decomposers that promote the decay of once-living matter. As we mentioned back in Module #1, the role of the decomposers in nature is very important. In a single autumn, the average elm tree will drop as much as 400 pounds of leaves on the ground. If it were not for fungi and other decomposers, those leaves would continue to pile up until the tree choked on its own dead leaves in just a few seasons! Because of the decomposers, however, the leaves will be broken down into chemicals that can then be re-used by the tree and other organisms in creation.

Although mushrooms are the most well-known fungi, there are many other organisms that make up this important kingdom. For example, the yeast that is used to make bread is a member of kingdom Fungi. The mold that later grows on that bread is also a fungus. There are some rather exotic creatures such as slime molds that belong to this important kingdom as well. Some fungi are beneficial to us beyond their roles as decomposers. We eat some fungi; other fungi are used in the production of cheese; others are used in baking; and some even produce important medicine. Still other members of kingdom Fungi are pathogenic. Diseases such as St. Anthony's fire, histoplasmosis, potato wart, and Dutch elm disease are all caused by fungi. A study of kingdom Fungi, then, should prove to be quite interesting.

NOTE: In order to perform Experiment 4.3, you will need to grow some mold on bread, jelly, and/or fruit. You need to start that process now. To do this, take a slice of bread (sprinkle a little water on it), a sample of jelly, and a piece of sliced fruit and set them out in the open. Putting them outside works best, but you will need to find a place where the birds and other animals cannot get to them. If you can't find such a place outside, put them somewhere like a garage or shed. They will attract fruit flies, so do not put them in the house! If you live in a very arid climate (like Arizona), put the samples in a plastic bag with some water. Keep the plastic bag open and on its side. This will keep the samples moist, which is necessary for the growth of molds, but it will also keep them exposed to air. As you read through the first part of this module, the mold should start to grow. By the time you reach Experiment 4.3, at least one of the samples should have mold on it. The experiment has non-microscope components, so you should prepare for it whether or not you have a microscope.

General Characteristics of Fungi

Although very diverse, the organisms in kingdom Fungi do not vary nearly as widely as those in kingdom Protista. As a result, we can identify a few general characteristics for the vast majority of fungi. For example, most fungi are heterotrophs. Most are saprophytic (feeding on the remains of dead organisms), although there are a few species that are parasitic (feeding on living hosts). Whether saprophytic or parasitic, however, fungi digest their food outside of their bodies. The fungus grows on and in its food, secreting a chemical onto the food that digests it *before* it is ingested. The digested food is then absorbed into the cells of the fungus, as a quick source of nutrition and energy. This

extracellular digestion can be beneficial to other organisms, which often absorb some of the nutrients before the fungus has a chance to absorb them.

Extracellular digestion – Digestion that takes place outside of the cell

If a fungus is harmful to other creatures, it is often the substances that the fungus excretes for extracellular digestion which are responsible for the harm that it inflicts.

Another trait common to all fungi is an aspect of reproduction. All fungi reproduce by making spores. Most fungi have other means of reproduction at their disposal as well, but they all have this mode of reproduction in common. If the fungus is multicellular (as most are), the fungus will form a specialized structure to produce the spores. Once the spores are produced, they are dispersed from the fungus and begin to develop on their own. In the next section, we will look at the reproduction of fungi in detail. First, however, we need to learn a few more general characteristics of fungi.

The vast majority of fungi are multicellular creatures. The largest part of their body is the part that performs extracellular digestion and absorbs the digested food. This part of the fungus, called the **mycelium** (my sell' ee uhm) is composed of many interwoven filaments called **hypha** (hi' fuh).

Mycelium (plural is mycelia) – The part of the fungus responsible for extracellular digestion and absorption of the digested food

Hypha (plural is hyphae) – A filament of fungal cells

Figure 4.1 is a sketch of a typical mushroom. Notice the mycelium that exists below the mushroom's stalk. It is not uncommon for a mushroom's mycelium to be ten or twenty times as large as its stalk.

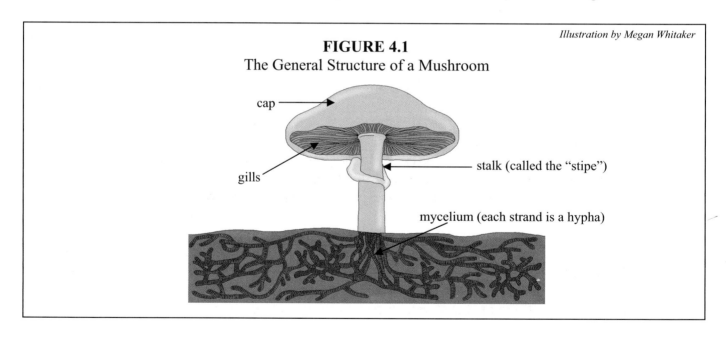

Illustration by Megan Whitaker

FIGURE 4.1
The General Structure of a Mushroom

cap

gills

stalk (called the "stipe")

mycelium (each strand is a hypha)

Now don't be fooled by this drawing. The mycelium is *not* a root system for the mushroom. Although the hyphae might look a lot like roots, there are *many* differences between a root system and the mycelium of a mushroom. A root system, for example, has one purpose: to pull nutrients and water from the soil so that they can be transported to the rest of the plant. The mycelium of the mushroom, on the other hand, is the *main part* of the mushroom. We will learn later that the stalk, cap, and gills of a mushroom exist only at a certain stage of the mushroom's life. The mycelium, however, exists throughout the entire life of the mushroom. Thus, whereas the root system is really just an extension of the tree, a mushroom's stalk, cap, and gills are really just an extension of the fungus's main body – the mycelium.

In some fungi, the hyphae are composed of individual cells separated from one another by cell walls. These hyphae are called **septate hypha**. Even though the individual cells are separated from one another by a cell wall, there is usually a hole or pore through which cytoplasm can be passed between the cells. Other fungi have hyphae that look like one big cell. There are no walls, and the nuclei are spread throughout the hypha. These hyphae are called **nonseptate hyphae**. The difference between these two types of hyphae is illustrated in the figure below.

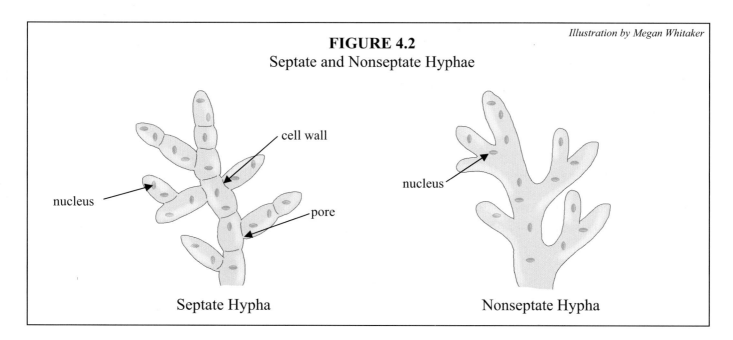

FIGURE 4.2
Septate and Nonseptate Hyphae

Illustration by Megan Whitaker

Septate Hypha

Nonseptate Hypha

Even when a fungus is composed of septate hypha, its cells are not completely separated from one another, because the pores in the cell walls allow cytoplasm to be passed between the cells. This is a characteristic that is unique to kingdom Fungi. Colonial protistans and monerans have cells that group together, but the cells do not exchange cytoplasm. Plants and animals are multicellular, but their individual cells are completely separate; they do not exchange cytoplasm. In kingdom Fungi, the cells are not completely separate.

Many fungi have hyphae that are designed to perform specific tasks. If a hypha is part of the mycelium, for example, it is called a **rhizoid** (rye' zoyd) **hypha**.

Rhizoid hypha – A hypha that is imbedded in the material on which the fungus grows

Rhizoid hyphae are responsible for supporting the fungus and digesting the food. As stated before, these hyphae are considered the main body of the fungus and are illustrated in Figure 4.3 on the next page.

Other types of hyphae can exist in a fungus, depending on the species and the circumstances. For example, an **aerial hypha** is not imbedded in the material on which the fungus grows.

Aerial hypha – A hypha that is not imbedded in the material upon which the fungus grows

As its name implies, an aerial hypha sticks up in the air. It can do one of three things: absorb oxygen from the air, produce spores, or asexually reproduce to form new filaments. If the hypha performs one of the latter two jobs, it is further specified as either a **sporophore** (spor' uh for) or a **stolon** (stoh' lun), respectively.

Sporophore – Specialized aerial hypha that produces spores

Stolon – An aerial hypha that asexually reproduces to make more filaments

Often, a sporophore will form its spores within an enclosure. When this happens, the sporophore is called a **sporangiophore** (spuh ran' jee uh for'). If no enclosure is made, the sporophore is called a **conidiophore** (kuh nid' ee uh for'). Sporophores are not the only way that fungi produce their spores. Likewise, stolons are not the only way a fungus asexually reproduces. In fact, many fungi do not form sporophores or stolons at all. Nevertheless, these two specialized hyphae are means by which some fungi reproduce.

In the case of a fungus that feeds on a living organism, a hypha can actually enter the cells of the living organism and draw nutrients directly from the cytoplasm of those cells. This kind of hypha is an extension of the mycelium and is called a **haustorium** (haw stor' ee uhm).

Haustorium – A hypha of a parasitic fungus that enters the host's cells, absorbing nutrition directly
    from the cytoplasm

Examples of these specialized hyphae are shown in Figure 4.3. As the figure implies, not all fungi have all of these structures. The only type of specialized hyphae that most fungi have is the rhizoid hyphae, because they make up the mycelium.

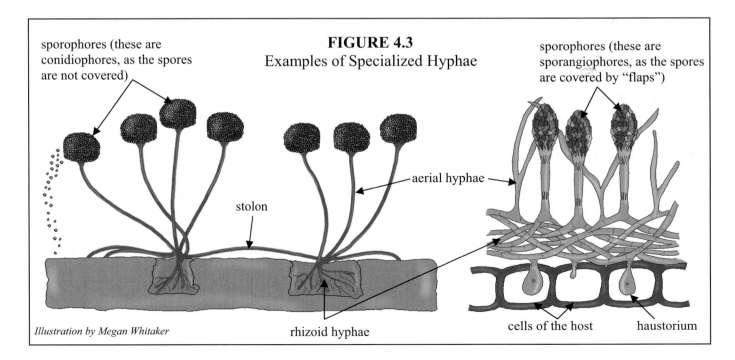

FIGURE 4.3
Examples of Specialized Hyphae

sporophores (these are conidiophores, as the spores are not covered)

sporophores (these are sporangiophores, as the spores are covered by "flaps")

aerial hyphae

stolon

rhizoid hyphae

cells of the host   haustorium

Illustration by Megan Whitaker

The final common characteristic of fungi refers to the cell wall. Most fungi have cell walls that contain **chitin** (kye' tin).

<u>Chitin</u> – A chemical that provides both toughness and flexibility

This chemical makes fungi hardy, because it provides protection, much like armor provided protection to the knights of old. Interestingly enough, this same chemical is found in the shells (called exoskeletons) of arthropods, such as spiders, beetles, ants, and lobsters.

ON YOUR OWN

4.1 An organism eats food and then digests it. Does this organism belong to kingdom Fungi? Why or why not?

4.2 A farmer tries to remove a patch of mushrooms from his field by pulling all of the stalks and caps in the patch out of the ground. Why has the farmer really not gotten rid of the fungus?

4.3 A fungus produces haustoria (plural of haustorium). Is it saprophytic or parasitic?

<u>Reproduction in Kingdom Fungi</u>

One reason fungi are so plentiful is that they have multiple means of reproduction at their disposal. As we mentioned previously, all fungi reproduce by making spores. Sometimes spores are formed asexually, but all fungi are assumed to have some sexual mode of spore formation as well. Asexual spore formation is accomplished by a hypha that becomes either a sporangiophore or a conidiophore. The spores are formed and then dispersed. Some fungi produce spores that have flagella for locomotion, but most spores are just carried by the wind until they land. If a spore finds itself in a suitable environment, it can grow into a hypha and form a new mycelium.

The sexual reproduction that usually occurs in fungi involves forming specialized structures called **fruiting bodies**. These fruiting bodies form as the result of sexual reproduction between compatible hyphae. Once the fruiting body is formed, it rises out of the mycelium and releases its spores. The cap and stalk that we normally call a mushroom are, in fact, just parts of the fruiting body of the mycelium of a fungus. In some fungi, sexual reproduction does not lead to a fruiting body. It just directly produces a new hypha, which will mature into a completely different mycelium.

Although spore formation is the reproductive mode common to all fungi, most can also reproduce by other asexual means. The hyphae cells in the mycelium can reproduce asexually, increasing the size of the mycelium. Also, when a stolon is formed, the cells within it will reproduce asexually, lengthening the stolon. After the stolon reaches a certain length, it will begin to reproduce into hyphae that will form the mycelium of a new fungus. Because the stolon is still attached to both the parent fungus and the new fungus, these fungi will always be attached by the stolon that gave rise to the new fungus. The new fungus can then form another stolon, which will give rise to yet another fungus. This process can be repeated over and over again, so it is not uncommon to find long chains of fungi, all linked together by stolons.

ON YOUR OWN

4.4 What job does the fruiting body of a fungus perform?

4.5 Spores of a fungus give rise to offspring that are identical in every way to the parent. Were the spores formed asexually or sexually?

Classification in Kingdom Fungi

Kingdom Fungi is divided into six phyla: **Basidiomycota** (buh sid' ee oh my' koh tuh), **Ascomycota** (ask' uh my' koh tuh), **Zygomycota** (zye' goh my' koh tuh), **Chytridiomycota** (kye' trid ee oh my' koh tuh), **Deuteromycota** (doo' ter oh my' koh tuh), and **Myxomycota** (myk so my' koh tuh). Organisms are placed within each phylum based mostly on their sexual reproduction. Members of phylum Basidiomycota form their spores on clublike structures called **basidia**, while the members of phylum Ascomycota form their spores in saclike structures called **asci** (as' kye). Phylum Zygomycota contains fungi that make their spores, called **zygospores**, in small structures that are formed where hyphae fuse with one another for the purpose of reproduction. The reproduction of fungi in phylum Chytridiomycota involves spores that have flagella and can therefore move on their own.

Phylum Deuteromycota is a phylum where we put fungi that have reproductive methods which are not completely characterized. As we have said before, all fungi reproduce by forming spores. Although some fungi have a means of forming spores asexually, all fungi in the other phyla have at least one mode of sexual reproduction that forms spores. We assume, therefore, that all fungi have at least one mode of sexual spore formation. There are some species of fungi, however, for which sexual reproduction is unknown. We place these fungi into phylum Deuteromycota, expecting that more research will eventually lead us to the discovery of a sexual, spore-forming mode of reproduction. If such a mode is found for a fungus in this phylum, it is reclassified into one of the other phyla. Of course, it is possible that some fungi have no sexual mode of reproduction. Thus, it is possible that some fungi will never move out of phylum Deuteromycota.

The last phylum in kingdom Fungi, Myxomycota, is rather controversial. This phylum contains the strange organisms known as **slime molds**. These interesting creatures grow in moist habitats on the decaying remains of living creatures (particularly trees) or in some cases exist as parasites that feed on living plants. They are typically brightly colored and slimy to the touch. In many ways, they resemble protozoa. Most of the time, they behave like a colony of single-celled creatures. However, at some point in their life cycle, many of them produce sporophores for reproduction. Since a sporophore is generally associated with a fungus, slime molds tend to resemble fungi during this stage of their life. Since the slime molds resemble protozoa during a part of their life cycle and fungi during another part of their life cycle, there is much controversy as to whether to place these creatures in kingdom Protista or kingdom Fungi. We choose to place them in kingdom Fungi for two reasons. First, since they reproduce by forming spores from specialized structures, they seem to fit in with the other fungi. Also, they are found in the same type of habitats as fungi, so it only seems natural to consider them a part of this kingdom. Be warned, however, that the next biology book you read might classify the slime molds as belonging to kingdom Protista! Table 4.1 sums up the classification of fungi into phyla.

TABLE 4.1
Phyla in Kingdom Fungi

Phylum	Characteristics
Basidiomycota	**Form sexual spores on clublike basidia**
Ascomycota	**Form sexual spores in saclike asci**
Zygomycota	**Form sexual spores where hyphae fuse**
Chytridiomycota	**Form spores with flagella**
Deuteromycota	**Fungi with no known method of sexual reproduction**
Myxomycota	**Fungi that look like protozoa for much of their lives**

Do not try to memorize the names of the phyla. They will always be given to you on tests. Instead, learn the characteristics that separate organisms into these phyla.

ON YOUR OWN

4.6 Construct a biological key that separates organisms into the phyla of kingdom Fungi. For this key, you can assume you already know that any organism you have is a part of kingdom Fungi. Thus, you just need to use the characteristics in Table 4.1 to construct the questions that will separate the organisms into the proper phyla.

Phylum Basidiomycota

Often referred to as the "club fungi," members of phylum Basidiomycota form spores (called **basidiospores**) on club-shaped cells known as basidia (singular is basidium). These spores are the result of sexual reproduction between mycelia. Mushrooms are the most common examples of the members of this phylum. As we pointed out in the previous sections, the cap and stipe of the mushroom actually make up only the fruiting body of a vast network of mycelia that exist below the surface upon which the mushroom grows. Example members of phylum Basidiomycota are shown in Figure 4.4. As you can see from the figure, most of the fungi with which you are familiar fall into this phylum. Most of these fungi are saprophytic, but a few (which we will discuss later) are parasitic.

Mushrooms photo © Toya Finn
Puffballs photo © Laurie Knight

Shelf fungi photo © Kristen Eckstein

FIGURE 4.4
Phylum Basidiomycota

Mushrooms   Puffballs   Shelf Fungi

The life cycle of a club fungus is rather interesting, and it is illustrative of the complex nature of most fungi. Therefore, we will study it in some detail. Begin by examining the figure below.

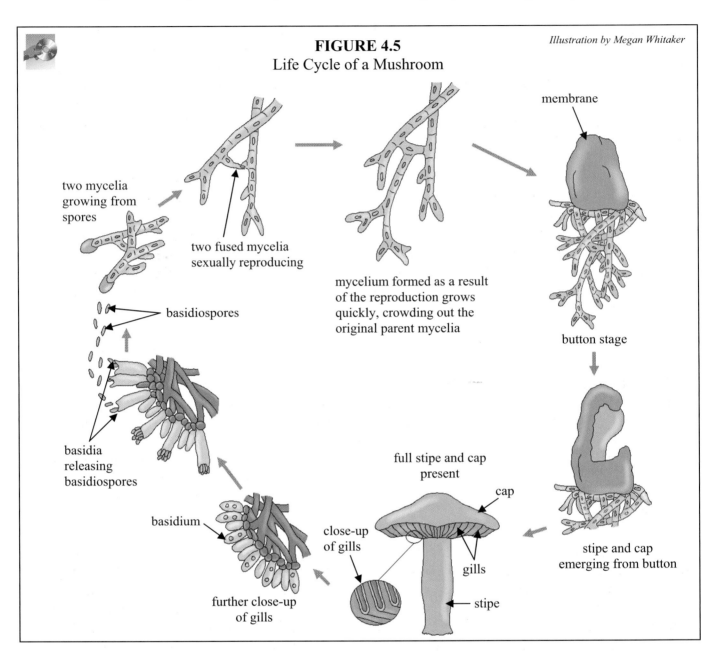

FIGURE 4.5
Life Cycle of a Mushroom

Illustration by Megan Whitaker

two mycelia growing from spores

two fused mycelia sexually reproducing

mycelium formed as a result of the reproduction grows quickly, crowding out the original parent mycelia

membrane

button stage

basidiospores

basidia releasing basidiospores

basidium

further close-up of gills

close-up of gills

full stipe and cap present

cap

gills

stipe

stipe and cap emerging from button

A mushroom begins life as a small mycelium that grows from spores which have come from another mushroom. As the mycelium begins to grow, it might encounter another mycelium with which it can mate. As the two mycelia begin to intertwine, their hyphae sexually reproduce. They accomplish this by aligning themselves parallel to each other and forming a small junction. At this point, we say that the hyphae are **fused**. Once fused, the hyphae exchange DNA and form a new mycelium. This new mycelium grows so quickly that it takes over, completely crowding out the original parent mycelia. Eventually, hyphae from this new mycelium will form a complex web and enclose themselves in a **membrane**.

Membrane – A thin covering of tissue

This membrane-enclosed web of hyphae (often called a "button") is the beginning of the mushroom's fruiting body.

When the hyphae are formed in the membrane, we say that the mushroom has reached the **button stage** of its existence. At this point, the hyphae begin filling with water quickly, and eventually the stipe and cap of the mushroom break through the membrane. The hyphae fill with water so quickly during the button stage that the stipe and cap of a mushroom can literally "pop up" out of the membrane overnight. Since many mushrooms are hard to see in their button stage, it gives the illusion that a mushroom formed itself overnight, even though the mushroom that you see is the result of many days' growth.

The stipe and cap of a mushroom form the fruiting body of a fungus, and that body's main function is to produce spores, which can grow into new mycelia. The cap is full of **gills**, small plates that are lined with basidia. The basidiospores contained on the basidia are released from the mushroom, where wind and water carry them to a new location, starting the process all over again. Once the fruiting body of the mushroom releases all of its spores, it withers and dies. Of course, the fungus is still very much alive, because the mycelium is still digesting and absorbing nutrients below the surface. Eventually, more mycelia will sexually reproduce, clump together, and form another fruiting body.

The fruiting body of the mushroom, of course, is the part that we eat. Although most mushrooms are tasty and nutritious, there are some that are quite toxic. The genus *Amanita* (ah mah nee' tuh), for example, contains mushrooms that are commonly called "destroying angel" mushrooms. These pure white mushrooms carry a poison that is deadly to humans. If you eat one of these mushrooms, it tastes quite normal. However, after eating one of those mushrooms, you can die in as little as 16 hours! There is virtually nothing that can be done to treat a person who has eaten these deadly mushrooms. Since there are rarely distinguishing marks that separate poisonous mushrooms from nonpoisonous ones, the only place that you should hunt mushrooms to eat is in the grocery store. Many people who try to hunt wild mushrooms end up in a hospital or a coffin because the mushrooms that look tasty are, instead, toxic.

Sometimes, you can find mushrooms that grow in an almost perfect circle (see Figure 4.6 on the next page). Inside or outside of this ring of mushrooms, no other mushrooms grow. These rings, often called **fairy rings**, are believed by some to have magical properties because of their unique appearance. Of course, there is no magic associated with a fairy ring. Instead, it is just a result of the saprophytic nature of the fungus. You see, when a fungal mycelium begins to grow in an area, it eats the remains of dead organisms. As it eats, it grows and reproduces. Eventually, the mycelium will spread out in all directions, making a relatively circular patch of hyphae. Once the hyphae in the center of the circle eat up all of the remains of dead organisms, there is no more food for them, and they die. The hyphae at the edge of the mycelium, however, still have food, because they haven't existed for as long as those at the center, and they therefore have not used up the food in their area. As a result, they continue to live, and the mycelium becomes ring-shaped. When it is time to reproduce, then, the ring produces fruiting bodies, forming a fairy ring.

As time goes on, the mycelium continues to grow outward, and the inner hyphae continue to die because they use up their food. As a result, the mycelium retains its ring shape, but the ring gets larger in diameter. When the reproduction cycle comes again, a new ring of stipes and caps are formed, and this ring is larger than the old one. This happens year after year. Some fairy rings have been found in which the ring of mushrooms is only a few inches wide, but it has a diameter of nearly 20 feet!

FIGURE 4.6
A Fairy Ring

Photo © Lauren Frederick

Other Members of Phylum Basidiomycota

Phylum Basidiomycota is also home to the **puffball** fungi (middle of Figure 4.4). Puffballs, which are saprophytic, produce their spores on basidia inside a membrane, rather than in the gills of a cap. When pressure is exerted on the membrane by a passing animal or a heavy wind, the spores are "puffed" out through a hole near the top of the puffball. The spores, which are as fine as dust, are often carried on the wind for several miles before they hit the ground. As a result, you are unlikely to find dense patches of puffballs.

The **shelf fungi** (right side of Figure 4.4) are generally found either on dead wood or on living trees. If you find them on dead wood, they are obviously one of the saprophytic species of shelf fungi, while those found on living trees are the parasitic species. Although parasitic fungi are uncommon, they do exist, and phylum Basidiomycota is home to some of them.

The spores of the shelf fungi are formed in the pores of the shelves. These spores are also very fine, so that when they are released, they can travel great distances to find another tree on which to grow. Once again, the shelves are just the fruiting bodies of these fungi. The mycelia are inside the wood of the tree. In fact, some parasitic shelf fungi actually add a new layer to the fruiting body each year, resulting in huge shelves growing out of the tree trunk.

Rusts are another form of fungi that are parasitic. They typically grow on living plants, reducing the plant's ability to grow and mature. If the rust happens to be living on a commercial crop, it can make the crop virtually useless as a source of food. One particularly bothersome form of rust is **wheat rust**. This fungus is well known for destroying tons of wheat crops over the course of history. Its life cycle is actually rather complex, because it requires two hosts, a **main host** and an **alternate host**. Both of them are necessary for the rust to complete its life cycle.

When wheat rust infects a wheat plant, it produces a red spore called a **uredospore** (yoo ree' duh spor). These red spores can look like rust, which is how the fungus gets its name. They can travel on the wind to other wheat plants and grow on them, destroying entire fields of wheat. As the wheat season ends, however, the wheat plants turn yellow, and the rusts form a different type of spore, called a **teliospore** (tee' lee uh spor). These spores survive the winter and then grow into basidia in the spring. The basidia produce basidiospores, but those spores cannot grow into fungi on wheat. Instead, they find their way to a barberry bush and grow on the underside of the leaves of this bush. They form tiny cups in which **aeciospores** (ee' see uh sporz) are produced. These spores can then find their way to wheat plants and grow into rust there. The rust, then, lives most of its life cycle on wheat, its main host, but in the spring it must spend a certain part of its life cycle on its alternate host, the barberry bush. One way to protect wheat crop from rust is to eliminate any barberry bushes surrounding the wheat field. Without nearby alternate hosts, the rust will not find it as easy to infect the wheat. The need for an alternate host is rather common among parasitic fungi.

Smuts are another group of parasitic fungi that belong in phylum Basidiomycota. These fungi also feed on crops such as wheat, barley, rye, and corn, resulting in millions of dollars worth of crop loss each year. Typically, farmers cannot control smuts or rusts chemically, because anything that kills the fungi also kills the plants upon which they live. Instead, agricultural scientists try to develop strains of wheat, barley, rye, and corn that are resistant to these fungi. Many advances have been made in such crop-related research and, as a result, the crops planted today are less likely (but not immune) to be ruined by fungi.

To finish our study of phylum Basidiomycota, perform the following experiment. Although this lab calls for a microscope, you can really perform all of the steps with just a magnifying glass. You won't see quite as much, but you will still learn a lot!

EXPERIMENT 4.1
Phylum Basidiomycota

Supplies:
- Microscope
- Magnifying glass
- Slides
- Coverslips
- Water
- Needle (or probe from your dissection kit)
- Mushrooms
- Puffballs
- Shelf fungi
- Gloves

Autumn, when leaves are dropping to the ground and there is abundant moisture, is a perfect time to find mushrooms, shelf fungi, and/or puffballs in yards and woods nearby. They decompose all of the dead organisms around them. If you cannot locate any specimens, purchase a few mushrooms from the store. **You only need one of the specimens in the list to do the experiment**. If you can find more, however, you will learn more.

Object: To observe fungi that are readily found in most areas and to understand how members of phylum Basidiomycota grow and reproduce

Procedure:

1. Go for a walk and look for mushrooms, puffballs, and shelf fungi in your area. A wooded area with fallen leaves or dead logs should have some fungi.
2. Put on the gloves and collect samples to take back home. Handle them with care. Do not breathe in the spores, especially the puffball spores. They can be an irritant to your respiratory system. Always wear gloves to handle the fungi, and wash your hands after you are done.
3. Observe each specimen you collected. Study it using a magnifying glass. Sketch what you see. If you collected a mushroom, label the stipe, cap, and gills.
4. Using a magnifying glass, look at the gills. Try to see the basidia and the basidiospores. The basidia are on the gills, and the basidiospores are in the powdery substance attached to the basidia. If you have shelf fungi, look for pores that hold the basidia.
5. If you have a mushroom, cut the stipe off at the base of the cap. Tap the cap lightly on a microscope slide to get some spores (which will look like powder) on the slide. If you have a puffball, look for a small opening in the membrane. Point that opening towards the slide and gently squeeze the membrane. Spores should "puff" out of the opening. **Be careful! Don't get the spores in your eyes while you are handling the puffball!** If you have a shelf fungus, just tap the sample against the slide. Spores should fall onto the slide. This may not work if your fungus isn't dry or has come from a store. If you cannot get any spores this way, you might try scraping the fungus with a knife or the scalpel from your dissection kit.
6. Place a drop of water and coverslip over the powder. Observe the sample at all magnifications. If you do not have a microscope, tap the mushroom against a small plate, add water, and then observe the spores with a magnifying glass. You should see oval-shaped basidiospores. Sketch a few.
7. If you have a mushroom, cut the cap vertically.
8. Use your needle or probe to pull a gill off the cap.
9. Place the gill on a slide.
10. Add a drop of water and coverslip. Observe on all magnifications. Once again, if you don't have a microscope, use a small plate and a magnifying glass. Note the basidia and the darker basidiospores. Sketch and label what you see.
11. Wash and dry all slides and eyedroppers.
12. Dispose of all fungi. They do not keep; do not try to save them.
13. **WASH YOUR HANDS!**
14. Clean the microscope lenses with lens paper and return all supplies to the proper places.

ON YOUR OWN

4.7 A mushroom is in its button stage. Has it released its spores yet?

4.8 One major characteristic that separates the members of phylum Basidiomycota into different subgroups is the structure in which the fungi form their basidia. Where are the basidia formed in mushrooms? What about in puffballs? Where are they formed in shelf fungi?

Phylum Ascomycota

The members of phylum Ascomycota are both single-celled creatures and multicellular organisms. They are generically referred to as **sac fungi**, because they form their spores in protective membranes (sacs) shaped like globes, flasks, or dishes. These sacs are called asci, and the spores inside are called **ascospores** (as' kuh sporz). The single-celled members of this phylum are generically called **yeasts**. The yeast that we use in cooking is an example of a single-celled member of phylum Ascomycota.

Cup fungi photo © Bob Ford/Nature Portfolio
Morel photo © Corbis, Inc.

FIGURE 4.7
Members of Phylum Ascomycota

Yeast photo by Kathleen J. Wile

Cup Fungi   Morel   Yeast (Magnified 400x)

Yeasts

Since yeasts are probably the most well-known members of phylum Ascomycota, we will begin there. Most yeasts are saprophytic, although there are examples of parasitic yeasts as well. When forming spores, they reproduce sexually, producing ascospores. Most yeasts, however, have a form of asexual reproduction, called **budding**, at their disposal as well. When a yeast buds, the nucleus of the cell reproduces inside a single cell. A section of the cell wall and plasma membrane then swell to form a pouch, into which the nucleus and some cytoplasm flow. This pouch with its nucleus is called a bud. The bud continues to grow until it is about the same size as the parent cell, and then the two cells separate. Budding is distinct from the asexual reproduction with which you are already familiar (such as that of the bacteria studied in Module #2), because the daughter cell remains attached to the parent cell as it grows.

Yeasts are typically egg-shaped cells that are only somewhat larger than bacteria. Besides a nucleus, about the only organelle in a yeast cell is a vacuole that stores food substances and certain chemicals that the yeast needs. Certain species of yeast store substances useful to humans in these vacuoles. For example, there are many species of yeasts that tend to store vitamins in their vacuoles. Some people eat these yeasts in a ground-up, powdered form to obtain the vitamins.

The yeast with which you are most familiar, however, is the type used in baking. Active dry baker's yeast that you can buy at the grocery store contains *Saccharomyces* (sak air' oh my seez) *cerevisiae* (sehr uh vuh say' ee) spores. When mixed with water, these spores mature into yeast cells that can carry on a process called **fermentation** (fur men tay' shun).

<u>Fermentation</u> – The anaerobic breakdown of sugars into smaller molecules

Fermentation is responsible for making bread dough rise. The yeast mixed with the bread dough feed on the sugars in the bread, breaking them down into alcohol and carbon dioxide. Since carbon dioxide is a gas, it pushes its way out of the dough, causing the dough to rise. When the dough is put into the oven, the yeasts are killed and the alcohol evaporates.

In fact, the nice smell that you associate with baking bread is a mixture of two things: alcohol and another substance called ozone. In large quantities, ozone is poisonous to humans. In small quantities, however, it simply has a distinct odor. When bread is in the oven, the heat causes the alcohol made by the yeast to evaporate. The alcohol, as it evaporates, can chemically react with other substances in the air to make ozone. The mixture of ozone and alcohol in the air makes the unique smell that we associate with bread baking.

Yeasts are also used in the manufacture of alcoholic beverages such as beer and wine. Since one of the products of fermentation is alcohol, yeasts are used to put the alcohol into the alcoholic beverages. Yeasts feed on the sugars in hops (a plant) and barley to perform the fermentation process in the manufacture of beer. In the manufacture of wine, they feed off of sugars in the grapes (or other fruits) that are used to make the wine. Interestingly enough, yeast cannot survive high concentrations of alcohol. Thus, as they continue the fermentation process, the increasing amount of alcohol actually ends up killing them. Wild forms of yeast can stand only a 4% level of alcohol before they begin to die off. Wineries and breweries, however, have bred strains of yeast that can survive levels of up to 12%. A mixture of wild yeast and specially-bred yeast is used in the making of most alcoholic beverages. Alcoholic drinks (such as whiskey) that have levels of alcohol much greater than 12% are made by taking a drink with 12% alcohol and boiling it. When a mixture of alcohol and water is boiled, the alcohol boils off at a lower temperature than the water. If you collect the vapors produced by boiling at this lower temperature and condense them, the result is a solution with a higher concentration of alcohol. This process is called **distillation**.

Learn more about yeast by performing the following experiment. Despite the fact that the supplies call for a microscope, the first five steps in the procedure should be done even if you do not have a microscope and the related equipment. Performing these steps will allow you to observe the fermentation process.

EXPERIMENT 4.2
Yeast and the Fermentation Process

<u>Supplies:</u>

- Packet of active dry yeast (can be purchased at a grocery store)
- Warm water
- Tablespoon
- Measuring cup
- Glass that holds at least 2 cups of water
- Sugar

- ♦ Microscope
- ♦ Eyedropper
- ♦ Slides and coverslips
- ♦ Methylene blue

Object: To observe the fermentation process and how yeast reproduce through budding

<u>Procedure:</u>

1. Mix one packet of yeast with two cups of warm water in the glass.
2. Add a tablespoon of sugar.
3. Stir gently.
4. Let the mixture stand at least five minutes. Active dry yeast contains the spores of baker's yeast for use in baking. Adding them to water and sugar causes them to begin growing.
5. As the mixture stands, you should observe bubbles beginning to form. Most likely, the bubbles formed will be very small. The best way to observe them is to watch the top of the mixture through the side of the glass. A layer of foam caused by the bubbles will appear and grow thicker as time goes on. The bubbles are the result of carbon dioxide produced in the fermentation process. There should be a familiar odor for those who bake bread. The odor is partially caused by the alcohol produced in the fermentation process.
6. Place a drop of the yeast solution on a slide and add a coverslip.
7. Observe the slide under low, medium, and high powers. Look for oval-shaped cells that have another cell attached to them, such as what is shown in the picture on the top right. The attached cell is a bud from the budding process.
8. Sketch an example of budding that you see in your specimen.
9. On another slide, place a drop of yeast solution and then a drop of methylene blue. The methylene blue is a stain. It will stain the yeast cells, making their features stand out. Observe on all three powers.
10. Sketch both an individual cell and a cell that is budding.
11. Let the yeast sit for thirty minutes or more and observe a sample (both stained and unstained) under the microscope again. You might see some budding that is forming chains, as shown in the picture on the bottom right.
12. Clean up: wash and dry all slides, coverslips, and eyedroppers. Wipe the lenses clean with lens paper. Put all of the supplies away in their proper place.

*Photos by
Kathleen J. Wile*

Other Members of Phylum Ascomycota

Many of the tasty, edible fungi also belong in phylum Ascomycota. Morels, whose fruiting bodies look like sponges, are one of the most sought-after forms of edible fungus. A morel is shown in the middle of Figure 4.7. The ascospores of these fungi are formed within the holes that make up the spongelike fruiting body. Wind and rain release the spores, allowing them to travel. Once again, however, just because a fungus looks like a sponge, it is not necessarily edible. Many amateur fungus-hunters have been tricked into eating toxic fungi because the fruiting body happened to look like a sponge! Cup fungi (shown on the left in Figure 4.7), whose fruiting bodies look like small cups, form their ascospores on the inside of the cup. When raindrops hit the cup, the force of impact releases the ascospores.

Many of the fungi that cause disease are in this phylum as well. *Claviceps* (kluh vye' seps) *purpurea* (per per ee' uh), better known as **ergot of rye**, can be deadly to humans. As its popular name implies, it feeds on rye grain. If rye bread made with rye that has *Claviceps purpurea* in it is eaten, it is often deadly. History tells us that Peter the Great was thwarted in his efforts to conquer the known world because his troops were fed rye bread that contained this fungus. In addition, many historians believe that the calamities that plagued the early settlers in New England were caused by rye that contained *Claviceps purpurea*. Unfortunately, the settlers at that time had no idea about this fungus, so they blamed it on witches and started the famous Salem witch trials in colonial Massachusetts.

Have you heard about **Dutch elm disease**? What about **chestnut blight**? Both of these diseases, caused by fungi in phylum Ascomycota, affect trees. The American chestnut was once one the most important sources of hardwood lumber in the United States. The fungus that causes chestnut blight, *Cryphonectria* (cry' fohn ek tree' uh) *parasitica* (perh' uh sit' ik uh), spread so quickly across the United States, however, that the American chestnut was completely wiped out! Likewise, many regions have lost their elm trees due to the fungus that causes Dutch elm disease, *Ophiostoma* (oh' fee oh stoh' muh) *ulmi* (uhl' me).

ON YOUR OWN

4.9 A single-celled organism asexually reproduces by duplicating its nucleus, causing a bulge to form in its plasma membrane, transferring the copied nucleus and some cytoplasm to the bulge, and then separating the bulge into a small cell. The small cell grows to the size of the parent in a day or so. How does this compare to the budding that takes place in yeasts?

4.10 Bread rises because of the fermentation process. Since this process produces both alcohol and carbon dioxide, why don't you get drunk when you eat bread?

Phylum Zygomycota

The next phylum we want to discuss contains those fungi that form **zygospores** (zye go' sporz).

Zygospore – A zygote surrounded by a hard, protective covering

Of course, this definition does little good if you do not know what a **zygote** (zye' goht) is. We will study this in detail later. However, for right now think of it this way: a zygote forms as a result of sexual reproduction when each parent contributes only half of the DNA necessary to form the offspring. When those two halves join together, a full set of DNA is formed and the offspring can begin development.

Zygote – The result of sexual reproduction when each parent contributes half of the DNA necessary for the offspring

If this definition confuses you, don't worry about it. When we cover reproduction in detail, you will understand it much better. For right now, just think of a zygote as a certain product of sexual reproduction. Later we will learn much more about what this product is and how it is different from other products of reproduction.

You have probably had some experience with at least a few of the members of this phylum. For example, the mold that grows on old bread is a member of phylum Zygomycota. Because the members of this phylum have a variety of reproduction methods at their disposal, it is useful to study an example of their life cycle in detail. Begin by examining Figure 4.8, an illustration of the bread mold's life cycle.

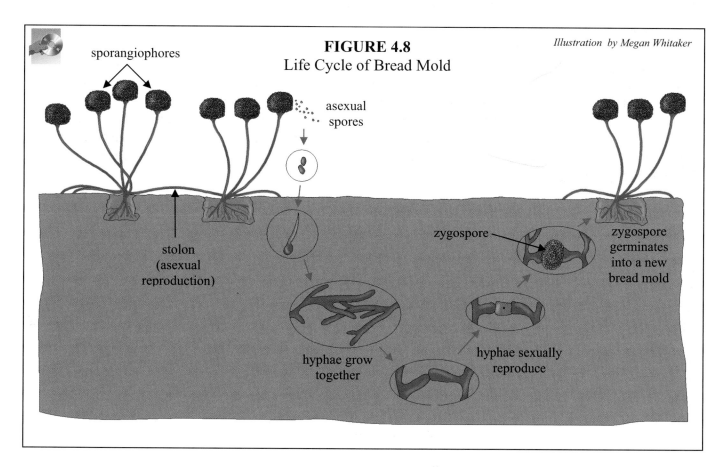

FIGURE 4.8

Life Cycle of Bread Mold

Illustration by Megan Whitaker

sporangiophores

asexual spores

stolon (asexual reproduction)

zygospore

zygospore germinates into a new bread mold

hyphae grow together

hyphae sexually reproduce

As you can see from the figure, there are three ways that these molds can reproduce. They can asexually reproduce when a stolon lengthens and forms new filaments. The new filaments become a new mycelium and thus a new fungus. Another form of asexual reproduction involves the production of sporangia (from aerial hyphae) that release spores. Finally, hyphae can fuse together and sexually reproduce to form a zygospore that can then mature into another fungus. Although the latter form of reproduction is what separates these fungi from the ones in the other phyla, all three means of reproduction are used.

The most well-known members of this phylum come from genus *Rhizopus*, which contains most of the common bread molds. Because these fungi have so many reproductive modes at their disposal, their spores are in the air virtually everywhere. If you leave bread out in the open, bread mold spores will eventually land on it and, within a matter of days, the growing mold will be noticeable. Most molds that grow on bread and other baked goods are harmless if consumed in small quantities. The molds that grow on fruit are typically members of this phylum as well. To become more familiar with these organisms, perform Experiment 4.3. Although this experiment calls for a microscope, you can perform the first two steps with just a magnifying glass. Thus, you should perform this experiment even if you do not have a microscope.

EXPERIMENT 4.3
Molds

Supplies:

♦ Bread, jelly, and/or fruit mold grown earlier (Only one specimen is necessary, but if you observe more than one specimen, you will learn more!)
♦ Magnifying glass
♦ Knife
♦ Needle (or probe from your dissection kit)
♦ Microscope
♦ Slides and coverslips
♦ Water
♦ Eyedropper

Object: To observe various molds and the differences in how they look both macroscopically and microscopically

Procedure:

1. Observe the mold or molds that you grew. For each specimen, note the source and describe the mold by color, shape, size and texture (fuzzy, smooth, flat, raised, etc.).
2. Observe each specimen with a magnifying glass. Sketch the magnified image, noting the source of each specimen.
3. Using a knife, needle, or probe, scrape off a section of the mold and place a tiny amount on a slide.
4. Observe the dry specimen on all magnifications.
5. Sketch an example of your observations, noting the source of each specimen.
6. Add a drop of water to the specimen on the slide and then cover it with a coverslip.
7. Observe the specimen again on all magnifications and sketch an example, noting the source.
8. If you were able to observe both a bread mold and a mold from fruit, try to note the differences between the two.
9. Clean up: Wash the slides, coverslips, and eyedropper; dry and put everything away. Wipe the microscope lenses with lens paper. Return the microscope to a safe place. Dispose of all molds.

ON YOUR OWN

4.11 A bread mold forms a stolon for reproduction. Is it reproducing sexually or asexually?

4.12 A fungus forms a fruiting body. Is it likely to be a bread mold?

Phylum Chytridiomycota

Phylum Chytridiomycota contains the single-celled fungi called **chytrids** (kye' trids). Chytrids inhabit muddy or aquatic areas. They are typically saprophytic, feeding on decaying water plants. Some species of chytrids, however, are parasitic. One well-known parasitic chytrid is *Synchytrium* (sin kye' tree uhm) *endobioticum* (en' doh by ah' tih kum), which causes potato wart. This fungus has been responsible for destroying many potato crops over the years. However, most commercial potatoes grown today are resistant to this fungus.

The characteristic that sets the members of this phylum apart from the rest of the fungi is the fact that their spores have flagella. Remember that the spores of the other fungi must be carried by wind or some other method in order to be dispersed, because they cannot move on their own. The spores of the fungi in this phylum, however, need no help in being dispersed, as they can use their flagella to move on their own.

Phylum Deuteromycota: The Imperfect Fungi

If a fungus is studied and scientists cannot determine a sexual reproductive phase in its life cycle, it is placed in this "phylum" until it can be better classified. The reason we put phylum in quotation marks here is that many biologists don't really consider this classification group a true phylum. Instead, they consider it a "holding area" until more can be learned about the fungus. Many biologists think that every fungus has a phase of sexual reproduction. Since some fungi are rather hard to study in detail, however, there are some whose mode of sexual reproduction eludes us. As a result, we place them in this phylum until the sexual reproduction method can be found. Once that happens, the fungus can be placed in one of the other phyla in kingdom Fungi. Because we do not fully understand the fungi in this phylum, we often call them **imperfect fungi**.

Since many biologists consider this classification group a temporary holding area, they often say that this phylum "has no taxonomic status." Remember, taxonomy is the science of classification. This statement, then, is equivalent to saying that phylum Deuteromycota does not really exist. It is simply a place for us to stick fungi that are not yet fully understood. We can't overemphasize, however, that the reason this classification has no taxonomic status is because biologists *assume* that all fungi have a sexual mode of reproduction. This, of course, could very well be an incorrect assumption, and perhaps this should be a true phylum. The fact that some of the fungi in this group have been extensively studied for many years without finding a sexual mode of reproduction might be considered strong evidence that the assumption is, indeed, wrong!

One of the most useful imperfect fungi comes from the genus *Penicillium*. These fungi produce the drug penicillin. In 1928, Alexander Fleming, an English physician, discovered this wonderful drug quite by accident. He had taken a short vacation, leaving a bacteria culture open to the air. When he returned, he saw that the culture was overrun by bacteria, except in a certain place where a blue mold (from the genus *Penicillium*) was forming. The blue mold seemed to be producing a substance that killed the bacteria. Fleming isolated that chemical and called it **penicillin**. With the help of two other scientists, he demonstrated that this substance can kill bacteria associated with many human sicknesses. Thus, the first **antibiotic** (an tie by ah' tik) was discovered.

Antibiotic – A chemical secreted by a living organism that kills or reduces the reproduction rate of other organisms

Because penicillin and other antibiotics have been so successful in treating many forms of sickness, these three scientists shared the Nobel Prize in Medicine in 1945.

One interesting fact about antibiotics is that although they can be very effective at killing many bacteria that cause disease, they can lose their effectiveness after a while. You see, God created His creatures with the ability to adapt and change to better fit their environment. It turns out (we will dwell on this in a later module) that God has made it possible for bacteria (and other organisms) to change significantly over the course of several generations. A colony of bacteria that is being destroyed by an antibiotic can, under certain conditions, produce offspring that are not at all affected by the antibiotic. We say that these bacteria are **immune** to the antibiotic's effect. As an antibiotic is used more and more, strains of bacteria develop that are immune to the antibiotic. To counter this, medical scientists must find new antibiotics to which these bacteria are not immune. This is a constant struggle. Bacteria adapt to become immune to an antibiotic, and then humans adapt and find another antibiotic. Paradoxically, the more antibiotics are used to cure disease, the more ineffective they become, because the very act of using an antibiotic can give rise to a strain of bacteria immune to it!

Other members of genus *Penicillium* are useful to us because they flavor certain cheeses. Camembert and Roquefort cheese are both flavored by the growth of *Penicillium* species. Although it might at first sound strange that fungi are grown on cheese to give them flavor, it is no different from putting mushrooms in a sauce to change its flavor. Although the term "fungus" often carries a bad connotation, many fungi are completely edible and, to some, quite tasty!

If you would like, perform the following experiment to observe some of these cheese molds. Note that these molds do not produce penicillin, because they are not the same species as the one analyzed by Dr. Fleming. They are just within the genus *Penicillium*, so they are similar to Dr. Fleming's fungus, but they will not produce penicillin.

OPTIONAL EXPERIMENT 4.4
Imperfect Fungi

(Perform this experiment only if you are interested and if you can find at least one sample of Camembert or Roquefort cheese.)

Supplies:

- Camembert cheese (available at large supermarkets)
- Roquefort cheese (available at large supermarkets)
- Microscope
- Slides and coverslips
- Eyedropper
- Knife
- Water

Object: To observe various molds and the differences in how they look both macroscopically and microscopically

Procedure:

1. Refrigerate the cheese or cheeses until about 6 hours before performing the experiment.
2. Take the cheese or cheeses out of the refrigerator and let them sit for about 6 hours.
3. The mold on the Camembert cheese (*Penicillium camemberti*) generally looks like white masses that form the mycelium of the fungus. Scratch a small amount of mold loose with a knife and place on a slide.
4. Add a drop of water and a coverslip to the sample on the slide.
5. Observe the fungus on all magnifications. Sketch and label what you see.
6. The blue areas on the Roquefort cheese are samples of the mold *Penicillium roqueforti*. Because of the color of the fungus, this is often called "blue cheese." Scratch off some of this blue with the knife and place on a slide.
7. Add a drop of water and a coverslip to the sample on the slide.
8. Using the microscope, try to find chains of the tiny, blue-green fungal spores. You might only see them as individuals, but they are usually found in chains.
9. Sketch and label this fungus.
10. Clean up: Wash the slides, coverslips, and eyedropper; dry and put everything away. Wipe the microscope lenses with lens paper. Return microscope to a safe place. Dispose of all molds.

ON YOUR OWN

4.13 A biology teacher once said, "The only thing imperfect about the imperfect fungi is our knowledge of them." What does the biology teacher mean?

4.14 In medical journals these days, there is a lot of concern about the overuse of antibiotics. Doctors think that since antibiotics are so effective, they are prescribed far too often for patients. Why are doctors worried about overuse of antibiotics?

Phylum Myxomycota

As we mentioned previously, placing this phylum in kingdom Fungi is rather controversial. Of course, placing it in Protista, the other kingdom in which you might find it, can be controversial as well. The reason is quite simple: the members of this phylum, typically called **slime molds**, behave like fungi when they reproduce, and they behave like colonial protists when they feed. As a result, there are arguments for placing them in either kingdom. Some biologists have even considered making a sixth kingdom in which to place them; however, there are really not enough species of slime mold to justify such a drastic step.

Although modern textbooks do tend to place phylum Myxomycota in kingdom Protista, we choose to leave it in its traditional place, as a part of kingdom Fungi. This is mainly because we consider reproduction to be a more fundamental aspect of life than feeding. After all, many species have several feeding options open to them. If you look at a population of wild cats, for example, you will find that they are all carnivores, but their choice of diet and means of finding food are quite different depending on their environment. Their mode of reproduction, however, will always be the same. Also, the definition of species that you learned in Module #1 references reproduction, so it is obviously a very fundamental means of classification. As a result, we will place the slime molds in the kingdom which contains other organisms that reproduce in a similar fashion. Another reason to classify slime molds with fungi is that you find them in habitats similar to those of other fungi.

Realize, however, that the reasons we just presented simply reflect our preconceived notions. Despite what you might think, scientists are not very objective. We all approach science with an inherent bias. As a result, the conclusions that we come to are often affected by this bias. This is clear in the means of classification that we employ. It is also clear in the general outlook that we have. For example, if you were to pick up a biology textbook written by a scientist who does not believe in God, the entire book would be quite different from this one. Atheistic scientists (and even some scientists who believe in God) tend to teach biology within the framework of the theory of evolution. Since this theory attempts (quite unsuccessfully) to explain the formation of life without reference to a Creator, these scientists must look at life through those blinders. As a result, their entire view of biology is quite different from that of this book. In a later module, we will study evolution in great detail, showing you why it ought to lose its status as a theory because it really is, at best, an unconfirmed hypothesis. The point we are trying to make now is that you should never be fooled into thinking that scientists are objective. We cannot be. Our preconceptions and our world views will always color the way we see and do science.

Now that we have that out of the way, we can start learning about the interesting creatures known as slime molds. Most of the time, you will see slime molds in their feeding stage, when they resemble colonial protozoa. They usually can be found on the bark of decaying logs or between the layers of leaves on the floor of a forest. Despite the fact that slime molds seem rather exotic, they are quite common. If you ever lay tree bark or dead leaves down as mulch for a garden, you can almost always turn the mulch over during the middle of the growing season and find a slime mold, providing that the mulch has been kept moist. Figure 4.9 shows two slime molds: one in its feeding stage, where it resembles a colonial protozoan, and the other in its reproductive stage, where it resembles a fungus.

Fuligo photo © Bob Ford/Nature Portfolio

FIGURE 4.9
Slime Molds

Didymium photo © J. L. Carson/Custom Medical Stock Photo

Slime mold from genus *Fuligo* in its feeding stage

Slime mold from genus *Didymium* in its reproductive stage

Slime molds are usually white, red, orange, or yellow. They exist in their feeding stage as a mass of living matter called a **plasmodium**. This is rather confusing, however, since there is a genus in kingdom Protista that is given that same name. The term plasmodium when applied to slime molds, however, has no connection to the genus *Plasmodium* found in kingdom Protista! As their name implies, most slime molds are slimy to the touch. Despite the fact that their name and appearance are rather disagreeable, most slime molds are not harmful to living plants or animals, as they are almost all saprophytic. The nonsaprophytic slime molds tend to feed on bacteria, although there are some parasitic slime molds.

Slime molds can move about as a unit in search of food. When the food supply is exhausted, however, or when unfavorable conditions occur, the slime mold will produce fruiting bodies that are best described as sporophores (see the right side of Figure 4.9). These sporophores contain spores that are released to find new sources of food. If they find a new source of food, they will grow and reproduce to form a new slime mold. Some slime molds produce motile spores (spores that can move on their own), while others produce nonmotile spores (spores that cannot move on their own).

Probably the main feature of a slime mold's habitat is water. Wherever dead trees, bark, or leaves are kept moist, slime molds are almost certain to grow. To get rid of a slime mold, you simply need to dry the area. Gardeners often dislike slime molds because of their appearance. To rid themselves of the slime molds, they need only to rake the surface of their gardens daily. This will bring the moist mulch to the surface, allowing it to dry.

ON YOUR OWN

4.15 A biologist observes a slime mold only during its feeding stage. In what kingdom will the biologist most likely classify it?

Symbiosis in Kingdom Fungi

Before we leave our discussion of kingdom Fungi, we must mention two different forms of symbiosis in which its members participate. The most well-known form of symbiosis in which you will find a fungus is the **lichen** (lie' kun). Lichens are produced by a mutualistic relationship between a fungus (usually of phylum Ascomycota) and an alga (usually of phylum Chlorophyta). The alga in the relationship produces food for itself and the fungus by means of photosynthesis, while the fungus supports and protects the alga. As a result of this mutualistic arrangement, you can find lichens where other organisms cannot survive. They are commonly found growing on dry rocks, brick walls, fences, and trees. In certain cold, snowy regions, lichens grow so large that they can cover a few square miles!

Despite the fact that the algae which make up the known species of lichens can live independently, the fungi that make them up cannot. This should make sense to you, because the alga produces its own food. Thus, it can live with or without the fungus. The fungus just makes its survival easier. The fungus, on the other hand, cannot live without the food that the alga produces.

Since the fungus of a lichen cannot live without the alga, you might wonder how lichens reproduce. Actually, it is rather fascinating. Most lichens reproduce by releasing a dustlike substance called a **soredium** (suh ree' dee uhm). The soredium contains spores of *both* the alga and the fungus in a protective case. Thus, the soredium is like a spore that contains two different spores. Wherever the soredium lands, then, both the fungus and the alga can grow. Isn't that amazing? The two separate species work together not only to survive, but also to reproduce! There is still a lot that science does not understand about this fascinating mutualistic relationship.

A more prevalent (but less well-known) symbiotic relationship in which fungi participate is called a **mycorrhiza** (my' kuh rye' zuh), or a "fungus root." Nearly 80% of all plants with root systems participate in this mutualistic symbiotic relationship with a fungus. In a mycorrhiza, the fungus forms haustoria that penetrate the cell walls of the root system's cells. The fungus absorbs nutrients from the roots as they are transported to the plant. In return, the fungus gives the plant certain needed chemicals, called minerals, that it cannot absorb efficiently from the soil.

You see, in order to absorb minerals effectively from the soil, an organism must be wide and thin. This is how the mycelium of a fungus grows. This is not, however, the way that a root system grows. Roots usually grow thick and long, trying to go deep into the soil. Since the mycelium of the fungus has the ideal structure for absorbing minerals, it does so in exchange for food. In laboratories, it has even been demonstrated that fungi absorb the minerals when they are plentiful and store the excess. These excess minerals are then released slowly into the roots of the plant when the minerals are scarce in the soil!

ON YOUR OWN

4.16 Suppose a biologist were to separate the fungus spores from the alga spores in a soredium. Could the fungal spores develop into a free-living fungus? Could the spores from the alga develop into a free-living alga?

4.17 There are some scientists who have studied the effect of air pollution on fungi. They conclude that air pollution destroys fungi at a much higher rate than it destroys other organisms. These same scientists say that if air pollution kills too many fungi, trees and other plants will begin to die as well. Why?

Summing Up Kingdom Fungi

As you can see, we still have a lot to learn about this fascinating kingdom. Are imperfect fungi simply not completely understood, or are they really fungi that have no means of sexual reproduction? How do we classify slime molds? How do lichens form soredia? How do the fungi in mycorrhizae (the plural of mycorrhiza) know to store up excess minerals and release them when their hosts need them most?

Of course, despite the fact that there are things we don't know about kingdom Fungi, there are many things that we do know. We know that the saprophytic nature of most fungi is critical to the balance that exists in nature. We also know that many fungi are useful to humans as food, flavoring, and medicine, while others are deadly or toxic. Finally, we know that fungi are another wonderfully interesting part of God's creation, bringing glory and honor to Him.

ANSWERS TO THE "ON YOUR OWN" PROBLEMS

4.1 <u>No, the organism is not a part of kingdom Fungi.</u> Fungi digest their food before eating it. This process, called "extracellular digestion," is common to all fungi.

4.2 <u>The stalks and the caps are not the main parts of the mushroom. The mycelium, which is underneath the ground, is the main body of the fungus.</u> Even though the farmer removed the stalks and caps, the mycelium is still there. It will produce more stalks and caps later on.

4.3 <u>Only parasitic fungi produce haustoria, so the fungus must be parasitic.</u> Since a haustorium's job is to invade a living cell and draw nutrients from it, there is no reason a saprophytic fungus would produce a haustorium.

4.4 <u>The fruiting body holds and releases the spores for reproduction.</u> Once all spores are released, the fruiting body withers and dies.

4.5 <u>The spores result from asexual reproduction,</u> because asexual reproduction results in offspring that are identical to parents.

4.6 Your key can be ordered differently and can even have more steps in it. However, it should separate fungi into phyla based on the criteria in Table 4.1

1. Organism in feeding stage resembles protozoa .... *phylum Myxomycota*
   Organism never resembles protozoa **2**
2. Fungus produces spores on basidia ....*phylum Basidiomycota*
   Fungus does not produce spores on basidia **3**
3. Fungus produces spores in asci .... *phylum Ascomycota*
   Fungus does not produce spores in asci **4**
4. Fungus produces spores where hyphae fuse .... *phylum Zygomycota*
   Fungus does not produce spores where hyphae fuse **5**
5. Fungus produces cells with flagella for sexual reproduction .... *phylum Chytridiomycota*
   Fungus does not produce cells with flagella for sexual reproduction .... *phylum Deuteromycota*

Please note that in step 5, there is no need to ask about whether or not the sexual mode of reproduction is known in order to determine that the fungus must be in phylum Deuteromycota. After all, if it is not in any of the other five phyla, it must be in that one.

4.7 <u>No.</u> The button stage comes before the stipe and cap are formed. Since the spores are released from the gills in the cap, a mushroom cannot release its spores until the stipe and cap are formed.

4.8 <u>In mushrooms, basidia form on the gills of the cap. In puffballs, they form inside the membrane of the fruiting body. In shelf fungi, they form in the pores of the fruiting body.</u>

4.9 <u>It is very similar, but not identical.</u> For yeasts, the bud typically does not detach itself until after it is fully grown.

4.10 Baked bread does not have alcohol in it because the heat of the baking process evaporates the alcohol. Thus, even though alcohol is formed during the making of bread, it is removed by the heat of the baking process.

4.11 Stolons are an asexual means of reproduction, so the mold is reproducing asexually.

4.12 No, bread molds do not form a fruiting body. In sexual reproduction, they form an underground zygospore that grows into a new mold. They do have sporangiophores that form asexual spores, but those are not fruiting bodies; they are just specialized aerial hyphae.

4.13 The fact that a fungus is called imperfect means that we simply *do not know about its sexual reproduction*. Since we do not know about an aspect of its life, our knowledge of it is imperfect.

4.14 The more an antibiotic is used, the more likely the chance of bacteria (or other pathogen) strains immune to the antibiotic will form. If that strain reproduces and is spread, a new antibiotic must be made to destroy that strain of bacteria (or other pathogen).

4.15 Since slime molds in their feeding stage resemble protozoa, the biologist will most likely classify it in kingdom Protista.

4.16 The spores from the fungus could never grow into a free-living fungus, because the fungus in a lichen (that's where a soredium comes from) has no food supply without the alga. The alga would be able to live on its own, because the fungus simply gives it support and protection. The alga can live without that. It will not be as prolific as it could be with the fungus, but it can live on its own.

4.17 Because nearly 80% of plants have a symbiotic relationship with the mycelia of fungi, if fungi die off, the trees will no longer be able to participate in the symbiotic relationship. The fungi help trees absorb vital minerals from the soil. Without the aid of the fungi, the trees will not be able to absorb enough minerals, and they will begin to die.

STUDY GUIDE FOR MODULE #4

1. Define the following terms:

a. Extracellular digestion	h. Haustorium
b. Mycelium	i. Chitin
c. Hypha	j. Membrane
d. Rhizoid hypha	k. Fermentation
e. Aerial hypha	l. Zygospore
f. Sporophore	m. Zygote
g. Stolon	n. Antibiotic

2. Which of the following characteristics or structures exist for the vast majority of fungi? Which are present in only a few species?

extracellular digestion	sporangiophores	motile spores
stolons	mycelia	septate hyphae
chitin	hyphae	cells
caps and stalks	haustoria	rhizoid hyphae

3. Some biologists say that a mushroom is much like an iceberg, because only about 10% of an iceberg is visible from the surface of the ocean. What do they mean?

4. What is the difference between septate and nonseptate hyphae?

5. What is the function of the following specialized hyphae?

> rhizoid hyphae   stolon   sporophore   haustorium

6. Of the hyphae listed in question 5, which are aerial?

7. What is the difference between a sporangiophore and a conidiophore?

8. Give the main characteristic associated with each of the phyla of kingdom Fungi: Basidiomycota, Ascomycota, Zygomycota, Chytridiomycota, Deuteromycota, and Myxomycota.

9. Describe each of the stages (in chronological order) associated with the life cycle of a mushroom, starting with the formation of a mycelium.

10. What is the main difference between shelf fungi, puffballs, and mushrooms?

11. What is an alternate host? List a type of fungus that uses one.

12. What type of fungus is best known for fermentation? To which phylum does it belong?

13. How is budding different from the asexual reproduction in bacteria?

14. Name at least two pathogenic fungi and the maladies that they cause.

15. Describe the three ways a bread mold can reproduce. In each case, specify whether the reproduction is sexual or asexual.

16. What puts a fungus into phylum Deuteromycota?

17. What can happen when an antibiotic is used too much?

18. Name the genus of the fungus that produces penicillin.

19. When a slime mold is a plasmodium, it resembles organisms from what kingdom?

20. What is the easiest way to get rid of slime molds?

21. What are the two major forms of mutualism in which fungi participate? Describe each relationship and the job of each participant in that relationship.

22. What is a soredium?

"Why won't you go out with me? I'm a fun guy."

Cartoon by Speartoons

MODULE #5: The Chemistry of Life

Introduction

In the past three modules, we introduced three of the five kingdoms in God's creation. You've learned a lot about the organisms within these kingdoms, and hopefully you've begun to develop a keen appreciation for the grandeur and complexity of the Creator's work. Now it's time to step back and look at biology from a different angle. Rather than studying life itself, in this module we will be studying the chemistry that helps make life possible.

In the future, you should spend an entire year studying chemistry. After that year, you will have received a basic introduction to this vast field. Thus, what we cover in this module will barely scratch the surface of the science called chemistry. Nevertheless, without this brief introduction, you will probably be lost in later modules.

Atoms: The Basic Building Blocks of Matter

In chemistry, we study **matter**. A basic definition of matter is given below.

<u>Matter</u> – Anything that has mass and takes up space

Of course, this definition does us no good unless we know what **mass** is. Unfortunately, mass is a difficult term to define; thus, we will not try to define it in this course. Instead, we will say this: if something has mass, it will also have weight. As a result, we could say that matter is anything that has weight and takes up space. Now if you think about it, almost everything has weight and takes up space. In chemistry, then, we really study just about everything!

As you learned in Module #1, cells are the basic building blocks of life. Indeed, the type of cell or cells that make up an organism is the first thing used to start the process of classifying it. Organisms made of prokaryotic cells belong in kingdom Monera, whereas those made of eukaryotic cells belong in one of the other four kingdoms. Well, just as cells are the basic building blocks of life, **atoms** are the basic building blocks of matter. Everything from the tiniest speck of dust to the biggest mountain in the world is made up of atoms.

Although atoms make up matter, we cannot see them because they are quite small. Indeed, in the dot of an "i" on this page, there are approximately a billion (1,000,000,000) atoms! As a result, atoms are simply too small to see. Even using the most powerful microscope in the world, we cannot see them. Nevertheless, scientists have performed detailed experiments that provide ample evidence for their existence. Using the right kind of equipment, for example, scientists have been able to get computers to draw "maps" of the surfaces of a metal. These computer-generated maps show peaks and valleys on what appears to be a very smooth surface. Those peaks and valleys are most likely the result of the atoms that make up the metal. Although such maps are not really *pictures* of atoms, they provide a great deal of evidence that atoms do, indeed, exist. As a result, the existence of atoms has become a foundational principle that guides our understanding of chemistry.

Believe it or not, atoms are made up of even smaller things called protons, neutrons, and electrons. You will learn a lot more about protons, neutrons, and electrons when you study chemistry.

For right now, you can think of them as small, spherical particles that are arranged in a very specific way to form an atom. For example, a simplistic drawing of an atom is shown in the figure below.

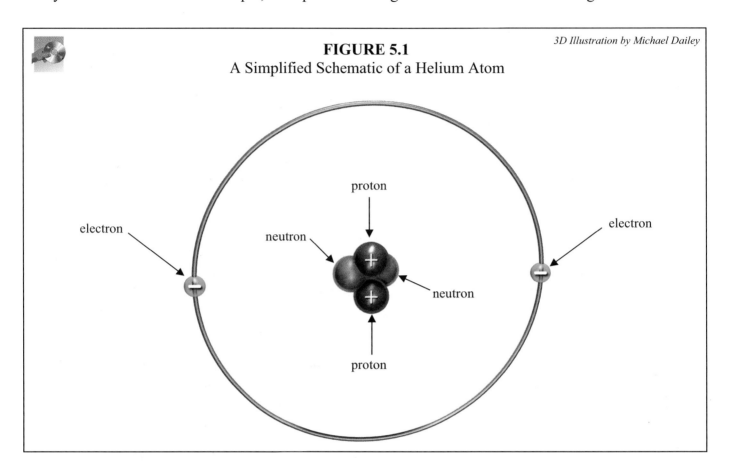

FIGURE 5.1
A Simplified Schematic of a Helium Atom

3D Illustration by Michael Dailey

As we note in the title of the figure, this is an illustration of a helium atom. When you fill a balloon with helium so that it will float, you are filling it with a huge number of these atoms. Notice that the neutrons and protons clump together in the center of the atom, which is called the **nucleus**. Now don't get confused here. The term nucleus is also used to identify the organelle that holds the DNA in a eukaryotic cell. When discussing atoms, however, the nucleus refers to the center of the atom where the protons and neutrons clump together. Whirling around in an orbit outside of the nucleus, we find the electrons.

You might wonder how in the world we know what atoms look like if we cannot see them. The fact is, we don't really know what they look like. However, scientists have done many experiments designed to help us understand the structure of the atom and, as far as we can tell, the drawing in Figure 5.1 is consistent with most of those experiments. As a result, we say that this is a good **model** of the atom.

Model – An explanation or representation of something that cannot be seen

The fact that a model is consistent with experiments does not necessarily mean that it is an accurate representation of the object being studied. Later on, someone might come up with experiments that contradict the model, or someone else might come up with a different model that is also consistent with all of the experiments done. Thus, the drawing in Figure 5.1 is instructive in that it gives us a picture of what an atom might look like, but please realize that it is not an actual drawing of an atom. It is merely a representation of what an atom *might* look like.

In fact, we already know that the model presented in Figure 5.1 is not really accurate. There are certain experiments that clearly show this model (called the "Bohr model") to be in error. As a result, a new model of the atom, called the "quantum mechanical model," is currently the model believed by most scientists. However, that model is far too complex to be shown here. Indeed, to truly understand the quantum mechanical model, you need to have several years of mathematics beyond calculus! Thus, even though we know that the Bohr model is not exactly correct, we will use it. Although you might not like the idea of using an incorrect model to learn about the atom, don't worry. Most of the concepts that you learn using the Bohr model will help you to understand the quantum mechanical model when you get to it. When you take chemistry, you will get a brief introduction into the quantum mechanical model of the atom.

With all of this in mind, let's take a look at the Bohr model of the atom. Protons and neutrons are packed into the nucleus, which is at the center of the atom. Electrons whirl around the nucleus in circular orbits. The number of protons, neutrons, and electrons in an atom determines all of the properties of that atom. As we already mentioned, the atom pictured in Figure 5.1 is a helium atom. You already know that helium is a gas that is lighter than air. That's why a balloon full of helium will float. Well, the fact that helium is a gas and the fact that it is lighter than air are both a result of the number of protons, electrons, and neutrons in the atom. If you change the number of protons, electrons, and neutrons in an atom, you change its properties. For example, an atom that contains 6 protons, 6 neutrons, and 6 electrons is a type of carbon atom. Unlike helium, carbon is not a gas; it is a solid. It is black and brittle. These properties (as well as all of the other properties of the atom) are once again the result of the number of protons, electrons, and neutrons in the atom.

No matter how many protons, electrons, and neutrons make up an atom, there is one general principle that applies to them all. All atoms have equal numbers of protons and electrons. Thus, if an atom has 3 protons, you know that it has 3 electrons. If it has 17 electrons, you know that it has 17 protons. There is, of course, a reason for this. Protons and electrons each have electrical charge. Now we won't even try to explain what electrical charge is; we will just tell you that it comes in two types: positive and negative. Electrons have a negative charge, while protons have a positive charge. Well, it turns out that an atom cannot have any net electrical charge, so the number of positive charges (protons) in the atom must always balance out the number of negative charges (electrons) in the atom. That way, the total charge is zero. Neutrons, by the way, have no electrical charge.

Here's the really interesting part: the vast majority of an atom's properties are determined by the number of electrons that it has. Although this might sound a little weird, it actually makes a great deal of sense. You see, based on the Bohr model of the atom, the protons and neutrons are tucked away in the center of the atom (the nucleus). Since the electrons orbit around the nucleus, they are, in effect, what makes up the "outer layer" of the atom. Thus, if two atoms were to come close to one another, their electrons would be the first things to interact. As a result, the electrons determine the vast majority of an atom's properties.

To confuse this issue a little bit, remember that all atoms have the same number of protons and electrons. Thus, even though it is in reality the *number of electrons* that determine an atom's properties, since the number of protons is always equivalent to the number of electrons, we could also say that the *number of protons* can be used to determine an atom's properties. Now don't let this confuse you! The electrons are what really determine an atom's properties; however, because of some complicated history, we track the properties of atoms by their number of protons. Although this is a bit confusing, there is nothing wrong with it, since the number of protons and electrons in an atom are always the same.

In summary, then, if two atoms have the same number of protons, they will have the same number of electrons. Thus, even if the number of neutrons in the first atom is different from the number of neutrons in the second, the vast majority of the two atoms' properties will be the same, because they have the same number of electrons.

ON YOUR OWN

5.1 The Bohr model of the atom is sometimes called the "planetary model" of the atom. Why?

5.2 What determines all of an atom's properties? What determines the vast majority of an atom's properties?

5.3 An atom has 13 electrons. How many protons does it have?

Elements

Since the number of electrons (and therefore the number of protons) is the most important factor in determining an atom's properties, we tend to separate atoms by their number of protons. Atoms that have the same number of protons (regardless of their number of neutrons) are said to belong to the same **element**.

Element – A collection of atoms that all have the same number of protons

Helium, for example, is an element. One of the atoms that makes up this element is pictured in Figure 5.1. It has 2 protons, 2 electrons, and 2 neutrons. However, there is another atom that is a part of the element helium. It has 2 protons, 2 electrons, and 1 neutron.   Since both of these atoms have the same number of protons and electrons, they have the same basic properties. Thus, when you fill a balloon with helium, you are actually filling it with two kinds of atoms: one that has 2 protons, 2 electrons, and 2 neutrons, and one that has 2 protons, 2 electrons, and 1 neutron.

How do we identify these atoms? Well, first and foremost, we call them by the element to which they belong. Thus, both types of atoms are called helium atoms. To distinguish between the two different atoms that make up helium, we add their protons and neutrons together. Thus, the atom with 2 protons, 2 electrons, and 1 neutron is called "helium-3," because 2 protons and 1 neutron add to 3. The atom picture in Figure 5.1 is called "helium-4," because the 2 protons and 2 neutrons add to 4.

It is often convenient for us to abbreviate the name of elements so as to reduce the amount of time it takes to write them. Usually, an element is abbreviated by the first one or two letters of its name. Helium, for example, is abbreviated "He," while carbon is abbreviated "C." Unfortunately, some elements are also abbreviated by the first one or two letters of their Latin names. Sodium, for example, is abbreviated as "Na" because its Latin name is "natrium." Even if an element has two letters in its abbreviation, only the first is capitalized. There is no real way to learn these abbreviations short of memorization, so you will need to learn the abbreviations of the following biologically-important elements.

TABLE 5.1
Some of the Biologically Important Elements

Element Name	Abbreviation
carbon	C
hydrogen	H
oxygen	O
nitrogen	N
phosphorus	P
sulfur	S

Although you are not required to remember this, you might be interested in the makeup of these elements. Carbon contains all atoms that have 6 protons and 6 electrons. Hydrogen is made up of all atoms that have 1 proton and 1 electron. Oxygen contains all atoms that have 8 protons and 8 electrons, while nitrogen is made up of the atoms that have 7 protons and 7 electrons. Finally, phosphorus contains those atoms with 15 protons and 15 electrons, while sulfur is made up of the atoms that have 16 protons and 16 electrons.

Now, of course, there are many, many more elements that you will not really deal with until you take chemistry. In fact, there are currently 116 known elements in God's creation. Most likely, more will be discovered. For right now, however, we will concentrate mostly on the elements listed in Table 5.1.

Before we move on to the next section, step back and review what you have learned. Atoms are made up of protons, neutrons, and electrons. No matter how many protons an atom has, it will have the same number of electrons. When different atoms have the same number of protons (and therefore the same number of electrons), they are said to belong to the same element, because they have the same basic properties. To name an atom, we call it by the element to which it belongs, followed by the sum of its protons and neutrons. Thus, when you see a name like sulfur-32 (or S-32 or sometimes ^{32}S), you know that we are talking about a *particular* atom, the one that has 16 protons (that's what makes it sulfur) and 16 neutrons (that's how to get a 32 when you add the protons and neutrons together). On the other hand, if you see just the name sulfur (or S), then you know that we are talking about an element, which probably contains more than one type of atom. After all, there are several atoms that all have 16 protons and thus belong to the element sulfur but have different numbers of neutrons.

Now it turns out that for the vast majority of situations in chemistry, the number of neutrons in an atom is completely irrelevant. This is (once again) because the major properties of an atom are determined by its number of electrons (and therefore its number of protons). As a result, the element to which an atom belongs is, by far, the most important aspect of identifying an atom. Thus, you will usually see just the element (like sulfur) instead of seeing a particular atom (sulfur-32) when you study chemistry. You always need to remember, however, that an element (such as sulfur) is composed of many individual atoms (such as sulfur-32, sulfur-33, sulfur-34, and sulfur-36).

ON YOUR OWN

5.4 Two atoms have slightly different properties, but they belong to the same element. What is different about them: their numbers of protons, neutrons, or electrons?

5.5 The element carbon is composed of all atoms that have 6 protons. One of the atoms in the element carbon is carbon-13. How many protons, neutrons, and electrons are in a carbon-13 atom?

<u>Molecules</u>

If there really are only 116 known elements in God's creation, you might think that there are only 116 different types of matter. After all, if matter is made up of atoms, and each element contains atoms that have the same basic properties, then 116 different elements make 116 different types of matter, right? Wrong! You see, atoms are only the basic building blocks of matter. In order to provide the chemical diversity necessary for life, God has designed atoms to link together much like the pieces of a jigsaw puzzle. When atoms link together, they form **molecules**.

<u>Molecules</u> – Chemicals that result from atoms linking together

For example, carbon dioxide is a gas that humans (and many organisms) produce as a part of their metabolism. This gas is formed when 1 carbon and 2 oxygens link together.

Molecules have abbreviations, just like elements. We call these abbreviations **chemical formulas**. Carbon dioxide, for example, is abbreviated as "CO_2." Where does this abbreviation come from? Well, carbon dioxide is made up of carbon and oxygen. The "C" stands for carbon and the "O" stands for oxygen. Numbers that appear as subscripts indicate how many of those atoms are in the molecule. The "2" that appears as a subscript after the "O" tells us that there are 2 oxygen atoms in a carbon dioxide molecule. If no subscript appears after an element, we know that there is only 1 atom of that type in the molecule. Thus, since there is no subscript after the "C," we know that there is only 1 carbon in a molecule of carbon dioxide.

Just to make sure you really understand this, we want to give you two more examples. Another important molecule in the chemistry of life is methane. This molecule is abbreviated as CH_4, so you should be able to tell that there is 1 carbon atom linked to 4 hydrogen atoms in a methane molecule. Glucose, which is the basic food substance that most autotrophic organisms produce via photosynthesis, is abbreviated as $C_6H_{12}O_6$. This should tell you that a molecule of glucose has 6 carbon atoms, 12 hydrogen atoms, and 6 oxygen atoms all linked together.

The important thing to realize about molecules is that the properties of a molecule are determined by the type and number of atoms that link together, as well as the way that they link together. For example, as we already mentioned, carbon dioxide is formed when one carbon atom links to 2 oxygen atoms. However, this is not the only molecule that can be formed by carbon and oxygen atoms. If 1 carbon atom links to 1 oxygen atom, the result is carbon monoxide (CO). Carbon monoxide and carbon dioxide are two completely different molecules! Carbon dioxide, for example, is a gas that is generally harmless to humans unless it exists in very large quantities. In fact, humans produce it as a result of their metabolism. Specifically, we produce it in the catabolic process in which simple sugars such as glucose are burned for energy. Carbon monoxide, however, is poisonous to humans. If we breathe too much carbon monoxide, we will die of suffocation! Thus, not only the type but also the number of atoms that link together determine the properties of a molecule. Two molecules can be made up of the same exact atoms, but if the number of even one type of atom is different between the two molecules, the molecules will have completely different properties.

To make things just a little more confusing, atoms of the same type can join together as a molecule just like atoms of different types can. For example, oxygen is an element. However, when we breathe in oxygen from the air around us, we are not breathing individual oxygen atoms. Instead, we are breathing in the molecule O_2. This molecule is formed when 2 oxygen atoms link together. Even though we are lazy and tend to refer to this as "oxygen," it is not the same as an individual atom of oxygen. Instead, it is a molecule formed by 2 oxygens linking up. This gives it very different properties. If you were to breathe in a substantial amount of individual oxygen atoms, you would quickly die. However, breathing in oxygen molecules is absolutely necessary in order for you to live. Other examples of molecules formed by the same types of atoms linking up are: ozone (O_3), nitrogen gas (N_2), and hydrogen gas (H_2). As you can see from their chemical formulas, ozone is a molecule made up of 3 oxygen atoms linked together, nitrogen gas is a molecule formed by 2 nitrogen atoms linking up, and hydrogen gas results when 2 hydrogen atoms link together.

We see, then, that molecules are formed when atoms link together. Different atoms can link together (as is the case with methane, CH_4) or atoms of the same type can link together (as is the case with oxygen gas, O_2). Both the type of atoms that link together and the number of each type of atom determine the properties of the molecule. As you will learn in more detail momentarily, the *way* in which atoms link up to form a molecule also affects the properties of the molecule.

It is important for you to understand the distinctions between atoms, elements, and molecules. Atoms are the basic building blocks of matter, and their properties are determined by the numbers of protons, neutrons, and electrons that make them up. A group of atoms that have the same number of protons all belong to the same element. When you fill a balloon with the element helium, you are filling it with a bunch of helium atoms, some of which will be helium-3 and some of which will be helium-4. When atoms link together, they form molecules. The type of and number of atoms that link together determine the properties of the molecule. When you blow into a balloon to fill it up, for example, you are filling it with a lot of nitrogen molecules (N_2), oxygen molecules (O_2), and carbon dioxide molecules (CO_2). These molecules have different properties because the number of and type of atoms that make them up are different.

ON YOUR OWN

5.6 Identify each of the following as either an atom, element, or molecule:

   a. NH_3 b. P   c. carbon-14   d. S   e. P_4

5.7 A student is told to study the chemicals nitrogen monoxide (NO) and nitrogen dioxide (NO_2) and determine their differences. The student reports back that there are no differences between the molecules because they are made up of the same elements. Is the student right or wrong? Why?

5.8 Name each element and the number of atoms of that element in one molecule of acetic acid ($C_2H_4O_2$), which is the active ingredient of vinegar.

<u>Changes in Matter</u>

As we all know, God's creation is constantly changing. Seasons change; weather changes; organisms grow and mature, etc. On a smaller level, however, there are countless changes occurring in the elements and molecules that make up life and the world that contains it. These changes fall into one of two broad categories: **physical change** or **chemical change**.

<u>Physical change</u> – A change that affects the appearance but not the chemical makeup of a substance

<u>Chemical change</u> – A change that alters the makeup of the elements or molecules of a substance

These definitions probably seem a little cryptic right now, but in the next couple of paragraphs, their meanings should become clear.

Let's start with physical change. Suppose you were to mix sugar and water. What would happen? The sugar would slowly "disappear" into the water, right? Actually, it wouldn't disappear. It would just dissolve in the water. You would not be able to see it, but it would still be there. You know that because the water has a sugary taste. Now before you mixed the sugar in the water, you had sugar in a spoon and water in a glass. What do you have once you mix the two? You still have sugar and water. They are now mixed, but they are still the same substances you started with. Their molecules have not changed; they have just been mixed together. That's a physical change.

Chemical change is something entirely different from physical change. Suppose you were to take a piece of paper and light it with a match. What would happen? The paper would begin to burn, making flames and smoke appear. In the end, you would be left with a small pile of ashes where there was once a piece of paper. This is an example of a chemical change. The molecules that made up the paper have been completely changed. They were changed into gases (principally carbon dioxide and water vapor) and ash. One substance (paper) changed into other substances (carbon dioxide, water vapor, and ash). That's a chemical change. As with physical change, don't worry if you are a little confused on this point; we will be studying chemical change in great detail soon!

One idea that is often helpful to students in determining whether a change in matter is physical or chemical is as follows: **physical changes are generally reversible; chemical changes are generally not reversible**. For example, you can reverse the process of dissolving sugar in water. All you would have to do is boil the mixture. Once the water was boiled off, you would have sugar left in the pan. Additionally, if you collected the steam as the water boiled and then cooled it, you would get the water back as well. Thus, by boiling the sugar water, you can reverse what you did in making it. On the other hand, you cannot reverse the burning of paper. Once it has burned, there is no way to reverse the process and get the paper back. It has been changed forever. Thus, dissolving sugar in water is a physical change because it can be reversed, and burning paper is a chemical change because it cannot be reversed. Now please realize that this is just a helpful tool in distinguishing chemical changes from physical changes. When you take chemistry later on in your high school experience, you will actually learn that chemical changes *can* be reversed, at least to some degree and under the right conditions. Thus, this is not a "hard and fast" rule of nature. Nevertheless, the idea that chemical changes are not reversible and physical changes are reversible can be a great help in determining whether a change is physical or chemical.

ON YOUR OWN

5.9 Identify the following changes as chemical or physical:

   a. putting milk on cereal   b. baking bread   c. boiling salt water

Physical Change

Now that you have at least some understanding of the definition of physical change, it is time to study a few specific types of physical change in detail. The first important type of physical change involves the **phases of matter**.

   Phase – One of three forms – solid, liquid, or gas – which every substance is capable of attaining

When we change a substance from one phase to another, it is a physical change. For example, when you put water in the freezer, it turns from its liquid phase (water) into its solid phase (ice). Despite the change in appearance, the molecules in the ice are still water molecules; they are simply in a different phase. Thus, phase changes are physical changes, as they do not change the chemical makeup of the substance undergoing the change.

   In general, we all know that to change something from its liquid phase to its solid phase, we must freeze it. This involves cooling it down. When you cool a substance down, you actually remove energy from the molecules of the substance. Thus, freezing a substance is really just a matter of removing energy from its molecules. We also know that to change a substance from its solid phase to its liquid phase, we simply have to melt it. This involves heating it up. When you heat a substance up, you are, in fact, adding energy to its molecules. Thus, changing the phase of a substance is really just a matter of modifying the energy that the molecules of the substance have. If you take energy away from the molecules of a substance, the phase of the substance changes from gas to liquid to solid. If you add energy to a substance, the phase of the substance changes from solid to liquid to gas. This relationship can be illustrated schematically as follows:

$$\textbf{SOLID} \underset{\substack{\text{TAKE HEAT}\\ \text{AWAY}}}{\overset{\text{ADD HEAT}}{\rightleftarrows}} \textbf{LIQUID} \underset{\substack{\text{TAKE HEAT}\\ \text{AWAY}}}{\overset{\text{ADD HEAT}}{\rightleftarrows}} \textbf{GAS}$$

Once again, it is important to realize that since a phase change involves only removing or adding energy to the molecules of a substance, the chemical makeup of those molecules does not change. As a result, phase changes are physical changes.

   One type of physical change that is very important in biology occurs when one substance is dissolved in another. In the physical change example we gave you previously, we discussed what happens when sugar is dissolved in water. Although the sugar seems to disappear, it does not. Instead, it is simply mixed throughout the water. As a point of terminology, when one substance is dissolved in another, it is a physical change, and the result is called a **solution**. The substance being dissolved in the liquid is called the **solute**, and the liquid is called the **solvent**. In the case of salt being dissolved in water, then, the solute is salt, the solvent is water, and the resulting solution is salt water.

A lot of the chemistry of life involves solutions and the changes that occur in them. It is therefore important to study two aspects of how molecules behave in solution with the following experiments. Note that once you set it up, the first experiment has to sit for two hours, so plan your school day accordingly! The second experiment takes several days to complete, so there is not much you can do about that.

EXPERIMENT 5.1
Diffusion

Supplies

♦ Sugar
♦ Tablespoon
♦ Water
♦ A small glass
♦ A paper napkin
♦ Cellophane tape
♦ Plastic wrap

Object: To observe and learn about the process of diffusion

Procedure:

1. Measure out a tablespoon of sugar and dump it into the center of an unfolded napkin.
2. Fold the napkin around the lump of sugar so that it forms a nice "package," completely surrounding the sugar. Tape the napkin so that it does not unfold. DO NOT cover the entire napkin with tape! Tape it with only a few strips of tape just so it does not unfold. In the end, you should be able to pick up the napkin without any sugar spilling out of it.
3. Put the napkin/sugar "package" into the small glass and fill the glass with water. The napkin will probably float up to the top of the water. That's okay.
4. Cover the glass with plastic wrap and let it sit for at least 2 hours.
5. Once at least 2 hours are up (if you wait longer your results will be better), remove the plastic wrap and then the napkin from the glass. Swirl the water in the glass and then take a small sip of the water. What does it taste like? **NOTE: You should never, ever taste the results of an experiment unless you are instructed to do so.** You can taste the results of this experiment because I know that the solution is harmless. However, unless someone who really knows chemistry tells you to do so, never taste the results of an experiment.
6. Clean up your mess.

What did the water taste like in step #5? If everything went well in your experiment, the water tasted sweet. Why? Well, sugar obviously found its way out of the napkin and into the water. It did so by a process known as **diffusion**.

<u>Diffusion</u> – The random motion of molecules from an area of high concentration to an area of low concentration

Of course, this definition does not do you much good unless you know the definition of concentration.

<u>Concentration</u> – A measurement of how much solute exists within a certain volume of solvent

Now that you have those two definitions under your belt, you can understand what really happened in the experiment. You see, when you placed the napkin in the water, there was at first no water in the napkin or in the sugar. Thus, the concentration of water inside the napkin was zero. Since there are many holes in a napkin, water molecules were able to travel through those holes and start filling up the napkin. Thus, water molecules began moving from an area of high concentration (the water in the glass) to an area of low concentration (the napkin). You probably noticed this right away because the napkin got wet.

What you probably did not notice was that the sugar was doing the same thing. Inside the napkin, there was a *lot* of sugar. Outside the napkin, there was none. When in its solid form, the sugar was too large to get through the tiny holes in the napkin, so the sugar stayed where it was. However, as soon as the water got inside the napkin, some of the sugar dissolved. When dissolved, the sugar molecules could individually get through the tiny holes in the napkin. Thus, the sugar moved from inside the napkin (where the concentration of sugar was high) into the water outside of the napkin (where the concentration was low).

It's important for you to realize that the motion of the molecules in this experiment was not directed by some mysterious force. Instead, molecules that are in their liquid phase and molecules that are dissolved in liquid tend to move about randomly as a matter of course. That's why the definition of the term diffusion includes the word "random." The water molecules simply were moving about randomly. Some of them happened to find themselves in the napkin as a result of their random motion. Likewise, once the water began to dissolve the sugar molecules, the dissolved sugar molecules began randomly moving. Since they could fit through the holes of the napkin once they were dissolved in the water, some of the sugar molecules found themselves in the water outside of the napkin. Thus, the sugar molecules did not suddenly "decide" that they had to get to an area of low sugar concentration. Instead, they just randomly ended up there. Given enough time, this random motion would have evenly mixed the sugar and water throughout the glass. Now perform the next experiment, which explores another kind of behavior that evens out the concentration of solutes in a solution.

EXPERIMENT 5.2
Osmosis

<u>Supplies</u>:

♦ Three coffee mugs
♦ One fresh, raw egg
♦ A measuring cup for liquids
♦ A tape measure
♦ White vinegar
♦ Clear sugar syrup (like Karo® syrup)
♦ Distilled water (You can purchase this at any large supermarket.)

Object: To observe the process of osmosis. Note: this experiment takes three or four days to complete.

Procedure:

1. Use the tape measure to measure the distance around the widest part of the egg's middle.
2. Using the measuring cup, measure ¾ cup of vinegar and pour it into a coffee mug.
3. Place the egg in the mug. The egg should be almost totally covered with the vinegar. Since the egg probably floats, there will be a small part of the egg that is not covered. That's okay. If there is not enough vinegar to cover the egg except for the tiny part at the top, add more vinegar.
4. Let the egg soak in the vinegar for 24 hours.
5. Very carefully remove the egg from the vinegar. You should notice that it is soft to the touch because the shell has been removed by the vinegar. If the shell has not been removed, dump the old vinegar out and replace it with fresh vinegar. Leave the egg in the new vinegar for another 24 hours.
6. Once the shell has been removed, measure ¾ cup of the syrup and pour it into the second mug.
7. Very carefully rinse the egg with water and note its appearance.
8. Use the tape measure to measure the distance around the widest part of the egg's middle.
9. Place the egg in the mug of syrup.
10. Let the egg soak in the syrup for 24 hours.
11. After 24 hours, measure ¾ cup of distilled water and pour it into the third mug.
12. Very carefully remove the egg from the syrup and rinse it.
13. Note its appearance and use the tape measure to measure the distance around the widest part of its middle.
14. Put the egg in the water and let it soak for 24 hours.
15. After 24 hours, very carefully remove the egg from the water. Note its appearance and use the tape measure to measure the distance around the widest part of its middle.
16. Throw the egg away.
17. Clean up your mess.

What did you see in the experiment? While it was in the vinegar, the egg's hard calcium shell reacted with the acetic acid in the vinegar. This turned the shell into a soluble compound (calcium acetate) that dissolved, which left the membrane under the shell exposed. That membrane is a **semipermeable membrane**.

Semipermeable membrane – A membrane that allows some molecules to pass through but does not allow other molecules to pass through

Thus, by the time you reached step #6, the egg was no longer surrounded by its shell, but it was surrounded by only a semipermeable membrane.

Did your measurement of the distance around the egg's middle indicate that the egg was larger after it had soaked in the vinegar than before you put it in the vinegar? That's because of **osmosis**.

Osmosis – The tendency of a solvent to travel across a semipermeable membrane into areas of higher solute concentration

You see, inside the egg there are lots of chemicals and only a little water. Thus, the solute concentration inside the egg is large. The concentration of solute in the vinegar solution is much lower. Now, if the solute molecules inside the egg could diffuse, a lot of them would leave the egg and enter the vinegar. However, they could not, because the semipermeable membrane would not allow them to pass through. However, water *can* pass through the membrane. Thus, once the hard shell was dissolved, water from the less-concentrated vinegar solution moved to the higher-concentrated interior of the egg. This caused the egg to swell.

What happened when the egg was soaked in the syrup? In that situation, the concentration of dissolved substances *outside* the egg (in the syrup) was higher than the concentration of dissolved substances *inside*. As a result, osmosis dictated that water had to move out of the egg and into the mug. This is why the egg appeared dimpled in step #13. This is also why the distance around the egg's middle decreased as compared to when you took the egg out of the vinegar. Since water went from the egg into the syrup, the egg got smaller.

When the egg soaked in water (step #14), the concentration of dissolved substances inside the egg was once again higher than the concentration of dissolved substances in the water, so the water moved back into the egg, once again making it larger. In this experiment, then, you saw the effects of osmosis several times. Osmosis caused water in the vinegar to travel into the egg when the egg soaked in vinegar. Then, it caused water in the egg to travel out of the egg when the egg soaked in syrup. Finally, it caused water to move back into the egg when the egg soaked in plain water.

Now please understand that had the membrane around the egg not been semipermeable, this would not have happened. Had the membrane allowed solutes to travel through, the solute would have randomly traveled back and forth across the membrane until the concentration of solutes was the same throughout both the egg and the solution in the mug. However, that could not happen because of the semipermeable membrane. The membrane allowed only water to pass through it. Thus, in an attempt to even out the concentration of solutes, the water had to travel across the membrane from an area of low solute concentration to an area of high solute concentration.

To fully understand the difference between osmosis and diffusion, examine the figure below.

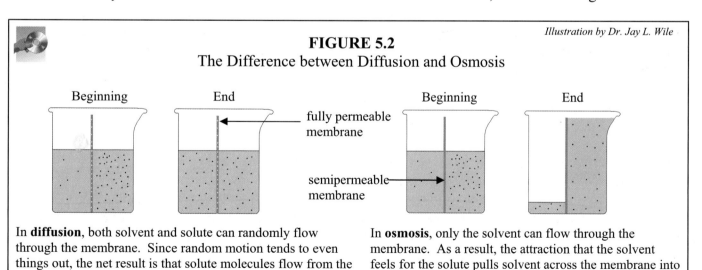

Illustration by Dr. Jay L. Wile

FIGURE 5.2
The Difference between Diffusion and Osmosis

Beginning   End   fully permeable membrane   Beginning   End

semipermeable membrane

In **diffusion**, both solvent and solute can randomly flow through the membrane. Since random motion tends to even things out, the net result is that solute molecules flow from the side of high concentration to the side of low concentration.

In **osmosis**, only the solvent can flow through the membrane. As a result, the attraction that the solvent feels for the solute pulls solvent across the membrane into the side with the most solute. This evens out the concentration of solute, but in a completely different way.

Do you see the difference between osmosis and diffusion? In diffusion, molecules move randomly so that solute and solvent spread out evenly throughout the solution. That's what happened in the first experiment. In osmosis, the travel of the solute is restricted by a semipermeable membrane. As a result, only the solvent can move. Thus, the solvent moves across the semipermeable membrane into the area with the highest concentration of solute. This is what happened in the second experiment.

In "Rime of the Ancient Mariner," a famous poem by Samuel Taylor Coleridge about a ship lost at sea, the mariner laments:

> "Water, water, every where,
> And all the boards did shrink;
> Water, water, every where,
> Nor any drop to drink."

This is a reference to a common problem in early sea travel. Many crews died of thirst because their voyage took longer than expected and not enough fresh water had been taken along. Despite the fact that the ship was surrounded by water, none of it could be drunk. You see, ocean water has a high concentration of salt in it. If you drink this salt water, it will come into contact with the cells of your body, which are full of water that has a *much lower* concentration of salt. Since the cells in your body are covered with a semipermeable membrane, osmosis causes the water in your cells to leave the cell (where the solute concentration is low) and enter the salt water that you drank (where the solute concentration is high). This causes your cells to lose water, which in turn causes them to shrivel up and die. Thus, if you drink a lot of salt water, you will kill yourself one cell at a time! That's what was frustrating to the mariner in the poem. His crew was dying of thirst, despite the fact that there was ocean water all around!

It turns out that osmosis and diffusion play rather critical roles in many biological processes. Do you remember the contractile vacuoles you learned about in Module #3? A lot of the organisms from kingdom Protista have them. Do you remember their job? They must remove excess water from the cell. How does the excess water get into the cell? It gets there by osmosis. Consider an amoeba that lives in fresh water. The concentration of solutes dissolved inside the cell is much higher than the concentration of solutes dissolved in the fresh water. What happens as a result? Water travels into the cell by osmosis. If that were the end of the story, the cell would eventually explode because it absorbed too much water. However, God has designed such organisms with contractile vacuoles, which get rid of the water as it is being absorbed. That should give you an idea of how important osmosis is in biology. Many members of kingdom Protista have specific organelles designed to fight the effects of osmosis.

ON YOUR OWN

5.10   A semipermeable membrane is placed in a beaker. Equal amounts of salt water solution are placed on each side of the membrane, but the solution on one side is twice as concentrated with salt as is the solution on the other side. After one hour, the water level of the solution on the right increases, and the water level of the solution on the left decreases. Which solution (the one on the left or the one on the right) started out with the higher salt concentration?

Chemical Change

Now that we've spent some time on physical change, we need to study chemical change. Remember, chemical change occurs when the molecules in a substance change their chemical makeup. For example, most people have seen a natural gas stove or furnace flame. It is a pleasant blue flame that can be used to cook food or heat a home. Well, in order to make that flame, natural gas is burned. The main component of natural gas is methane, and it has a chemical formula of CH_4. Thus, a molecule of methane is 1 carbon atom linked to 4 hydrogen atoms. When methane burns, the oxygen in the air interacts with the methane molecules, changing them into water molecules and carbon dioxide molecules. The resulting chemical change causes energy to be released, which we detect as heat and light from the flame.

If we wanted to write out the chemical change that occurs when methane burns, we could do so as follows:

methane and oxygen interact to make water and carbon dioxide

This is a bulky way of informing someone what happened, so we will abbreviate it. We will use the chemical formulas of the substances in the reaction instead of their names; we will replace "and" with a plus sign; and we will replace "interact to make" with an arrow. The result is:

$$CH_4 + O_2 \rightarrow H_2O + CO_2$$

Now it turns out that if we were able to watch this happen on a molecular level, we would see that it takes 2 oxygen molecules to interact with 1 methane molecule, and the result is 2 water molecules and 1 carbon dioxide molecule. To really abbreviate what happened, then, we need to add this information:

$$CH_4 + 2O_2 \rightarrow 2H_2O + CO_2$$

What we did was put numbers to the left of the chemical formulas. These numbers indicate how many of each molecule it took to make the chemical change. If there was only 1 molecule, no number was written. This is called a **balanced chemical equation**. The chemical change that occurs is often called a **chemical reaction**. Thus, we could say that the balanced chemical equation above describes the chemical reaction that takes place between methane and oxygen.

It is important for you to understand that when a chemical reaction occurs, the molecules on the left side of the arrow are destroyed, and the molecules on the right side of the arrow are produced. Thus, in the burning of methane, we start out with 1 methane molecule (CH_4) and 2 oxygen molecules (O_2). These molecules interact, exchanging atoms so that the methane and oxygen molecules are destroyed. In the process, 2 water molecules (H_2O) and 1 carbon dioxide molecule (CO_2) are made. In chemical terminology, we call the molecules on the left side of the arrow **reactants**, because they react with one another and are destroyed. The molecules on the right side of the arrow are called **products**, because they are produced as a result of the chemical reaction.

When looking at a chemical equation, students often get confused between the numbers that are subscripts and the numbers that are placed to the left of the chemical formulas. You need to remember that the subscripted numbers refer to how many atoms of a certain type are in the molecule. If a subscripted number does not exist after an element's symbol, it means that only 1 such atom is in the molecule. Thus, the subscript "2" in H_2O tells us that there are 2 hydrogen atoms in the water molecule. The fact that there is no subscript after the "O" tells us that there is 1 oxygen atom in the water molecule. Thus, the formulas with the subscripted numbers (the chemical formulas) tell us *the types of molecules* that are in the reaction. On the other hand, the "2" that is written to the left of the water molecule's abbreviation tells us that 2 water molecules are produced in the chemical reaction. The numbers to the left of the chemical formulas, then, tell us *how many of each molecule* is used or made in the reaction. Make sure you understand this distinction!

ON YOUR OWN

5.11 One of the main chemical reactions used to run an automobile engine is the combustion of octane (C_8H_{18}):

$$2C_8H_{18} + 25O_2 \rightarrow 16CO_2 + 18H_2O$$

a. Write the chemical formulas of the reactants in this equation.

b. Write the chemical formulas of the products in this equation.

c. How many molecules of C_8H_{18} are used in the reaction?

d. How many molecules of H_2O are made in the reaction?

Photosynthesis

Now that you understand how to interpret chemical equations, let's talk about photosynthesis, a subject we have already discussed, but not in detail. The most prevalent means by which autotrophic organisms create their own food is photosynthesis. As you already learned, this process converts carbon dioxide and water into oxygen and a simple sugar called glucose. The balanced chemical equation is:

$$6CO_2 + 6H_2O \rightarrow C_6H_{12}O_6 + 6O_2$$

In photosynthesis, then, 6 carbon dioxide (CO_2) molecules interact with 6 water (H_2O) molecules. Once they interact, 1 glucose molecule ($C_6H_{12}O_6$) and 6 oxygen (O_2) molecules are formed.

Think for a moment about what you've already learned regarding photosynthesis, and you will probably realize that something is missing from this chemical equation. What is it? Well, you've already learned that photosynthesis requires sunlight. Where is the sunlight in this equation? Sunlight isn't in the equation, but it is still necessary in photosynthesis. You see, if a bunch of carbon dioxide and water were put into a container, they would not interact to form glucose and oxygen. In order for them to interact, the carbon dioxide and water molecules must be pushed together. This takes energy, which the plant gets from sunlight.

Okay, then, in order for plants to make glucose and oxygen, they just need carbon dioxide, water, and energy, right? Well, no. Even after supplying a lot of energy to a container full of carbon dioxide and water, you would not produce much glucose. You see, even with a lot of supplied energy, the interaction between carbon dioxide and water is very slow. Thus, the reaction would have to continue for a long, long time before even a small amount of glucose would be produced. This is unacceptable to an autotrophic organism, because it must produce a lot of food for itself. Thus, the organism must speed up the reaction, making it produce glucose much faster than it normally would.

How does an organism speed up the photosynthesis reaction? It turns out that there are two ways to speed up a chemical reaction. First, the reaction could be performed at a higher temperature. Higher temperatures increase the speed of most chemical reactions. This isn't a viable option, however, because most autotrophic organisms cannot survive unusually high temperatures. Thus, autotrophic organisms use another method. They employ a **catalyst** (kat' uh list).

Catalyst – A substance that alters the speed of a chemical reaction but is not used up in the process

Most catalysts speed up chemical reactions, but there are a few which are known to slow down chemical reactions.

Now look at that definition for a moment. Catalysts alter the speed of a chemical reaction, but they are not used up in the process. If you think about it, this is really an amazing feat. Somehow, the catalyst *affects* the reaction, but it isn't *used up by* the reaction. Pretty nifty, huh? How does a catalyst accomplish this feat? Unfortunately, you will have to wait until chemistry class to learn the answer to that question.

So, autotrophic organisms speed up the photosynthesis reaction by using a catalyst. What is the catalyst? For most autotrophic organisms, it is **chlorophyll**. Back in Module #3 we studied green algae in kingdom Protista, and we briefly discussed chlorophyll, telling you that it was necessary for photosynthesis in these organisms. Now you know why. Chlorophyll speeds up the photosynthesis reaction so that an autotrophic organism can produce a reasonable amount of glucose for its food supply. Green plants use chlorophyll as a catalyst in photosynthesis as well; however, chlorophyll is not the only catalyst that can be used in photosynthesis. Blue-green algae, for example, use a class of substances called blue phycobilins (fye' koh bil uhns) in addition to chlorophyll. Members of phylum Rhodophyta (the red algae) use red phycobilins as their catalysts in photosynthesis.

At this point, your head might be swimming a bit because you've taken in a lot of information. Let's slow down and do a little review. Most autotrophic organisms use photosynthesis to create their own food. This process is a chemical change that takes 6 carbon dioxide molecules and 6 water molecules to make 1 glucose molecule and 6 oxygen molecules. The glucose is food for the organism, and the oxygen goes back into the atmosphere for organisms to use. In order for this chemical reaction to occur, however, the organism needs more than just carbon dioxide and water. It needs energy, which it gets from sunlight, and it needs a catalyst, which is usually (but not always) chlorophyll.

Now if you think about this, it is an incredibly complex system! It's amazing enough that an organism can actually produce its own food. Once you learn the details of how this is accomplished, however, you should be overcome with a deep sense of awe. God has designed His creation so intricately that even a single-celled organism like the euglena knows how to feed itself by collecting carbon dioxide and water, producing a catalyst that will speed up their reaction, and trapping energy from the sun to make the reaction occur. If it hasn't already become clear to you, this should drive a fundamental point home: **there is no such thing as a simple life form**. All of God's creation is intricate and complex, and as you learn more about it, your appreciation of His power should increase dramatically!

ON YOUR OWN

5.12 A plant loses all of its chlorophyll. Will it be able to produce any glucose at all?

5.13 A chemist is trying to speed up a chemical reaction. If the chemist does not have a catalyst, what other means can be used?

Organic Chemistry

The most important chemicals in the study of life fall into a broad category called **organic molecules**.

Organic molecule – A molecule that contains only carbon and any of the following: hydrogen, oxygen, nitrogen, sulfur, and/or phosphorous

This is actually a simplistic definition of organic molecules, but it fits more than 99% of them, so for right now it will do. By this definition, molecules like CH_4, CO_2, and C_2H_4O are organic molecules (CH_4 contains only carbon and hydrogen, CO_2 contains only carbon and oxygen, and C_2H_4O contains only carbon, hydrogen, and oxygen), but $CaCO_3$ and H_2O are not organic molecules (the element Ca is not on the list, and H_2O contains no carbon).

Many organic molecules are actually produced by living organisms. When a living organism takes small molecules and makes larger molecules, we call it **biosynthesis** (bye oh sin' thuh sis).

Biosynthesis – The process by which living organisms produce larger molecules from smaller ones

Photosynthesis is an example of biosynthesis. Living organisms (plants, for example) take small molecules like CO_2 and H_2O and make a large molecule like glucose. That's biosynthesis. If you remember our discussion of anabolism and catabolism from Module #1, you should recognize biosynthesis as an example of anabolism (building large molecules out of simple ones). The products of biosynthesis are many and varied, but they fall into a few broad categories, most of which we will discuss in the next few sections.

Carbohydrates

Carbohydrates are organic molecules that contain only carbon, hydrogen, and oxygen. In addition, they have the same ratio of hydrogen atoms to oxygen atoms as does water. For example, one of the simplest carbohydrates is glucose, $C_6H_{12}O_6$. Notice that it has 12 hydrogen atoms and 6 oxygen atoms. In other words, there are twice as many hydrogen atoms as there are oxygen atoms. This is the same as water (H_2O). In fact, that's where the term "carbohydrate" originates. "Carbo" stands for the carbon in the molecule, and "hydrate," which means "to add water," stands for the fact that there are twice as many hydrogen atoms as oxygen atoms, just like water.

Now although giving you the chemical formula of glucose is instructive, there is an even more instructive means of describing glucose to you. The **structural formula** of a molecule gives you the type and number of atoms in the molecule, but it also tells you something else. It tells you which atoms are linked to which. For example, one structural formula for glucose is:

Illustration by Megan Whitaker

If you simply count the atoms in the structural formula, you will see that there are 6 carbon atoms, 12 hydrogen atoms, and 6 oxygen atoms. Thus, as we knew already, the chemical formula of glucose is $C_6H_{12}O_6$. By looking at the structural formula, however, we can see a lot more. The lines in the structural formula represent **chemical bonds**, which link atoms together in a molecule. Thus, we can see that the 6 carbon atoms are all linked to one another in a straight line, which we call a **carbon chain**. The first carbon in the chain (the one on the far left) has 2 H's attached to it, an O attached to it, and another C attached to it. This is exactly the way we would see the atoms arranged in the molecule, if we were able to see a molecule of glucose.

Looking at the structural formula of glucose a little more carefully, you will see two strange things. First, you will notice that there are two lines which link the carbon on the far right of the carbon chain to the oxygen at the end of the structural formula. This is called a **double bond**, and it is approximately twice as strong as the bonds represented by the single lines in the structural formula. For biology, it is not critical that you understand what a double bond really is. You will learn that next year in chemistry. For right now, just realize that it is a chemical bond that is roughly twice as strong as most chemical bonds. There are, in fact, **triple bonds** that exist in some molecules. In structural formulas, triple bonds are represented by three lines and are roughly three times as strong as most chemical bonds.

The second strange thing you will see in the structural formula of glucose is the fact that there are five times in which an oxygen atom (O) and a hydrogen atom (H) are drawn next to each other but there is no line linking them together. Does that mean they are not linked to one another? No, not really. It turns out that when an oxygen atom and a hydrogen atom are linked together in a molecule, biologists rarely draw a line to represent the bond that exists between them. This is because the combination of an O and an H linked together is fundamentally important in biology, so to represent this, we purposely do not draw the bond. This makes that part of the molecule stand out. Thus, even though it is not drawn, you need to realize that there is a bond linking the O to the H in an OH group.

Now that you have some understanding of structural formulas, it is time to throw a curveball at you. **Many molecules have more than one structural formula!** In fact, this is the case for glucose. Although some glucose molecules look like the one drawn above, other glucose molecules look like this:

Illustration by Megan Whitaker

What's the difference? Well, the chemical formula is still the same. If you count the atoms, you will find 6 C's, 12 H's, and 6 O's. The main difference is that the double bond is gone between the oxygen and the carbon and, instead, the atoms have arranged themselves in a ring. Not surprisingly, the first structural formula of glucose is called the **chain structure**, while this one is called the **ring structure**. In biology, the ring structure is the most prevalent form of glucose.

Are you ready for another curveball? We hope so! You see, not only can the same molecule have different structural formulas, but different molecules can also have the same chemical formula! When two different molecules have the same chemical formula, they are called **isomers**:

<u>Isomers</u> – Two different molecules that have the same chemical formula

Now before we move on to *why* two different molecules can have the same chemical formula, it is imperative that you know what we mean when we say "different." Have you ever noticed that both lettuce and fruit taste sweet, but they taste sweet in different ways? Well, the reason for this is that there are two different types of sugar molecules in them. Lettuce has glucose, the molecule that we have been discussing. This molecule gives lettuce its sweet taste. On the other hand, fruit has a sugar called fructose. It tastes different from glucose, so the sweetness of fruit is different from the sweetness of lettuce. These two sugars taste different because they are different molecules. They have different chemical characteristics, one of which is the way we taste them. That's what we mean when we say "different" molecules. Different molecules have different chemical properties.

What's the chemical formula of fructose? It's $C_6H_{12}O_6$, the same as glucose! If these two molecules have the same chemical formula, how can they be different molecules? Well, they have the same *chemical formula*, but they have different *structural formulas*. The ring and chain structural formulas of both glucose and fructose are shown in the figure below.

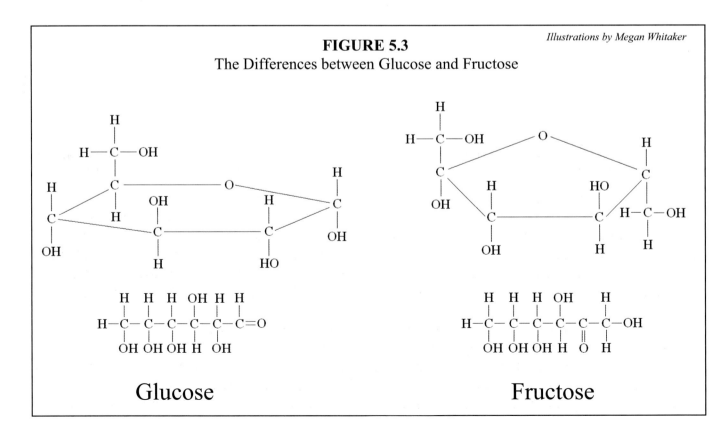

FIGURE 5.3

Illustrations by Megan Whitaker

The Differences between Glucose and Fructose

Glucose   Fructose

These different structural formulas are the reasons behind the taste difference and other chemical differences between glucose and fructose.

Glucose and fructose belong to a class of compounds known as **monosaccharides** (mahn uh sak' uh rides), which are also called **simple sugars**.

Monosaccharides – Simple carbohydrates that contain 3 to 10 carbon atoms

The reason that these are called monosaccharides is that they form the basic building blocks of more complex carbohydrates called **disaccharides** (dye sak' uh rides) and **polysaccharides** (pahl ee sak' uh rides).

Disaccharides – Carbohydrates that are made up of two monosaccharides

Polysaccharides – Carbohydrates that are made up of more than two monosaccharides

These more complex carbohydrates form the basis of much of the food that we eat.

For example, table sugar is a disaccharide called **sucrose** (soo' krohs). It is formed when glucose and fructose chemically react in a process known as a **dehydration** (dee hye dray' shun) **reaction**.

Dehydration reaction – A chemical reaction in which molecules combine by removing water

The dehydration reaction that makes sucrose is shown in the figure below.

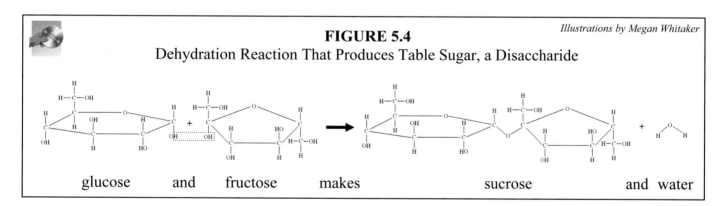

FIGURE 5.4
Dehydration Reaction That Produces Table Sugar, a Disaccharide

Illustrations by Megan Whitaker

glucose   and   fructose   makes   sucrose   and water

Do you see now why we call this a dehydration reaction? When you dehydrate something, you remove water. Well, in this reaction, a hydrogen from the glucose and an OH from the fructose (the atoms surrounded by the red rectangle) combine to make water, "removing water" from the two molecules. The remaining oxygen atom then links the two molecules together, making a disaccharide. Many other disaccharides make up the sugars that sweeten our foods. Glucose and galactose, for example, combine in a dehydration reaction to make lactose, the sugar that gives milk its sweetness.

When several monosaccharides link together, the result is a polysaccharide. Typically, polysaccharides are not sweet, because the sweetness of the monosaccharides is lost when several combine. Nevertheless, polysaccharides are an important part of our diet. The polysaccharides known as starches, for example, are found in most plants. When a plant has extra monosaccharides, it will store them as polysaccharides by running many dehydration reactions that link the monosaccharides together. Potato starch, corn starch, and starch from wheat, rice, and other grains are a major source of food for people. People and animals make their own starch, **glycogen** (glye' koh jen), when they have excess carbohydrates to store. The nondigestible part of our diet (often called "roughage") is made up of the polysaccharide known as cellulose. Although most organisms cannot digest cellulose, God has created a few protozoa, bacteria, and fungi that use it as their major source of food.

As you might have already learned by studying nutrition, carbohydrates are one of the principal sources of food energy for most animals. Interestingly enough, however, most animals can only use monosaccharides for energy. Thus, when an organism eats disaccharides or polysaccharides, it must first break them down into their individual monosaccharide components. How is this done? Well, monosaccharides combine to form disaccharides and polysaccharides by dehydration. To break these complex molecules back down into their monosaccharide components, all you have to do is add water to them. This process is called **hydrolysis** (hye drah' luh sis).

Hydrolysis – Breaking down complex molecules by the chemical addition of water

Hydrolysis is essentially the reverse of dehydration. We will discuss it in more detail when we discuss proteins and enzymes.

ON YOUR OWN

5.14 What is the chemical formula of the molecule with the following structural formula?

$$H-C=C-H$$
$$\;\;\;\;|\;\;\;\;\;|$$
$$\;\;\;\;H\;\;\;\;H$$

5.15 A chemist takes a polysaccharide and turns it into many disaccharides. Has the chemist used dehydration or hydrolysis?

Organic Acids and Bases

Two classes of molecules that you will spend a lot more time learning about in chemistry are **acids** and **bases**. In general, acids are substances that taste sour, while bases are substances that tend to taste bitter. Many of the drinks that we enjoy (soda pop and fruit juice, for example) contain acids, and many of the substances we use for cleaning (soap and spray-cleaner, for example) contain bases. When acids and bases react together, they typically form water and another class of molecule called "salts." Table salt is just one example of the general chemical class known as salts.

Organic acids, in general, contain the following pattern of atoms bonded together:

$$O$$
$$\|$$
$$-C-OH$$

When a molecule has a section that looks like this, it is generally considered an organic acid. Not surprisingly, then, this is often called an **acid group**. Figure 5.5 shows the structural formula of three organic acids. Notice that even though they look quite different, they each have one thing in common. They each have at least one acid group.

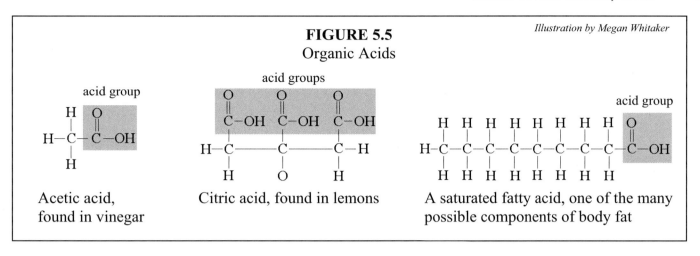

FIGURE 5.5
Organic Acids

Illustration by Megan Whitaker

Acetic acid, found in vinegar

Citric acid, found in lemons

A saturated fatty acid, one of the many possible components of body fat

Organic bases also have a group of atoms in common, which is often called the **amine group**. This group is a little harder to spot than the acid group, so we will not concentrate on the structural formulas of organic bases.

In most of the chemical reactions that make life possible, the amount of acid or base present has a profound effect on the speed and effectiveness of the reaction. As a result, tracking the level of acid or base is quite important. Chemists have a scale called the "pH" scale to do just this. The pH scale runs from 0 to 14. When a solution has a pH of 7, it is considered **neutral**, having no net acid or base characteristics. Solutions with pH from 0 to just under 7 are **acidic**. The lower the pH, the stronger the acidic nature of the solution. Solutions with pH from just above 7 to 14 are called **alkaline** and have the characteristics of a base. The higher the pH, the more like a base the solution becomes. Figure 5.6 illustrates this scale, along with some common substances and where they fall within it.

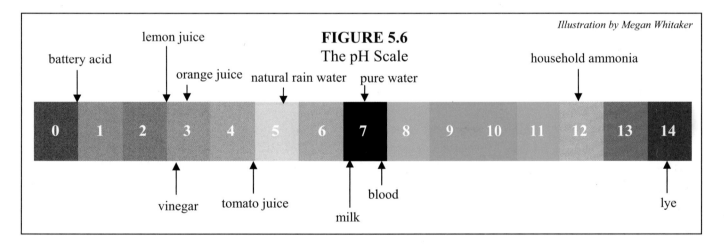

FIGURE 5.6
The pH Scale

Illustration by Megan Whitaker

ON YOUR OWN

A chemist measures the pH of several solutions. The results are: Solution A: 8.1, Solution B: 1.1, Solution C: 5.5, Solution D: 13.2.

5.16 Which solution is the most alkaline?

5.17 Which solution is the most acidic?

Lipids

Lipids, also known as fats, are complex molecules formed when three fatty acids, like the one shown in Figure 5.5, link to a substance known as glycerol in a dehydration reaction. A sample dehydration reaction that forms a lipid is shown in the figure below.

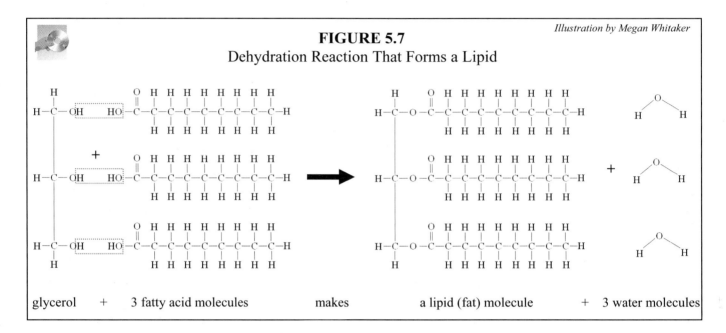

FIGURE 5.7
Dehydration Reaction That Forms a Lipid

Illustration by Megan Whitaker

glycerol   +   3 fatty acid molecules   makes   a lipid (fat) molecule   +   3 water molecules

In general, lipids cannot be dissolved in water. Cooking oil, for example, is made of lipids. What happens when you try to mix water and cooking oil? They don't mix well, do they? This is because the structure and components of lipid molecules make them **hydrophobic**.

Hydrophobic – Lacking any affinity to water

Since lipids are not attracted to water, they simply cannot be dissolved in it.

Lipids are important biologically because most animals store most of their excess food as lipids. Even though animals can take their excess carbohydrates and turn them into glycogen (as we learned just a few pages ago), lipids are the preferred method of storing excess food, because more than twice as much energy can be stored in an ounce of fat than in an ounce of carbohydrates. As a result, animals make fats whenever they have excess food. Later, when food is scarce, their bodies can digest the fat for energy. For most humans in the United States, of course, there isn't a lot of food scarcity. Thus, the fat just builds up in our bodies, making us overweight.

The fact that more than twice as much energy can be stored in an ounce of fat as can be stored in an ounce of carbohydrates has implications for our diets. People tend to like the taste of fats, and in the United States, we have a lot of food that is rich in fat. Since a lot of energy can be stored in fat, the higher your diet is in fat, the more excess energy you tend to intake. After all, most people tend to eat until their stomachs are full. Well, if you fill your stomach with foods that are mostly fat, you have taken in twice as much energy as you would have if you had filled your stomach with foods that are mostly carbohydrates. As a result, your body will have a lot more excess energy, and it will make more fat to store that excess energy. Diets high in fat, then, can lead to being overweight.

Nowadays, nutritionists also tell us that **saturated fats** tend to lead to heart troubles while **unsaturated fats** are less likely to. As a result, they suggest that we eat unsaturated fats. What's the difference? When a fat is made from fatty acids that have no double bonds between the carbons, it forms a saturated fat. We call the fat "saturated" because it has all of the hydrogens it can take. As a result, it is "saturated" with hydrogen. If a fatty acid has one or more double bonds between the carbons, however, it has fewer hydrogens than it possibly could. As a result, we call it "unsaturated."

Saturated fat – A lipid made from fatty acids that have no double bonds between carbon atoms

Unsaturated fat – A lipid made from fatty acids that have at least one double bond between carbon
                            atoms

Look back at Figure 5.7. The fatty acid molecules that make the fat drawn in that diagram are saturated, as there are no double bonds between the carbon atoms. When saturated fatty acid molecules attach to the glycerol, they make a saturated fat. If, on the other hand, there are one or more double bonds between the carbons in the fatty acid molecules, when they link to the glycerol, they make an unsaturated fat.

How do you know whether a fat is saturated or unsaturated? Well, one way is to look at the structural formula of the fat. If there are double bonds between the carbons in the fatty acids, it is an unsaturated fat. Otherwise, it is saturated. There is another way to distinguish these two types of fats, however. Generally, you can look at their phase. At room temperature, saturated fats tend to be solid while unsaturated fats tend to be liquid. Thus, the solid lard that you use for baking is a saturated fat. The liquid oils that you use for frying, however, are unsaturated fats.

ON YOUR OWN

5.18 If plenty of glycerol is available, how many fat molecules can be made from 15 fatty acid molecules?

Proteins and Enzymes

Proteins make up one of the most important groups of molecules in the chemistry of life. They are involved in virtually every chemical reaction that supports life. As you might expect, they come in many different shapes and sizes and are very complex. They are mostly made from carbon, nitrogen, hydrogen, and oxygen, but some proteins have phosphorous and sulfur in them as well.

Like polysaccharides, proteins have some basic building blocks. The basic building blocks of a protein are called **amino** (uh mee' no) **acids**. There are about 20 different amino acids in the proteins that make up life, and the type of amino acid along with the order in which the amino acids are linked up determine the shape and function of a protein. As you might expect, amino acids link up using dehydration reactions, just like monosaccharides do when they link up to form polysaccharides. As the amino acids link together, ejecting water, a bond called a **peptide bond** forms.

Peptide bond – A bond that links amino acids together in a protein

Now unfortunately, it is impossible to give you a structural formula for a protein. You see, the average amino acid has about 20-40 atoms, and even the simplest protein of life contains 124 amino acids linked together. Obviously, then, drawing the structural formula of a protein would be quite a job. The situation would be even worse for the "average" protein in the chemistry of life, which contains several thousand amino acids!

Even though it is impossible to give you a structural formula for a protein, it is possible to give you some idea of what one looks like. Figure 5.8 is a schematic representation of a protein called ribonuclease (rye boh new' klee ays).

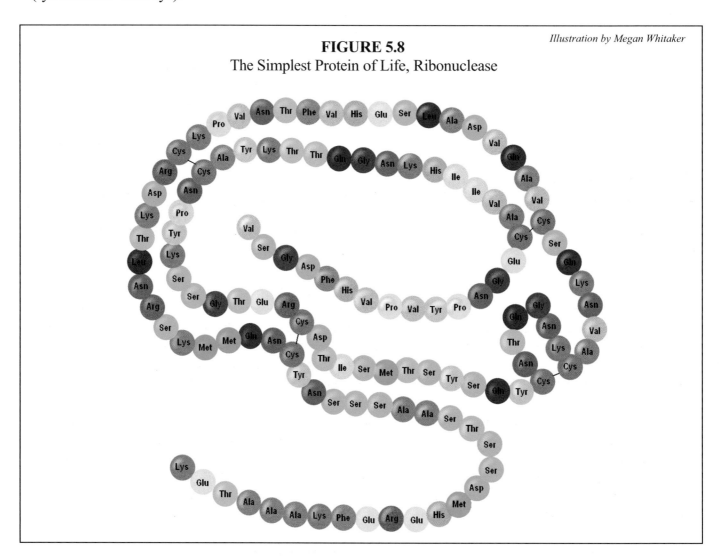

FIGURE 5.8
The Simplest Protein of Life, Ribonuclease

Illustration by Megan Whitaker

What does this schematic mean? Well, each three-letter abbreviation stands for an amino acid. The letters "Lys," for example, stand for the amino acid called "lysine," while the letters "Glu" stand for glutamic acid. This schematic, then, tells you the type of amino acids and the order in which they appear in the protein ribonuclease. In addition, the shape that the amino acids are lined up in is roughly equivalent to the shape of ribonuclease. Of course, since the paper is only two-dimensional, it is impossible to draw the entire three-dimensional shape, but what is drawn here is a reasonable approximation.

Remember back in Module #1 when we promised to tell you why we think that abiogenesis (the formation of life from nonliving chemicals) is impossible? Well, proteins are a part of the reason. In order for abiogenesis to work, proteins would have to be able to form from random chemical reactions. Without this happening, life could never appear because proteins are such a fundamental component of the chemical reactions that make life possible. It is our contention that the formation of proteins from random chemical reactions is impossible. You see, when assembled as shown in Figure 5.8, those 124 amino acids form a protein, ribonuclease, that performs a vital task in the chemistry of life. If the amino acids were to link up in the wrong order, or if even one of the amino acids were the wrong type, then the resulting protein would be *unable to perform ribonuclease's job*. Thus, if the first amino acid were "Glu" instead of "Lys," the protein would no longer be ribonuclease and would be unable to do the job that ribonuclease must do. Think, for a moment, about the probability of such a protein forming by chance. Is it possible for such a molecule to form by chance from a mixture of lots of amino acids?

Let's make it easy on ourselves and assume that the only amino acids in the mixture are the 17 types needed to make this particular protein. In fact, there are about 20 amino acids that are a part of the chemistry of life, but adding more amino acid types would significantly reduce our chance of forming ribonuclease. So, in order to make the outcome more likely, we will restrict ourselves to using only the 17 different types of amino acids that make up this molecule. Making this assumption, we can say that the possibility of forming a protein that has "Lys" as its first amino acid is 1 in 17. Those aren't bad odds at all. However, the chance of forming a protein with "Lys" as its first amino acid and "Glu" as its second amino acid is 1 in 17 *times* 1 in 17, or 1 in 289. Suddenly the odds are looking less and less favorable.

The probability of forming a protein whose first three amino acids are "Lys," "Glu," and "Thr" in that order are 1 in 17 *times* 1 in 17 *times* 1 in 17, or 1 in 4,913. If you were to complete this calculation, you would find that the odds for making this protein by chance from a mixture of the proper amino acids is approximately 1 in 10^{152} (a 1 followed by 152 zeros). In order to illustrate just how ridiculously low these odds are, the probability for forming ribonuclease by chance is roughly equivalent to the probability of a poker player drawing a royal flush (the most unlikely hand you can draw) *19 times in a row without ever exchanging cards!* Remember, ribonuclease is a "simple" protein. There are proteins in our bodies that contain more than 10,000 amino acids! Clearly the idea that these proteins could form by chance is absurd, which makes the whole idea of abiogenesis very hard for a scientist to believe.

Enzymes are a special class of proteins that act as catalysts for many of the chemical reactions that support life. For example, remember from our discussion of carbohydrates that in order to digest disaccharides and polysaccharides, animals must break them down into monosaccharides using hydrolysis. Well, it turns out that hydrolysis reactions like these are incredibly slow, and animals would die long before they could digest polysaccharides if it weren't for enzymes. These proteins are catalysts for the hydrolysis reactions, making them fast enough to allow the proper digestion of food.

It turns out that most enzymes do their job based on the shape that the enzyme molecule has. The shape of a hydrolysis enzyme, for example, complements the disaccharide or polysaccharide that it is trying to break down. As a result, the enzyme can attach itself to the molecule it is breaking down and help force the water between the monosaccharides, speeding up the hydrolysis reaction. This process is illustrated in Figure 5.9 for the specific case of sucrase (soo' krayz), the enzyme that catalyzes the hydrolysis reaction that breaks down table sugar (sucrose) into its monosaccharides (glucose and fructose).

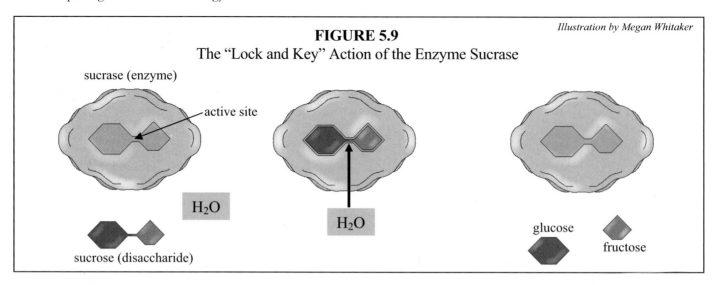

FIGURE 5.9
The "Lock and Key" Action of the Enzyme Sucrase

Illustration by Megan Whitaker

As we mentioned before, when you eat sucrose, your body must break it down into its two monosaccharides (glucose and fructose) before it can be used. Of course, the body uses a hydrolysis reaction to do this. To speed up the hydrolysis reaction, the body produces an enzyme called sucrase. This enzyme has an **active site** that complements the shape of the sucrose molecule. As a result, the sucrose molecule fits snugly into the active site of sucrase, and sucrase can then make it easier for the water molecule to react with the sucrose. Once that happens, the enzyme has not changed, but the sucrose has been broken down into glucose and fructose. The sucrase, since it has not been changed, is free to catalyze more hydrolysis reactions so that more sucrose molecules can be broken down.

Because the sucrose fits into the sucrase like a key fits into a lock, this view of how an enzyme works is often called the **lock and key theory of enzyme action**. Of course, if you think about it, a key can only fit one lock. Thus, sucrase is best used to catalyze the hydrolysis of sucrose, not some other disaccharide. After all, the shape of another disaccharide will be different, and it will not fit into the active site of sucrase as well as sucrose does. Thus, you often find that a given chemical reaction in the body is associated with a given enzyme. If a person does not have a particular enzyme, he or she might be able to digest most carbohydrates fine, but there will be one carbohydrate that he or she cannot digest well. This is the case with people who are lactose intolerant. These people cannot digest milk or milk-based products. As a result, they can get sick by eating or drinking dairy products. This happens because certain cells in their bodies are defective and cannot manufacture the enzyme necessary to catalyze the hydrolysis of lactose (the disaccharide in milk). Thus, they can digest many carbohydrates just fine, but they cannot digest lactose. As a result, they cannot drink milk or eat many dairy products.

Now it turns out that enzymes, because they are so complex and because their function is shape-dependent, are quite fragile. Enzymes are easy to destroy, as can be illustrated in the following experiment.

EXPERIMENT 5.3
The Fragility of an Enzyme

Supplies:

◆ Part of a *fresh* pineapple (It cannot be canned. It must be fresh.)
◆ A blender or fine cheese grater
◆ Three small bowls
◆ A small box of Jell-O® gelatin mix - any flavor (Generic brands work just as well.)

- Pot
- Stove
- Refrigerator
- Two tablespoons

Object: To see how easily enzyme function can be destroyed

Procedure:

1. Cut the pineapple to remove any skin. Mix it in a blender or grate it with the cheese grater and then mash the gratings so that in the end you get a thick, pulpy mixture of fresh pineapple. You need about a cup of this mixture.
2. Prepare the Jell-O as described in the directions on the box.
3. As you are boiling the water for the Jell-O, take a tablespoon of the thick, pulpy pineapple mixture and pour it into one of the three small bowls. Save that tablespoon for use with that bowl *only*. Mark the bowl as "room-temperature pineapple juice."
4. Take the rest of the thick, pulpy mixture and pour it into a pot. You will eventually heat it, but *do not* do that now.
5. When you have finished preparing the Jell-O up to the point where you stick it in the refrigerator, pour one-third of it into each of the three bowls. Stir the bowl that has Jell-0 and the thick, pulpy pineapple mixture with the tablespoon you saved in step (3).
6. Take the pot of thick, pulpy pineapple mixture and heat it on high for five minutes. Keep stirring it constantly, in order to distribute the heat evenly. The thick, pulpy mixture may boil. That's fine; just keep stirring.
7. After five minutes of heat, take one tablespoon of the hot pulpy mixture and pour it into one of the two bowls that have only Jell-O in them. Use a different tablespoon than the one you used in step (3). Stir vigorously. Label that bowl as "heated pineapple juice."
8. Put all three bowls in the refrigerator and wait for the amount of time described on the Jell-O box.
9. Examine the three bowls of Jell-O. What happened?
10. Clean up your mess.

You should have seen that the Jell-O in the bowl with nothing added gelled as you would expect, as did the Jell-O in the bowl that had heated pineapple juice added to it. However, the Jell-O in the bowl that had room-temperature pineapple juice in it should not have gelled. If it did, you did not use fresh pineapple. Why did this happen?

Pineapple contains an enzyme that stops the reaction which causes Jell-O to gel. As a result, when pineapple juice is added to Jell-O, the Jell-O cannot gel. As is the case with most enzymes, however, this enzyme is very fragile. The heat that you added to the thick, pulpy pineapple mixture in the pot was enough to destroy the enzyme, and that's why the Jell-O in the bowl with the heated pineapple juice was still able to gel. This should give you an idea of how fragile enzymes are. In fact, enzymes are so fragile that most food processing destroys them. That's why the experiment called for a fresh pineapple. In processed pineapple juice (canned or frozen), the enzyme used in this experiment has been destroyed by the processing. Interestingly enough, pineapple also contains enzymes which can break down some of the tissue in your gums. That's why your gums may bleed when you eat a fresh pineapple.

Because enzymes are so fragile, they break down soon after they are formed. Even in the fresh pineapple that you used in the experiment, much of the relevant enzyme had already broken down before you used it. However, even a tiny amount of enzyme is enough to see the effect that you saw in the experiment, so even though a lot of the enzyme in the pineapple was gone, enough was left to make the experiment work. Since enzymes break down soon after they are formed, your body must continually produce more and more enzymes, just to replace the ones that are breaking down. That's why it's important to eat protein in your diet. The protein that you eat gets broken down into its constituent amino acids, and those amino acids are shipped to the cells in your body so that they can produce the enzymes that your body needs.

ON YOUR OWN

5.19 Since dehydration reactions link amino acids in order to form proteins, you can probably guess that hydrolysis reactions break them down. Why don't proteins quickly break down into their amino acids when they are mixed with water?

DNA

Our discussion of the chemistry of life would not be complete, of course, without a brief description of DNA, the molecule which forms the basis of life. If you thought proteins were complex, you haven't seen anything until you have studied DNA! To begin your study of DNA, examine the schematic representations given below.

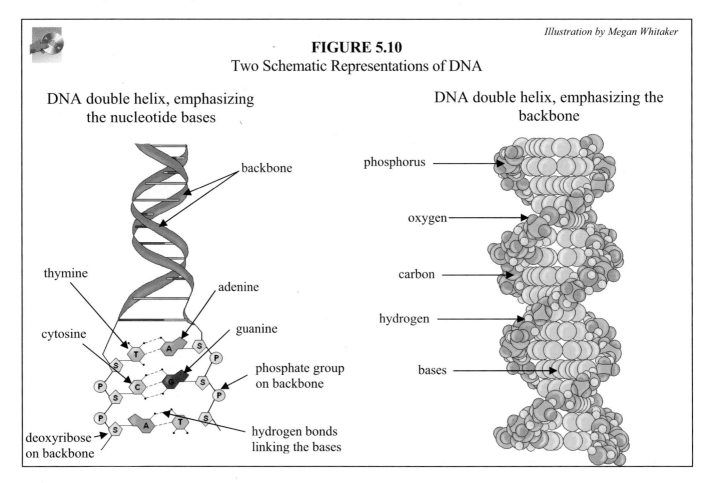

Illustration by Megan Whitaker

FIGURE 5.10
Two Schematic Representations of DNA

DNA double helix, emphasizing the nucleotide bases

DNA double helix, emphasizing the backbone

Deoxyribonucleic (dee' ox ee rye boh noo klay' ik) acid, or DNA, is a double chain of chemical units known as **nucleotides** (noo' klee uh tides). These two chains twist around one another in the double helix that is so familiar to most people who have studied any amount of biology. The nucleotides that make up these two chains are composed of three basic constituents: **deoxyribose** (a simple sugar that contains 5 carbons), a **phosphate group** (an arrangement of phosphorous, hydrogen, and oxygen atoms), and a **nucleotide base**. A nucleotide's base can be one of four different types: **adenine** (ad' uh neen), **thymine** (thye' meen), **guanine** (gwah' neen), or **cytosine** (sye' toh seen).

In the lower part of the left side of the figure, you can see that the phosphate groups link to the deoxyribose units, which support the bases. The two nucleotide chains are held together because the bases link together in a process known as **hydrogen bonding**.

Hydrogen bond – A strong attraction between hydrogen atoms and certain other atoms (usually oxygen or nitrogen) in specific molecules

Hydrogen bonding is actually a very complex process that you will learn much more about when you study chemistry. For right now, you just need to realize that the attraction between the atoms in hydrogen bonding is about 15% as strong as the attraction between two atoms that have a true chemical bond linking them. Thus, the hydrogen bonds in DNA are strong enough to keep the two chains together in a double helix, but they are significantly weaker than a true chemical bond. Since they are weaker than a true chemical bond, it is rather easy for the two helixes in DNA to unravel. That, as we will learn in the next chapter, is *very* fortunate.

Hydrogen bonding is highly dependent on the molecules involved, and as a result, only certain nucleotide bases can link together using hydrogen bonds. The nucleotide base adenine can only hydrogen bond to thymine. It cannot hydrogen bond to cytosine or guanine. In the same way, thymine can only hydrogen bond to adenine. Likewise, cytosine can only hydrogen bond to guanine and guanine to cytosine. As a result, DNA is often pictured in a very simple way, as shown below.

FIGURE 5.11
A Simplified Sketch of DNA   *Illustration by Megan Whitaker*

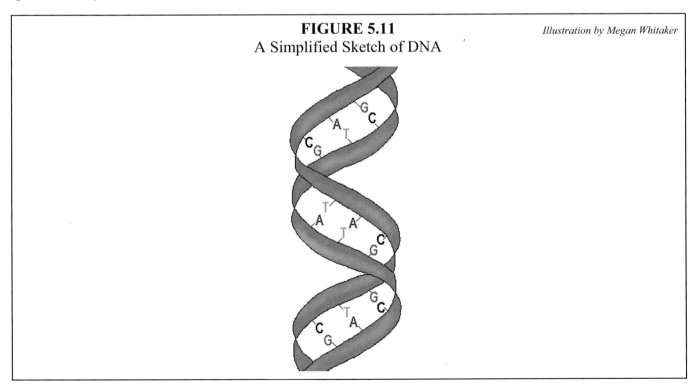

In this figure, "T" represents thymine, "A" adenine, "G" guanine, and "C" cytosine. The ribbons in the figure represent the backbone of the DNA. Notice that T's are only linked to A's, and C's are only linked to G's, and vice-versa. This represents the fact that only adenines and thymines can link together and that only cytosines and guanines can link together.

Now you can finally learn how DNA stores all of the marvelous information that it uses as the instructions for making life. Just as the entire English language can be reduced to sequences of dots and dashes in Morse code, all of the information necessary for life can be reduced to sequences of nucleotide bases in a DNA molecule. Cells, as we will learn in the next module, have chemical machinery that decodes the sequences of nucleotide bases into instructions for what structures need to be built and where to build them. Isn't that marvelous?

Of course, that's not the end of the story. One of the most interesting questions you can ask about DNA is *how much* information can be stored there. After all, modern technology has gotten pretty good at storing information. We can now make handheld computers that hold more information than a whole shelf full of books. How does that compare to the amount of information that DNA can hold? *It doesn't even come close.* Calculations indicate that if we were to write out all of the information that can be stored in a sample of DNA the size of a pinhead, we would end up with a pile of books that is *500 times higher than the distance from the earth to the moon*! This should tell you something about the incredible power and wisdom of the One who designed DNA!

ON YOUR OWN

5.20 Suppose you have just one strand of a portion of DNA. This strand has the following order of nucleotide bases:

adenine, cytosine, thymine, guanine

If you could find the other strand of DNA that connects to this one to form the double helix, what would the order of nucleotide bases be?

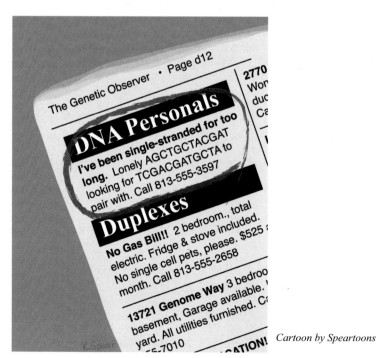

Cartoon by Speartoons

ANSWERS TO THE "ON YOUR OWN" PROBLEMS

5.1 <u>Since the electrons orbit the nucleus, they look like planets orbiting a sun.</u>

5.2 <u>All of an atom's properties are determined by the number of electrons, protons, and neutrons that it contains. The vast majority of an atom's properties, however, are determined by the number of electrons (or protons) that an atom has.</u>

5.3 Since atoms have the same number of protons and electrons, this atom must have <u>13 protons</u>.

5.4 Atoms that belong to the same element have the same number of protons and therefore the same number of electrons as well. Thus, <u>their numbers of neutrons are different</u>.

5.5 Since carbon-13 is a member of the element carbon, it has <u>6 protons and 6 electrons</u>. The "13" in carbon-13 represents the sum of all protons and neutrons. Thus, there must be <u>7 neutrons</u>, because you must get 13 when you add the protons and neutrons together. Since the number of protons must be 6, the only possible number of neutrons is 7, as $6 + 7 = 13$.

5.6 a. <u>This is a molecule</u>, because it has several atoms (1 nitrogen and 3 hydrogens) linked together.

b. <u>This is an element</u>, because it is a single abbreviation. With no other atoms linked to it, it is not a molecule. Also, since there is no number following it, we do not know its sum of neutrons and protons, so it is not a specific atom. It represents a group of atoms that all have the same number of protons.

c. <u>This is an atom</u>, because it has the element name as well as the total number of neutrons and protons.

d. Like (b), <u>this is an element</u>.

e. Even though this has only one element abbreviation, the subscript indicates that there are 4 phosphorus atoms linked together. Thus, this is a <u>molecule</u>.

5.7 <u>The student is wrong. Even though these molecules are comprised of the same elements, there are different numbers of oxygen atoms. Different numbers of atoms result in molecules with different properties.</u>

5.8 The subscript after the "C" tells us that there are <u>2 carbons</u>. The subscript after the "H" tells us that there are <u>4 hydrogens</u>. The subscript after the "O" tells us that there are <u>2 oxygens</u>.

5.9 a. This is a <u>physical change</u>, because it is reversible. You could pick the cereal out of the milk and then dry the cereal.

b. This is a <u>chemical change</u>, because there is no way to "unbake" bread. Thus, it is not reversible.

c. This is a <u>physical change</u>. Remember we gave you the example of boiling a mixture of sugar and water to separate the two. Since salt water is just salt dissolved in water, the same would apply here.

5.10 The water level of the solution on the right increased, while the water level of the solution on the left decreased. Since we are dealing with a semipermeable membrane, this is osmosis. In osmosis, the solvent travels from an area of low solute concentration to one of high solute concentration. Since the water levels indicate that the solvent traveled from left to right, the <u>right side had a higher salt concentration</u>.

5.11 a. Reactants appear on the left side of the arrow. Also, the numbers to the left of the formulas are not part of the formulas. They simply tell us how many of each molecule is in the reaction. Thus, we do not include them in our answer. The reactants, then, are <u>C_8H_{18} and O_2</u>.

b. Products are on the right side of the arrow and, once again, the numbers to the left of the formulas are not a part of the formulas. The products, then, are <u>CO_2 and H_2O</u>.

c. The number to the left of the chemical formula for C_8H_{18} tells us how many octane molecules are used in the reaction. Thus, the answer is <u>2</u>.

d. The number to the left of the chemical formula for H_2O tells us how many water molecules are made in the reaction. Thus, the answer is <u>18</u>.

5.12 <u>It can produce *some* glucose</u>, but without the catalyst, the rate will be far too slow for the plant to sustain itself.

5.13 <u>The chemist can increase temperature</u>. This speeds up most reactions.

5.14 In the structural formula, 2 C's are pictured along with 4 H's. The chemical formula, then, is <u>C_2H_4</u>.

5.15 To turn a polysaccharide into a disaccharide, you have to break it down. <u>Hydrolysis</u> reactions break polysaccharides down.

5.16 The pH scale says below 7 is acidic and above 7 is alkaline. In addition, the higher the pH, the more alkaline the solution. Based on that, then, <u>Solution D is the most alkaline</u>.

5.17 The pH scale says below 7 is acidic and above 7 is alkaline. In addition, the lower the pH, the more acidic the solution. Based on that, then, <u>Solution B is the most acidic</u>.

5.18 Each fat molecule takes 3 fatty acids along with a glycerol. As long as we have plenty of glycerol, then, we can make <u>5 fat molecules</u> with 15 fatty acid molecules.

5.19 As we discussed with regards to enzymes, in order for hydrolysis reactions to work with any worthwhile rate, they need a catalyst. <u>If proteins are mixed with just water, there are no enzymes to catalyze the hydrolysis reaction, so they will not break down into amino acids very quickly</u>.

5.20 Remember that adenine can only link to thymine and that cytosine can only link to guanine. Thus, if one strand has adenine, the other must have thymine. If one strand has cytosine, the other must have guanine. This tells us that the order on the other strand must be <u>thymine, guanine, adenine, cytosine</u>.

STUDY GUIDE FOR MODULE #5

1. Define the following terms:

a. Matter
b. Model
c. Element
d. Molecules
e. Physical change
f. Chemical change
g. Phase
h. Diffusion
i. Concentration
j. Semipermeable membrane
k. Osmosis
l. Catalyst
m. Organic molecule

n. Biosynthesis
o. Isomers
p. Monosaccharides
q. Disaccharides
r. Polysaccharides
s. Dehydration reaction
t. Hydrolysis
u. Hydrophobic
v. Saturated fat
w. Unsaturated fat
x. Peptide bond
y. Hydrogen bond

2. Describe where the protons, neutrons, and electrons are in an atom.

3. What determines the vast majority of characteristics in an atom?

4. What does the number after an atom's name signify?

5. What is the difference between an element and an atom?

6. How many electrons are in an atom that has 32 protons?

7. How many atoms (total) are in a molecule of C_3H_8O? What atoms are present and how many of each atom?

8. Identify the following as an atom, element, or molecule:

a. H_2CO_3 b. nitrogen-14   c. P

9. If you add energy to the molecules of a liquid, will it turn into a gas or a solid?

10. A chemist wants to study diffusion. Should a semipermeable membrane be used?

11. Two solutions of different solute concentration are separated by a membrane. After a while, the water levels of the two solutions change. Has osmosis or diffusion taken place? What kind of membrane is being used?

12. Consider the following chemical reaction:

$$N_2 + 3H_2 \rightarrow 2NH_3$$

a. What are the reactants?
b. What are the products?
c. How many molecules of H_2 are used in the reaction?

13. What is the chemical equation for photosynthesis? What 4 things are necessary for a plant to carry out photosynthesis?

14. Other than using a catalyst, how can a reaction be sped up?

15. Which of the following is a carbohydrate?

    a. NH_3 b. CO_2 c. C_2H_4O d. $C_5H_{10}O_5$ e. $C_3H_8O_3$

16. What kind of reaction is used for building disaccharides, polysaccharides, fats, and proteins? What kind of reaction can break these substances down?

17. Which of the following is an acid?

18. Describe the pH scale and what it measures.

19. What are the basic building blocks of proteins, lipids, and polysaccharides?

20. If two proteins contain the same type and number of amino acids, but the order in which they link up is different, are the properties of the two proteins the same?

21. What are enzymes, and for what purpose are they usually used?

22. What is the "lock and key" theory of enzyme action?

23. What are the basic parts of a nucleotide?

24. How does DNA store information?

25. What holds the two helixes in a DNA molecule together?

MODULE #6: The Cell

Introduction

In Module #1 we started talking about the cell, and we really haven't stopped talking about it since. Indeed, in Modules #2 and #3, you learned quite a bit about the cell structures that exist in kingdoms Protista and Monera. Well, believe it or not, we have barely scratched the surface of all that can be learned about the cell. That's why we are now going to devote an entire module to learning about its structure and functions. Since we have already studied kingdom Monera, however, we will not spend any time on prokaryotic cells. This module will concentrate entirely on eukaryotic cells. Take heed that there is a lot of information in this module. You will want to spend an extra week or so studying it, in order to be able to absorb all of the information that it contains. Now please understand what we mean by that. Don't take three weeks reading the module. It is not a lot longer than other modules in this course. Instead, plan to spend an extra week just *studying* the information in this module, especially making sure to learn all of the vocabulary words. Vocabulary is an important part of biology, and in this module, it is *especially* important.

Since cells are the basic building blocks of life, it's not surprising that we need to study them in detail. What might surprise you, however, is that you have already studied the most diverse eukaryotic cells that exist – those in kingdom Protista, and, to a lesser extent, those in kingdom Fungi. In the last two kingdoms (Animalia and Plantae), there is not nearly as much diversity among cells. All of the incredibly diverse organisms in these two kingdoms are built from cells that have very similar blueprints!

Cellular Functions

In the process of life, cells must carry on several different functions. First of all, they must maintain enough energy to live and perform the duties for which God has designed them. To maintain energy, cells must perform three basic functions:

Absorption –The transport of dissolved substances into cells

Digestion – The breakdown of absorbed substances

Respiration – The breakdown of food molecules with a release of energy

In **absorption**, dissolved substances must enter the cell from the outside. This is actually more complicated than it sounds, because cells cannot let just anything inside them. After all, some substances are poisons that would immediately kill the cell if they were allowed to enter to any great extent. As a result, absorption is a complicated process in which the cell "recognizes" the substances that are trying to enter. Useful substances are allowed in, while harmful substances are not. Of course, this system is not perfect, so some toxic substances do get inside the cell. Nevertheless, the plasma membrane has a lot of chemical machinery designed to stop many toxic substances from entering.

Although their definitions seem rather similar, there is a big difference between **digestion** and **respiration**. In digestion, large molecules are broken down into smaller ones. For example, when polysaccharides are broken down into monosaccharides, we call this digestion. As we mentioned in Module #5, this has to happen, because most organisms can only use energy from the breakdown of monosaccharides. When a cell actually breaks down the monosaccharides and produces energy, however, we call that **respiration**. There are two reasons that we cannot lump digestion and respiration together. First, they usually occur in different places within an organism or a cell. Second, the products of digestion can be used for other processes besides just respiration. For example, when proteins are digested, they are broken down into their constituent amino acids. Instead of being used for respiration, these amino acids can be used by the cells to produce more proteins. In this case, digestion is used to provide the building blocks of **biosynthesis**, a cellular function we discussed in the previous module. Going back to the terminology you learned in Module #1, digestion and respiration are both a part of the cell's catabolism, while biosynthesis is a part of the cell's anabolism.

Since a cell absorbs substances from the outside environment, it makes sense that, at some point, it must eliminate excess substances as well. There are three methods by which cells eliminate substances:

<u>Excretion</u> – The removal of soluble waste materials

<u>Egestion</u> – The removal of nonsoluble waste materials

<u>Secretion</u> – The release of biosynthesized substances

Once again, even though these definitions look similar, they describe completely different processes. If a substance can be dissolved in a fluid, it is called "soluble." Thus, **excretion** involves the removal of substances that can be dissolved in the fluids of a cell. On the other hand, **egestion** involves the removal of substances that cannot be dissolved in the fluids of a cell. Egestion is a much more difficult process than excretion. Why? Well, think of Experiment 5.1. When you put solid (undissolved) sugar in a napkin, the sugar could not leave the napkin. Once water seeped into the napkin and dissolved the sugar, however, the sugar molecules could diffuse right through the napkin. Thus, transporting soluble substances is simpler than transporting nonsoluble ones. Finally, **secretion** does not involve removal of wastes at all. Instead, it involves the removal of useful substances that the cell has manufactured for other cells.

Cells also must perform functions of **movement** and **irritability**. When we say "movement," we might mean the actual locomotion of a cell from one point to another, or we might mean movement of things *within* a cell. Also, when we say "irritability," we do not mean that the cell must get cranky from time to time. Instead, biologists use this term to mean sensing and responding to changes in the surroundings.

In order for a cell to continue its existence, there are two other functions that it must perform:

<u>Homeostasis</u> – Maintaining the status quo

<u>Reproduction</u> – Producing more cells

In order to survive, the cell must make sure that all of its organelles are functioning properly, that all organelles are supplied with the substances that they need, and that everything within the cell is running according to God's design. This is called **homeostasis** (ho mee' oh stay' sis). In addition, all cells die. Thus, in order to maintain life, cells must produce other cells by **reproduction**.

In fact, the last sentence in the previous paragraph is another reason why we think that abiogenesis could never happen. You see, scientists have been studying cells in one way or another since the 1600s. In the last century, especially, an enormous amount of scientific resources have been devoted to studying cells. In fact, the science of studying cells has become so fundamental that we have a name for it: **cytology** (sye tahl' uh jee).

<u>Cytology</u> – The study of cells

Well, in the entire history of cytology, scientists have seen cells produced in only one way: from other cells. Never have chemicals or any other nonliving substances produced cells. In fact, scientists cannot even produce cells in the lab unless they have a living cell to start with. Even the process of cloning starts with a living cell. Without that living cell, cloning would not work. If, in the history of cytology, scientists have only seen cells produced by other cells, why in the world would some scientists want to believe that cells could have been produced in a different way at some time in the past?

Summing up, then, cells must perform at least 11 main functions in order to support and maintain life: absorption, digestion, respiration, biosynthesis, excretion, egestion, secretion, movement, irritability, homeostasis, and reproduction. Interestingly enough, in the single-celled organisms that you studied in kingdom Protista, all of these functions were performed by a single cell. In the multicellular life forms that you studied in kingdom Fungi and that you will study in the next semester of this course, these functions are performed by different groups of cells. Some cells have been designed to specialize in certain functions, while others have been designed to specialize in other functions. As a result, in multicellular organisms, most of the cells do not perform all of the processes listed. Instead, the functions have been assigned to certain groups of cells, and the individual groups of cells work together to make sure all of the above-listed functions are performed.

This brings up an interesting point that we have already emphasized but cannot emphasize enough. **There is no such thing as a simple life form**. Biologists are fond of calling the organisms in kingdoms Protista and Monera "simple," because they mostly consist of only a single cell. However, in some ways, these organisms are more complex than those in the other kingdoms! Certainly, having several cells working together and specializing in different life functions adds a level of complexity to the organism. However, since the cells need not perform all functions associated with life, they need not be as sophisticated as a single cell that must perform *all* of life's functions. Thus, multicellular organisms are certainly more complex when you look at them as a whole, but when you focus in on one cell, single-celled organisms could be considered more complex. In the end, then, there is nothing "simple" about life, not even when it is composed of a single cell!

ON YOUR OWN

6.1 A cell makes proteins in an organelle that is near the center of the cell. It then transports the proteins to the edge of the cell and sends them into the surroundings to be used by other cells. What three of the basic life functions are employed to accomplish this series of tasks?

6.2 A cell takes in a polysaccharide and sends it to an organelle to be digested. The digestion products are then used to produce energy. The soluble waste products are eliminated. What five of the basic life functions are performed in this procedure?

Cell Structure

Once we get through the cells in kingdoms Protista and Monera, we generally lump the remaining cells into two distinct categories: **plant cells** and **animal cells**. Schematic, idealized examples of a plant cell and an animal cell are shown in Figure 6.1.

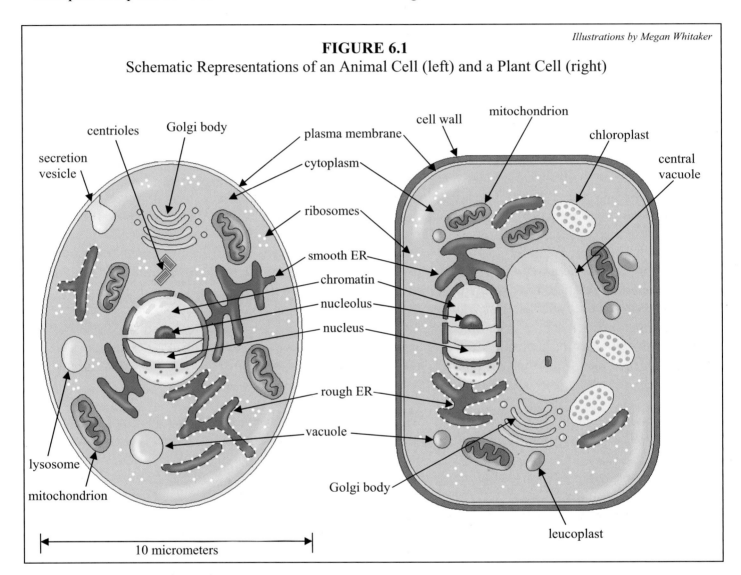

FIGURE 6.1
Schematic Representations of an Animal Cell (left) and a Plant Cell (right)

Illustrations by Megan Whitaker

Before we discuss the individual components of a cell, we need to discuss size. As the legend in Figure 6.1 indicates, most cells are just a little less than 10 micrometers across. As we mentioned in Module #2, a micrometer is one millionth of a meter. Now if you aren't familiar with the metric system, don't worry about it. When you study chemistry, you will learn it in detail. For right now, just realize that about 2,000 cells can fit across your fingernail. That should give you an idea of how small a cell is.

It turns out that cells are small for a reason. You see, the volume of materials in a cell increases with the cube of the radius of the cell. If you haven't had geometry yet, that phrase might be a little mystifying to you. Don't worry. What it means is that when the distance across the cell doubles, the cell's volume (the amount it holds) goes up by a factor of 8! Well, when a cell's volume increases, it must absorb more nutrients to survive. The amount of nutrients it must absorb actually depends on its volume. Thus, when the distance across a cell doubles, its absorption must increase by a factor of 8. Since the amount of absorption that a cell needs is so dramatically dependent on the cell's size, there is a fundamental size limit for cells. After that point, they cannot grow any larger. That's why cells are so small.

The Cell Wall

One of the major distinctions between plant cells and animal cells can be found on the outside of the cells. Plant cells usually have a **cell wall**, and animal cells do not.

Cell wall – A rigid structure on the outside of certain cells, usually plant and bacteria cells

The cell wall is typically made of cellulose and **pectin**, a carbohydrate that hardens cellulose. These materials are secreted by organelles in the cell. Although the main function of the cell wall is to protect the cell from its surroundings, the cell wall is full of small holes called **pores**. These pores allow substances (like nutrients) to diffuse through the cell wall and into the cell. They also allow waste products from the cell to diffuse into the cell's surroundings.

As a plant cell grows, the cell wall must grow along with it. Thus, as a plant cell is maturing, the cell wall needs to be rather flexible so as to allow for the growth. Once a plant cell has matured and stopped growing, however, there is no longer any need for the cell wall to be flexible. Therefore, at that time, the cell starts producing **secondary cell walls**. These walls are formed on the inside of the original cell wall and are much more rigid, providing better protection for the contents of the cell. Once the secondary cell walls are formed, the original cell wall is usually referred to as the **primary cell wall**.

A single plant, of course, has many, many cells. These cells are usually arranged adjacent to one another, separated by a thin film called the **middle lamella** (luh mel' uh).

Middle lamella – The thin film between the cell walls of adjacent plant cells

The middle lamella is made primarily of pectin.

Besides plant cells, many bacteria and algae have cell walls. It is important to note, however, that the cell wall of a bacterium is chemically different from what we have discussed here. Thus, this discussion is specifically for plant cells. In addition, some cells have structures that substitute or add to the cell wall. For example, you learned in Module #2 that in addition to a cell wall, some bacteria form a capsule that surrounds the cell wall and adds protection. A few bacteria have a capsule and no cell wall. Also, some bacteria have neither cell wall nor capsule. Finally, in Module #3 you learned that a euglena has a pellicle, which functions much like a cell wall but is thinner and more flexible.

The Plasma Membrane

The **plasma membrane** is inside the cell wall, if the cell has one. In the case of animal cells (which do not have a cell wall), the plasma membrane is all that separates the cell from its surroundings.

Plasma membrane – The semipermeable membrane between the cell contents and either the cell wall
        or the cell's surroundings

As the definition states, the plasma membrane is semipermeable. It allows certain substances
(nutrients, water, and oxygen, for example) to pass through and enter the cell, but it does not allow
other substances (certain toxins, for example) in. Likewise, the plasma membrane allows water and
waste products to leave the cell, but it does not allow the necessary contents of the cell to leave. How
does it accomplish this feat? To understand this, we need to understand a bit more about how the
plasma membrane is built, so we will leave that to another section. For right now, just realize that the
plasma membrane is semipermeable, allowing certain substances, and not others, to pass through it.

    Now remember, in animal cells, there is no cell wall. That means the plasma membrane is the
animal cell's only protection, right? Well, not really. It turns out that cells within animals are
protected by other cells. In animals, there are special cells (white blood cells, for example) that protect
the other cells of the body, keeping their surroundings free of major contaminants. This is one of the
things that allows animal cells to exist without cell walls.

The Cytoplasm

    Cytoplasm (sye' tuh plaz' uhm) is a jellylike fluid in which all of the cell organelles are
suspended.

    Cytoplasm – A jellylike fluid inside the cell in which the organelles are suspended

The cytoplasm is made up of a mixture of several different compounds, including water, small organic
molecules, proteins, fats, and carbohydrates. In addition, there are substances called **ions** in the
cytoplasm.

    Ions – Substances in which at least one atom has an imbalance of protons and electrons

As you already learned, atoms must have the same number of protons and electrons in order to remain
electrically balanced. Well, if an atom loses or gains an electron, it is suddenly thrown out of electrical
balance. If it gains electrons (which are negatively charged), it ends up with an overall negative
charge. If it loses electrons, it ends up with an overall positive charge. Once this happens, it ceases to
be an atom and is called an ion. Since ions have electrical charge, they respond to electrical stimuli.

    The ions in the cytoplasm are responsible for a very important process called **cytoplasmic
streaming**.

    Cytoplasmic streaming – The motion of cytoplasm in a cell that results in a coordinated movement of
        the cell's contents

 The multimedia CD has video of cytoplasmic streaming in the cells of plant roots.

You see, it is important for a cell to be able to move its organelles in order to respond to change. In
addition, the cell must be able to move substances from one place to another within the cell. To

accomplish these tasks, the ions in the cytoplasm cause certain parts of the cytoplasm to become more watery and less jellylike. This causes other portions of the cytoplasm that remain jellylike to flow into the watery areas, in essence moving that portion of the cytoplasm along with anything contained within it. The ions in the cytoplasm are also responsible for other chemical aspects of cell function that go beyond the scope of this course.

The Mitochondrion

One of the most important organelles in a cell is the **mitochondrion** (my tuh kahn' dree uhn). The plural is mitochondria.

Mitochondria – The organelles in which nutrients are converted to energy

These bean-shaped organelles are often referred to as the "powerhouses of the cell" because they are responsible for respiration in the cell. Because of this, cells that need a lot of energy tend to have a lot of mitochondria. Muscle cells, for example, have a lot more mitochondria than skin cells, because muscle cells need more energy than skin cells. We will spend a great deal of time on the functions of the mitochondria in an upcoming section of this module.

The Lysosome

Another difference between plant and animal cells is the presence of an organelle called the **lysosome** (lye' soh sohm).

Lysosome - The organelle in animal cells responsible for hydrolysis reactions that break down proteins, polysaccharides, disaccharides, and some lipids

Remember, in order to use polysaccharides or disaccharides, a cell must first break them down into monosaccharides using hydrolysis reactions. In addition, cells must break certain proteins and lipids down into their constituent parts (amino acids and fatty acids, respectively) in order to use them for biosynthesis. All of the necessary hydrolysis reactions take place in the lysosome. Now remember, hydrolysis reactions do not occur on their own. They need enzyme catalysts in order to make them happen. Thus, the lysosome is full of enzyme catalysts for the hydrolysis reactions that it needs to perform.

Interestingly enough, when the lysosome was first discovered, it was called the "suicide sac," because prior to the death of certain cells, the lysosomes would release chemicals that destroyed the cell. Later on, scientists realized that as the cell dies, the membrane that encloses the lysosome begins to deteriorate. When that happens, the contents of the lysosome start to spill into the cytoplasm of the cell. Why does this kill the cell? Think about it. The cytoplasm is full of proteins, lipids, and carbohydrates. What do the contents of the lysosomes do? They break down these molecules. Thus, when the contents of the lysosome are released into the cell, they begin breaking down the contents of the cell. That spells the death of the cell.

This brings us to one last thing that lysosomes do. Because lysosomes can digest the contents of the cell, they can also act as a "cleaning crew," breaking down old, worn-out organelles as well as any debris that collects in the cytoplasm. This keeps the cytoplasm clear of chemicals and structures that might hinder cytoplasmic streaming and other functions of the cytoplasm.

Ribosomes

Because they control respiration and digestion, the mitochondria and the lysosomes are primarily responsible for breaking down and destroying molecules. As you should recall from Module #1, these processes are a part of what we call catabolism. In order to live, however, cells must also synthesize molecules. This is referred to as the cell's anabolism, and there are several organelles involved in anabolic processes. The **ribosomes** (rye' buh sohms) are non-membrane-bounded organelles that are found in both eukaryotic and prokaryotic cells. These organelles are responsible for synthesis of proteins in the cell.

Ribosomes – Non-membrane-bounded organelles responsible for protein synthesis

Since protein synthesis is an integral part of DNA's job, we will discuss this process heavily in the next module.

The Endoplasmic Reticulum

The **endoplasmic** (en doh plaz' mik) **reticulum** (rih tik' yuh luhm), commonly abbreviated as ER, is composed of an extensive network of folded membranes.

Endoplasmic reticulum – An organelle composed of an extensive network of folded membranes that performs several tasks within a cell

The ER typically runs throughout the cytoplasm. In animal cells, it helps to maintain the shape of the cell, since there is no cell wall to perform this task. In addition, the ER aids in the transport of complex molecules through both animal and plant cells. There are two types of ER, **rough ER** and **smooth ER**.

Rough ER – ER that is dotted with ribosomes

Smooth ER – ER that has no ribosomes

Since it has ribosomes, rough ER is a part of protein synthesis. Typically, specialized proteins that are secreted by certain cells are produced here. Thus, cells that are designed to specialize in the secretion of proteins are filled with rough ER. Although no protein synthesis occurs in smooth ER, many other chemicals can be produced in this organelle. Lipids are made in order to store excess energy, and hormones and steroids, which affect a range of cell functions, can be made here as well. In addition, smooth ER inactivates certain harmful byproducts of digestion and respiration and then sends them to the plasma membrane to be ejected.

The Plastids

Another organelle involved in biosynthesis is found only in plant cells, algae cells, and some protozoa. These organelles, called **plastids**, are generally grouped into two categories: **leucoplasts** (loo' kuh plasts) and **chromoplasts** (kroh' muh plasts).

Leucoplasts – Organelles that store starches or oils

<u>Chromoplasts</u> – Organelles that contain pigments used in photosynthesis

If you recall from our discussion of starch, when plants have excess monosaccharides from photosynthesis, they are typically linked together in a long polysaccharide called starch. This allows the plant to store them for future use. The leucoplasts are where these starches are stored. Potatoes, for example, are full of starch because that's where a potato plant has a large fraction of its leucoplast-containing cells.

The chromoplast with which you are most familiar is the **chloroplast**. This structure contains the pigment chlorophyll, which you have already learned is a catalyst for the photosynthesis process. Photosynthesis actually takes place in the fluid that fills the chloroplast. This fluid is called the **stroma**.

Vacuoles and Vesicles

In general, a **vacuole** is a membrane-bounded "sac." In Module #3, we learned about **food vacuoles** and **contractile vacuoles**. In plant and animal cells, however, there are other types of vacuoles to consider. For example, most plant cells have a **central vacuole**.

<u>Central vacuole</u> – A large vacuole that rests at the center of most plant cells and is filled with a solution that contains a high concentration of solutes

Because of the high concentration of solutes in the solution that fills the central vacuole, water tends to enter the central vacuole by osmosis. This makes the central vacuole bigger and bigger, causing it to push the cytoplasm and all of the organelles against the cell wall. This causes the cell to be pressurized, much like a balloon. This pressure, called **turgor pressure**, helps keep a plant rigid. When a plant begins to wilt, it is because a lack of water has resulted in a lack of turgor pressure inside the plant's cells. Fresh lettuce, for example, is crisp because its cells are highly pressurized. As the lettuce gets old, however, it loses water, decreasing the turgor pressure and resulting in wilted leaves.

Nondigestible, nontoxic waste products are contained in **waste vacuoles**.

<u>Waste vacuoles</u> – Vacuoles that contain the waste products of digestion

Remember, toxic waste products are typically handled by the smooth ER. Nontoxic waste products, however, are typically enclosed in waste vacuoles. These vacuoles then move by cytoplasmic streaming to the plasma membrane, where the waste products can be eliminated.

Another type of vacuole is the **phagocytic** (faj uh sih' tik) **vacuole**. When some cells come across a food substance that is too large to be pulled through the plasma membrane, they simply surround the food substance and engulf it. Once a food substance is engulfed, a vacuole is formed around it so that digestion can begin. You should recall from Module #3 that this is how an amoeba feeds. This process is often called **phagocytosis** (faj' uh sye toh' sis), and the vacuole formed is called a phagocytic vacuole.

<u>Phagocytosis</u> – The process by which a cell engulfs foreign substances or other cells

<u>Phagocytic vacuole</u> – A vacuole that holds the matter which a cell engulfs

In animals, the white blood cells that protect other cells feed in this way. When they recognize a foreign cell (such as a bacterium), they destroy it by phagocytosis. This process is shown in the figure below.

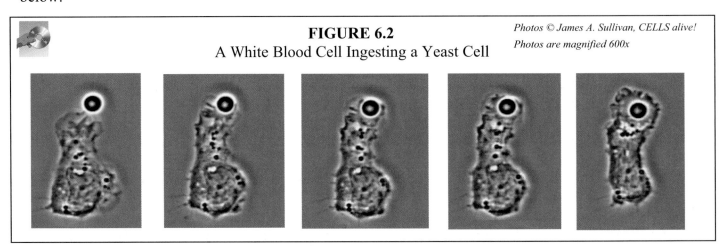

FIGURE 6.2
A White Blood Cell Ingesting a Yeast Cell

Photos © James A. Sullivan, CELLS alive!
Photos are magnified 600x

When a vacuole is small, it is typically called a vesicle. There are two important types of vesicles: **pinocytic** (pin uh sih' tik) **vesicles** and **secretion vesicles**.

Pinocytic vesicle – Vesicle formed at the plasma membrane to allow the absorption of large molecules

Secretion vesicle – Vesicle that holds secretion products so that they can be transported to the plasma membrane and released

Pinocytic and secretion vesicles are, in some ways, polar opposites. When a cell needs to absorb a molecule that is large (but not so large that it needs to be engulfed), a tiny pocket is formed in the plasma membrane. That pocket is then pinched off, forming a vacuole. This process, which is similar in some ways to phagocytosis, is called **pinocytosis** (pin uh sye toh' sis) and is illustrated in the figure below.

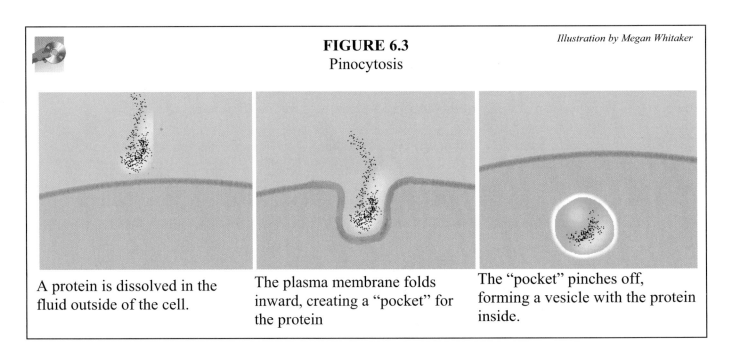

FIGURE 6.3
Pinocytosis

Illustration by Megan Whitaker

A protein is dissolved in the fluid outside of the cell.

The plasma membrane folds inward, creating a "pocket" for the protein

The "pocket" pinches off, forming a vesicle with the protein inside.

We will discuss secretion in the context of the Golgi bodies, which are integral to that process.

Golgi Bodies

Golgi (gohl' jee) **bodies** look like a stack of pancakes. In fact, they are comprised of flattened, interconnected membrane sacs that store proteins and lipids.

Golgi bodies – The organelles where proteins and lipids are stored and then modified to suit the needs of the cell

When the cell produces proteins and lipids, they are transferred to the Golgi bodies, where they are sorted and stored until needed. When a protein or lipid is needed by the cell, the Golgi bodies package the molecule so that it can be transported to the place where it is needed. The "packaging" that goes on in the Golgi bodies is typically some process of adding a small molecule to the protein or lipid that is being processed. This chemical then can be used by the cell as a marker, telling the cell where to transport the molecule. Thus, the Golgi bodies function much like a mailing service. They take in molecules, package and address them, and then send them to where they are needed.

As we have already mentioned, in multicellular organisms, specific cells are designed to perform specific tasks. One of those tasks is secretion, which we have also discussed. Golgi bodies are abundant in cells that are designed for secretion, because they are heavily involved in the process. In secretion, chemicals that are made by the cell are transported to the Golgi bodies, and once the Golgi bodies are finished processing them, a secretion vesicle forms. That vesicle then travels to the plasma membrane where it can secrete its contents into the cell's surrounding. This process is illustrated in the figure below.

FIGURE 6.4
Secretion

Illustration by Megan Whitaker

The Golgi bodies form a bulge with the appropriate substance inside the bulge.

The bulge pinches off, becoming a vesicle that travels to the membrane.

The vesicle opens the membrane, releasing the chemical.

Centrioles

Centrioles (sen' tree ohlz) are interesting organelles found mostly in animal cells. Some plant cells have them, but most do not. Centrioles have two rather different functions. First, they form flagella or cilia in cells that have them (euglena and paramecia, for example). This is because centrioles are composed of structures called **microtubules**.

Microtubules – Spiral strands of protein molecules that form a tubelike structure

When centrioles contact the plasma membrane of the cell, the microtubules are lengthened by the addition of more secreted proteins. The result is either a cilium or a flagellum, depending on the cell involved. Once the centriole has formed a cilium, it stays at the base of the cilium and is called a **basal body**.

Amazingly enough, however, centrioles have a completely different function as well. In most animal cells, centrioles can appear in pairs. Each individual centriole is oriented at a 90 degree angle to its partner, like two connecting sides of a square. These centrioles are involved in the asexual reproduction of cells. This process, called "mitosis," is very important, and we are going to discuss it extensively in the next module. As a result, we will put off any further discussion of centrioles until that time.

The Nucleus

Earlier, we told you that the mitochondrion was one of the most important organelles in the cell. Well, the **nucleus** is just as important as, if not more so than, the mitochondrion. Often called the "control center of the cell," the nucleus holds the cell's main DNA, which tells the cell everything it needs to know about its structure and functions. Since it is stored in the nucleus, this DNA is often called **nuclear DNA**, and it is a copy of the nuclear DNA that was in the parent cell. Although the nucleus can make additional nuclear DNA, the new nuclear DNA will once again be a copy of the old nuclear DNA. Thus, there is no way for the nucleus to make DNA that has different information than the DNA that it already has. Therefore, a cell cannot generate any new information about structure and function. It can only copy the information that is currently available.

In the previous module, you learned that information is stored in DNA as a sequence of nucleotide bases, much like Morse code. However, you probably still have no idea how that information is *used*. We will deal with this more completely in the next module, but for right now, we want you to know that the main function of DNA, as far as biologists can tell, is to produce proteins. The information stored in DNA, then, is really in the form of recipes. These recipes tell the cell how to make the proteins that it needs in order to function.

The nucleus of the cell has its own structure, starting with the **nuclear membrane**. This membrane has large pores in it (see Figure 6.5 on p. 174), allowing the diffusion of materials between the nucleus and the cytoplasm.

Nuclear membrane – A highly-porous membrane that separates the nucleus from the cytoplasm

Inside the nuclear membrane, you will find clusters of DNA surrounded by a rich concentration of proteins and a substance called **ribonucleic** (rye' buh noo klay' ik) **acid**, usually referred to as RNA.

We will learn a lot about RNA in the next module. These clusters of DNA, RNA, and proteins are called **chromatin** (kroh' muh tun).

Chromatin – Clusters of DNA, RNA, and proteins in the nucleus of a cell

Also inside the nuclear membrane, you will find the **nucleolus** (noo klee' uh lus). This is the center where RNA is made. Ribosomes are also assembled in the nucleolus and then transported out of the nucleus.

It is important to note that while the major portion of a cell's DNA is stored in the nucleus, there is actually some DNA stored in the mitochondrion. This DNA is, not surprisingly, called **mitochondrial DNA**. It is involved in producing the proteins that the cell needs specifically for respiration. Interestingly enough, however, mitochondrial DNA cannot do its job independently. The proteins produced by the mitochondrial DNA must contain units that are formed by the nuclear DNA in order to work. Thus, mitochondrial DNA and nuclear DNA work together in some way that is not completely understood at this time.

The Cytoskeleton

Before we leave our discussion of cellular organelles, we need to mention the **cytoskeleton** (sye' toh skel' uh tuhn). This is not a single organelle. Instead, it is a network of fibers that holds the cell together, helps the cell keep its shape, and even aids in movement.

Cytoskeleton – A network of fibers that holds the cell together, helps the cell to keep its shape, and aids in movement

The cytoskeleton is made up of three basic kinds of fibers: **microfilaments**, **intermediate filaments**, and **microtubules**. You already learned about microtubules when we discussed the centrioles, but the other two types of filaments are new to you.

Microfilaments – Fine, threadlike proteins found in the cell's cytoskeleton

Intermediate filaments – Threadlike proteins in the cell's cytoskeleton that are roughly twice as thick as microfilaments

Each of these types of fibers makes its own contribution to the cytoskeleton.

The microfilaments are mostly associated with movement. For example, microfilaments are responsible for the ponderous motion exhibited by amoeba. They also can cause certain cells to contract. Muscle cells, for example, do their job by contracting and relaxing. The microfilaments in the muscle cells take care of this function. Finally, in some cells, the microfilaments also generate cytoplasmic streaming. Microtubules also participate in cell movement. As you have already learned, they are used to form cilia and flagella in the cells that have those structures. They also provide a "track" upon which organelles and vesicles can travel as they move throughout the cell. The intermediate filaments are mostly responsible for strengthening and supporting the cell, which allows it to keep its shape. Another important role of the cytoskeleton is to keep the various organelles of the cell in their proper positions. Cellular organelles are not free to "float around" the cell, because the internal cellular structure is just too complex to allow for that. Thus, the cytoskeleton holds this complex structure together by keeping each organelle in its proper position.

The makeup of an individual cell's cytoskeleton is characteristic of the task for which the cell has been designed. For example, skin cells must be very rigid. As a result, the cytoskeleton of a skin cell contains a lot of intermediate filaments so that it can hold its shape well. A muscle cell does a lot of contracting and relaxing, so it has a lot of microfilaments in its cytoskeleton. A paramecium needs lots of cilia, so it has a high concentration of microtubules. Although you might not think about it, there are cells in your body that have cilia as well. The bronchial tubes that bring air to your lungs are lined with cells that have cilia. These cilia filter dust and debris from the air that you breathe and then beat back and forth to sweep the dust and debris away from your lungs.

As If This Isn't Already Complicated Enough!

Now we know that we've just thrown a *lot* of information at you. Unfortunately, we have to do this, because the cell is just so complex. That's why we say you need three weeks for this module. There is just a lot of information to absorb. Before we move on, however, we need to let you know something about Figure 6.1. In order to make it easy to see the organelles that we wanted to discuss, we *radically simplified* the structure of the plant and animal cells in that figure. For example, the cells are drawn as flat, even though they are not. Cells (and the organelles inside) are three-dimensional, having length, width *and* height. Also, whereas the cells in Figure 6.1 contained at most a few of each organelle, there are, in reality, several of each organelle (except the nucleus) present in most cells. To give you an idea of the complexity of a cell, then, the figure below shows you a less simplified drawing of an animal cell.

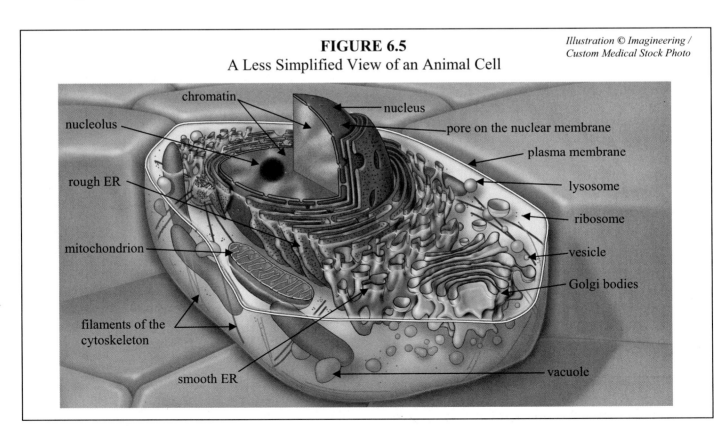

FIGURE 6.5
A Less Simplified View of an Animal Cell

Illustration © Imagineering / Custom Medical Stock Photo

ON YOUR OWN

6.3 In the first section of this module, we discussed the 11 main functions of life that a cell has to perform (absorption, digestion, respiration, biosynthesis, excretion, egestion, secretion, movement, irritability, homeostasis, and reproduction). Take the items listed in Figure 6.1 (with the exception of the nucleolus and the chromatin) and indicate which of the 11 functions the organelle helps the cell to perform. For example, you would list <u>respiration</u> for the mitochondrion, because it breaks down molecules with a release of energy. Some organelles might participate in more than one of the 11 basic life functions, so be sure to list multiple functions when appropriate. Please note that you will need to spend some serious time reviewing what you have read in order to answer this question. That's the purpose of the question. Please take it seriously and work hard, as you will need this kind of review to be able to assimilate all of the information you were given so far.

6.4 As we mentioned in Module #5, lactose-intolerant people cannot digest the disaccharide lactose due to the lack of an enzyme. Which organelle in the cells of a lactose-intolerant person does not have what it needs to get its job done?

6.5 List the organelles that are commonly found in plant cells but not in animal cells. List those organelles commonly found in animal cells that are not found in plant cells.

6.6 Even though ribosomes are considered organelles, they exist in both prokaryotic and eukaryotic cells. A student claims that this contradicts the definition of a prokaryotic cell, because he says that a prokaryotic cell cannot have organelles. Why is the student wrong?

To give you an idea of how complex cell structure can be, perform the following experiment if you have a microscope.

EXPERIMENT 6.1
Cell Structure I

<u>Supplies:</u>

♦ Microscope
♦ Lens paper
♦ Slides
♦ Coverslips
♦ Eyedroppers
♦ Water
♦ Iodine
♦ Onion
♦ Cork (Any cork item can be used.)
♦ Knife or scalpel
♦ Prepared slide: Hydra
♦ Prepared slide: *Ranunculus* root
♦ Prepared slide: *Zea mays* root

Object: To study the internal structures of various cells

Procedure:

A. Examine a wet mount of cork cells:
1. Secure the cork so that it will not slip. With a sharp knife or a scalpel, cut a very thin slice of cork. **BE CAREFUL!** Do not cut yourself!
2. Place the cork on a clean slide.
3. Add a drop of water and then a coverslip.
4. Place the slide on the microscope and observe on all three powers. You will not be able to see the organelles of the cells, but you will see their basic shape. Note how rectangular they are and that the cell wall is very thick and easy to see.
5. In your notebook, sketch what you see.

B. Examine a wet mount of onion epidermis cells:
1. Quarter an onion and peel the outside layer off one quarter.
2. With the point of your knife, carefully go under the skin (epidermis) on the inside of the layer that you just peeled off. Remove a paper-thin sample of the epidermis.
3. Place the epidermis sample on a clean slide
4. Add a drop of iodine and then a coverslip. The iodine will stain the cells to make the nuclei easier to see.
5. Observe the onion epidermis on all three powers. Note the cell walls and the bricklike configuration of the cells. Also, note that the nuclei are at different positions in different cells.
6. Sketch in your notebook what you have observed.

C. Compare plant and animal cells:
1. Observe each of the following prepared slides on all three magnifications: Hydra, *Ranunculus* root, and *Zea mays* root.
2. Draw a sample of a cell from each, and note the differences between the cells for each slide.
3. Can you tell just by looking at the cells whether you have a plant or an animal? The hydra is an animal, while the other two are plants. Probably the easiest thing that allows you to tell them apart is the boundary between cells. Note how thick the boundaries between cells are in the plant specimens as opposed to those in the animal specimen. This is because the plant cells have cell walls.

D. Clean up:
1. Wash and dry all slides and coverslips.
2. Return all supplies to their storage area.
3. Clean the microscope and put it away.

How Substances Travel In and Out of Cells

As we learned in the previous section, the plasma membrane is responsible for substances moving in and out of a cell. If a substance is to leave the cell or enter it, it must first pass through the plasma membrane. To see how the plasma membrane works, however, we first need to see how it is made. A plasma membrane is usually constructed of proteins, cholesterol, and **phospholipids**.

Phospholipid – A lipid in which one of the fatty acid molecules has been replaced by a molecule that contains a phosphate group

Remember from Module #5 that lipids are molecules that contain three fatty acids linked together on a glycerol molecule. Well, a phospholipid has, instead of the third fatty acid molecule, a small molecule that contains a phosphate group, which is composed of phosphorus, hydrogen, and oxygen. The figure below shows the structural formulas for a lipid and a phospholipid so that you can see the difference. The bottom part of the figure also shows you the typical symbols used for lipids and phospholipids in biological drawings.

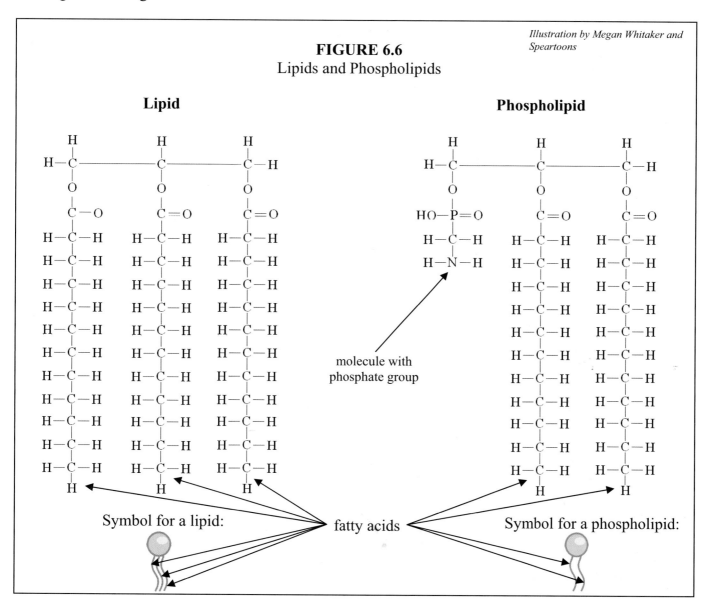

FIGURE 6.6

Lipids and Phospholipids

Illustration by Megan Whitaker and Speartoons

Lipid

Phospholipid

molecule with phosphate group

Symbol for a lipid:

fatty acids

Symbol for a phospholipid:

The interesting thing about a phospholipid is that the phosphate group gives the molecule a slight affinity to water. It turns out, however, that since the two fatty acid components are so long and since the molecule containing the phosphate group is so short, the affinity to water exists only on one end of the molecule – the end that contains the glycerol. The other end of the molecule still has no affinity for water. Thus, we say that a phospholipid has a **hydrophilic end** ("hydrophilic" means "water-loving") and a **hydrophobic end** ("hydrophobic" means "water-fearing"). In the little symbol used for a phospholipid, then, the top part of the molecule (the end with the circle) is attracted to water, while the bottom part (at the end of the wavy lines) is repelled by water.

Now that you understand what a phospholipid looks like, you can learn how a plasma membrane is constructed. A plasma membrane contains proteins, cholesterol, phospholipids, and other molecules arranged as shown below.

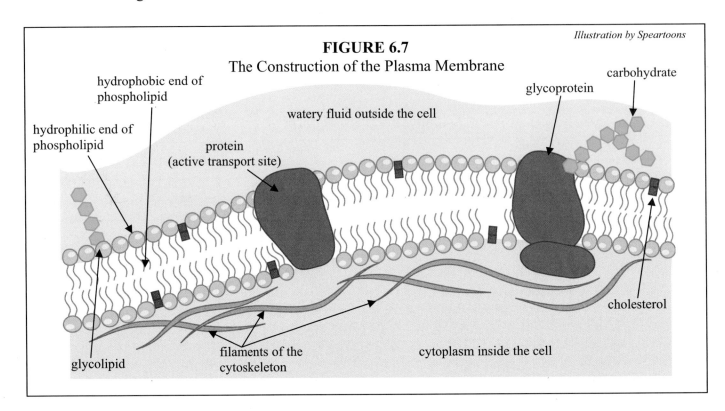

FIGURE 6.7
The Construction of the Plasma Membrane

Illustration by Speartoons

In the plasma membrane, phospholipids are arranged in a double layer (called the **phospholipid bilayer)** with the hydrophobic ends pointing towards each other and the hydrophilic ends pointing out towards the surroundings and in towards the cell. This should make sense, because water is both outside the cell in the surroundings and inside the cell in the cytoplasm. Thus, the hydrophilic ends will point towards the water, and the hydrophobic ends will point away from the water. Proteins and cholesterol are interspersed throughout the phospholipid bilayer. Some proteins are near the top of the bilayer, and others are near the bottom. In addition, some proteins span the entire width of the plasma membrane, leaving no room for phospholipids. This protein "bridge" can be an **active transport site**, which we will discuss in a moment.

In addition to what we have already discussed, there are carbohydrate molecules attached to the outside of the plasma membrane. When a carbohydrate is attached to a protein, the protein is called a **glycoprotein**, and when it is attached to a phospholipid, the result is called a **glycolipid**. The prefix "glyco" refers to glucose, which you should recall is a monosaccharide that can link together to form large carbohydrate molecules. The amount and type of carbohydrates on the plasma membrane varies from one cell type to another as well as from one individual to another. As a result, they serve as a means of cell-to-cell recognition. For example, people have four blood types (A, B, AB, and O). These blood types are distinguished from one another by the arrangement of carbohydrates on the plasma membrane of a person's red blood cells.

One neat feature of the plasma membrane is that it can self-reassemble when it is broken. When a secretion vesicle, for example, reaches the plasma membrane, the membrane actually breaks to allow for the release of the secretion vesicle's contents. Afterwards, the plasma membrane reassembles by itself. This self-reassembly is due primarily to the fact that the hydrophobic ends of the phospholipids are attracted to one another and the hydrophilic ends are attracted to the water on the inside and outside of the cell. Thus, the phospholipids always end up "pointing" in the right direction, once they have time to regroup.

There are two basic ways that substances enter or exit a cell: **passive transport** or **active transport**.

Passive transport – Movement of molecules through the plasma membrane according to the dictates of osmosis or diffusion

Active transport – Movement of molecules through the plasma membrane (typically opposite the dictates of osmosis or diffusion) aided by a process that requires energy

Now remember, you have already learned about osmosis and diffusion. If you have forgotten what they are, you should go back to Module #5 and review those concepts.

In passive transport, no cellular energy is expended as a result of the substance passing through the plasma membrane. Water, for example, is a very small molecule and thus passes easily through the spaces between the proteins in the plasma membrane. As a result, it will travel according to the dictates of osmosis. Thus, if the concentration of solutes in the solution surrounding the cell is essentially equal to the concentration of solutes inside the cell, water will aimlessly wander back and forth across the plasma membrane, resulting in no net change in the amount of water inside or outside the cell. When cells are in a solution whose solutes are essentially equal in concentration to those of the inside of the cell, the cell is said to be in an **isotonic** (eye' suh tahn' ik) **solution**.

Isotonic solution – A solution in which the concentration of solutes is essentially equal to that of the cell which resides in the solution

The red blood cells in your bloodstream, for example, are in an isotonic solution. Your kidneys get rid of excess solutes in the blood to make sure that the bloodstream stays isotonic with your red blood cells.

Why do your kidneys need to make sure that your bloodstream stays isotonic with your red blood cells? Well, remember the story about "The Rime of the Ancient Mariner" from Module #5? If cells are placed in a solution that has a significantly higher concentration of solutes than that of the cell, osmosis will drive water out of the cell and into the surroundings. This will cause the cell to lose too much water, killing it! Solutions that have a higher concentration of solute than the inside of a cell are called **hypertonic** (hi' pur tahn' ik) **solutions**. Although a hypertonic solution has the potential to kill any cell, the implosion that results from placing cells that have cell walls in a hypertonic solution is given a specific name – **plasmolysis** (plaz mahl' uh sis).

Hypertonic solution – A solution in which the concentration of solutes is greater than that of the cell that resides in the solution

Plasmolysis – Collapse of a walled cell's cytoplasm due to a lack of water

This is why you should not *overfertilize* a plant. If you put too much fertilizer in the ground near a plant, when the fertilizer mixes with the water in the soil, a hypertonic solution could be formed. This will result in plasmolysis of the plant's cells, killing the very plant that you were trying to fertilize!

Osmosis can also cause **cytolysis** (sye tahl' us sis), when a cell is placed in a **hypotonic** (hi' puh tahn' ik) **solution**.

Cytolysis – The rupturing of a cell due to excess internal pressure

Hypotonic solution – A solution in which the concentration of solutes is less than that of the cell that resides in the solution

In a hypotonic solution, there is greater solute concentration on the inside of the cell than on the outside. As a result, water flows by osmosis into the cell until the cell bursts from too much water pressure.

Since many organisms live in freshwater habitats, a lot of cells must deal with living in hypotonic solutions. After all, fresh water has few solutes dissolved in it; thus, it is a hypotonic solution for cells. Plant cells typically deal with living in a hypotonic solution by using their central vacuole and their cell wall. As discussed in the previous section, as water flows into a plant cell, it collects in the central vacuole. This expands the vacuole, pushing the cytoplasm against the rigid cell wall. This results in an increase in pressure, which we called turgor pressure. Once the pressure gets large enough, it counteracts the effect of osmosis, and no more water can get into the cell. Even though the water "wants" to get into the cell due to osmosis, the turgor pressure keeps the water out. This is what we mean when we say that turgor pressure counteracts the effects of osmosis, keeping the cell from rupturing.

Plant cells can do this, of course, because they have a rigid cell wall against which the cytoplasm can push, causing turgor pressure. What about cells that must live in freshwater environments but do not have cell walls? Well, if you think back to Module #3, you will remember that certain protozoa have contractile vacuoles that constantly pump excess water out of the cell. Animal cells that live in hypotonic solutions also have those vacuoles. Thus, they cannot stop water from entering the cell as plant cells can, but they do get rid of it right away, so excess pressure does not build up. Actually, when scientists first observed contractile vacuoles, they thought that they were looking at hearts, because the contractile vacuoles pumped constantly, just like hearts. What they were doing, however, was just forcing water to leave the cell.

When the contractile vacuoles force water out of the cell as discussed above, they are engaging in active transport, because the cell is expending energy to cause a substance (water) to travel across the plasma membrane against the dictates of osmosis. Another example of active transport involves the proteins we discussed earlier. When a protein spans the width of the plasma membrane, it often has been fashioned to recognize a certain type of molecule. When it recognizes that molecule, it goes through a series of shape changes that pulls the molecule across the plasma membrane and into (or out of) the cell. That's why the area pointed out in Figure 6.7 is called an active transport site, because the protein actively works to transport molecules, with the help of energy provided by the cell. An example of this can be found in potassium. Many cells take in potassium, but the concentration of potassium in a cell's surroundings is usually low compared to inside the cell itself. As a result, the cell must expend energy to pull potassium across a protein in the plasma membrane and into the cell.

ON YOUR OWN

6.7 If the plasma membrane were made out of regular lipids (not phospholipids) it could never self-reassemble. Why?

6.8 A scientist observes a cell and watches as it explodes. Was the cell in an isotonic, hypertonic, or hypotonic solution?

6.9 A cell's mitochondria cease to function, and the cell has no more energy. Will all transport across the plasma membrane stop? Why or why not?

If you have a microscope, perform the following experiment before you go on to the next section. It will give you the opportunity to see some of the processes that we have been discussing.

EXPERIMENT 6.2
Cell Structure II

Supplies:

> **Note:** If you have trouble seeing the plasmolysis you are supposed to see in this experiment, look at the course website that is discussed in the "Student Notes" section at the beginning of this book. It has links to pictures of plasmolysis. They may help you.

- Microscope
- Lens paper
- Banana
- Water
- Eyedropper
- Slides and coverslips
- Anacharis leaves (Anacharis is a generic term for aquatic plants that come from genus *Elodea* and genus *Egeria*. Sometimes called "water weeds," they can be purchased at aquarium stores. If you live in a state that has outlawed the sale of these plants due to their tendency to take over an ecosystem, ask the salesperson at the aquarium shop what they are selling in place of Anacharis. You could also use thin leaves from another plant like Impatiens. The main point is that the leaves need to be alive and very thin.)
- Iodine
- Cotton swab
- Salt water (1 tablespoon of salt in about $\frac{1}{8}$ cup of water)

Object: To see how cells vary in structure and to observe plasmolysis

Procedure:

A. Study a living leaf:
  1. Select a young or new leaf for your observation.
  2. Place the leaf on a slide, add a drop of water and place a coverslip on top. Be careful not to crush the leaf.
  3. Observe under all three powers.
     a. Some kinds of leaves will have distinct veins. These will appear to be long and thin.
     b. Look for cells of different shapes as you look at various areas of the leaf.

4.   At the edge of the leaf, you should be able to see some clear cells. If you look closely (lower the brightness of your light source), you should see the nuclei. In your notebook, sketch some of the different kinds of cells.

5.   Note the green (usually somewhat oval) objects in the cells. These are chloroplasts. In some of your leaves you might see movement of these chloroplasts in what looks like a "stream." This is the cytoplasmic streaming you learned about earlier. Sometimes putting the leaf close to a warm bright light makes the streaming begin.

6.   Place a drop of salt water on the slide next to the coverslip. Take a tissue and place it against the other side of the coverslip. As the tissue absorbs the water that was originally on the leaf, the salt water should replace it. You should be watching through the eyepiece as you do this.

7.   If your leaf is thin enough, you should see some cells begin to lose their cytoplasm. You can see this by watching the chloroplasts in the cells. They will start out evenly distributed throughout the cell and then slowly concentrate in one region of the cell as the cytoplasm leaks out. This is an example of plasmolysis.

B. Study banana cells:
1.   With a cotton swab or the end of a knife, smear a small amount of banana on a slide.
2.   Observe the slide on all three powers and sketch some cells in your notebook.
3.   Add a drop of iodine and a coverslip over the smear of banana. The iodine will stain the starch inside the cells, making it easier to see.
4.   Observe again and sketch in your notebook. Note the leucoplasts that contain the starch. They should appear dark now that the sample has been stained with iodine.
5.   Clean up and put away all equipment. Make sure your microscope is clean.

How Cells Get Their Energy

Cells produce useable energy in a chemical reaction that, at first glance, seems rather simple:

$$C_6H_{12}O_6 + 6O_2 \rightarrow 6CO_2 + 6H_2O + \text{energy}$$

This chemical equation represents what happens when a molecule of glucose is burned. Six molecules of oxygen from the air interact with a glucose molecule. The atoms rearrange, forming 6 carbon dioxide molecules and 6 water molecules. As is the case when anything burns, energy is also released. This is, in essence, how a cell produces the energy that it uses to sustain all of the processes it must run. Of course, we say "in essence" because the actual process of burning monosaccharides in cells is actually quite complicated.

Although you've probably never tried it (and you really shouldn't – it makes a mess), sugar does burn. If you strike a match and light some sugar, the sugar will burn, producing water, carbon dioxide, heat, and smoke. The heat is a form of energy. Although you expended some energy striking the match, once you get enough sugar burning, the heat that it generates will be much, much more than the energy you spent using the match to light the sugar. You see, the chemical reaction described by the equation above needs a bit of a "push" to get it started. This push, called **activation energy**, is supplied by the match.

Activation energy – Energy necessary to get a chemical reaction going

Once the chemical reaction starts, however, it produces lots of energy. Some will be used as activation energy to start more reactions and thus keep the fire going, while the rest will be used to produce heat and light in the form of flames.

Now it turns out that all chemical reactions require some amount of activation energy in order to start. Reactions such as the burning of glucose, however, end up producing a lot more energy than that which is required to start them. The net result is lots of energy that can be used for other things.

If a mitochondrion were simply to burn sugar in the way that we just described, however, the results would be disastrous! The reaction of glucose and oxygen is very fast once it gets going. It produces a large amount of energy in a short amount of time. This large concentration of energy in a short time interval would be so violent that it would destroy the mitochondrion! So, the cell simply slows the reaction down, right? Wrong! You see, the cell needs a lot of energy. If the cell were to slow the burning of glucose down to a point where the mitochondria could withstand the heat, there would not be enough energy produced to allow the cell to run all of the processes that it needs to run. Thus, something else must happen.

What happens is really quite complicated. We will give you a brief introduction to it here, but please realize that the actual processes involved are quite a bit more complicated than anything that we will do here. This is simply a short, sketchy introduction into one of the most complicated topics in biology: **cellular respiration**. Cellular respiration can occur aerobically or anaerobically. In this course, we will concentrate on aerobic respiration.

Aerobic cellular respiration is accomplished through a four-stage process that begins in the cytoplasm. When the cell wants to "burn" a monosaccharide for energy, the monosaccharide first goes into the cytoplasm. There, enzymes catalyze a reaction that causes the monosaccharide molecule to lose some of its hydrogen atoms and form 2 molecules of a substance called **pyruvic acid** ($C_3H_4O_3$). A "bare-bones" chemical equation that describes this process is:

$$C_6H_{12}O_6 \rightarrow 2C_3H_4O_3 + 4H + energy$$
glucose   pyruvic acid

The cell has to provide the activation energy required to get this to happen, but that's okay. Much like striking a match to start something burning, the amount of energy needed to begin this reaction is half of that which is released when the reaction finishes. This is the first stage of cellular respiration and is called **glycolysis** (glye kahl' uh sis).

Once 2 molecules of pyruvic acid are made, they head to a mitochondrion. As they cross into the mitochondrion, the next stage, called **formation of acetyl** (uh see' tuhl) **coenzyme** (koh en' zyme) **A**, takes place. This stage is also called the "oxidation of pyruvic acid" in some biology texts. In this process, the pyruvic acid is broken down, and a protein (coenzyme A) is attached to the remains. A "bare-bones" chemical equation for this process can be written as follows:

$$2C_3H_4O_3 + 2(coenzyme\ A) \rightarrow 2C_2H_3O\text{-}(coenzyme\ A) + 2CO_2 + 2H$$
pyruvic acid   acetyl coenzyme A

Now don't get put off by the equation. It's not that bad. It just takes the 2 molecules of pyruvic acid produced in glycolysis and attaches 2 molecules of coenzyme A to them. Since coenzyme A is a big molecule, we do not give its chemical formula. We just call it by its name in the equation. Now, notice what happens in the reaction. Pyruvic acid has 3 carbons in its chemical formula. When the coenzyme A reacts with the pyruvic acid, a carbon is removed. You can see this by looking at the chemical formula of the first product, acetyl coenzyme A. It has 2 carbons in its chemical formula instead of the 3 that pyruvic acid has. Where did the carbon go? It went into the second product, CO_2. The pyruvic acid also lost a hydrogen atom and 2 oxygen atoms. The oxygen atoms are used to make the CO_2, and the hydrogen atom is the final product in the equation.

You can therefore think of what happens in the formation of acetyl coenzyme A this way: Each pyruvic acid molecule loses a carbon, 2 oxygens, and a hydrogen. The carbon and oxygen atoms become CO_2, and the hydrogen atom is a product in the reaction. In order for this to happen, however, a large molecule called coenzyme A had to be added to the pyruvic acid. Eventually, that molecule will leave the picture, but we are getting ahead of ourselves.

Once the acetyl coenzyme A is formed, the next stage (called the **Krebs cycle**), begins. This stage is sometimes called the "citric acid cycle," and it also takes place in the mitochondrion. It takes the 2 molecules of acetyl coenzyme A and reacts them with oxygen to make hydrogen, carbon dioxide, and coenzyme A again. This is actually a long, complicated procedure that involves many reactions which are carefully controlled by enzymes. The overall chemical reaction that occurs as a result of this process, however, is fairly simple:

$$2C_2H_3O\text{-(coenzyme A)} + 3O_2 \rightarrow 6H + 4CO_2 + 2\text{(coenzyme A)} + \text{energy}$$
acetyl coenzyme A

Notice what happens here. The 2 molecules of acetyl coenzyme A that were produced in the previous step are reacted with oxygen to make hydrogen atoms, carbon dioxides, and coenzyme As. This reaction also produces some energy.

Now think about what has happened so far. We originally had a molecule of glucose, which has 6 carbon atoms, 6 oxygen atoms, and 12 hydrogen atoms. From glycolysis to the end of the Krebs cycle, the glucose molecule has been broken down. The carbon chain has been destroyed, and the 6 carbons that were a part of that chain are now a part of 6 individual CO_2 molecules. Two of those CO_2 molecules were formed in the formation of acetyl coenzyme A (the second stage), and 4 of them were formed in the Krebs cycle (the third stage). Half of the oxygens needed to form those CO_2 molecules came from the glucose molecule itself, and the other half came from oxygen that was added during the Krebs cycle. Thus, the carbons and oxygens in the glucose molecule have all been accounted for.

So far, the only thing that has been actually added to the glucose is the 3 oxygen molecules that were a part of the Krebs cycle (the third stage). You might think that coenzyme A was added as well, but it really wasn't. After all, the formation of acetyl coenzyme A (the second stage) required the addition of 2 molecules of coenzyme A, but what happened at the end of the Krebs cycle? Two molecules of coenzyme A were produced. Thus, the second stage used 2 molecules of coenzyme A, but the third stage produced 2 molecules of coenzyme A. The net effect, then, is that the amount of coenzyme A never changes. It is just used and then remade over and over again.

Alright then. So far, the glucose molecule has been broken down into 6 CO_2 molecules and 12 hydrogen atoms. In order to accomplish this, 3 molecules of oxygen were used. What's left? Well, the last stage of cellular respiration is called the **electron transport system**. This stage produces the most energy of all four stages. Just like the Krebs cycle, it takes place in the mitochondrion. In this stage, the hydrogen that was produced in the previous three stages is carefully reacted with oxygen to make water. Now think about it. Glycolysis produces 4 hydrogen atoms, the formation of acetyl coenzyme A produces 2 more, and the Krebs cycle produces 6 hydrogen atoms. How many total do we have now? We have all 12 hydrogens that were originally in the glucose molecule. They react with 3 oxygen molecules to make water.

$$12H + 3O_2 \rightarrow 6H_2O + \text{energy}$$

Once again, like the Krebs cycle, this is actually done through a complicated series of enzyme-controlled reactions.

We've gone through a lot of information here, and your head might be swimming with all of this new information. As a result, we want to step back and summarize what you have learned so far. In the first stage of aerobic cellular respiration (called glycolysis), glucose is broken down into 2 smaller molecules as well as 4 hydrogen atoms. This produces a small amount of energy. In the second stage (the formation of acetyl coenzyme A), a carbon, 2 oxygens, and a hydrogen are removed from each of the 2 molecules made in glycolysis, and 2 molecules of coenzyme A are attached to the remains. This results in the production of 2 acetyl coenzyme A molecules, 2 carbon dioxide molecules, and 2 hydrogen atoms. In the third stage (the Krebs cycle), the 2 molecules of acetyl coenzyme A are reacted with 3 oxygen atoms. This produces 6 hydrogen atoms, 4 carbon dioxide molecules, 2 molecules of coenzyme A, and some energy. Finally, in the fourth stage (the electron transport system), the 12 hydrogen atoms that were produced in the previous three stages are reacted with 3 oxygen molecules to make 6 water molecules and a lot of energy.

Okay, so what's the overall effect of these four stages? Well, we know that 1 molecule of glucose is used. We also know that 3 molecules of oxygen are used in the Krebs cycle and another 3 are used in the electron transport system. As a result, a total of 6 oxygen molecules are used. Two molecules of coenzyme A are used in the formation of acetyl coenzyme A, but they are made again in the Krebs cycle. Thus, there is no net use or production of coenzyme A. What is produced, then? Well, lots of things are produced, but most get used up again. All of the hydrogen atoms produced in the first three stages, for example, are used in the fourth stage (the electron transport system). All of the other products of glycolysis are used. All of the other products of the formation of acetyl coenzyme A are used except for the 2 carbon dioxide molecules. All of the other products of the Krebs cycle except for the 4 CO_2 molecules are used. Thus, a total of 6 carbon dioxide molecules are made. In addition, the electron transport system produces 6 water molecules. In addition, all stages but the second produce energy. Thus, after everything is said and done, 1 glucose molecule is reacted with 6 oxygen molecules to make 6 carbon dioxide molecules, 6 water molecules, and energy. What is that reaction?

$$C_6H_{12}O_6 + 6O_2 \rightarrow 6CO_2 + 6H_2O + \text{energy}$$

That's the reaction we showed at the beginning of this section. In the end, then, the overall chemical equation that tells us how cells get their energy is very simple. The process by which that reaction occurs, however, is incredibly complex!

ON YOUR OWN

6.10 Make a table that describes what you have learned so far. The table should have one row for each stage of aerobic cellular respiration, and it should have four columns. In the first column, indicate the name of the stage. In the second column, indicate what is used during that stage. In the third column, indicate what is produced. In the fourth column, indicate where the stage happens. Finally, make a list at the bottom that details what is used and made after all four stages are complete.

6.11 What stages in the process of aerobic respiration make it an aerobic process?

ATP and ADP

Now, as if what you have learned weren't complicated enough, there is one more twist we have to add. To make things a little more complicated, the energy produced in cellular respiration must be stored so that it can be transported to different parts of the cell and used when the cell needs it. Energy can be stored in nearly any molecule, and when that molecule is broken down, the energy will be released. In a cell, however, the energy must be stored in such a way as to allow for a *gentle* release of energy. If the energy is not released gently enough, it will destroy a part of the cell rather than help it perform its tasks. Thus, the energy produced in cellular respiration must be stored in a molecule that, when broken down, releases only a small amount of energy. That molecule is **adenosine** (uh den' uh seen) **triphosphate** (try fahs' fate), which is usually abbreviated ATP.

ATP is composed of a substance called adenosine linked with 3 phosphate groups. When 1 of the phosphate groups is broken away from the rest of the molecule, a gentle release of energy occurs. This energy is sufficient to accomplish most tasks in a cell, but gentle enough so that it will not harm the cell. When the phosphate group breaks off from the rest of the molecule, an adenosine attached to 2 phosphate groups is left over. This molecule is called **adenosine diphosphate** (ADP). This reaction is usually represented as follows:

$$ATP \rightarrow ADP + P + energy$$

The products of this reaction (ADP and the lone phosphate group, P) find their way back into either the cytoplasm or a mitochondrion so that they can be reassembled into ATP to store more energy.

Now it's really important that you sit back and understand what's going on here. In three of the four stages of cellular respiration, energy is released. This energy cannot be transported to the parts of the cell that need it unless it is "packaged" in some way. It must be packaged so that it can be gently released directly in the place in which it is needed. That's the job of ATP. Thus, the energy produced in each stage of cellular respiration is used to make ATP from ADP and a phosphate group (P). The ATP is then transported to the place where energy is needed and, at that point, enzymes break the ATP down into ADP and a phosphate group. The resulting gentle release of energy is used by the cell, and the ADP and phosphate are returned so that more ATP can be produced.

Since each stage of cellular respiration releases different amounts of energy, you might be interested to know how much energy is released in each stage. The first stage, glycolysis, actually requires 2 ATPs in order to get going. Thus, when a monosaccharide needs to be broken down, 2 ATPs are broken down, and the resulting energy released supplies the activation energy required to start glycolysis. The process of glycolysis, however releases enough energy to make 4 ATP molecules from the ADP and phosphates that are in the cytoplasm. Thus, 2 ATPs are used in glycolysis but 4 are produced, for a net gain of 2 ATPs. The Krebs cycle releases enough energy to produce 2 more ATPs from the ADP and phosphate in the mitochondrion. Finally, the electron transport system produces enough energy to make 32 ATPs from the ADP and phosphates in the mitochondrion. Adding it all up, a single glucose molecule produces enough energy in aerobic cellular respiration to make 38 ATP molecules. The process requires 2 ATPs to get going, however, so the net gain is 36 ATPs. These 36 ATP molecules can then be transported to where the cell needs energy. Once there, enzymes will break a phosphate group off of the ATP, providing a gentle release of energy for the cell. To sum up, cellular respiration can be illustrated by the schematic in the figure below.

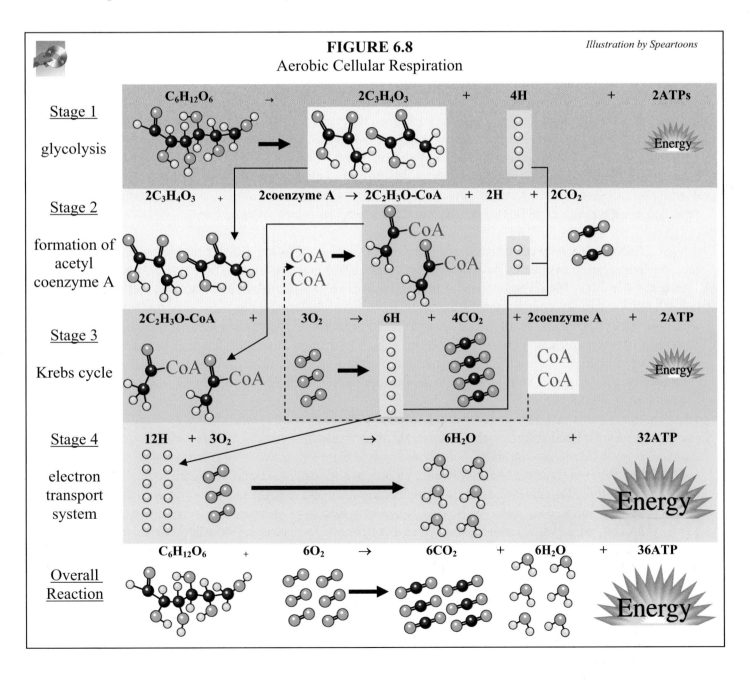

FIGURE 6.8
Aerobic Cellular Respiration

Illustration by Speartoons

We know that the figure is complex, but that's because it describes a complex process. Please be patient and read our description of the figure, flipping back from time to time to look at it. First, notice that there are four bands of color in the figure. Each band denotes a stage of aerobic cellular respiration. The white band at the very bottom is the overall result of the four-stage process.

Let's go through each band so that you can once more review what is going on. The first band (pink) represents stage one, which is glycolysis. In this stage, a glucose molecule ($C_6H_{12}O_6$) is broken down into 2 pyruvic acid molecules ($2C_3H_4O_3$), and 4 hydrogen atoms (4H). This results in enough energy to make 2 ATPs. Now notice that the pyruvic acid molecules are highlighted in yellow, and the hydrogen atoms are highlighted in blue. Why? Because the pyruvic acid molecules will be used in the next stage (which is represented by the yellow band), and the hydrogen atoms will be used in the final stage (which is represented by the blue band). The only product in this stage that is not used in a later stage is the energy produced.

The second band (yellow) represents the formation of acetyl coenzyme A. In this stage, the 2 pyruvic acid molecules that were produced in glycolysis are reacted with 2 coenzyme A molecules. This results in 2 acetyl coenzyme A molecules ($2C_2H_3O$-CoA), 2 hydrogen atoms (2H), and 2 carbon dioxide molecules ($2CO_2$). No ATPs are produced in this step. Notice that the acetyl coenzyme A molecules are highlighted in green. That's because they will be used in the next stage, which is represented by the green band. The hydrogen atoms are highlighted in blue because they will be used in the fourth stage, which is represented by the blue band. The carbon dioxide molecules are not highlighted, because they are not used again. So far, then, glycolysis produced 2 ATPs that are not used by a later stage, and this stage produced 2 carbon dioxide molecules that will not be used by a later stage. We could consider these the "final products" of the process so far.

The third band (green) represents the Krebs cycle. In this stage, the 2 acetyl coenzyme A molecules that were formed in stage two are reacted with 3 oxygen molecules ($3O_2$). This makes 2 coenzyme A molecules, 6 hydrogen atoms (6H), 4 carbon dioxide molecules ($4CO_2$), and another 2 ATPs. Notice that the coenzyme A molecules are highlighted in yellow. That's because they will be used in stage two of a *later* aerobic cellular respiration process. Thus, they will be used up when another glucose undergoes aerobic respiration. The 6 hydrogen atoms are highlighted in blue because they are used in stage four, which is represented by the blue band. The carbon dioxide molecules and ATPs are not highlighted, as they are not used again. Thus, they add to the "final products" we had from stages one and two.

The fourth band (blue) represents the electron transport system. In this stage, the 12 hydrogen atoms that have been made so far (4 from stage one, 2 from stage two, and 6 from stage three) are reacted with three oxygen molecules. This makes 6 water molecules and 32 ATPs. Neither of these products is used for anything else, so they are "final products" as well.

The last band (white) represents the results of these four stages. First, let's think of what had to be put into the four stages to get them to work. Glycolysis (stage one) required the input of 1 glucose molecule. The Krebs cycle (stage three) required the input of 3 oxygen molecules, and the electron transport system (stage four) required the input of an additional 3 oxygen molecules. That's it. Thus, this whole process requires 1 glucose molecule and 6 oxygen molecules. Remember, coenzyme A (used in stage two) is not a part of the required input for this process, since it comes from the aerobic respiration of a *previous* glucose molecule. What do we get out of the process? Well, glycolysis (stage one) gives us 2 ATPs, the formation of acetyl coenzyme A (stage two) gives us 2 carbon dioxide molecules, the Krebs cycle (stage three) gives us 4 more carbon dioxide molecules and 2 more ATPs, and the electron transport system (stage four) gives us 6 water molecules and 32 more ATPs. Anything else produced in those four stages is used up later on. As a result, the final products are 6 carbon dioxide molecules, 6 water molecules, and 36 ATPs. That's what's shown in the final band.

When a single molecule goes through all four stages of aerobic cellular respiration, then, the cell gains 36 molecules of ATP. Now although that sounds like a great trade (one glucose for 36 ATPs), it actually isn't as good as it first sounds. Remember, each ATP holds only a small amount of energy. This is necessary so that energy is released to the cell gently. Well, 36 ATPs actually store only about 55% of the energy contained in a glucose molecule. Where does the other 45% go? It makes heat. Although the heat warms the cell, it cannot be used for things like active transport, so in effect, the cell can use only 55% of the energy contained in the glucose. Even though this suddenly doesn't sound very good, it is still great compared to our modern technology! For example, a top-of-the-line automobile is only about 20% efficient in converting the energy of its fuel into energy of motion. Thus, although an efficiency of 55% sounds bad, it is significantly better than the best automobile that we can design! This, of course, should not surprise you. After all, the process of cellular respiration was designed and implemented by God. Anything that He produces will always be superior to the best that humankind has to offer!

We want to add one final note regarding cellular respiration. Notice that the first stage of the process (glycolysis) uses no oxygen. Thus, it can happen regardless of whether or not oxygen is present. If oxygen levels are very low, glycolysis can at least produce 2 ATPs from the glucose molecule. At that point, however, the pyruvic acid molecules produced in glycolysis cannot continue in the process we described. Even though the formation of acetyl coenzyme A does not require oxygen, it also does not produce energy. As a result, there is no benefit to the cell if the pyruvic acid molecules continue to that stage. Instead, in anaerobic conditions, the cell eventually converts the pyruvic acid into alcohol or lactic acid. This is called **cellular fermentation**, and in anaerobic conditions, it is the only option available to cells that are designed for aerobic respiration.

Now think about what you have learned for a moment. *Every cell in nature* must undergo either aerobic respiration, anaerobic respiration (which we have not covered here), or cellular fermentation. Obviously, cellular respiration is an incredibly complex process, as evidenced by the fact that it is so difficult to understand. This process, however, is just one of the many processes that occur constantly in a cell. This should reemphasize the point that there is no such thing as a "simple" organism. Even a single cell is a marvel of complexity, because it was designed by a marvelous God.

ON YOUR OWN

6.12 Despite the fact that a cell has all of the enzymes necessary for cellular respiration, plenty of glucose, functioning mitochondria, and a plentiful supply of oxygen, it cannot produce energy that is useable by the cell. What is it missing?

6.13 How many glucose molecules would have to undergo respiration in anaerobic conditions in order to produce the same amount of energy that 1 glucose molecule produces in aerobic conditions?

6.14 After a hard workout, your muscles often ache. One reason for this is the buildup of lactic acid in your muscles. Where does that lactic acid come from, and why does the process that produces lactic acid occur in your muscle cells during a hard workout?

ANSWERS TO THE "ON YOUR OWN" PROBLEMS

6.1 In making the proteins, the cell engaged in <u>biosynthesis</u>. In transporting the proteins to the edges of the cell, it engaged in <u>movement</u> (movement includes moving things *inside* the cell, not just moving the cell itself). Finally, in sending the materials out into the surroundings, it engaged in <u>secretion</u>. Secretion is different from excretion and egestion in the sense that secretion sends *useful* things outside the cell, whereas the others send out waste products.

6.2 In taking in the polysaccharide, the cell engaged in <u>absorption</u>. A polysaccharide cannot be used directly in respiration, however, because it first must be digested. Thus, the cell also performed <u>digestion</u>. Once the cell broke the polysaccharide down into monosaccharides, it performed <u>respiration</u> on the monosaccharides. In order to move all of these things to the proper organelles, as well as to move the waste products to the edge of the cell, it also engaged in <u>movement</u>. Finally, since the waste materials are soluble, the cell performed <u>excretion</u> to get rid of them.

6.3 Before you look at the answers, do not be too concerned if you did not list all of the life functions with each organelle. It's tough to think of them all. In addition, <u>homeostasis</u> could probably be added to every organelle's function, because, after all, these organelles all help the cell maintain the status quo.

<u>Organelle</u>	<u>Function(s)</u>
Cell wall	<u>Absorption, secretion, excretion, egestion</u>
Plasma membrane	<u>Absorption, secretion, excretion, egestion</u>
Cytoplasm	Really, the cytoplasm allows <u>all functions</u>, but <u>movement</u> is probably the most important.
Mitochondrion	<u>Respiration</u>
Lysosome	<u>Digestion</u>
Ribosome	<u>Biosynthesis</u>
Smooth endoplasmic reticulum	<u>Biosynthesis, movement, excretion, egestion</u>
Rough endoplasmic reticulum	<u>Biosynthesis, movement</u>
Golgi body	<u>Movement, biosynthesis, secretion</u>
Chloroplast	<u>Biosynthesis</u>. You could also include <u>absorption</u> here, since the chloroplasts absorb energy from sunlight.
Leucoplast	<u>Homeostasis</u>. By storing food for later use, this organelle helps to maintain the status quo. You could also include <u>biosynthesis</u> here, because without the leucoplasts, there would be nowhere to put the products of biosynthesis.
Central vacuole	<u>Homeostasis</u>. By maintaining turgor pressure, this helps maintain the status quo.
Vacuole	<u>Homeostasis</u>
Secretion vesicle	<u>Secretion, movement</u>
Centrioles	<u>Movement, reproduction</u>
Nucleus	Since the nucleus really holds all of the information regarding cell structure and function, it allows <u>all functions</u> of life.

6.4 The <u>lysosome</u> is where digestion of disaccharides and polysaccharides occurs. If a person lacks the enzyme to digest lactose (a disaccharide), it must be the lysosomes that don't have it.

6.5 <u>The cell wall, chromoplasts, leucoplasts, and the central vacuole are in plant cells but not animal</u> <u>cells. The lysosome and centrioles are in animal cells but not plant cells.</u> Some plant cells have centrioles, but most do not. You could also list chloroplast instead of chromoplasts, but remember that a chloroplast is really just one example of a chromoplast. It is the most common example, but there are other types of chromoplasts.

6.6 <u>A prokaryotic cell cannot have *membrane-bounded* organelles. Ribosomes, however, are not</u> <u>membrane-bounded</u>. Thus, they can exist in prokaryotic cells.

6.7 The reason that the plasma membrane can self-reassemble is that the phospholipids in the plasma membrane have both a hydrophobic end and a hydrophilic end. <u>Since regular lipids have no</u> <u>hydrophilic end, they do not have the ability to self-reassemble.</u>

6.8 <u>The cell was in a hypotonic solution.</u> In hypotonic solutions, the solute concentration is lower in the solution than in the cell. Thus, osmosis forces the water in the solution to flow into the cell. This builds up cell pressure, causing the cell to explode.

6.9 <u>Not all transport will cease. Active transport will cease, because it requires energy. However,</u> <u>passive transport requires no energy from the cell</u>. Thus, passive transport will continue.

6.10   The table should look something like this:

Name	Used	Makes	Takes Place In
glycolysis	$C_6H_{12}O_6$	$2\ C_3H_4O_3$ 4 H energy	cytoplasm
formation of acetyl coenzyme A	$2\ C_3H_4O_3$ 2 coenzyme A	2 acetyl coenzyme A $2\ CO_2$ 2 H	mitochondrion
Krebs cycle	2 acetyl coenzyme A $3\ O_2$	6 H $4\ CO_2$ 2 coenzyme A energy	mitochondrion
electron transport system	12 H $3\ O_2$	$6\ H_2O$ energy	mitochondrion

If you look at what is made and used, two $C_3H_4O_3$ are made in glycolysis but then used in the formation of acetyl coenzyme A. This uses two coenzyme A, but they are produced again in the Krebs cycle. The formation of acetyl coenzyme A also produces two acetyl coenzyme A, but they are used in the Krebs cycle. The first three stages also make a total of twelve H's, but they are all used in the electron transport system. In the end, then, only <u>1 $C_6H_{12}O_6$ and 6 O_2 are used</u>. In addition, only <u>6</u> <u>CO_2, 6 H_2O, and energy are produced</u>.

6.11 Only the <u>Krebs cycle and the electron transport system</u> use oxygen, so technically, this is what makes the entire process aerobic. However, the formation of acetyl coenzyme A will not happen unless the Krebs cycle can proceed, so in actuality, the final three stages are all considered aerobic.

6.12 <u>The cell is missing ADP and/or phosphate</u>. In order to make energy that is useable by the cell, respiration must make ATP from ADP and phosphate. If all of the components necessary to make energy are there, respiration must be making energy, but without ADP and phosphate to make ATP, the cell cannot use the energy.

6.13 One molecule of glucose in aerobic conditions allows the cell to gain 36 molecules of ATP. In anaerobic conditions, 1 molecule of glucose allows the cell to gain only 2 molecules of ATP, because the cell can only perform glycolysis. Thus, <u>18 molecules of glucose</u> must undergo respiration in anaerobic conditions to make as much energy as 1 molecule of glucose does in aerobic conditions.

6.14 <u>The lactic acid comes from cellular fermentation</u>. Remember, after glycolysis, if there is no oxygen, the pyruvic acid molecules are converted to either alcohol or lactic acid. In human muscle cells, it is converted to lactic acid, which causes muscles to ache. <u>Cellular fermentation occurs because the demand that muscle cells have for oxygen during intense workouts is more than the bloodstream can supply. As a result, the oxygen supply becomes low during a tough workout, and the cells are faced with anaerobic conditions. Thus, the only option for the cells is to perform glycolysis and then covert the resulting pyruvic acid into lactic acid.</u> That's cellular fermentation.

"I'm tired of living my life under a microscope."

Cartoon by Speartoons

STUDY GUIDE FOR MODULE #6

1. Define the following terms:

a. Absorption
b. Digestion
c. Respiration
d. Excretion
e. Egestion
f. Secretion
g. Homeostasis
h. Reproduction
i. Cytology
j. Cell wall
k. Middle lamella
l. Plasma membrane
m. Cytoplasm
n. Ions
o. Cytoplasmic streaming

p. Mitochondria
q. Lysosome
r. Ribosomes
s. Endoplasmic reticulum
t. Rough ER
u. Smooth ER
v. Golgi bodies
w. Leucoplasts
x. Chromoplasts
y. Central vacuole
z. Waste vacuoles
aa. Phagocytosis
bb. Phagocytic vacuole
cc. Pinocytic vesicle
dd. Secretion vesicle

ee. Microtubules
ff. Nuclear membrane
gg. Chromatin
hh. Cytoskeleton
ii. Microfilaments
jj. Intermediate filaments
kk. Phospholipid
ll. Passive transport
mm. Active transport
nn. Isotonic solution
oo. Hypertonic solution
pp. Plasmolysis
qq. Cytolysis
rr. Hypotonic solution
ss. Activation energy

2. Name the organelles that play a role in biosynthesis.

3. What helps the cell hold its shape if it does not have a cell wall?

4. How does a plant cell fight osmosis in a hypotonic solution?

5. A cell contains centrioles and lysosomes. Is it a plant or animal cell?

6. What organelles are involved in secretion? Don't worry about the biosynthesis necessary to get the secretion product. Just deal with the process *after* the product is made.

7. What things in the cell (not just organelles, but anything we have studied) deal with the cellular movement?

8. What are the 11 major tasks cells must perform?

9. What is the plasma membrane made of?

10. What is the difference between a phospholipid and a regular lipid?

11. What makes it possible for the plasma membrane to self-assemble?

12. A cell begins running low on food, and its energy output decreases by 20%. What kind of plasma membrane transport (active or passive) is affected?

13. Identify the structures pointed out below:

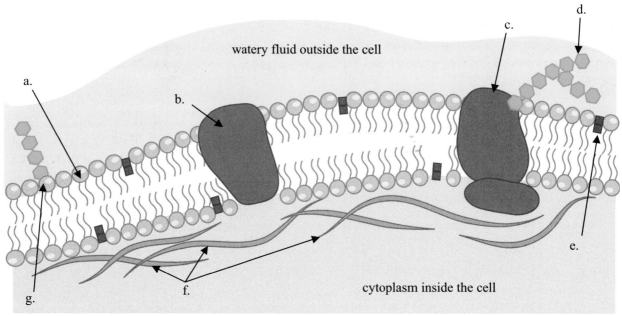

Illustration by Speartoons

14. If a cell dies by implosion, what kind of solution (isotonic, hypertonic, or hypotonic) was it in?

15. List the four stages of aerobic cellular respiration in the order in which they occur. In addition, note the net number of ATPs that are made in each step.

16. What is ATP's purpose in the cell?

17. If a cell has no oxygen, what stage(s) of aerobic cellular respiration can still run? How many ATPs can it make from a molecule of glucose?

18. A scientist determines a means to extract all ADP from a cell. Why will this kill the cell?

19. What organelle is responsible for breaking polysaccharides into monosaccharides?

MODULE #7: Cellular Reproduction and DNA

Introduction

As we learned way back in Module #1, all living organisms reproduce. Indeed, the ability to reproduce is one of the four criteria for life. It makes sense, then, that we should study this process in great detail. As is the case with many of the concepts in biology, however, reproduction is a difficult and detailed process to study. Thus, we will spend both this and the next module exploring some of the facets of reproduction.

The process of reproduction has always fascinated scientists. Throughout the history of biological inquiry, scientists have tried to understand how an organism's offspring receive the traits and characteristics that make them what they are. Sometimes, an organism's offspring might look identical or nearly identical to its parent or parents. In other cases, the offspring and parents look so different that you can hardly believe the offspring came from those parents.

When we look at a child, for example, we usually try to find the characteristics that the parents and child have in common. When we see that a boy and his father both have thick, brown hair, we say that the boy has his father's hair. When we see that he and his mother share vividly green eyes, we say that the boy has his mother's eyes. Sometimes, however, we have a hard time finding such similar traits. For example, two brown-haired, brown-eyed parents can have a child with blonde hair and blue eyes. How does that happen?

Well, the explanation for these facts eluded scientists for centuries. To be sure, there have always been theories that attempted to explain how characteristics passed from parent to offspring, but many were quite ridiculous. For example, there were those that thought characteristics like the height, strength, and weight of an offspring were determined by the activities of the parents. These scientists, called Lamarkian biologists, thought that if parents spent a lot of time in strenuous activity, their children would be born stronger. Somehow, they thought that the parents' activities would send "messages" to their reproductive systems, telling them what kind of offspring to produce.

The beginning of a real explanation for how traits are passed on from generation to generation was developed by a monk name Gregor Mendel in the mid 1800s. He did careful experiments with peas, determining how pea plants reproduced and passed on characteristics to their offspring. Unfortunately, this magnificent work was tucked away in a monastery for more than 50 years! When it was finally discovered, it laid the groundwork for all of our modern studies of **genetics**.

Genetics – The science that studies how characteristics get passed from parent to offspring

Genes, Chromosomes, and DNA

Building on the work of Gregor Mendel, scientists today have determined that each organism contains a storehouse of information that governs its traits and characteristics. As we have already mentioned, this storehouse is called DNA, and its main function is to tell the cell what proteins to make. Although it sounds hard to believe, most of your traits and characteristics are governed by what kinds of proteins your cells make. Your eye color, for example, is completely dependent upon what proteins are produced in some of the cells in your eyes. Thus, by coding for the production of certain proteins in your eyes, your DNA determines your eye color.

Now, of course, not all of your traits and characteristics are completely determined by your genetic makeup. For example, if you lift weights and work out, you will develop strong muscles. You might think, then, that your muscle strength is not determined by your DNA. Well, this is only partially correct. Even though you can increase your strength by working out, there is a fundamental limit to how strong you can become. For example, a person might lift weights and work out every day but never become as strong as another individual who really does not exercise seriously at all. In the case of muscle strength, your DNA determines a general range of how strong your muscles can become. If you do not exercise your muscles, you will stay on the weak side of the muscle strength range determined by your DNA. If, on the other hand, you work out seriously, you can increase your muscle strength to the strong side of the range your DNA has determined. You might never become as strong as you want to become, however, because there is some inherent limit set up by your DNA.

It turns out that the majority of characteristics in your body are determined in this way. The DNA sets a range of possible characteristics (sometimes called your **genetic tendency),** and your activities determine what portion of that range actually manifests itself in your body. Some people have a genetic tendency to be overweight. If they eat just a little more than they need, they will immediately gain weight. Others, however, have a genetic tendency to stay slim. They can eat almost as much as they like and never gain a pound. Although such tendencies exist, they can nevertheless be controlled. Even if you have a genetic tendency to be overweight, you can keep your weight down with strict diet and exercise. Thus, your DNA sets a general tendency, but you can work to fight that tendency, if you wish.

You might have read or heard that scientists have discovered that alcoholism is genetic. This is, in fact, partially true. A large study in Sweden, for example, looked at alcohol use in 3,000 children who were adopted and raised by nonrelatives. The incidence of alcoholism was 2½ times higher among those children who had at least one biological parent who was an alcoholic. What does this tell us? Well, consider an alcoholic whose father was also an alcoholic. You could argue that he inherited his alcoholism from his father through his DNA, or you could argue that he became an alcoholic because his father's actions "trained" him to be an alcoholic. In this study, however, the parents who had the children did not raise them. Thus, if alcoholism wasn't at least partly genetic, you would expect the rate of alcoholism to be independent of the child's biological parents. It wasn't. That tells us alcoholism has a genetic component. However, there are plenty of people whose parents are alcoholic who are not alcoholics themselves, and there are plenty of people who are alcoholics but have no family history of alcoholism. Thus, the data seem to indicate that DNA might set up a tendency towards or away from alcoholism, but that people can break that tendency.

The main point to this discussion, then, is that your DNA alone does not determine who you are or what you will become. There are some traits (blood type, for example) that are completely determined by your DNA. For many of your characteristics, however, your DNA just sets up a general framework. Within that general framework, what you do with yourself will ultimately decide what you become. Since DNA is not the only factor in determining what kind of person you will be, we generally say that there are three factors in determining the characteristics of a person: **genetic factors, environmental factors,** and **spiritual factors.**

Genetic factors – The general guideline of traits determined by a person's DNA

Environmental factors – Those "nonbiological" factors that are involved in a person's surroundings such as the nature of the person's parents, the person's friends, and the person's behavioral choices

Spiritual factors – The factors in a person's life that are determined by the quality of his or her relationship with God

These three factors each work together to determine what kind of person you are. There is no real consensus in the scientific community as to which of these three factors is the most important, if any. Needless to say, however, from the Christian point of view, spiritual factors are of the utmost importance. Since spiritual factors do not really apply to any organism other than humans, we generally say that the characteristics of animals and plants are determined by genetic and environmental factors.

Even though we do not know much about the relative importance of genetic factors and environmental factors, we can make one statement for sure. The genetic factors are laid down first. From a scientific point of view, these are the factors most easily studied, so we will concentrate on them in this course. The information in an organism's DNA is split up into little groups known as **genes** (jeens). Although a firm definition for a gene is rather hard to pin down, the following will work for the purposes of this course:

Gene – A section of DNA that codes for the production of a protein or a portion of protein, thereby causing a trait

As we mentioned before, your genetic traits or tendencies are determined by what proteins are produced in your cells. Thus, by coding for the production of a certain protein, a gene is actually coding for a particular genetic trait or tendency. How does a gene code for the production of a protein? You will learn about that in the next section. For right now, we want you to have a bit of experience with DNA by performing the following experiment.

EXPERIMENT 7.1
DNA Extraction

Supplies:

♦ Blender
♦ Toothpick
♦ Clear liquid hand soap or dish soap (The liquid hand soap tends to work just a bit better, and colorless will work a bit better than soap that is tinted with a color.)
♦ Salt
♦ Water
♦ Strainer
♦ Small glass
♦ Meat tenderizer (Make sure it has been bought within the last year or so.)
♦ Rubbing alcohol

- ♦ ½ cup of split peas
- ♦ Measuring cups and spoons
- ♦ Flashlight

Object: To extract DNA from peas so that you can see what DNA looks like on a macroscopic scale

Procedure:

1. Dissolve ¼ teaspoon salt in one cup of water.
2. Place the peas and salt water into the blender and blend for 20 seconds. The result should have the consistency of thin pea soup. The solution you have is a mixture of water, salt, and pea cells.
3. Hold the strainer over the small glass and slowly pour the contents of the blender into the strainer, allowing the liquid to drip into the small glass.
4. Add two tablespoons of the soap to the solution in the glass.
5. Slowly mix the liquid and the soap by swirling, and then let it sit for at least 10 minutes. The soap will dissolve the phospholipids in the plasma membranes of the pea cells. This will destroy the plasma membranes, opening the cells. The contents of the cells will then flow into the solution.
6. Add ¼ teaspoon of meat tenderizer and mix *very gently*. The meat tenderizer has enzymes that destroy proteins. This is what tenderizes meat. In this experiment, the meat tenderizer will destroy the proteins that coat the DNA, exposing the strands of DNA.
7. Slowly add as much rubbing alcohol as you have liquid. Do not stir! You want the alcohol to sit on top of the liquid. Wait for a few minutes, and you should see white strands appear in the layer of alcohol. To aid you in seeing the strands, shine the flashlight down onto the surface of the solution and look at the solution from the side of the glass. These strands are the peas' DNA!
8. Depending on your experimental technique, the strands might be long enough for you to twist them onto a toothpick and remove them from the solution.
9. Clean up your mess.

Now you know what the strands of DNA look like on a macroscopic scale. If all of the DNA from a single human cell were extracted in this way, you would have about two yards of strands.

ON YOUR OWN

7.1 Two identical twins have exactly the same set of genes. They are separated at birth and grow up in different households. If a scientist were to study the twins as adults, would he find them to be identical in every way, since they have the same genes? Why or why not?

Protein Synthesis – Part 1: Transcription

You might be wondering how instructing cells to make proteins can result in the traits that DNA gives us. If you are wondering that, perhaps you don't understand how important proteins are. Remember, in multicellular organisms, specific cells have specific tasks which they must accomplish so that the organism will survive. In your body, for example, you have skin cells that shield the inside of your body from contaminants. You also have cells in the retina of your eye that detect light and send electronic messages to your brain based on the light that they detect. How in the world do these cells perform such radically different tasks? The answer is simple: they produce different types of proteins. By and large, the tasks that a cell can complete are dependent on the proteins that it produces. If it produces one set of proteins, a cell functions as a skin cell. If it produces another set of proteins, it acts as a cell in the retina. So, how does a cell know what proteins to produce? That's the function of a gene, and that's what you will learn in this section.

Before we can show you how a cell uses DNA to produce proteins, we first need to give you a little more background on an important chemical of life, ribonucleic acid (RNA). We briefly mentioned this molecule in the previous module, but now we need to spend a little more time on it. As you might have already guessed by virtue of its name, RNA must have some relationship to DNA. Well, it does. In fact, there are a lot of similarities between them. RNA is also made up of nucleotides, but the individual structure of the nucleotides is a bit different. First of all, the sugar that makes up the foundation of the nucleotides is ribose, not deoxyribose (as is the case with DNA).

Like DNA, the nucleotides of RNA join together in long strands. Unlike DNA, however, they do not form a double helix. Instead, RNA is usually a single strand of joined nucleotides. Like DNA, RNA has four nucleotide bases. While three of them are the same as DNA's (adenine, cytosine, and guanine), one of them is different (uracil). Uracil performs the same tasks in RNA that thymine does in DNA. Thus, RNA has uracil in place of thymine. This can be confusing to students. Don't let it stump you. When you are thinking about RNA, just remember that uracil takes the place of thymine. It turns out that DNA is much more chemically stable than RNA because DNA uses thymine and deoxyribose instead of uracil and ribose. This means it is unlikely to undergo major changes with time, making it the ideal molecule for storing information from generation to generation, so that each organism reproduces after its kind (Genesis 1:11).

So, RNA has a lot of similarities to DNA but also some differences. Why is it important in the process of how the cell makes proteins? Well, one kind of RNA acts a lot like a camera. You see, the main portion of a cell's DNA is in its nucleus, but proteins are made in the ribosomes, which are outside the nucleus. In order to get information from the nuclear DNA to the ribosomes, one type of RNA makes a "snapshot" of the DNA and takes that information out of the nucleus to the ribosome.

How does the RNA do this? Well, first you need to remember something from Module #5. When we told you about DNA in that module, we told you that certain nucleotide bases can only link to other nucleotide bases. For example, cytosine can link only to guanine, and thymine can link only to adenine. Well, RNA has nucleotide bases as well. It has cytosine, guanine, and adenine just like DNA, and it also has uracil. Now remember, uracil in RNA acts like thymine in DNA. Thus, the uracil in RNA can link to the adenine in DNA. How does this help RNA take a "snapshot" of DNA? Think about it. The RNA can link its nucleotide bases to the bases on the DNA. The uracil in RNA can link to the adenine in DNA, the adenine in RNA can link to the thymine in DNA, the cytosine in RNA can link to the guanine in DNA, and the guanine in RNA can link to the cytosine in DNA.

What does all of this accomplish? When RNA does its job, it produces a "negative" of the DNA section that it is copying. Everywhere the DNA has a cytosine, the RNA will have a guanine. Everywhere the DNA has a guanine, the RNA will have a cytosine. Everywhere that DNA has a thymine, the RNA will have an adenine, and everywhere DNA has an adenine, the RNA will have a uracil. This is much like what happens when film is developed. During the developing process, the first thing to appear is a negative. This negative is dark everywhere that the picture is supposed to be light, and it is light everywhere that the picture is supposed to be dark. Based on this negative, then, a picture can be made. Based on the "negative" produced by RNA, a protein can be made. Before we show you how a protein can be made from this negative, we want to make sure you understand what we have discussed so far by showing you an illustration of it.

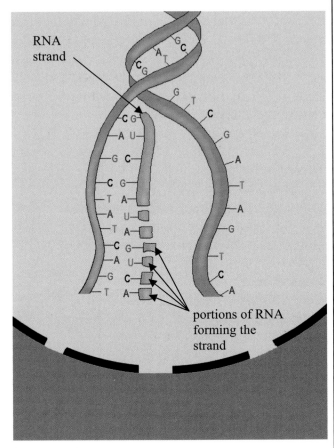

FIGURE 7.1
How RNA Takes a Snapshot of a DNA Section

Illustration by Megan Whitaker

When it is time for the cell to manufacture a protein, DNA "unwinds" the section of the double helix that codes for the production of that protein.

Individual RNA nucleotides begin bonding to the exposed DNA nucleotides. They bond with their partners: cytosine with guanine and uracil with adenine. This builds a strand of RNA.

When the RNA strand is built, then, it is a negative image of the DNA strand to which it linked. This process is called **transcription**, and the negative image produced by transcription can be used by the ribosome to make a protein. Before we show you how that is done, however, make sure you understand the process of transcription by answering the following questions.

ON YOUR OWN

7.2 An RNA strand has the following sequence of nucleotides:

uracil, adenine, adenine, guanine, cytosine, cytosine

What was the nucleotide sequence in the DNA that it transcribed?

7.3 A DNA strand has the following sequence of nucleotides:

thymine, thymine, thymine, adenine, guanine, cytosine

What will the RNA sequence be when this DNA section is transcribed?

Protein Synthesis – Part 2: Translation

After transcription, the RNA leaves the nucleus and moves to the ribosome. This carries the negative image of the gene that was transcribed to the organelle that produces the protein. Because the RNA that performs transcription is essentially a messenger (sending instructions from the nuclear DNA to the ribosome), we usually call it **messenger RNA (mRNA)**.

Messenger RNA – The RNA that performs transcription

Now if you're getting a bit confused at this point, don't worry; we'll sum it all up in a nice figure at the end. For right now, just understand that in order to get the DNA's instructions from the nucleus to the ribosome, RNA produces a negative image of the DNA's nucleotide sequence and takes it to the ribosome.

The ribosome is surrounded by amino acids, enzymes, and a different kind of RNA called **transfer RNA (tRNA)**. Transfer RNA is a big molecule that contains a special sequence of three nucleotides called an **anticodon**.

Anticodon – A three-nucleotide base sequence on tRNA

In addition to this anticodon, tRNA is bonded to an amino acid. The amino acid bonded to the tRNA is determined by the three nucleotide bases in the anticodon. For example, when the anticodon is made up of uracil, uracil, and cytosine (in that order), the tRNA is bonded to the amino acid lysine. If, instead, the anticodon is made up of uracil, cytosine, and cytosine, the tRNA is bonded to the amino acid arginine. This bond means that everywhere the tRNA goes, the amino acid goes as well.

When an mRNA molecule comes to the ribosome, the tRNA strands are attracted to the three nucleotide bases that will bond with the three nucleotide bases in their anticodon. Thus, if a tRNA strand has guanine, cytosine, and adenine, it will be attracted to a section of the mRNA that has cytosine, guanine, and uracil in succession. If such a sequence exists, the tRNA will go and link up with it, carrying the amino acid to which it is bonded the whole way. The three-nucleotide base sequence that attracts the tRNA anticodon is called, reasonably enough, a **codon**. Since a codon attracts a tRNA with a particular amino acid on it, we say that a codon refers to a specific amino acid.

Codon – A sequence of three nucleotide bases on mRNA that refers to a specific amino acid

Now don't get lost in all of this terminology. A strand of mRNA can be thought of as a bunch of three-nucleotide sequences. Each three-nucleotide base sequence is called a codon. A strand of tRNA contains a three-nucleotide base sequence called an anticodon. A certain anticodon on tRNA results in a certain amino acid bonded to the tRNA. Since the tRNA anticodons are attracted by the mRNA codons, the net result is that a codon on mRNA attracts a specific amino acid. The figure below illustrates this process.

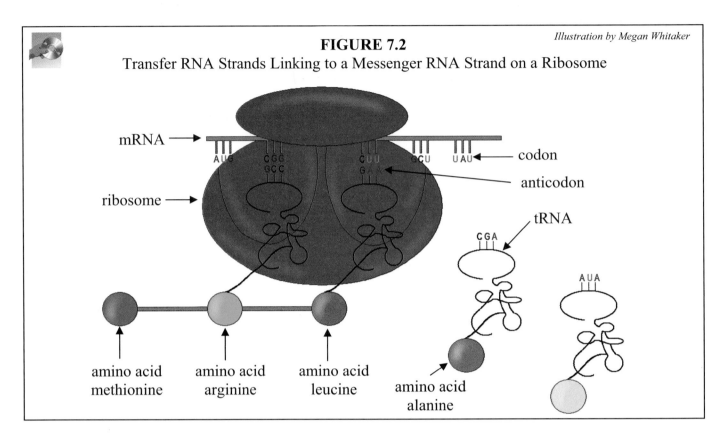

FIGURE 7.2
Illustration by Megan Whitaker
Transfer RNA Strands Linking to a Messenger RNA Strand on a Ribosome

Now look at what happens. When the tRNA strands have linked up to the mRNA, there are amino acids lined right up next to each other. What, by the way, is a protein? It is a bunch of amino acids linked together in a particular sequence. Well, at this point, the amino acids sitting right next to each other link up. This happens again and again and again, so that many, many amino acids link up together. When all the amino acids called for by the codons on mRNA are linked together, the result is a protein! The entire process shown in Figure 7.2 is called **translation**.

Now that we have been through the process once, let's summarize it so that you can review what you have learned. When a cell needs to make a protein, its DNA has the plan for making that protein in a long series of three-nucleotide base sequences. Messenger RNA reads this sequence and makes its "negative image" of the relevant portion of DNA. It then takes this series of nucleotide base sequences out to the ribosome. Once at the ribosome, each codon (set of three nucleotide bases) on the mRNA will attract a particular anticodon (set of three nucleotide bases) on tRNA. The tRNA that is attracted to the codon has a particular amino acid riding on it. This amino acid is determined by the anticodon on the tRNA. Thus, since each codon on mRNA attracts a particular anticodon, the codons, in effect, attract a particular amino acid. As the amino acids line up, they link together, eventually forming the protein.

The process we just reviewed is split into two phases: **transcription** and **translation**. In transcription, mRNA makes its negative image in order to copy the nucleotide sequence in DNA. This is much like a transcriptionist "copying" a conversation by writing everything down. In the next step, translation, the mRNA leaves the nucleus and goes to the ribosome, where tRNA strands carry amino

acids to the mRNA and line them up in the order determined by the sequence of nucleotides. The amino acids then bond together, making a protein. This step is called translation because the "language" of nucleotide base sequences in RNA is translated into the "language" of amino acid sequences in a protein. Thus, as a bare-bones description of how a cell makes a protein, we could say:

$$DNA \xrightarrow{\text{transcription}} RNA \xrightarrow{\text{translation}} \textbf{Protein}$$

Now of course, this doesn't show you the details of transcription and translation. Those are summarized in the figure below.

Illustrations by Megan Whitaker

FIGURE 7.3
A Schematic Describing Protein Synthesis in Cells

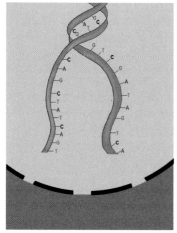

In the nucleus of the cell, DNA unwinds.

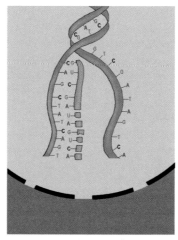

RNA nucleotides in the nucleus bond to the exposed DNA nucleotides, forming a strand of RNA.

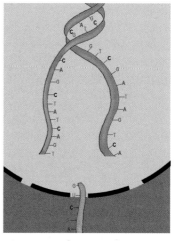

The mRNA then leaves the nucleus through a nuclear pore and goes to a ribosome, where there are plenty of tRNA molecules and the associated amino acids.

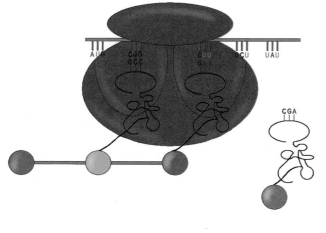

In the ribosome, tRNA strands are attracted to mRNA sections that have a codon with which their anticodon can bond. They bond to that section of the mRNA, dragging their amino acids along with them. This results in amino acids sitting next to each other. The amino acids bond, and after this happens many, many times, a protein is formed.

Now before we end this section, we want you to be aware that we have simplified this process a great deal in order to make it easier for you to learn. There are a lot of things that make this process more complicated, but we did not want that to interfere with your getting a basic understanding of how a gene is translated into a protein. However, we do want to illustrate the level of complexity by just mentioning two things in brief. First, a given amino acid can be called for by several different codons. For example, the amino acid cysteine can be called for by a sequence of uracil, guanine, and cytosine or by a sequence of uracil, guanine, and uracil. Of course, a single codon cannot call for more than one amino acid, but a single amino acid can have many codons that call for it.

In addition, there are certain things that have to happen between transcription and translation. For example, in eukaryotic cells, the mRNA that comes from the nucleus actually has too much information on it. You see, a gene in a eukaryotic cell contains sections called **exons** and other sections called **introns**. When the mRNA makes its negative image, it copies everything on the gene. However, only the exons are a part of the instructions for making a protein. The introns are sequences of nucleotide bases that separate the exons, almost like "spacers" that separate one set of books from another set of books on a shelf. In order for the ribosome to have the proper instructions, the mRNA must be "processed" before it gets to the ribosome. During this "processing," the introns are removed and the exons are spliced together, so that by the time it gets to the ribosome, the mRNA has the information necessary to make the protein (the exons) and nothing more (no introns).

If the introns are removed from the mRNA before the protein is made, what is their purpose? The short answer is that we are not sure. Since they are not a part of the instructions for the protein that the gene produces, some scientists call them "junk DNA." However, we know that God doesn't make junk, so introns must have a function. Indeed, current research is pointing to that fact already. For example, research indicates that different cells can use the same gene to produce different proteins, because they can keep certain exons from the gene and discard other exons. Thus, because the introns split the protein-making instructions into pieces, one gene can actually code for two (or more) proteins, based on what pieces are used by the cell. This makes the genetic information even more compact. Isn't God's creation incredible?

ON YOUR OWN

7.4 A scientist is studying a nucleic acid, but her notes are sketchy. You do not know whether she is studying DNA or RNA. You can make out the following nucleotide sequence, however:

guanine, cytosine, cytosine, uracil, guanine, adenine

Is the scientist studying DNA or RNA?

7.5 A protein has alanine as its first amino acid. One codon that calls for alanine has the following sequence of nucleotide bases:

guanine, cytosine, adenine

a. If a portion of DNA instructs a cell to make this protein, what will be the first three nucleotide bases of that DNA portion?

b. What will be the sequence of nucleotide bases on the tRNA that responds to the mRNA codon given above?

Now that you know what a gene is and how it codes for the production of a protein, your next question should be, "How are genes passed on from one generation to another?" After all, we know that offspring receive traits from their parents, and we know that traits are governed (to one extent or another) by genes. Thus, the genes of the parents must be transferred to the genes of the offspring. How does that happen? Well, to understand how that happens, you need to understand more about cellular reproduction. As a result, the rest of this module will be devoted to how eukaryotic cells asexually reproduce as well as the way in which eukaryotic cells are made for the purpose of sexual reproduction. Once you learn these details, you will finally be able to understand how genes are passed on from one generation to the next, and that will be the focus of Module #8.

Mitosis: Eukaryotic Asexual Reproduction

As you have already learned, the main component of a cell's DNA is stored in long, thin strands in the nucleus of the cell. If all of the DNA from one of your cells were strung together end to end, the string would dangle from your head to your toes, and still some would be on the floor. How does such a large amount of DNA get packed into a tiny cell? Examine the figure below.

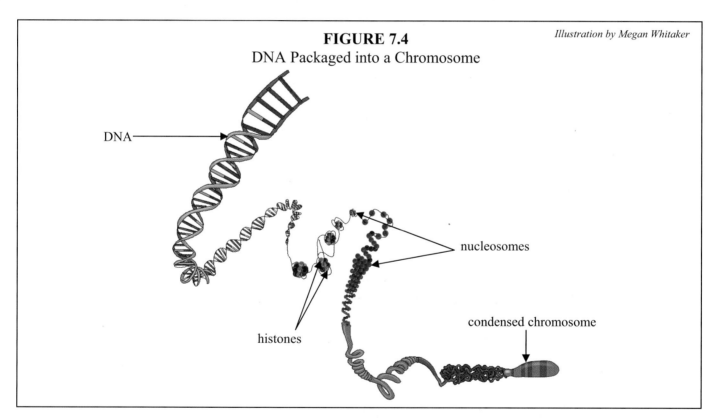

FIGURE 7.4
Illustration by Megan Whitaker
DNA Packaged into a Chromosome

DNA

nucleosomes

condensed chromosome

histones

The DNA of a eukaryotic cell is tightly bound together with a network of proteins. Certain proteins called **histones** act as spools which wind up small stretches of DNA. The DNA wrapped around these histones form what could be described as "beads on a string," which we call **nucleosomes** (noo' klee uh sohmz). Other proteins (including other histones) stabilize and support these spools, making a complex network of DNA coils and proteins. This network is called a **chromosome** (kroh' muh sohm).

Chromosome – DNA coiled around and supported by proteins, found in the nucleus of the cell

Most of the time, these chromosomes appear as long, fuzzy strands in the nucleus. As you learned in Module #6, we call this structure chromatin. When a cell is ready to reproduce, however, the coils form an even more compact version of the chromosome, which looks something like a dumbbell. This is often called a **condensed chromosome**, as it represents the smallest "package" in which nuclear DNA can be put.

Now as you might expect, it takes a lot of DNA to carry all of the genes necessary to code for the proteins of a living organism. Thus, there are several different chromosomes in the nucleus of the cell. What you might not expect, however, is that the number of chromosomes is dependent on the organism studied. For example, humans have 46 chromosomes in the nuclei of their cells, while a cat has only 38. Don't get the idea that more complex organisms have more chromosomes than less complex organisms, however. A horse has 64 chromosomes in the nuclei of its cells, and a crayfish (which you might call a crawdad) has 200!

As we have already learned, there are two types of reproduction in God's creation: sexual and asexual. You have already studied the asexual reproduction of prokaryotic cells, but now we want to look at the asexual reproduction of eukaryotic cells. We call this process **mitosis** (mye toh' sis).

<u>Mitosis</u> – A process of asexual reproduction in eukaryotic cells

Now remember, asexual reproduction results in offspring that have exactly the same genetic code as that of the parent. Since this is the case, we often call mitosis "cellular division," because, in the end, the process makes two duplicate cells from a single cell.

When a cell is busy with its normal life functions, it is said to be in a state of **interphase**.

<u>Interphase</u> – The time interval between cellular reproduction

During this time, the cell is simply maintaining the status quo. The chromosomes are in the nucleus of the cell, but not in their condensed form. From time to time, sections of the DNA on the chromosomes unwind to begin the process of protein synthesis. Otherwise, the DNA remains in its chromatin form. The chromosomes are not as tightly packed as they could be, but they are still pretty tightly packed. The combination of interphase and mitosis is often called the **cell cycle**, because the cell is either in one or the other. The cell spends most of its time in the interphase.

When it is time to for the cell to asexually reproduce, two things happen. First, the chromosomes must duplicate. After all, in order to make two cells, we need two sets of DNA. Thus, the chromosomes must make copies of themselves. Although this is a fascinating process, it goes beyond the scope of this course. If you are interested in learning about how this is done, visit the course website that we discussed in the "Student Notes" at the beginning of this text. It has links to websites that explain this fascinating process. After duplication, the chromosome and its duplicate are called **sister chromatids**. The second thing that happens to initiate mitosis is that the centrioles, which are outside the nucleus, duplicate themselves so that there are two sets of centrioles. Now that the DNA and the centrioles are duplicated, the cell is ready to begin mitosis. At this point, it is called a **mother cell**.

<u>Mother cell</u> – A cell ready to begin reproduction, containing duplicated DNA and centrioles

A mother cell is ready for **prophase**. In prophase, several things happen. The duplicated chromosomes coil into their condensed form, and they attach to each other at a point called the **centromere** (sen' truh meer).

<u>Centromere</u> – The region that joins two sister chromatids

As you can see in Figure 7.5, when the sister chromatids join together at the centromere, they form an "X" shape. This is the common representation of a chromosome. You have to understand, however, that chromosomes only look like this because they have been duplicated. The condensed form of an unduplicated chromosome is the dumbbell shape that we discussed earlier.

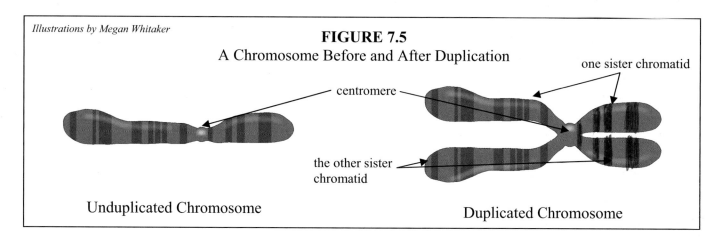

Illustrations by Megan Whitaker

FIGURE 7.5
A Chromosome Before and After Duplication

one sister chromatid

centromere

the other sister chromatid

Unduplicated Chromosome

Duplicated Chromosome

Also during prophase, each set of centrioles migrates towards one end of the cell. Microtubules extend from the centrioles in all directions, forming the **aster**. As the centrioles migrate, microtubules from the aster grow, forming **spindle fibers**, which make up the **mitotic spindle**. As the mitotic spindle is forming and the centrioles are moving, the nuclear membrane breaks apart, exposing the chromosomes to the rest of the cell. Microtubules from each centriole attach themselves to the chromosomes and move them to the center of the nucleus. They line up the chromosomes along an imaginary line called the **equatorial plane**, which runs down the center of the cell, equidistant from the two centrioles.

The brief time in which the chromosomes are lined up along the equatorial plane is called **metaphase**. At the conclusion of metaphase, the microtubules on the spindle pull the chromosomes towards the centrioles. Since each of the sister chromatids has a microtubule pulling on it from opposite centrioles, the sister chromatids separate from one another. One gets pulled to one side of the cell, and the other gets pulled to the opposite side of the cell. This phase of mitosis is called **anaphase**.

When the chromosomes are pulled far from one another, the last phase of mitosis, **telophase**, begins. During this time, the spindle begins to disintegrate, and the plasma membrane begins to constrict along the equatorial plane, eventually forming two cells where there was once only one. A nuclear membrane begins to form around each group of chromosomes, and the chromosomes uncoil from their condensed form, forming the chromatin material of the nucleus. At this point, there are two **daughter cells** where there was just one mother cell before. Please note that this phase is slightly different in plants, as we will see in a moment.

The process of mitosis is rather long and complicated, so it certainly deserves some review. Examine the figure below, which summarizes the process of mitosis.

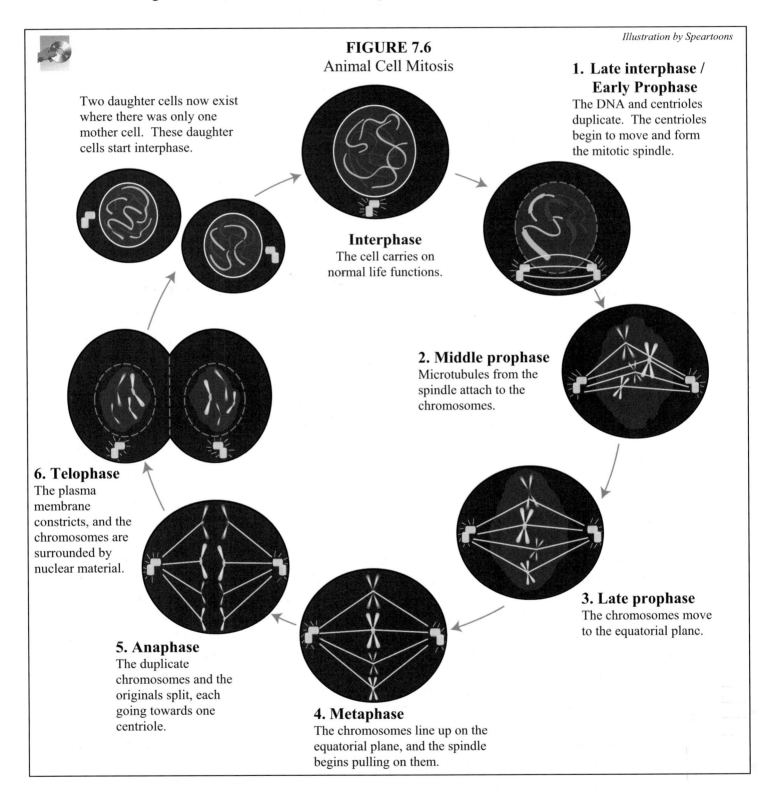

FIGURE 7.6
Animal Cell Mitosis

Illustration by Speartoons

Two daughter cells now exist where there was only one mother cell. These daughter cells start interphase.

Interphase
The cell carries on normal life functions.

1. Late interphase / Early Prophase
The DNA and centrioles duplicate. The centrioles begin to move and form the mitotic spindle.

2. Middle prophase
Microtubules from the spindle attach to the chromosomes.

3. Late prophase
The chromosomes move to the equatorial plane.

4. Metaphase
The chromosomes line up on the equatorial plane, and the spindle begins pulling on them.

5. Anaphase
The duplicate chromosomes and the originals split, each going towards one centriole.

6. Telophase
The plasma membrane constricts, and the chromosomes are surrounded by nuclear material.

Now you might wonder what happened to all of the other organelles in the cell during mitosis. Well, they are all still present while mitosis is occurring, we simply chose to ignore them in our discussion, because we wanted to concentrate on just those organelles that are important to mitosis. It turns out that the division of organelles between the two daughter cells is regulated during the process of mitosis. Each daughter gets at least one of each organelle. If there was only one of a given organelle prior to mitosis, it is divided between the two daughters.

But wait, how does each daughter get along with only half of the organelles or with only *part* of an organelle? Well, remember that the DNA in each of the cells has all of the information necessary to construct any organelles that the cell needs. Thus, the cells can build more of each organelle, and they can also complete any partial organelles that they have. In fact, since mitochondria have their own DNA, they can actually replicate themselves. As a result, if there is only one mitochondrion in a cell during interphase, it can actually replicate during mitosis so that each daughter gets one.

Figure 7.6 might have been easy to follow since it was an idealized, schematic drawing. In real life, of course, things are a little more difficult to see. Examine the figure below to see what mitosis looks like in animal cells when they are viewed under a microscope.

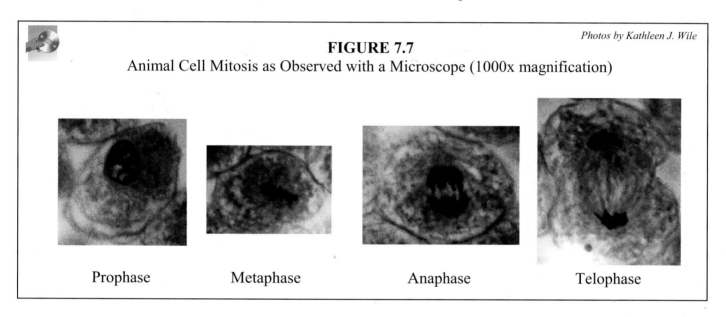

Photos by Kathleen J. Wile

FIGURE 7.7
Animal Cell Mitosis as Observed with a Microscope (1000x magnification)

Prophase   Metaphase   Anaphase   Telophase

When you think of asexual reproduction, you probably think of organisms like bacteria, protozoa, and perhaps fungi. Most of the organisms with which you are familiar (dogs, cats, humans, etc.), however, reproduce sexually. This means mitosis is irrelevant in the study of these organisms, right? Wrong! It turns out that nearly every eukaryotic organism uses mitosis. After all, when you are born, you are tiny. As time goes on, you grow. How does this happen? Well, your body produces more cells. How are those cells produced? By mitosis! Also, even after you are fully grown, your cells must continually go through mitosis in order to replace cells that have died. For example, millions of red blood cells die every minute in your body. If they weren't replaced, you would soon run out of them. How are they replaced? By mitosis, of course.

So you see that mitosis is relevant to nearly every creature in God's creation. Now the mitosis you just learned about occurs in animals. A slightly different kind of mitosis occurs in plant cells. The major difference between plant and animal mitosis is found in telophase. Because of the rigid cell wall in a plant cell, the plasma membrane cannot just constrict as we learned in animal cell mitosis. Instead, once the chromosomes have been pulled to the centrioles, vesicles containing the building blocks for the plasma membrane line up along the equatorial plane. These vesicles begin fusing together, forming a new plasma membrane. As the new plasma membrane forms, cellulose is formed in the middle, forming a new cell wall that splits the plasma membrane in two. Eventually, even a middle lamella is formed, resulting in two cell walls, two plasma membranes, and thus, two cells.

If you have a microscope, you can get more experience identifying the stages of mitosis by performing the following experiment. If you do not have a microscope, you can skip the experiment.

EXPERIMENT 7.2
Mitosis

Supplies:
♦ Microscope
♦ Prepared slide of *Allium* (onion) root tip
♦ Prepared slide of *Ascaris* Mitosis

Object: To observe how a cell divides during mitosis as well as the difference between plant and animal mitosis

Procedure:

A. **Animal Cell Mitosis:**

1. Observe the *Ascaris* Mitosis slide under low power. Look for cells that have stained (dark) objects in the cell. Those dark objects are the condensed chromosomes. Since condensed chromosomes exist only during reproduction, you know that a cell with visible chromosomes is in a stage of mitosis.
2. As you find a cell with visible chromosomes, increase the magnification so that you can see the pattern that the chromosomes form in the cell.
3. Using Figure 7.7 as a guide, determine which stage of mitosis the cell was in when it was stained.
4. As you scan the slide, you should find examples of prophase, metaphase, anaphase, and telophase. Draw an example of each.

B. **Plant Cell Mitosis:**

1. Observe the prepared slide of the onion root tip. Once again, begin under low power.
2. Most of the mitosis in an onion root occurs near the tip. Scan the slide until you can find the tapered end of the onion root. The cells of interest will be above that tapered end.
3. Look for cells with visible chromosomes, indicating that they are in one of the stages of mitosis.
4. When you find such a cell, increase the magnification so that you can see the pattern that the chromosomes form in the cell, and once again, determine the stage of mitosis.
5. Look for each of the stages of mitosis as you did with the *Ascaris* slide, and draw an example of each in your notebook.
6. Note that telophase is different in the onion root tip compared to what you saw in the *Ascaris* slide. You will distinguish telophase in this slide by looking for two nuclei that have split apart but have little or no cell wall between them.
7. Clean up and return all equipment to the proper place.

ON YOUR OWN

7.6 The phases of a cell's life are listed below. Which one is not a part of mitosis? Take the remaining phases and order them according to when they occur in the mitosis process.

anaphase, prophase, interphase, telophase, metaphase

7.7 A cell uses vesicles to build the plasma membrane during the telophase of mitosis. Is it a plant cell or an animal cell?

7.8 In which phase of mitosis are the chromosomes separated from their duplicates?

Diploid and Haploid Cells

As we mentioned before, humans have 46 chromosomes. What we failed to mention, however, is that these chromosomes come in pairs. Thus, in a human cell, there are 23 chromosome pairs, for a total of 46 chromosomes. The two members of a chromosome pair are very similar, but they are not identical. We will come back to this point in a little while. Because the members of a pair are similar but not identical, the pairs are called **homologous** (ho mahl' uh gus) **chromosome pairs**. The term "homologous" means similar but not identical. Each member of a homologous chromosome pair is called, reasonably enough, a **homologue** (hahm' uh lawg).

Now this can get a little confusing, so we will try to illustrate it with a figure. Suppose we were to take the chromosomes from a human cell during the metaphase of mitosis. Remember, at that point, the chromosomes are attached to their copies at their centromeres. This gives them an "X"-shape. If we were to look at the chromosomes at that point, we would get something that looks like the figure below.

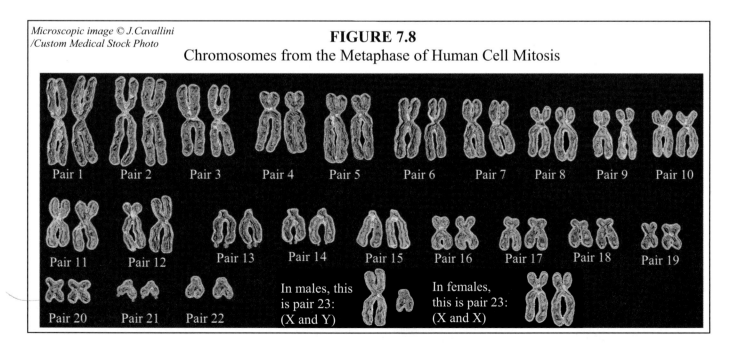

Microscopic image © J.Cavallini /Custom Medical Stock Photo

FIGURE 7.8
Chromosomes from the Metaphase of Human Cell Mitosis

Pair 1   Pair 2   Pair 3   Pair 4   Pair 5   Pair 6   Pair 7   Pair 8   Pair 9   Pair 10

Pair 11   Pair 12   Pair 13   Pair 14   Pair 15   Pair 16   Pair 17   Pair 18   Pair 19

Pair 20   Pair 21   Pair 22   In males, this is pair 23: (X and Y)   In females, this is pair 23: (X and X)

Notice that the chromosomes are grouped in pairs. The two chromosomes in each pair look incredibly similar, but this is not because the chromosomes have been duplicated for reproduction. The "X" shape of each chromosome is the result of the fact that the chromosome has been duplicated, because each chromosome is attached to its duplicate (its sister chromatid) at the centromere. The fact that the chromosomes come in pairs is simply a property of the nuclear DNA in a human cell. Thus, each cell has two of chromosome #1, two of chromosome #2, and so on. The only exception to this is found in males. As shown in the figure, if a cell has similar chromosomes in its 23rd pair, the cell must be from a female. If not, it must be from a male. Since this pair of chromosomes can be used to distinguish between the sexes, they are called **sex chromosomes**. By convention, the two sex chromosomes in a woman's cell are called **X chromosomes**, while the chromosome that distinguishes a male's cell is called the **Y chromosome**. As a result, we say that females have two X chromosomes, while males have an X and a Y chromosome. We will learn a great deal more about that in the next module.

As a point of terminology, a figure like Figure 7.8 is called a **karyotype** (kehr' ee uh type).

Karyotype – The figure produced when the chromosomes of a species during metaphase are arranged according to their homologous pairs

In this particular karyotype, you can see that there are 23 pairs of chromosomes (chromosomes 1-22 and a pair of sex chromosomes). With the exception of the XY pair in males, each homologue within a pair has exactly the same number of genes as its partner. These genes also code for the same trait. For example, you will learn in the next module that whether your blood type is A, B, AB, or O depends on a specific gene. That gene is located on the chromosomes of pair #9. Each homologue in the pair has a gene for blood type, and you can find that gene in exactly the same spot on each chromosome. Now remember, these chromosomes are not identical; they are similar. Thus, they each have a gene for blood type, and that gene is found at the same place on each chromosome. However, one gene might be for type A blood and the other might be for type O. In the next module, we will learn the implications of this fact.

When a cell's chromosomes come in pairs, we say that it is a **diploid** (dih' ployd) cell.

Diploid cell – A cell with chromosomes that come in homologous pairs

Many species have diploid cells, and this results in a great deal of diversity within the species. Cells with chromosomes that do not come in homologous pairs are called **haploid** (hap' loyd) cells.

Haploid cell – A cell that has only one representative of each chromosome pair

As we will see in the next section, even species that have diploid cells will have some haploid cells.

Before we leave this section, we need to make sure you understand what we mean when we talk about the number of chromosomes in a cell. When we say that a horse has 64 chromosomes, this means that it has 32 pairs of chromosomes. The total count is 64, but they come in 32 pairs. Since this terminology can be a little confusing, we sometimes make it a little clearer by referring to the **diploid number** (sometimes abbreviated as "2n") or the **haploid number** (sometimes abbreviated as "n").

Diploid number (2n) – The total number of chromosomes in a diploid cell

Haploid number (n) – The number of homologous pairs in a diploid cell

Based on these definitions, we could say that the diploid number (2n) of a human cell is 46, while the haploid number (n) is 23. Now remember, a haploid cell has only one chromosome from each pair. Thus, the haploid number is also the number of chromosomes in a haploid cell.

ON YOUR OWN

7.9 A pea plant has seven pairs of homologous chromosomes. What is its haploid number? What is its diploid number?

7.10 In a scientist's notebook, you find notes regarding a new species that is being studied. The notes say that the species is diploid, with a chromosome number of 17. Is this the haploid or diploid number? If this is the haploid number, give the diploid number. If this is the diploid number, give the corresponding haploid number.

Meiosis: The Cellular Basis of Sexual Reproduction

Now that we have spent some time discussing asexual reproduction and chromosome number, it is time to move on to sexual reproduction. There are many facets of sexual reproduction, but we will concentrate on the cellular level, because it forms the basis of what happens during all of the other stages of sexual reproduction. Sexual reproduction begins with **meiosis** (my oh' sis).

Meiosis – The process by which a diploid (2n) cell forms gametes (n)

Now this definition might seem a little mystifying, especially since you do not know what a gamete is. That's okay. This definition will become clear to you as you continue to read.

You learned way back in Module #1 that in sexual reproduction, the offspring are not identical to the parents. Instead, the offspring might have some characteristics in common with the parents but other characteristics that are quite different from the parents. This happens because in sexual reproduction, each parent contributes DNA to the offspring. As a result, the offspring has DNA from both parents.

If a human has 46 chromosomes and each parent contributes to the DNA of the offspring, then it should make sense that each parent can contribute only 23 chromosomes. If the parents each contributed more than 23 chromosomes, the offspring would have more than 46 chromosomes. For reproduction, then, the DNA of the parents must get "whittled down" to 23 chromosomes. How does this happen? That's what the process of meiosis is all about. In meiosis, diploid cells get split into haploid cells called **gametes** (gam' eats).

Gametes – Haploid cells (n) produced by diploid cells (2n) for the purpose of sexual reproduction

In animals, the gamete produced by the mother is called an **egg** cell, or ovum, while the gamete produced by the father is called the **sperm** cell.

A human gamete produced by meiosis, then, will have only one of each chromosome pair. This makes 23 chromosomes. When a gamete from one parent joins up with a gamete from the other parent, the two sets of 23 chromosomes merge to form a diploid cell that has 23 *homologous pairs* of chromosomes, for a total of 46. One member of each homologous pair will come from the father, and the other member of each homologous pair will come from the mother. When gametes join together, the resulting diploid cell is called a **zygote**. We actually presented the definition of a zygote back in Module #4, but now at least you know how a zygote forms.

So, before sexual reproduction can begin, diploid cells must form gametes. This is accomplished through the process of meiosis, which is actually similar to mitosis in several ways. In fact, the phases of meiosis have the same names as the phases of mitosis. The details of these phases, however, are a bit different. In addition, meiosis involves a few more steps. In fact, the process of meiosis is split into two groups of phases, **meiosis I** and **meiosis II**. Each of these parts of meiosis involves prophase, metaphase, anaphase, and telophase.

In order to begin meiosis, a cell must duplicate its DNA, just as it must for mitosis. The centrioles must duplicate as well. At that point, the cell is ready for the first stage of meiosis, **prophase I**. During this phase, the centrioles move to opposite sides of the cell, just as in mitosis. Also, the mitotic spindle begins to form, just as is the case in mitosis.

By the time the cell has reached the next phase, **metaphase I**, the spindle has formed, and the microtubules are attached to the centromeres so that there is a single microtubule for each set of sister chromatids. This is different from what happens in the metaphase of mitosis, because in mitosis, one microtubule attaches to a chromosome and another microtubule from the other centriole attaches to the chromosome's duplicate. Thus, in mitosis, the microtubules attach so as to pull the chromosomes away from their duplicates. In meiosis, however, the microtubules attach so as to pull the homologous pairs apart while leaving the chromosomes attached to their duplicates.

In **anaphase I**, then, the homologous pairs are separated, being pulled towards the centrioles on the opposite sides of the cell. Now think about how this is different from mitosis. In the anaphase of mitosis, each chromosome is separated from its duplicate. Thus, the "X"-shape goes away, because the chromosome goes one way and its duplicate goes the other. Both chromosome homologues go one way while both duplicate homologues go the other. In anaphase I of meiosis, however, the chromosomes stay with their duplicates. Instead, the homologues are separated from one another, breaking up the homologous pairs.

In **telophase I**, the plasma membrane constricts along the equatorial plane, forming two cells. These cells are haploid, because the homologues have been separated. However, each chromosome still has its duplicate attached to it at the centromere. Thus, each of these cells needs to split those duplicates. Well, if you think about it, these haploid cells are ready for mitosis, because they have their chromosomes duplicated with the duplicates attached to the originals. Thus, each of the two haploid cells goes through a process very similar to mitosis, which is called meiosis II.

In meiosis II, both cells enter **prophase II** by having the centrioles duplicate and once again travel to opposite ends of the cell, forming a spindle. In **metaphase II**, the spindle is fully formed, the chromosomes are lined up along the equatorial plane, and the microtubules attach to each chromosome and its duplicate at the centromere. In **anaphase II**, the microtubules of the spindle pull the chromosomes away from their duplicates, with the originals being moved towards one centriole and the duplicates moving towards another. Finally, in **telophase II**, the plasma membrane in each cell constricts along the equatorial plane, forming two pairs of cells where there were originally just two cells. The four cells formed, then, are still haploid, because no homologous pairs are present. In addition, there are no chromosome duplicates in any of the cells. The process of meiosis is summarized in the figure below.

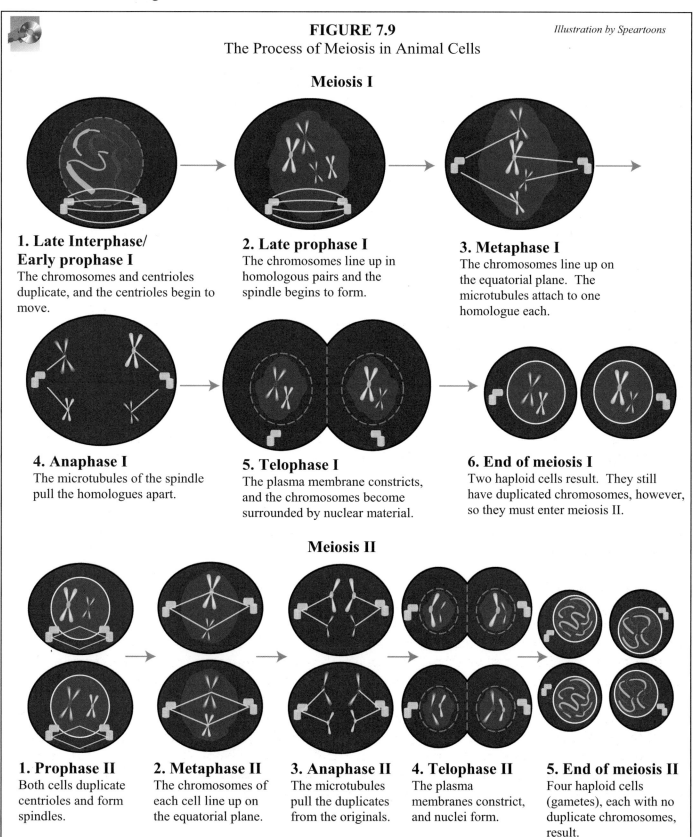

FIGURE 7.9
The Process of Meiosis in Animal Cells

Illustration by Speartoons

Meiosis I

1. Late Interphase/ Early prophase I
The chromosomes and centrioles duplicate, and the centrioles begin to move.

2. Late prophase I
The chromosomes line up in homologous pairs and the spindle begins to form.

3. Metaphase I
The chromosomes line up on the equatorial plane. The microtubules attach to one homologue each.

4. Anaphase I
The microtubules of the spindle pull the homologues apart.

5. Telophase I
The plasma membrane constricts, and the chromosomes become surrounded by nuclear material.

6. End of meiosis I
Two haploid cells result. They still have duplicated chromosomes, however, so they must enter meiosis II.

Meiosis II

1. Prophase II
Both cells duplicate centrioles and form spindles.

2. Metaphase II
The chromosomes of each cell line up on the equatorial plane.

3. Anaphase II
The microtubules pull the duplicates from the originals.

4. Telophase II
The plasma membranes constrict, and nuclei form.

5. End of meiosis II
Four haploid cells (gametes), each with no duplicate chromosomes, result.

So if we compare mitosis and meiosis, we can see several differences. Mitosis takes one cell and makes an exact duplicate, resulting in two cells. If the original cell is diploid, the copy will be diploid as well. In contrast, meiosis takes a single cell and produces four cells. The cells produced are quite different from the original. The original is diploid, but the four cells produced are haploid. These haploid cells are called gametes. In addition, meiosis has twice as many steps as mitosis. Finally, while the anaphase of mitosis keeps the homologous pairs of chromosomes together and separates the chromosomes from their duplicates, anaphase I of meiosis separates the homologous pairs, keeping the chromosomes and their duplicates together.

While studying cellular reproduction, there is one very important thing to remember: *two sister chromatids that form the "X" shape during mitosis and meiosis count as only one chromosome*. Thus, if you look at Figure 7.9, the two cells at the end of meiosis I have two chromosomes each. Both of those chromosomes are duplicated, but the chromosome and its duplicate count as only *one* chromosome, because they each carry *exactly* the same information. Since the duplicate adds no new information to the nuclear DNA, it does not count as a second chromosome.

Before we leave this section on meiosis, it is important for you to understand that although meiosis occurs in both the female and the male of an animal species, there are differences between the two. As we mentioned before, in males, the gametes produced by meiosis are called sperm, while in females, the gametes are called eggs. It turns out that these gametes are quite different. As a result, the process of meiosis II is different in males and females.

In male animals, once telophase II ends, the gametes produced grow flagella. This is accomplished by a centriole, which moves to the plasma membrane and grows a microtubule through the membrane and out into the surroundings. The sperm can then use these flagella to move about in search of an egg. A schematic of male meiosis is shown below.

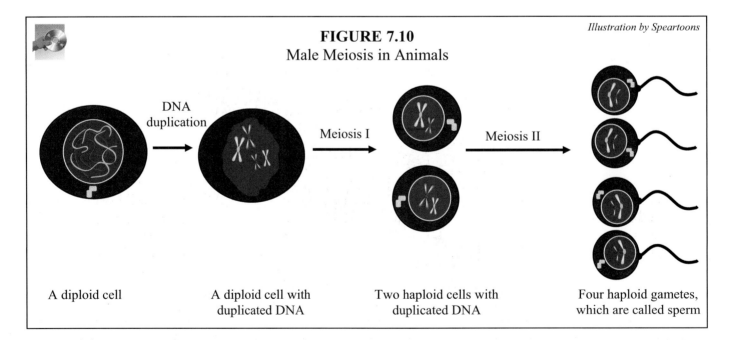

FIGURE 7.10
Male Meiosis in Animals

Illustration by Speartoons

DNA duplication

Meiosis I

Meiosis II

| A diploid cell | A diploid cell with duplicated DNA | Two haploid cells with duplicated DNA | Four haploid gametes, which are called sperm |

In female animals, something quite different happens. At the end of telophase I, one of the two cells that are produced takes most of the cytoplasm as well as most of the organelles. Thus, this cell is much bigger than the other. Both cells go through meiosis II. However, since the small cell is so small, the two gametes it produces are quite small. In addition, when the big cell reaches telophase II, one of the two gametes once again takes most of the cytoplasm and organelles. As a result, meiosis II ends up producing three tiny gametes and one large gamete. The three tiny gametes, often called **polar bodies**, are useless. If they are fertilized, the resulting zygote will quickly degenerate and die. Only the large gamete, called the egg, can produce a viable zygote through fertilization.

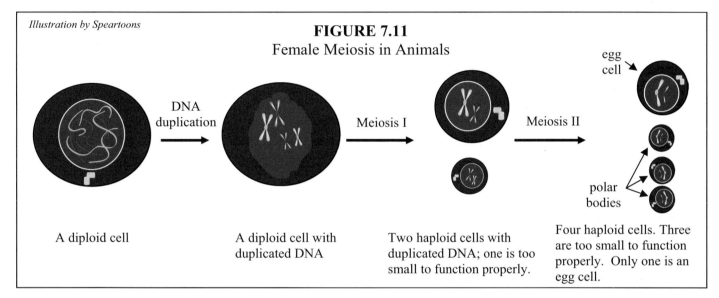

Illustration by Speartoons

FIGURE 7.11
Female Meiosis in Animals

A diploid cell

A diploid cell with duplicated DNA

Two haploid cells with duplicated DNA; one is too small to function properly.

Four haploid cells. Three are too small to function properly. Only one is an egg cell.

Since the functional egg produced by female meiosis has been formed by taking most of the cytoplasm and organelles during both meiosis I and meiosis II, it should make sense to you that the egg cell in a female is much bigger than the sperm cell in a male. In addition, since the sperm cells have a means of locomotion and the egg cell does not, the egg cell must sit still and wait for the sperm cell to travel to it. These facts are illustrated in the figure below.

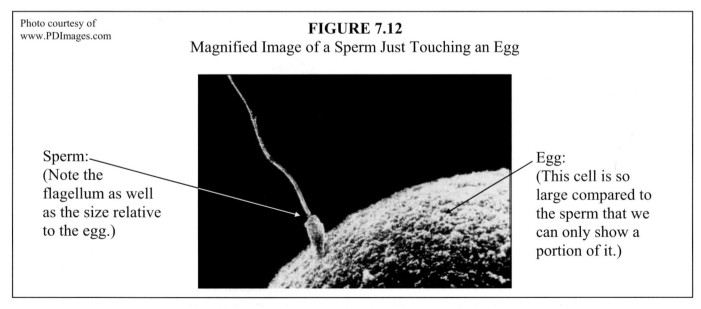

Photo courtesy of
www.PDImages.com

FIGURE 7.12
Magnified Image of a Sperm Just Touching an Egg

Sperm:
(Note the flagellum as well as the size relative to the egg.)

Egg:
(This cell is so large compared to the sperm that we can only show a portion of it.)

Once the egg and sperm meet, the sperm burrows into the egg, and the haploid cells fuse to form a diploid cell. This diploid cell, called the zygote, can then begin to develop and grow. To do this, the zygote must begin producing many cells, which it does by mitosis.

ON YOUR OWN

7.11 Which phases of meiosis are essentially the same as the corresponding phases of mitosis? Which are different?

7.12 A cellular reproduction process results in four diploid cells. Is this mitosis or meiosis? How many cells underwent this process?

7.13 A cellular reproduction process results in four haploid cells. Is this mitosis or meiosis? How many cells underwent this process?

7.14 A sperm cell finds a polar body and attempts to fuse with it. Will a viable zygote develop?

Viruses

In Modules #2 and #3, we spent considerable time discussing bacteria and protozoa that cause disease. Despite all of that discussion, however, we have not mentioned another common agent of disease: the **virus**. We decided to put off such a discussion until now. This is because viruses cause disease by overriding certain cellular processes, such as reproduction. As a result, it seems only natural that a discussion of viruses should occur in the module that discusses cellular reproduction.

The definition of a virus is rather specific.

> Virus – A non-cellular infectious agent that has two characteristics:
> (1) It has genetic material (RNA or DNA) inside a protective protein coat.
> (2) It cannot reproduce on its own.

Based on this definition, then, a virus is not alive. Since it cannot reproduce itself, it fails to meet one of life's basic criteria.

Well, if a virus is not alive, what is it? The best way to answer this question is to show you a couple of viruses.

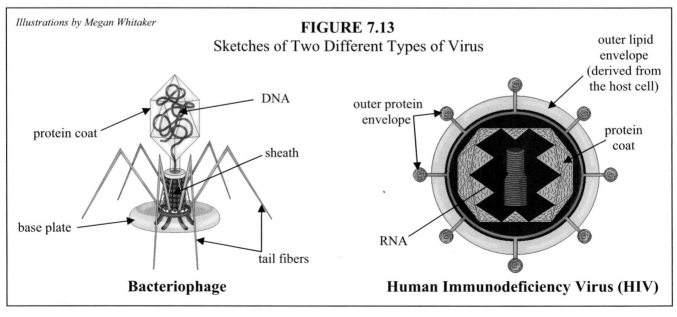

Illustrations by Megan Whitaker

FIGURE 7.13
Sketches of Two Different Types of Virus

Bacteriophage: DNA, protein coat, sheath, base plate, tail fibers

Human Immunodeficiency Virus (HIV): outer lipid envelope (derived from the host cell), outer protein envelope, protein coat, RNA

Bacteriophage   **Human Immunodeficiency Virus (HIV)**

Now don't let appearances deceive you here. Even though the bacteriophage looks like it has legs (the tail fibers), a body (the sheath), and a head (the protein coat), it is not a living organism. It is a highly-organized mixture of chemicals (mostly proteins and nucleotides) that can perform some pretty specific tasks, nothing more. Notice from the figure that while the genetic material in the bacteriophage is DNA, the genetic material in the HIV is RNA. This is the case with viruses. Some have RNA as their genetic material, and others have DNA.

The main thing you need to know about viruses is the way in which they infect their hosts. Since viruses cannot reproduce themselves, they rely on cells to do it for them. In order to do this, a virus will attach itself to a cell. The virus either enters the cell or injects its genetic material into the cell. The genetic material of the virus redirects the cell's reproductive machinery to reproduce the DNA or RNA of the virus, as well as the proteins that make up the virus. The cell's biosynthetic machinery is then directed to assemble these pieces into new viruses. This continues until there are so many viruses that the cell ruptures, destroying the cell and releasing new viruses to infect other cells. This process is called the **lytic** (lih' tik) **pathway**.

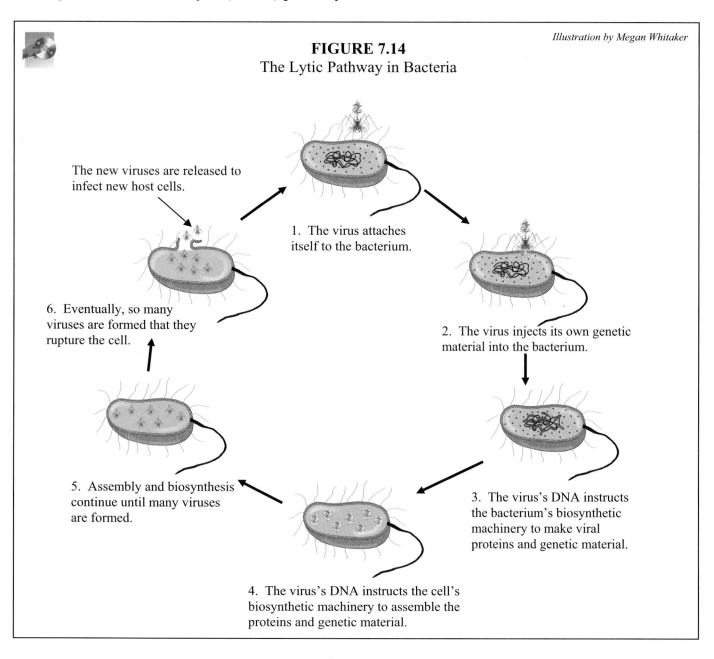

FIGURE 7.14
The Lytic Pathway in Bacteria

Illustration by Megan Whitaker

The new viruses are released to infect new host cells.

1. The virus attaches itself to the bacterium.

2. The virus injects its own genetic material into the bacterium.

3. The virus's DNA instructs the bacterium's biosynthetic machinery to make viral proteins and genetic material.

4. The virus's DNA instructs the cell's biosynthetic machinery to assemble the proteins and genetic material.

5. Assembly and biosynthesis continue until many viruses are formed.

6. Eventually, so many viruses are formed that they rupture the cell.

Although Figure 7.14 is drawn for the specific case of a bacteriophage infecting a bacterium, all viruses reproduce and infect via the lytic pathway. Some viruses, however, can inject their genetic material into a cell so that it lies dormant for as long as several years before beginning the lytic pathway. Thus, the time between a host receiving the virus and manifesting the symptoms of the malady caused by that virus can be several years. This is the case with HIV, the virus that causes AIDS (acquired immune deficiency syndrome).

Viruses cause many diseases and afflictions. Warts, chicken pox, the common cold, influenza, some forms of cancer, mumps, measles, AIDS, and many other diseases are caused by viruses. Of course, viruses affect other organisms as well as humans. Plants can be killed by viral infections. As illustrated in Figure 7.14, even bacteria are subject to viruses!

In most organisms, there are infection-fighting agents that can destroy viruses and other pathogens. People, for example, have a host of infection-fighting mechanisms in their bodies. In Module #6, you learned that the process of a cell engulfing a chemical or another cell is called phagocytosis. Well, there are certain cells that use this process to destroy pathogens, including viruses. These cells are called, reasonably enough, **phagocytic cells**.

Some white blood cells, for example, are phagocytic cells. They often circulate in the bloodstream, but when an infection strikes, they can leave the bloodstream and go to the point of infection in order to engulf the pathogen. Some phagocytic cells do not need to move, however. They reside in the **lymph** (limf) **nodes** of your body. Special vessels called lymph vessels carry fluids through the lymph nodes, and pathogens that are in that fluid get engulfed by the phagocytic cells that are there.

People have several defenses against pathogens. One very interesting defense is the ability to produce **antibodies** that ward off pathogens.

Antibodies – Specialized proteins that aid in destroying infectious agents

While some antibodies can help destroy a host of different pathogens, other antibodies are highly specific and can only aid in the destruction of one type of pathogen. When your body is infected by a pathogen, specialized cells work to produce antibodies that will help to destroy that pathogen. Once the cells are successful at producing such an antibody, other cells are produced that actually remember how to produce that antibody. If you happen to get infected by the same pathogen again (or one very similar to it), those cells will immediately help to produce the antibody that they know is successful against that pathogen. This increases the speed at which your body can fight off the pathogen, making you less likely to be overwhelmed by the pathogen. This makes it less likely that you will get sick from the infection.

This is the principle behind **vaccines**, the most common way that medical science can fight viruses.

Vaccine – A weakened or inactive version of a pathogen that stimulates the body's production of antibodies which can aid in destroying the pathogen

When a person is infected by many viruses, the only way to rid the person of the virus is for the body to make a specific antibody that will aid in destroying it. The body can produce antibodies against many, many different viruses. However, some viruses are so fast-acting that the body cannot produce antibodies before many viruses have completed the lytic pathway. By that time, the body is overwhelmed by the virus. As a result, the disease kills or permanently injures the person. When you are given a vaccine, a weakened or inactive version of the virus is injected into your body. This virus has been changed so that it cannot enter the lytic pathway very effectively. As a result, your body has plenty of time to figure out what antibodies to make in order to destroy it. Once your body produces the antibodies, it produces cells that remember how to do it. That way, when you are exposed to the real virus of the same type, those cells help the body produce the antibodies specific to the virus right away. This reduces the number of viruses that can enter the lytic pathway, making you less likely to get the disease that the virus causes.

So, in essence, a vaccine works by giving you a virus! Of course, since this altered form of the virus cannot enter the lytic pathway very easily, it is not as harmful as the real virus. However, your body doesn't know that, so it manufactures the antibodies to destroy the virus, thus protecting you if you later come into contact with the real thing. This is why vaccines must be given *before* you get infected by the virus. Once you get infected, your body begins producing antibodies. At that point, then, a vaccine will do you no good! Thus, vaccines act like a wall of defense. If you do not build the wall before the attackers come, you will be too late. One thing to note is that some vaccines don't even contain a virus at all. For some viruses, medical scientists have been able to construct a chemical "mimic" that the body thinks is a virus. This kind of vaccine is ideal, of course, because it doesn't contain a pathogen at all.

There is a movement afoot these days that says vaccines are bad for you. Some members of this movement say that vaccines aren't even effective at preventing disease. Others say that vaccines are effective at preventing disease but they have side effects that are not worth the protection that the vaccines give you. However, both of these ideas are just plain wrong.

There are many ways to demonstrate that vaccines are very effective at preventing disease. Consider, for example, the two graphs below. They show the rates of polio and measles versus the year for which the rates were measured.

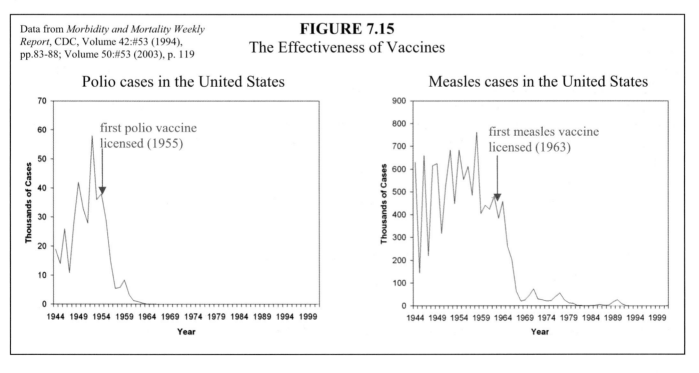

Data from *Morbidity and Mortality Weekly Report*, CDC, Volume 42:#53 (1994), pp.83-88; Volume 50:#53 (2003), p. 119

FIGURE 7.15
The Effectiveness of Vaccines

Notice that in the case of polio, the disease rate rose in a shaky but steady fashion from 1944 to 1952. Then, there was a slight (34%) decrease in the disease rate from 1953 to 1955. However, from 1955 to 1957, there was a *dramatic* decrease (80%) in the disease rate. What explains these drops in disease rate? Well, notice that the first polio vaccine was licensed in 1955. The dramatic decrease in disease rate, then, came right after the polio vaccine was used on the general public. What about the smaller decrease from 1953 to 1955? Well, the vaccine was developed in 1952 and tested shortly thereafter. For example, in 1954, it was tested in a double-blind study of 1.8 *million* children. The slight drop in disease rates prior to 1955, then, was most likely due to the testing of the polio vaccine.

Now look at the graph for measles. Once again, the story is similar. There is not nearly as much of a rise in the measles rate in the early years (1944-1958). However, once again, there is slight decrease in the disease rate just prior to the licensing of the vaccine (during the testing phase) and then a *dramatic* decrease in the disease rate after the vaccine began being used on the general public. These data dramatically illustrate the effectiveness of vaccines in preventing disease.

The safety of vaccines is also well established. Before a vaccine is approved for use in the United States, it must go through many tests on animals, and then it must go through three separate phases of testing on people. In each phase, the people that volunteer for the test are closely monitored for health problems. If the rate of health problems in the testing group is higher than that of those not getting the vaccine, the vaccine is not allowed to be used on the general public. By the end of this three-phase testing process, the vaccine has been tested on thousands of volunteers over a period of several years. If the vaccine were not safe for the vast majority of people, this three-phase testing system would demonstrate that, and the vaccine would not be approved for general use.

That's not the end of the story, however. Even after the vaccine is approved for general use, its effectiveness and safety are continually monitored by the Centers for Disease Control and Prevention (CDC). If a problem is found that is possibly associated with the vaccine, detailed studies are done to see if the vaccine does indeed cause the problem. If so, it is no longer allowed for general use. The stringent testing methods used on vaccines along with the constant monitoring done on them lead the American Academy of Pediatrics to state, "...vaccines are one of the safest forms of medicine ever developed" (Samuel L. Katz, representing the American Academy of Pediatrics in testimony before the Committee on Government Reform, U.S. House of Representatives, August 3, 1999).

Now if you think about it, the concept of a vaccine is, in fact, a testament to God's power and majesty. After thousands of years of medical science, people cannot fight viruses as efficiently as God's creation (the human body) does. Thus, to protect ourselves, we simply stimulate the body to do what God designed it to do in the first place. The vaccine gives the body a little push, and the body does the rest. If you would like to know more about vaccines, please visit the course website discussed in the "Student Notes" section at the beginning of this book. That website contains a link to a detailed discussion of the safety and efficacy of vaccines.

ON YOUR OWN

7.15 What is the principal difference between viruses and pathogenic bacteria?

7.16 The human body can produce the antibodies that destroy smallpox. If this is the case, why did so many people die from it? Why didn't their bodies just kill the virus?

ANSWERS TO THE "ON YOUR OWN" PROBLEMS

7.1 <u>No, they will not be identical in every way</u>. Because they have the same genes, they will have the same genetic tendencies and traits. Any characteristic that is determined completely by genetics will be the same in each twin. <u>However, any characteristics in which environmental and spiritual factors play a role (weight, personality, strength, etc.) will most likely be at least somewhat different in each, since they grew up in different environments.</u>

7.2 Guanine can bond to cytosine and vice-versa. Thus, when the RNA has guanine, the DNA must have had cytosine. When the RNA has cytosine, the DNA must have had guanine. Adenine bonds with uracil and vice-versa. Thus, when you see a uracil in RNA, there must have been an adenine in the DNA. Now, when you see an adenine in RNA, you might think that there was a uracil in the DNA. However, DNA has no uracil. It has thymine instead. Thus, when you see an adenine in RNA, there was a thymine in DNA. Therefore, the DNA sequence must have been:

<u>adenine, thymine, thymine, cytosine, guanine, guanine</u>

7.3 The same rules apply here, realizing that RNA has uracil. Thus, when DNA has cytosine, RNA will have guanine. When DNA has guanine, RNA will have cytosine. When DNA has thymine, RNA will have adenine. When DNA has adenine, RNA will have uracil.

<u>adenine, adenine, adenine, uracil, cytosine, guanine</u>

7.4 The scientist must be studying <u>RNA</u>, because only RNA has uracil.

7.5 a. Now remember, if this is a codon, then it must be on the mRNA. Thus, the sequence given is the sequence made by mRNA during transcription. The DNA, then, must have the complement of this sequence. We can therefore use the same reasoning that we did in question 7.2. Thus, the DNA must have had:

<u>cytosine, guanine, thymine</u>

b. The tRNA must also have the complement to this sequence. Unlike DNA, however, RNA uses uracil rather than thymine. Thus, the tRNA must have:

<u>cytosine, guanine, uracil</u>

7.6 <u>Interphase is not a part of mitosis</u>. This phase of a cell's life is defined as the time between reproduction cycles. The remaining phases occur in the following order: <u>prophase, metaphase, anaphase, telophase</u>.

7.7 <u>It must be a plant cell</u>. In animal cells, the two new cells are created when the plasma membrane constricts. This can't happen in a plant cell because of the rigid cell wall. Thus, plant cells use vesicles to carry the components of the plasma membrane and cell wall between the two nuclei and build them there.

7.8 <u>Chromosomes are separated from their duplicates during anaphase</u>.

7.9 The haploid number is the number of homologous pairs. <u>Thus, the haploid number is seven</u>. The diploid number is the total number of chromosomes in the nucleus. Since there are seven pairs and each pair has two, there are a total of 14 chromosomes in the nucleus. <u>The diploid number, therefore, is 14</u>.

7.10 Since the number is odd, it cannot be a diploid number. Remember, diploid cells have chromosomes that come in pairs. Since there are two chromosomes in a pair, then the total number of chromosomes (the diploid number) will always be even. By virtue of the fact that this is an odd number, then, we know that <u>it is a haploid number</u>. Since the haploid number is 17, the diploid number will be twice that. Thus, <u>the diploid number is 34</u>.

7.11 Since meiosis II is essentially mitosis that occurs on two cells, the stages of meiosis II are all essentially the same as mitosis. Thus, <u>prophase II, metaphase II, anaphase II, and telophase II are essentially the same as the corresponding stages of mitosis. In addition, prophase I is essentially the same as the prophase of mitosis. Metaphase I is different from the metaphase of mitosis</u>, because the microtubules attach so as to separate homologues in metaphase II. In the metaphase of mitosis, the microtubules attach so as to pull the chromosome duplicates apart from their originals. <u>Anaphase I is also different from the anaphase of mitosis</u>, because in anaphase I the homologues get pulled apart, whereas in the anaphase of mitosis, the chromosome duplicates get separated from their originals. Finally, <u>telophase I is different from the telophase of mitosis because the former results in haploid cells while the latter results in diploid cells</u>.

7.12 <u>This must be mitosis</u>. Meiosis always results in haploid cells. Since mitosis makes two cells for every one that goes through the process, <u>two cells must have gone through mitosis to produce four cells in the end</u>.

7.13 <u>This was meiosis</u>. Mitosis results in diploid cells. Since meiosis forms four cells for every one that undergoes the process, <u>one cell must have undergone meiosis to get four cells in the end</u>.

7.14 <u>A viable zygote will not develop</u>. Polar bodies are so small that they cannot function properly.

7.15 <u>A pathogenic bacterium is alive; a virus is not</u>.

7.16 <u>Even though the body can produce the antibodies, it cannot do so quickly enough to destroy the virus before many viruses reach the lytic pathway. As a result, the virus is reproduced so quickly that it overwhelms the body</u>. The smallpox vaccine defended so well against the virus, however, that the smallpox virus has been wiped out. It now only exists in laboratories and, as a result, this vaccine is no longer necessary!

STUDY GUIDE FOR MODULE #7

1. Define the following terms:

a. Genetics
b. Genetic factors
c. Environmental factors
d. Spiritual factors
e. Gene
f. Messenger RNA
g. Anticodon
h. Codon
i. Chromosome
j. Mitosis
k. Interphase
l. Centromere

m. Mother cell
n. Karyotype
o. Diploid cell
p. Haploid cell
q. Diploid number
r. Haploid number
s. Meiosis
t. Gametes
u. Virus
v. Antibodies
w. Vaccine

2. A DNA strand has the following sequence of nucleotides:

guanine, cytosine, adenine, adenine, thymine, guanine

a. What will the mRNA sequence be?

b. How many amino acids will the mRNA code for?

c. How many codons will the mRNA have?

d. What are the anticodons on the tRNAs that will bond to the mRNA?

3. Fill in the blanks:

a. _____   b. _____
DNA ⟶ RNA ⟶ Protein

4. An RNA strand has an anticodon. Is it tRNA or mRNA?

5. Protein synthesis is occurring in a ribosome. Is this a part of transcription or translation?

6. Suppose scientists determine that a set of genes is significantly more prevalent in murderers than in the population at large. Would that mean that murderers are not at fault for what they do? Why or why not?

7. If you look under a microscope and see distinct chromosomes in a cell, is the cell in interphase? Why or why not?

8. List (in order) the four stages of mitosis.

9. Identify the stage of mitosis for each of the following pictures:

a. 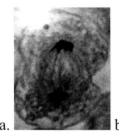   b. 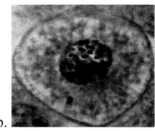   c. 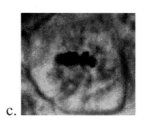   d.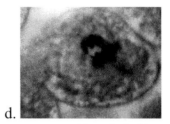

Photos by Kathleen J. Wile

10. The diploid number of a cell is 16. What is its haploid number?

11. The haploid number of a cell is 9. What is its diploid number?

12. What is the difference between a gamete and a regular animal cell?

13. List (in order) all of the stages of meiosis.

14. Which is closer to mitosis: meiosis I or meiosis II?

15. A single cell with seven pairs of homologous chromosomes goes through meiosis I. How many cells result at the end of meiosis I? How many chromosomes exist in each cell? Are the chromosomes in each cell duplicated or not?

16. Four cells that originally (prior to meiosis I) had seven pairs of homologous chromosomes go through meiosis II. How many cells result? How many (total) chromosomes exist in each cell? Are the chromosomes in each cell duplicated or not?

17. What are gametes produced in male animals called? What are gametes produced in female animals called?

18. How many useful gametes are produced in the meiosis of male animals? What about female animals?

19. What is the difference between a polar body and an egg?

20. Which gamete can move on its own: the male gamete or the female gamete?

21. What is the purpose of the lytic pathway?

22. If a virus uses DNA as its genetic material, is it alive? Why or why not?

23. A person decides to wait until he contracts measles before getting the vaccine. What is wrong with this strategy?

MODULE #8: Mendelian Genetics

Introduction

As we mentioned in the previous module, the means by which an offspring receives traits and characteristics from its parents has always intrigued scientists. Now that you have learned how the genetic material of a cell gets split into gametes for the purpose of sexual reproduction, you are ready to learn the mechanisms of this fascinating process. Before we get into the nuts and bolts of genetics, however, we need to spend some time discussing the founder of modern genetics: Gregor Mendel.

Gregor Mendel

Named Johann Mendel at birth, Gregor Mendel was born a peasant in Heinzendorf, Austria, in 1822. His father was a farmer who taught his son the techniques of animal breeding and plant grafting. These pursuits intrigued young Johann, and later in life, he spent a great deal of time studying them as no one had before. While still a child, his talent at learning impressed his school teacher so much that the teacher implored his parents to give him a higher education. At that time, a "higher education" really just meant our equivalent of high school. His parents agreed, but they were so poor that they could not easily afford an education. Johann struggled through his education, oftentimes nearly dying of starvation because he could not afford to eat. Eventually, his younger sister gave up her dowry so that he could finish his education.

Once he graduated, he spent some time trying to be a teacher. When he proved completely unsuccessful at that, he joined the Augustinian monastery of St. Thomas in Altbrünn. Back then, monasteries were centers for scholarly pursuits. This particular one had a great reputation for scientific achievement, so Mendel was happy to join. When a monk joins the Augustinian order, he must choose a new name. That was when Johann Mendel became Gregor Mendel.

While he was at the monastery, Gregor tried again to be a teacher. When he took the examinations to qualify, however, he failed. Nevertheless, one of his examiners, Andreas Baumgartner, was impressed by his originality of thought, and he pulled some strings to get Gregor admitted to the University of Vienna. While at the university, Mendel studied under Johann Christian Doppler, a physicist who became famous for his explanation of a physical effect now known as the "Doppler effect." When you take physics, you will learn all about that. Doppler taught Mendel the proper way to conduct experiments, and that's all that Mendel needed.

After finishing his studies at the university, he once again tried to take the examination to qualify for a teaching position. His examiners failed him a second time, however. This time, he was failed for showing *too much original thought*. Giving up the idea of being a teacher, he began his work as a scientist. For eight years, he conducted experiments on the subject that first intrigued him: breeding. He raised thousands of pea plants and carefully documented the results of breeding and crossbreeding them. You will learn much about these experiments in the next section. At the end of those eight years, he published a paper in which he presented a series of four conclusions that are the basis of what we call **Mendelian genetics**.

Unfortunately, Mendel's groundbreaking paper went largely unnoticed among the scientific community. In addition, Mendel had to give up his scientific endeavors shortly after writing the paper because he became embroiled in a political controversy. The government decided that it would tax the monasteries, and Mendel thought that this was an attack on religious freedom. As a deeply committed Christian, he bitterly opposed the government's idea and spent most of the rest of his life fighting against the taxation. Since he was so involved in this political struggle, he had no time for his scientific work. When he died in 1884, no one knew the significance of his experiments. Nevertheless, by the 1930s, his work was well known throughout the scientific community, and today he is known as the father of modern genetics.

Gregor Mendel's life story is inspiring on several levels. It shows what can happen when a person has a true desire to learn. Mendel was willing to sacrifice *eating* in order to pursue an education. As a result, he unlocked one of the deep mysteries of God's creation. Mendel's story also shows that when you fail, you should not give up. Mendel failed his teaching examination twice; nevertheless, he taught the worldwide scientific community about a process that has fascinated biologists for hundreds of years. Finally, his willingness to put all of that away in order to defend the faith against an attack from the government shows that Mendel had the proper set of priorities. He put his faith in Christ above everything!

Mendel's Experiments

During his eight years of scientific work, Mendel studied the breeding of pea plants. He noticed that pea plants had certain definable characteristics that seemed to change from plant to plant. For example, some pea plants were tall (about 6 feet high) and others were short (about 1½ feet high). In some pea plants, the flowers grew along the sides of the plant (these are called "axial flowers"), while in others, flowers grew on top of the plant (these are called "terminal flowers"). Pea pods produced by some pea plants were green, while those produced by others were yellow. In the same way, some pea plants produced yellow peas, and some produced green ones. In addition, some plants produced wrinkled peas, while others produced smooth peas.

Mendel noticed that some plants bred so as to produce offspring with the same characteristic. For example, some tall plants would always give rise to other tall plants. When this happens, we say that the tall plant has **bred true**.

True breeding – If an organism has a certain characteristic that is always passed on to its offspring, we say that this organism bred true with respect to that characteristic.

Mendel noticed that not all plants bred true. Some tall plants would produce a short plant every now and again. Mendel decided that the tendency for some plants to breed true while others do not was the key to understanding the mysteries of reproduction.

With this thought in mind, Mendel came up with an ingenious set of experiments. He took a tall pea plant that always bred true and allowed it to sexually reproduce with a short pea plant that also always bred true. No matter how many times he did this, the offspring of such a union were always tall plants. Now most people might think that the sexual reproduction of a tall plant and a short plant would result in a medium-sized plant. This, however, was not the case. Mendel noticed that with the other definable characteristics listed above, the outcome was similar. Figure 8.1 compiles the results of this portion of Mendel's work.

FIGURE 8.1

Illustrations by Megan Whitaker

Mendel's Experiments with Plants that Bred True

Plants Bred	Result
 Tall plants bred with short plants	 All offspring were tall plants.
 Axial-flowered plants bred with terminal-flowered plants	 All offspring had axial flowers.
 Green-pod plants bred with yellow-pod plants	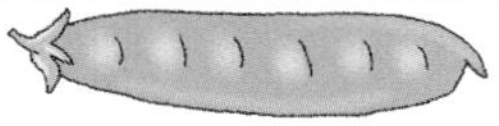 All offspring produced green pods.
 Yellow-pea plants bred with green-pea plants	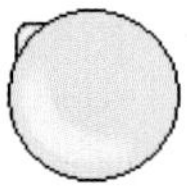 All offspring produced yellow peas.
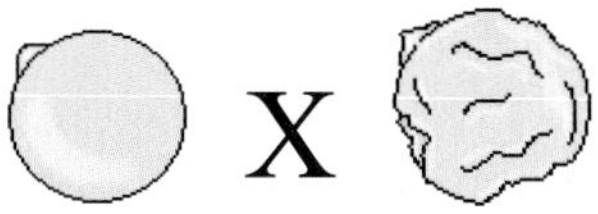 Smooth-pea plants bred with wrinkled-pea plants	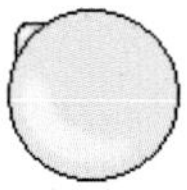 All offspring produced smooth peas.

Notice that in each experiment shown in Figure 8.1, the definable traits did not mix. A short true-breeding plant bred with a tall true-breeding plant did not produce a medium-sized plant. Instead, a tall plant was always produced. If a true-breeding plant that produced smooth peas was bred with a true-breeding plant that produced wrinkled peas, the result was not a plant that produced partly-smooth, partly-wrinkled peas. Instead, such a union would always result in a plant that produced smooth peas.

One interesting fact about pea plants (and many other plants as well) is that they can be self-bred. In other words, a pea plant can actually sexually reproduce with itself! Now the details of how this happens will be left to a discussion of plants in a later module. For right now, just accept this startling revelation and realize that the process of a plant sexually reproducing with itself is called **self-pollination**. Mendel extended his studies by using this property of plants. He decided to see what would happen when the offspring produced in the experiments described in Figure 8.1 were self-pollinated. When he tried this, an amazing thing happened. The majority of the offspring produced had the same definable trait as the parent. However, a minority of the offspring had the other trait. Thus, if the tall plants produced in the first experiment were self-pollinated, 75% of the offspring produced were tall; however, 25% were short. These results are summarized in Figure 8.2.

Illustrations by Megan Whitaker

FIGURE 8.2

Mendel's Self-Pollination Experiments Using the Offspring of the Previously-Discussed Experiments

Offspring From First Experiment Self-Pollinated	Result
 X Self Tall offspring from the first experiment were self-pollinated.	 75% 25% 75% of the offspring were tall, but 25% of them were short.
 X Self Axial-flowered plants from the first experiment were self-pollinated.	 75% 25% 75% of the offspring had axial flowers, but 25% of them had terminal flowers.
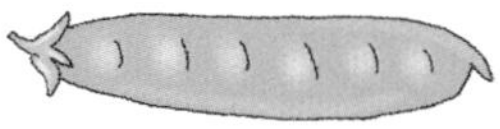 X Self Green-pod plants from the first experiment were self-pollinated.	 75% 25% 75% of the offspring produced green pods, but 25% of them produced yellow pods.
 X Self Yellow-pea plants from the first experiment were self-pollinated.	 75% 25% 75% of the offspring produced yellow peas, but 25% of them produced green peas.
 X Self Smooth-pea plants from the first experiment were self-pollinated.	75% 25% 75% of the offspring produced smooth peas, but 25% of them produced wrinkled peas.

These two sets of experiments led Mendel to develop four principles of genetics:

1. **The traits of an organism are determined by packets of information called "factors."**

2. **Each organism has not one, but two factors that determine its traits.**

3. **In sexual reproduction, each parent contributes ONLY ONE of its factors to the offspring.**

4. **In each definable trait, there is a dominant factor. If it exists in an organism, the trait determined by that dominant factor will be expressed.**

How do these principles help explain the data that Mendel collected? Well, Mendel assumed that if a pea plant always breeds true, it must have two factors that are identical. In other words, if a tall pea plant always produces tall offspring, it must have two factors that correspond to the trait of being tall. A short plant that always produces short plants must have two factors that correspond to the trait of being short. Since each parent has two of the same factor, they always contributed that factor to the offspring. Thus, the offspring produced when a true-breeding short plant was bred with a true-breeding tall plant would always have one factor that corresponded to being tall and one factor that corresponded to being short. By further assuming that the factor corresponding to being tall was dominant, each offspring would be tall, because each offspring had that dominant factor. This kind of reasoning explained all of Mendel's observations in his first set of experiments.

We are virtually certain that the previous paragraph was rather confusing to you, so we now want to explain it in a different way. Suppose we represented these factors with letters. Let's say that a capital "T" represents the factor that tells a plant it will grow tall. Further, let's say that a small "t" represents the factor that tells a plant it will grow short. What would the factors in their offspring look like? Study the figure below.

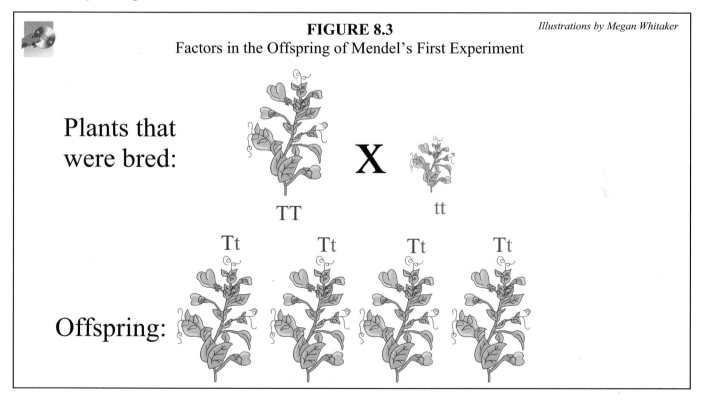

FIGURE 8.3

Illustrations by Megan Whitaker

Factors in the Offspring of Mendel's First Experiment

Plants that were bred: TT   X   tt

Offspring: Tt   Tt   Tt   Tt

Before we explain the figure, we want to tell you why we chose capital "T" to represent tall and a small "t" to represent short. We did that because we are going to assume that the factor for tallness is dominant. The capital "T" will help us keep that in mind. We could have used "s" for the factor that represented short, but we wanted to make it clear that these two factors affect the same trait. Thus, we kept the letters the same and just made the dominant one capital.

Now look at the labels in the figure. A true-breeding tall pea plant must have two factors that both tell it to grow tall. Thus, we represent its factors as "TT" in the figure. The true-breeding short plant must have two factors telling it to grow short. Thus, we represent its factors as "tt" in the figure. To make it easier to track where the factors come from, we have colored the tall parent's factors blue and the short parent's factors red. Now suppose these two plants were bred and produced four offspring. According to Mendel's principles, each parent will donate one factor to each offspring. Thus, each offspring must receive a "T" factor from the tall parent and a "t" factor from the short parent. Their factors, then, are all "Tt." Well, since the factor for tallness (T) is dominant, that trait will be expressed in every offspring that has it. Since each offspring has a dominant factor, each offspring will be tall. Now you should see why Mendel's principles explain the data in the first experiment.

What about the data in the second experiment? In that experiment, each of the offspring represented in Figure 8.3 was self-pollinated. If one of the offspring in Figure 8.3 is self-pollinated, it is like having two parents, each with the same set of factors ("Tt"). Thus, we now get the situation shown in the figure below.

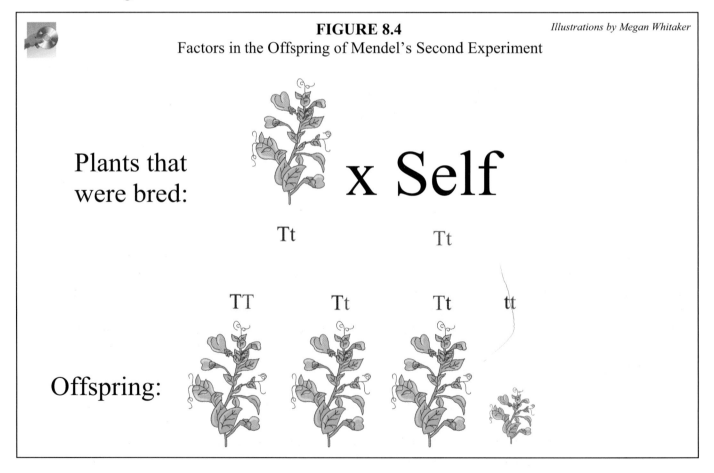

FIGURE 8.4
Factors in the Offspring of Mendel's Second Experiment

Illustrations by Megan Whitaker

In this figure, each parent's factors are once again drawn as either red or blue. You can see that the first parent (the one with the blue factors) gave a "T" to half of its offspring and a "t" to the other half. In the same way, the other parent gave a "T" to half of its offspring and a "t" to the other half.

As a result, one of their offspring has "TT" as its factors. Obviously, then, this plant will be tall. Two of the remaining offspring have "Tt" as their factors. Since the "T" is dominant, they will be tall as well. The final offspring, however, inherited a "t" from each parent. As a result, it has the factors "tt." Thus, this plant will be short, because it has only the factors that tell it to grow short, and no dominant factor telling it to grow tall.

Now do you see how Mendel's four principles can explain all of his experiments? Despite the fact that we've learned a great deal more about genetics since Mendel's time, these four principles are still the foundation of the science. This should give you a strong appreciation for Mendel's creativity and scientific ability. Without the aid of microscope, chemical analysis, or electronic instrumentation, he was able to devise and analyze experiments that helped him discover the foundation of a science that has been growing for more than one hundred years since his death!

ON YOUR OWN

8.1 The factor for producing smooth peas (we will call it "S") is dominant over the factor for producing wrinkled peas (which we will call "s"). Three plants (a, b, and c) have the following factors. Determine whether they will make smooth peas or wrinkled peas.

a. ss   b. Ss   c. SS

8.2 The factor for producing yellow peas ("Y") is dominant over the factor for producing green peas ("y"). Suppose a pea plant produces yellow peas. What possible combination(s) of factors can it have?

8.3 Given the information in the previous question, what possible combination(s) of factors can a pea plant that produces green peas have?

Updating the Terminology

As we said before, Mendel's work is still used today as the foundation of Mendelian genetics. Nevertheless, we have learned a great deal about this fascinating science since Mendel published his work, so we need to update the terminology a bit. For example, biologists do not use the term "factor" any more. We now know that the packets of information that Mendel envisioned are genes. In addition, since we know that animals have homologous pairs of chromosomes (as you learned in the previous module), we know that genes come in pairs, with one gene on each homologous chromosome. Each gene that makes up one of these pairs is called an **allele** (uh leel').

Allele – One of a pair of genes that occupies the same position on homologous chromosomes

Thus, Mendel's "factors" are now called alleles. Each definable trait has two alleles that help determine it, and they can each be represented by a letter.

When we put two alleles together ("Tt" for example), we say that we are describing a **genotype** (jee' nuh type).

<u>Genotype</u> – Two-letter set that represents the alleles an organism possesses for a certain trait

When we presented the offspring in Figure 8.3, the "Tt" we used to represent them is called their genotype. So by presenting the genotype of an organism, we are telling you what alleles an organism has on its homologous chromosomes.

We learned in Figure 8.4 that different genotypes can result in apparently identical organisms. After all, in Figure 8.4, one of the offspring had "TT" as its genotype and two others had "Tt" as their genotype. Nevertheless, all three of those plants were of exactly the same height. We determined that was the case because the "T" allele was dominant. Thus, if a plant has just one "T" in its genotype, it will be tall. Well, the *expression* of an organism's genotype is called the **phenotype** (fee' nuh type).

<u>Phenotype</u> – The observable expression of an organism's genes

Thus, there are two phenotypes possible in Figure 8.4: tall and short. That's how the genes studied in Figure 8.4 are expressed. The genotypes "TT" and "Tt" end up resulting in the same phenotype: tall. The genotype "tt" results in the other phenotype: short. We can therefore say that a phenotype is the outward expression of an organism's genotype.

Since the genotypes "TT" and "Tt" result in the same phenotype (tall), we have developed terminology to separate these two genotypes. When a genotype is composed of identical alleles ("TT" or "tt," for example), we say that the genotype is **homozygous** (ho muh zye' gus). When a genotype has mixed alleles ("Tt," for example), we say that it is **heterozygous** (het uh roh zye' gus).

<u>Homozygous genotype</u> – A genotype in which both alleles are identical

<u>Heterozygous genotype</u> – A genotype with two different alleles

We need to clarify two things before we go any further. First of all, there is no such thing as a "tT" genotype. This is because we have no way of knowing which chromosome is which in a homologous pair of chromosomes. Thus, in a heterozygous genotype, we always write the dominant allele first. Secondly, if we call the "T" allele dominant, what do we call the "t" allele? We call it the **recessive allele**.

<u>Dominant allele</u> – An allele that will determine phenotype if just one is present in the genotype

<u>Recessive allele</u> – An allele that will not determine the phenotype unless the genotype is homozygous in that allele

Now we realize that this is probably a lot of terminology to throw at you in one section, but it is important that you master it before you go any further. Therefore, look over the definitions again and make sure that you understand them. Then, to help you familiarize yourself with the terminology some more, read the set of statements on the next page, which restates Mendel's principles in modern terminology.

A RESTATEMENT OF MENDEL'S PRINCIPLES

1. **The traits of an organism are determined by its genes.**

2. **Each organism has two alleles that make up the genotype for a given trait.**

3. **In sexual reproduction, each parent contributes ONLY ONE of its alleles to its offspring.**

4. **In each genotype, there is a dominant allele. If it exists in an organism, the phenotype is determined by that allele.**

Now that you know Mendel's principles using updated terminology, you need not remember the ones using his older terminology. We will refer only to these from now on.

Before we finish this section, we need to make sure that you understand how the information you have learned in this module relates to meiosis, which you learned about in the previous module. Remember, the reason animals have two alleles for each genetic trait is because they have diploid cells. Thus, they have homologous pairs of chromosomes. In sexual reproduction, however, each parent contributes only one allele to the offspring. This is why meiosis takes diploid cells and makes them haploid. The process of meiosis separates the homologous pairs, separating the alleles from each other. Thus, each gamete produced has only one allele for each trait. When the male gamete (sperm) fuses with the female gamete (egg) and fertilization takes place, the resulting zygote then has two alleles: one from the father and one from the mother.

So the process of meiosis separates the alleles in the parents by separating the homologous chromosomes. That way, each parent contributes only one allele for each trait. If this is the case, then, how do we know which allele gets donated? Well, it all depends on which gamete fertilizes which. Remember, the male makes four functional sperm when one diploid cell undergoes meiosis. Thus, two of the sperm have one of the male's two alleles, and the other two have the other allele. Thus, if the male's genotype for a particular trait is "Tt," two sperm will have the "T" allele, and two will have the "t" allele. For the female, only one functional egg is produced. The allele that it has depends completely on which cell in meiosis I got the majority of the cytoplasm. For a heterozygous female, then, whether or not the single, functional egg has a "T" or "t" allele depends entirely on which cell got the most cytoplasm in meiosis I. The process that determines which cell gets the majority of cytoplasm is completely random; thus, half of all meiosis processes in a heterozygous female will produce an egg with a "T" allele, while the other half will produce an egg with a "t" allele.

In the end, then, the genotype of the offspring is determined by which sperm fertilizes which egg. If a "T" sperm fertilizes a "t" egg, the genotype of the offspring will be "Tt." Similarly, if a "t" sperm fertilizes a "T" egg, the genotype of the offspring will once again be "Tt." On the other hand, if a "T" sperm fertilizes a "T" egg, the genotype of the offspring will be "TT," and a "t" sperm fertilizing a "t" egg will result in a "tt" offspring. Thus, the fact that each parent contributes only one allele is a result of meiosis. The final result of which allele gets contributed, however, is determined by which gamete participates in fertilization.

ON YOUR OWN

8.4 A student repeats one of Mendel's experiments. He self-pollinates an "Ss" pea plant, where "S" is the dominant allele for smooth peas and "s" is the recessive allele for wrinkled peas. The result is one "SS" offspring, two "Ss" offspring, and one "ss" offspring. For each of these three sets of offspring, list the genotype, whether it is homozygous or heterozygous, and the phenotype.

<u>Punnett Squares</u>

Since the genotype of an offspring depends on which gamete from the mother and which gamete from the father end up participating in the fertilization process, the traits of the offspring can never be determined with absolute certainty. However, if we concentrate on a single, definable trait, we can determine the *percent chance* that the offspring has for possessing a certain genotype and corresponding phenotype. For example, if you look back at Figure 8.4, you will see that out of four offspring, one had the "TT" genotype (resulting in a tall phenotype), two had the "Tt" genotype (also resulting in a tall phenotype), and one had the "tt" genotype (resulting in the small phenotype). Thus, the chance of an offspring having the "TT" genotype was 25% (one in four). The chance of an offspring having a "Tt" genotype was 50% (two in four), and the chance for a "tt" genotype was 25% (one in four). Since these genotypes resulted in three tall plants and one small plant, the chance for a tall phenotype was 75% (three in four), and the chance for a small phenotype was 25% (one in four).

It turns out that for any single, definable trait, we can predict the likelihood that an offspring will have a given genotype and phenotype. We do this using **Punnett squares**. When we construct a Punnett square, we take the genotype of one parent and the genotype of the other and place them on the top and side of a grid. This allows us to predict all possible genotypes of the offspring. For example, another experiment that Mendel performed dealt with the pea plant's flower color. In the pea plant, the allele for producing purple flowers is dominant (we will call it "P"), and the allele for producing white flowers ("p") is recessive. In the first experiment, Mendel bred a homozygous purple-flowered pea plant ("PP") with a homozygous white-flowered pea plant ("pp"). Using Punnett squares, we can predict what possible genotypes and phenotypes will be produced.

EXAMPLE 8.1

A homozygous, purple-flowered pea plant is bred with a homozygous, white-flowered pea plant. If the allele for purple is dominant, what possible genotypes and phenotypes will be produced? What is the percent chance for each?

Since we are told that the allele for a purple flower is dominant, we can use "P" for that allele and "p" for the allele that causes white flowers. The genotypes of the purple-flowered parent, then, must be "PP." After all, the only way it could produce purple flowers is to have the "P" allele, and since it is homozygous, both its alleles must be the same. The white-flowered parent must have a genotype of "pp," because the only way it can express the recessive trait is if both alleles are recessive. For this parent, then, we don't even need to be told it is homozygous. There is no way it can express the recessive allele without being homozygous.

Now that we know the genotypes of the parents, we can construct a Punnett square. To do this, we take one parent's genotype and place each allele at the top of a 2x2 grid. We then take the other parent's genotype and place each allele to the left of the grid:

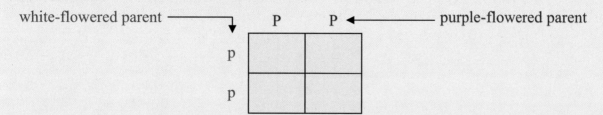

Now, since each parent can donate either allele to the offspring, all possible genotypes are determined by taking the allele to the left of a row and adding it to the allele on top of the column. That two-letter sequence goes in the appropriate box, giving you a possible genotype. In this case, the result is as follows:

	P	P
p	Pp	Pp
p	Pp	Pp

Since there are four boxes, there are four possible genotypes. In this case, they are all "Pp," resulting in a purple phenotype. Thus, the chance for genotype is <u>100% "Pp,"</u> and the phenotype chance is <u>100% purple flowers.</u>

Now that wasn't very hard, since all offspring had the same genotype. Therefore, let's look at a more complex example. In this example, we use a term that you will see a lot in genetics. When we breed two individuals, we often call it a "cross." Thus, we will sometimes say that we are "crossing" two individuals instead of "breeding" them.

EXAMPLE 8.2

A heterozygous purple-flowered pea plant is crossed with a heterozygous purple-flowered pea plant. What are the possible phenotypes and genotypes of the offspring, along with the percentage chance of each?

Since we have already been told that the allele for a purple flower is dominant, we can use "P" for that allele and "p" for the allele that causes white flowers. Since the parents are heterozygous, they must have one of each. Their genotypes, then, are both "Pp." The Punnett square we initially set up will therefore look like this:

	P	p
P		
p		

Now, since each parent càn donate either allele to the offspring, all possible genotypes are determined by taking the allele to the left of a row and adding it to the allele on top of the column. That two-letter sequence goes in the appropriate box, giving you a possible genotype. In this case, the result is:

	P	p
P	PP	Pp
p	Pp	pp

Based on this, then, we see three different genotypes. The "PP" genotype has a 25% chance of occurring, since it appears in one of the four boxes. The "Pp" genotype has a 50% chance of occurring, since it appears in two of the four boxes. The "pp" genotype has a 25% chance of occurring, since it appears in one of the four boxes. Since any offspring that has even one "P" allele will have purple flowers (that happens in three of the four boxes), the purple flower phenotype has a 75% chance of occurring. The only white-flowered pea plant will be the one with genotype "pp," which happens in one of the four boxes. The white-flowered phenotype, then, has a 25% chance of occurring.

ON YOUR OWN

8.5 Using a Punnett square, predict the possible genotypes and phenotypes, along with the percentage chance for each, when a heterozygous, purple-flowered pea plant is crossed with a white-flowered pea plant.

Pedigrees

Let's suppose we want to study an organism but we really don't know much about its genotype. Is there any way that we can find the genotype? Yes, it turns out that in many cases, you can determine the genotype of an organism if you construct a **pedigree**.

Pedigree – A diagram that follows a particular phenotype through several generations

Suppose, for example, we wanted to determine the genotype of a guinea pig in reference to its coat color. If we had enough information about its brothers and sisters, its parents, aunts, uncles, and grandparents, we could reach some fairly solid conclusions about its genotype with respect to coat color.

In a pedigree, we need to develop symbols that tell us everything we need to know about the animals that are being studied. In this course, we will represent males with squares and females with circles. In addition, whether or not a square or circle is filled in will tell us about the phenotype. In this particular example, a filled-in square or circle indicates that the guinea pig has a black coat. A hollow square or circle, on the other hand, indicates a white coat. When a circle and square are linked by a horizontal line, it indicates that the two individuals have been bred. The symbols that "hang" off that solid line are the offspring of the breeding. Examine Figure 8.5 on the next page to see what this looks like.

FIGURE 8.5
A Pedigree of Coat Color in Guinea Pigs

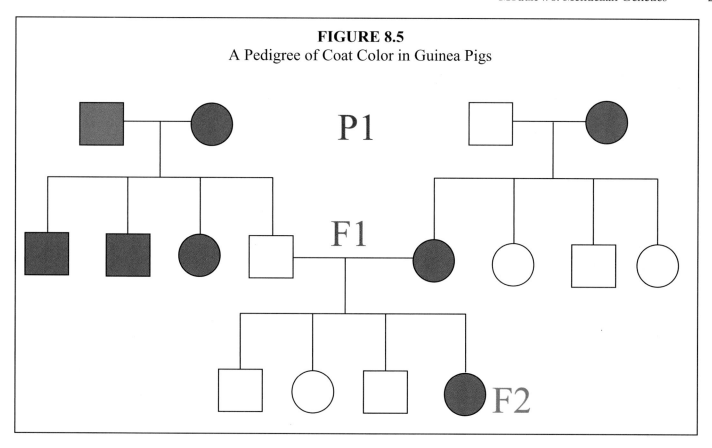

   In this figure, the two guinea pigs in the upper left corner are a black-coated male and a black-coated female. Since there is a line connecting them, they are bred and produce four offspring (the ones that "hang" off that line): two black-coated males, a black-coated female, and a white-coated male. The white-coated male was then bred with a black-coated female from another breeding. The letters "P1," "F1," and "F2" refer to the generations. The "P1" indicates the initial parents. "F1" represents their offspring, and "F2" represents their offspring's offspring.   In this pedigree, then, we are following the phenotype of coat color through three generations. If more generations were present, they would be labeled "F3," "F4," etc. That's what we mean when we say that a pedigree follows a particular phenotype through several generations.

   What can we learn through such a pedigree? The first thing we learn is that the allele which determines a black coat must be the dominant allele. Why? Think about it. In the cross between the black-coated male and black-coated female (the parents in the upper left corner of the figure), a white-coated male was produced. This means that the allele for a white coat *must* be present in the parents. However, the parents both have black coats. According to the principles of Mendelian genetics, the only way that a black-coated guinea pig can have an allele for a white coat is if the allele for a white coat is recessive. We will therefore call it "b." This means that the allele for black coat is dominant, and we will therefore call it "B." In addition, since we know that the allele for a white coat is recessive, each parent must have one. After all, for a recessive allele to be expressed, the organism must have two of that allele. Thus, since one of the offspring has two white-coat alleles, one must have come from each parent. Each parent, therefore, has an allele for a white coat. Nevertheless, each parent has a black coat. This means that each parent also must have an allele for a black coat. Thus, by looking at this pedigree, we can already say that the black-coated female and black-coated male in the first cross both have a genotype of "Bb." By just looking at the pedigree, we have determined the genotype of the first two parents!

We can also determine the genotypes of the other parents in P1 (the ones in the upper right corner of the figure). After all, the white-coated male must be "bb," because that's the only way it can be white. Also, since some of their offspring are white, the female parent must also have the recessive allele. This indicates that the black-coated female must be "Bb." Do you see how that works? By just following the phenotype through a generation, we have already determined the genotype of the original parents!

Now you should notice something that looks wrong in the figure. If the P1 parents on the right-hand side of the figure (the white-coated male and black-coated female) have genotypes "bb" and "Bb" as we just determined, you can work out the Punnett square to determine that the probability of a black-coated phenotype is 50%, and the probability of the white-coated phenotype is 50%. Thus, you might think that there should be two white-coated offspring and two black-coated offspring. Why are there three of one and one of the other?

Remember, just because the *probability* of having a white coat is two out of four (50%), it doesn't mean that two out of *every* four offspring will be white-coated. It only means that, if you have lots and lots of offspring, on average, two out of four will be white-coated. Nevertheless, for only a few offspring, the probabilities might not exactly work out. After all, if you flip a coin, the odds are 50% that you will get heads and 50% that you will get tails. This doesn't mean that *for every two times you flip it* one will give you heads and the next give you tails. Indeed, if you take a few minutes to flip a coin right now, you might get two or three heads in a row before getting a tail. If, however, you flipped the coin 1,000 times, you probably would get a total of 500 heads and 500 tails. The same thing holds true in pedigrees. You can't expect the probabilities to work out exactly each time!

EXAMPLE 8.3

The following pedigrees represent experiments involving the eye color of a certain animal. The only possible eye colors for this animal are blue and brown. A brown square or circle indicates the brown eye color, while a white circle or square indicates the blue eye color. What are the genotypes of individuals #1-4 in this pedigree?

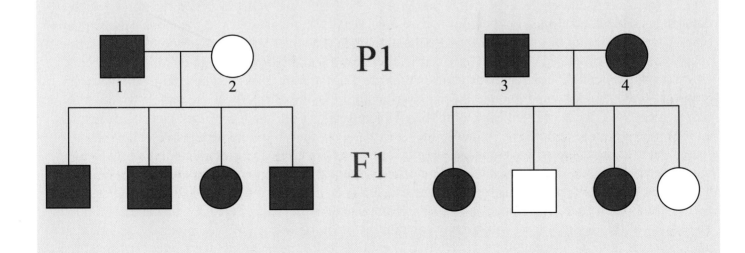

First of all, as we look at these pedigrees, we realize that blue eye color (white circles and squares) must be the recessive allele. We know that because even though #3 and #4 both have brown eyes, some of their offspring are blue-eyed. Therefore, they both must have a blue-eyed allele. Since they must have a blue-eyed allele but are nevertheless brown-eyed, the brown eye allele must be dominant. Therefore, we will call the brown eye allele "B" and the blue eye allele "b." Right away, then, we know that the genotype of #2 is "bb." After all, the only way you will express the phenotype of a recessive allele is if you have two of those alleles. Also, since #3 and #4 (who are both brown-eyed) must have a recessive allele to produce blue-eyed offspring, they must both be "Bb." What about #1? Well, when #1 and #2 breed, there are no blue-eyed offspring, despite the fact that #2 has two blue-eye alleles. This indicates that #1 must always give the dominant allele. Thus, it is probably "BB." In the end, then, <u>#1 is "BB," #2 is "bb," #3 is "Bb," and #4 is "Bb."</u>

ON YOUR OWN

8.6 In the following pedigrees, a biologist studies the presence of a tail on a certain species of animal. Some individuals have a tail (represented by the gray circles and squares) and others do not (represented by the white circles and squares). Which allele is dominant, the one for having a tail or the one for not having a tail? What are the genotypes of individuals #1-4?

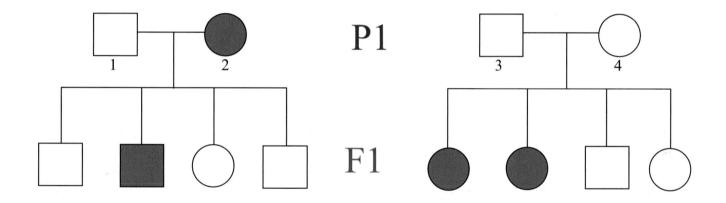

EXPERIMENT 8.1
Making Your Own Earlobe Pedigree

<u>Supplies</u>:

♦ Your parents, grandparents, aunts, uncles, and siblings (Even if your grandparents, aunts, and uncles are not living or are living far away, you might be able to find pictures of their ears, which is all that you need for the experiment. If you don't have many siblings or cannot determine the earlobe characteristics of your grandparents, aunts, and uncles, you might consider studying another family as well so that you can get even more information.)
♦ Mirror

Object: To interpret your own pedigree with respect to earlobe attachment

<u>Procedure</u>:

1. Examine the earlobes of each of your grandparents, parents, aunts, uncles, and siblings. There are typically two types of earlobes on people: free or attached. An attached earlobe is attached to the skin all of the way down, while a free earlobe has a portion that is not attached to the skin. Using the pictures below as a guide, determine whether the earlobes are free or attached for as many of your grandparents, parents, aunts, uncles, and siblings as you can.

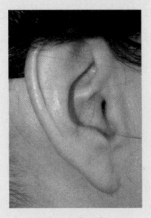 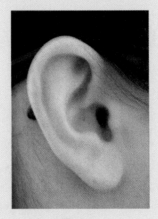

Photos by Kathleen J. Wile

          Attached   Free

2. Look in a mirror and determine whether your earlobes are free or attached.
3. Record all of your results in your laboratory notebook.
4. Draw a pedigree of your family. Circles should be used to represent females, while squares should be used to represent males. You can determine whether filled circles and squares represent attached or free earlobes.
5. Use the pedigree to determine whether free or attached earlobes are the dominant trait. It is possible that you will not be able to conclusively determine anything from your family's pedigree, but give it a try. At the end of the solutions to the study guide for this module, you can find a sample pedigree. The explanation of the pedigree will tell you which trait is dominant.
6. If you have the desire, do this kind of analysis on other families. The more families and generations you can include in your analysis, the more conclusively you can determine which trait is dominant and which is recessive.
7. See if you can determine the genotype of any of your parents or siblings based on your pedigree.

<u>More Complex Crosses</u>

The crosses that you have just done are relatively simple because they concentrate on only one trait. When we are looking at the sexual reproduction of two individuals and concentrate on only one trait, we say that we are doing a **monohybrid cross**.

<u>Monohybrid cross</u> – A cross between two individuals, concentrating on only one definable trait

More interesting situations develop, however, when we look at a **dihybrid cross**.

<u>Dihybrid cross</u> – A cross between two individuals, concentrating on two definable traits

For example, in the pea plants that we discussed in Examples 8.1 and 8.2, we looked only at flower color. As we know, however, a pea plant has many more traits. For example, it can be either short or tall, it can produce smooth peas or wrinkled peas, it can produce green or yellow peas, etc. What if we were to look at two of those traits simultaneously? With a little more work, we can predict the possible results when a tall, purple-flowered pea plant is crossed with a short, white-flowered pea plant.

How do we do this? Well, it involves making a Punnett square, but the size of that Punnett square depends on the genotypes present in each parent. You see, when we make a Punnett square, we need to know the alleles that will be present in all gametes produced by each parent. The alleles in each possible gamete from one parent will be listed at the top of the columns in the square, and the alleles in each possible gamete from the other parent will be listed at the left of rows in the square. In order to determine the gametes from each parent, however, we need to think a little more about meiosis.

In meiosis, each allele is separated. Half of the gametes end up with one allele, and half end up with the other. When we are talking about an individual trait (flower color, for example), the possibilities of what allele goes into what gamete are limited. If an organism has the "PP" genotype, for example, there is only one possible allele for each gamete, the "P" allele. If an organism has the "Pp" genotype, half of the gametes will have the "P" allele and the other half will have the "p" allele. Rather simple, huh?

When we start considering more traits, however, determining all possible gametes gets more complicated. Suppose, for example, a pea plant has the "Pp" genotype for flower color and the "Tt" genotype for height. What are the possible allele combinations for the gametes produced by meiosis? Well, when the alleles split in meiosis, the "P" allele for flower color might go into the same gamete as the "T" allele for height. That gamete would be *PT*. We italicize this combination of letters to make it clear that this is not a genotype. It is simply a list of two of the alleles that exist in a gamete. If the "P" allele for flower color went to the gamete that got the "t" allele for height, then the resulting gamete would be *Pt*. Since which allele goes where is completely random in meiosis, both of these combinations are possible. Furthermore, the "p" allele for flower color might go into the gamete that got the "T" allele for height. This would result in a gamete that has *pT*. Finally, if the "p" allele for flower color went to the gamete that got the "t" allele for height, the result would be a *pt* gamete.

Now we realize that this might be a little confusing, but stick with us here. Since all of these combinations of alleles are possible in a gamete produced by this plant, we must include all of them in our Punnett square. Thus, the Punnett square will have the following columns: *PT, Pt, pT, pt*. Notice that when writing the alleles in a gamete, we always keep the allele for one trait first and the allele for the other trait second. Unlike genotypes, we do not worry about whether an allele is dominant or recessive when determining where to put it. In the case of writing out the alleles in a gamete, we always keep them ordered by the trait that they affect.

In the end, then, when we are considering more than one trait, we have to determine all possible allele combinations in the gametes produced by each parent. That will help us build the Punnett square. Study the next example and then do the "On Your Own" problem that follows so that you can be prepared to do the Punnett squares that come next.

EXAMPLE 8.4

A tall, white-flowered pea plant is heterozygous in the gene that determines height and homozygous in the gene that determines flower color. What are the possible allele combinations for the gametes produced by this plant?

Since the plant is heterozygous in height, its genotype is "Tt." Note that we didn't really have to tell you that it is homozygous in flower color. The *only* way a pea plant can make white flowers is if its genotype is "pp." So, given these two genotypes, how many possible allele combinations are there? Well, the "T" allele could go with a "p" allele, making a *pT* gamete. Also, the "t" allele could go with a "p" allele, making a *pt* gamete. Hmm. That's all of the possible combinations. Since the plant is homozygous in one trait, the number of possible combinations is relatively low. Thus, the possible gametes formed are just *pT* and *pt*.

ON YOUR OWN

8.7 Suppose we are going to set up a Punnett square that concentrates on the smooth/wrinkled nature of the peas produced in a pea plant ("S" for smooth, "s" for wrinkled) and the color of the pea produced ("Y" for yellow and "y" for green). What are all possible allele combinations for the gametes produced by a plant that is heterozygous in both traits?

Now that you have experience thinking about the possible allele combinations in gametes when more than one trait is considered, it is time to study an example of making the Punnett square for a dihybrid cross.

EXAMPLE 8.5

A pea plant that is tall and purple flowered is homozygous in both corresponding genotypes. What are the possible genotypes and phenotypes, along with their percentage chances for occurring, when this pea plant is crossed with a short, white-flowered pea plant?

Since the tall, purple-flowered pea plant is homozygous in both corresponding genotypes, it has the "TT" and "PP" genotypes. What about the other plant? We are not told whether it is homozygous or heterozygous. Of course, we do not need to be told. As we already know, the allele for shortness is recessive. Thus, to be short, the plant must have the "tt" genotype. Similarly, since the white flower trait is recessive, it must have the "pp" genotype.

Now, to make a Punnett square for this cross, we need to think about what possible gametes will be produced by meiosis. To do that, we need to come up with all possible combinations of alleles in each parent. Well, for the "TT," "PP" parent, there is only one possible combination. No matter what, a "P" allele will be mixed with a "T" allele. The only possible gamete, then is *PT*. Likewise, since the other parent is homozygous in both traits, the only possible combination is *pt*. Thus, our Punnett square is rather simple:

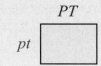

We can then determine the genotype of the offspring by just putting the alleles that affect the same trait together:

PT

pt | PpTt |

Since this is the only genotype possible, <u>100% of the genotypes will be "PpTt," with a resulting phenotype of tall with purple flowers.</u>

Now that example probably didn't seem too bad. Since both parents were homozygous, there was only one possible combination of alleles for each of their gametes. Thus, the Punnett square was rather simple. Notice that in the previous Punnett squares you drew, there were always two rows and two columns. This resulted in four boxes. We call such a Punnett square a 2x2 Punnett square, because there are two rows and two columns. In the example you just studied, there was only one row and one column, so we would call it a 1x1 Punnett square. In the next example, things are a bit more complicated, resulting in a 4x4 Punnett square.

EXAMPLE 8.6

A tall pea plant that produces purple flowers is heterozygous in both of the relevant genotypes. What are the genotypes and phenotypes, along with their percentages, for the offspring produced when this plant is self-pollinated?

Since this plant is heterozygous in both genotypes, we know that it is "Pp" and "Tt." Self-pollination is just like having two parents, each with the same genotype. So, we need to determine all possible combinations of alleles for the gametes produced by this genotype. Well, the "P" allele might go with the "T" allele, making a gamete that is *"PT."* Of course, the "P" allele might also go with the "t" allele, making a *Pt* gamete. The "p" allele could go with the "T" allele for a *"pT"* gamete, or the "p" allele could go with the "t" allele, making a *"pt"* gamete. Thus, there are four possible gametes. Since self pollination is the same as having two parents with identical genotypes, the Punnett square we draw will have these four gametes both on top of the columns as well as to the left of the rows, making a 4x4 Punnett square:

	PT	*Pt*	*pT*	*pt*
PT	PPTT	PPTt	PpTT	PpTt
Pt	PPTt	PPtt	PpTt	Pptt
pT	PpTT	PpTt	ppTT	ppTt
pt	PpTt	Pptt	ppTt	pptt

Since there are 16 boxes, we have 16 possibilities. <u>The genotype "PPTT" appears once.</u> Remember, to calculate the percentage chance of something happening, we take the number of times it occurs and then divide by the total number possible. After that, we multiply by 100 to make it a percentage.

Thus, <u>the percentage chance of the "PPTT" genotype occurring is (1/16)x100 = 6.25%.</u> The genotype "PPTt" happens twice, making its percentage chance (2/16)x100 = 12.5%. The genotype "PPtt" happens once, making its percentage chance (1/16)x100 = 6.25%. The genotype "PpTT" happens twice, making its percentage chance (2/16)x100 = 12.5%. The genotype "ppTT" happens once, making its percentage chance (1/16)x100 = 6.25%. The genotype "ppTt" happens twice, making its percentage chance (2/16)x100 = 12.5%. The genotype "PpTt" happens four times, making its percentage chance (4/16)x100 = 25%. The genotype "Pptt" happens twice, making its percentage chance (2/16)x100 = 12.5%. Finally, the genotype "pptt" happens once, making its percentage chance (1/16)x100 = 6.25%.

Now remember, many genotypes will result in the same phenotype. The genotype "PPTT" will result in a tall pea plant with purple flowers, but so will "PpTt," "PPTt," and PpTT." Adding the percentages of each genotype together, then, <u>the probability of producing a tall pea plant with purple flowers is 6.25% + 25% + 12.5% + 12.5% = 56.25%.</u> The genotype "ppTT" results in a tall pea plant with white flowers, but so does "ppTt." <u>The probability for producing a tall pea plant with white flowers is therefore 6.25% + 12.5% = 18.75%.</u> A short pea plant with purple flowers will result from genotypes "PPtt" and "Pptt." <u>The percentage chance of making a short pea plant with purple flowers is 6.25% + 12.5% = 18.75%.</u> Finally, the genotypes "pptt" will result in a <u>short pea plant with white flowers, giving a probability of 6.25%.</u>

That was pretty difficult, so we need to go back and review what we did. First of all, we recognized that we were dealing with two distinct traits. Thus, we knew that we had a dihybrid cross on our hands. We therefore determined all possible combinations of the four alleles present, to determine all of the different gametes that the parents could produce. We found four different gametes, so we made a 4x4 Punnett square. After filling in the boxes, we counted the number of times that each genotype occurred, and then we divided by the total number of boxes (16) and multiplied by 100. This gave us the percentage chance of making each genotype. Then, we determined all of the genotypes that would lead to the same phenotype and added their percentages, giving us the total percent chance of producing that phenotype.

This is tough stuff; no question about it. Nevertheless, it is important for you to get some practice doing it yourself. Thus, perform the following "experiment." The answers to the experiment are given after the solutions to the study guide for this module.

"EXPERIMENT" 8.2
A Dihybrid Cross

<u>Supplies:</u>
♦ Lab notebook

Object: To help understand how multiple traits are passed from one generation to another

Background: The ability of some people to taste PTC (a bitter substance to some people and tasteless to others) and the ability to roll their tongues into a "U" shape when extended from the mouth are both dominant traits. We will designate them with a "T" and an "R," respectively. Persons who have these dominant alleles are called "tasters" and "rollers." The inability to taste PTC or to roll the tongue is caused by being homozygous in the recessive alleles, which we will designate as "t" and "r," respectively. Those who are homozygous with the recessive alleles are called "nontasters" and "nonrollers."

Procedure:

1. Draw a 1x1 Punnett square.
2. Fill in the square with the following cross: a man that is a homozygous taster/roller and a woman that is a nontaster/nonroller.
3. What are the genotype and the phenotype of the children?
4. Now make a 4x4 Punnett square.
5. Cross a heterozygous taster/roller with another of the same genotype.
6. Indicate the percentage chance of producing each of the following: roller/taster, roller/nontaster, nonroller/taster and nonroller/nontaster.

Sex and Sex-Linked Genetic Traits

As we learned back in Module #7, humans have 23 homologous pairs of chromosomes. The last pair, however, is a little different from the other 22. Thus, we call the first 22 pairs of chromosomes **autosomes** (aw' toh sohms) and the last pair **sex chromosomes**, because the last pair of chromosomes determines the sex of an individual.

Autosomes – Chromosomes that do not determine the sex of an individual

Sex chromosomes – Chromosomes that determine the sex of an individual

If a person's sex chromosomes are perfectly homologous, they are called an XX pair, and the person is female. In males, the sex chromosomes (called an XY pair) are not perfectly homologous. There are fewer genes on the second chromosome (called the "Y" chromosome) than there are on the X. This causes an interesting effect called **sex-linked genetic traits**, which we will discuss in a moment.

Before we do that, however, we just need to briefly make it clear how the sex of offspring is determined using these chromosomes. In meiosis I, the homologous pairs of chromosomes are split up. Thus, since females have two X chromosomes, each cell at the end of meiosis I has an X chromosome. As a result, all gametes produced by a female have an X chromosome. At the end of meiosis I in the male, however, one cell has an X chromosome and the other has a Y. As a result, half of the gametes produced by the male will have an X chromosome and half will have a Y. In the end, a Punnett square for the sex chromosomes would look like this:

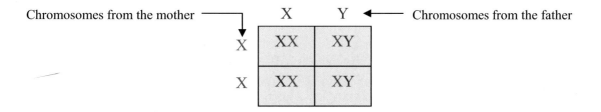

So you see that half of the offspring will have XX chromosomes and therefore be female, and the other half will have XY chromosomes and will therefore be male. As a result, half of the babies born are male, and half are female. Once again, in a given family, half of the children may not be male, because we are only discussing probabilities here. If you count *enough* children, half will be male and half will be female. It is not unusual, however, for a single family to "beat the odds" and have significantly more of one sex than the other.

Sex-linked genetic traits exist because the Y chromosome does not carry as many alleles as does the X chromosome. As a result, there are some traits for which alleles only exist on the X chromosome and not the Y chromosome. In essence, this means that for certain genetic traits that reside on the sex chromosomes, males have only one allele instead of two. As a result, expressing the recessive phenotype for such a trait is much more likely for males. After all, if you have two alleles for a given genetic trait, both recessive alleles must be present for you to express the recessive phenotype. For sex-linked traits, however, the male has only one allele. Thus, the male needs only one recessive allele to express the recessive phenotype!

Consider, for example, the fruit fly. If you leave fruit exposed for very long, you will most likely attract tiny flies from the genus *Drosophila*. If you examine them under a microscope, you will see that the vast majority have red eyes. A few, however, have white eyes, and that phenotype is much more likely among male *Drosophila* than female *Drosophila*. The reason for this is that the eye color allele for *Drosophila* exists on the X chromosome but not on the Y chromosome. Thus, to have white eyes, a female must have two recessive alleles. A male, however, has only one allele to begin with. If that allele is recessive, the male's eyes will be white. When a trait is much more prevalent among males than females, it is often governed by a sex-linked, recessive trait. Medical scientists often use this as a means of trying to understand diseases and other health problems. Autism, for example, is a neurological disorder that affects four times as many males as it does females. Because of this fact, medical researchers think that autism has a genetic component that is sex-linked. This has caused some medical researchers to concentrate on studying the differences in X chromosomes between autistic and nonautistic people, hoping that such analysis will help them find the genetic component to autism.

In analyzing sex-linked characteristics, we can still use a Punnett square. The notation has to be a little more involved, however. Instead of just representing an allele with a capital letter when it is dominant and a small letter when it is recessive, we have to note the fact that it exists only on the X chromosome. We do this by using an X with a superscript. In the case of *Drosophila*, for example, we could say that the allele for a red eye can be noted with an "X^R." The X refers to the fact that it is carried on the X chromosome, and the capital "R" indicates that it is the dominant allele. The recessive white allele, then, would be noted as "X^r." The Y chromosome from the male has no allele, so it will just be noted as a "Y." Let's see how this notation works in an example.

EXAMPLE 8.7

A heterozygous, red-eyed female *Drosophila* is crossed with a red-eyed male *Drosophila*. Will any of the offspring be white-eyed? What is the percentage chance of getting a white-eyed female? What is the percentage chance of getting a white-eyed male?

If the female is heterozygous, it has one of each allele. Thus, its gametes will be abbreviated as "X^R" and "X^r." The male has the allele only on the X chromosome, and, since he is red-eyed, it is the dominant allele. Thus, the male's gametes will be labeled as "X^R" and "Y." Remember, since the Y chromosome does not have this allele, it has no superscript. Now that we have the gametes produced by the male and female, we can make a Punnett square:

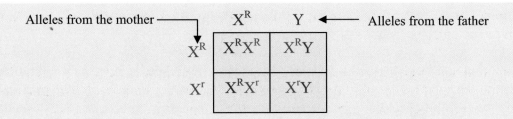

The two female offspring (represented by the boxes with two X chromosomes) will both end up red-eyed, since they each have at least one of the dominant allele. The male offspring (represented by the boxes with XY chromosomes), however, show a different story. One of the males (the one in the upper right corner) will be red-eyed, because he got the dominant allele from his heterozygous mother. The other one, however, will be white-eyed, since his only allele is the recessive one he got from his mother. Thus, there will be white-eyed offspring. The percent chance of a female offspring being white-eyed is 0%, but 50% (one out of two) of all males will be white-eyed.

Do you see the reasoning behind sex-linked characteristics? If we build the Punnett square showing that the allele exists on the X chromosome but not on the Y chromosome, predicting the sex and phenotype of the offspring is a piece of cake! Try your hand at it with the following "experiment." Once again, the answers to the experiment can be found after the solutions to the study guide for this module.

"EXPERIMENT" 8.3
Sex-linked Genetic Traits

Supplies:
♦ Lab notebook

Object: To help understand how sex-linked traits are passed from parents to offspring

Background: You should have read in your history books about the English and Russian royalty's problems with hemophilia, a disease that inhibits the blood's ability to form clots. People with this disease are prone to excessive bleeding. In fact, it is very possible for hemophiliacs to bleed to death as a result of minor cuts. This devastating disease is a sex-linked, recessive characteristic. Queen Victoria of England was heterozygous in this trait. Thus, she carried the recessive allele for hemophilia, but since she had the dominant allele on her other X-chromosome, she did not actually have the disease.

Procedure:
1. Since this trait is sex-linked, we designate the allele as X^h. The "h" is small because it is recessive.
2. If Queen Victoria was heterozygous, then she must have been $X^H X^h$. Queen Victoria's husband did not have hemophilia, so his genotype would have to have been $X^H Y$.
3. Draw a Punnett square that tells what the children's genotype might be. Also state all the possible phenotypes.
4. Their son Leopold was a hemophiliac. What was his genotype?
5. Hemophilia can also be present in females. Write a Punnett square to show how it is possible to produce hemophiliac females. Use whatever genotypes you need for the parents in order to get a female offspring with hemophilia.

A More Complete Understanding of Genetics

Although Mendel's principles form the basis of understanding how traits are passed from one generation to another, we have learned a lot more about genetics since Mendel's time, and we have found that Mendel's principles are only the first step in understanding this fascinating process. Although many traits follow Mendel's principles quite well, there are (not surprisingly) several exceptions to Mendel's principles, and we want to briefly mention them here.

Although many traits are caused by a specific gene, many traits are also caused by the interaction of *several* genes. This is called **polygenetic inheritance**, and it is very common in creation. In fact, the majority of observable traits in people are determined by the interaction of several genes. Eye color, for example, is governed by the interaction of at least three genes in human beings. Two of the genes are found on chromosome pair 15, and the other one is found on chromosome pair 19. If you are interested in learning the current theory for how these three genes interact to affect eye color, you should go to the course website that we mentioned in the "Student Notes" section at the beginning of this book. There is a link to a website that discusses this topic in detail. Height, skin color, and weight are other examples of polygenetic inheritance.

Some traits are controlled by alleles that exhibit **incomplete dominance**. For example, snapdragon flowers can be several colors, including white, red, or pink. This is because there is an allele for red flower color and another for white flower color, but neither is dominant. If a snapdragon is homozygous in the allele for red, its flowers will be red. If it is homozygous in the allele for white, it will have white flowers. If a snapdragon is heterozygous, however, its flowers will be pink, because neither allele is completely dominant. Instead, the two flower colors "mix," forming a color between the two.

In some cases, one set of alleles might affect how another set of alleles is expressed. This is called **epistasis** (ep' uh stay' sis), and a good example can be found in mice. There is one set of alleles that determines the coat color in mice. The allele for black coats (B) is dominant over the allele for brown coats (b). However, there is another set of alleles that determines whether or not this color is actually present. This set of alleles has a dominant form (C) which tells the brown/black set of alleles to "turn on." However, the recessive allele (c), tells the gene to "turn off." Thus, if the mouse has a genotype "CC" or "Cc," the coat will be either brown or black, depending on the other set of alleles. However, if the mouse has a "cc" genotype, it doesn't matter what alleles it has for the other set of genes. Those genes are "turned off," and the coat is white.

Sometimes, a single gene can affect multiple observable traits (multiple phenotypes). This is called **pleiotrophy** (plee' uh troh fee), and sickle-cell anemia is a commonly-cited example. In sickle-cell anemia, a set of alleles on chromosome 11 determines a particular aspect about the shape of a person's red blood cells. If a person is homozygous in the recessive allele, his red blood cells will have a different shape from normal red blood cells. This different shape causes the red blood cells to pile up in the bloodstream, causing blockages and damaging vital organs. Anemia, pneumonia, heart failure, and severe pain episodes are symptoms that can be produced by sickle-cell anemia. Thus, all of those observable traits (phenotypes) can be caused by being homozygous in just one set of alleles.

Two more exceptions to Mendel's principles are best illustrated by discussing the genetics of **blood type**. A person's blood type can be A, B, AB, or O. There is also a positive or negative associated with blood type, but we don't want to go into that right now. Whether a person has A, B,

AB, or O blood type is determined by one set of alleles, but unlike the alleles that we have been studying so far, there are three possibilities. This is an example of having **multiple alleles** for a gene. In the case of blood type, there is an allele for blood type A, another for blood type B, and a third for blood type O. Now, of course, a person has only two homologous chromosomes, so she can have only two alleles. There are just three possibilities for those two alleles.

The allele for blood type O is recessive to both A and B. Thus, a person with genotype "AO" will have type A blood, while a person with genotype "BO" will have type "B" blood. However, the A and B alleles are not dominant over each other. This is an example of **codominance**. The A and B alleles are both dominant over the O allele, but they are not dominant over each other. Thus, if a person has the genotype "AB," she does not have type A or type B blood, she has type AB blood. How does this all work when it comes to parents having children? Study the following example to find out.

EXAMPLE 8.8

A woman with type AB blood marries a man with type A blood. However, the man has the recessive O allele as well. What are the possible blood types for their children as well as the percentage chances for each?

If the woman is type AB, her genotype is "AB." The man is type A but has the recessive O allele. The only way that can happen is if his genotype is "AO." Thus, the Punnett square looks like this:

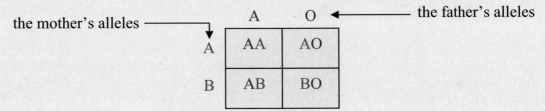

The possible blood types are A (50 % - both AA and AO make type A blood), AB (25%), and B (25%).

Now that you know how blood type is inherited, you might want to know why it is important. Well, the plasma membrane of a person's red blood cells can contain **antigens**.

Antigen – A protein that, when introduced in the blood, triggers the production of an antibody

Remember, your body produces antibodies to help fight off invaders. Type A blood is blood whose red blood cells have type A antigens on their surfaces. As a result, a person with type A blood does not produce the antibodies against the type A antigen. Otherwise, the body would kill its own blood cells. The red blood cells in type B blood have type B antigens, so the body does not produce antibodies against the type B antigen. The red blood cells in type O blood have neither A nor B antigens on their plasma membranes, so a person with type O blood produces the antibodies against both the A and the B antigens. A person with type AB blood has both A and B antigens on his red blood cells, so his body cannot produce either of the antibodies we have been discussing.

That's the difference in the blood types. Why does it matter? Well, your blood produces antibodies to fight infection. If you have type A blood, your body does not produce antibodies against A antigens. However, it *does* produce antibodies against B antigens. Thus, if you have type A blood and are unfortunate enough to get type B blood injected into you, the antibodies you have against the B antigens will help your body destroy the type B red blood cells. They do this by binding to the antigens and building "bridges" from one red blood cell to another. As a result, the type B red blood cells start clumping together, and that's bad. This clump can block blood vessels, cutting off blood flow to organs and tissues! Thus, you cannot receive blood that is incompatible with the antibodies which are already in your blood.

Although the blood types A, B, AB, and O are important, there is one other important thing to consider: the **Rh factor**. The "Rh" stands for "Rhesus monkey," because that's where the factor was first identified. It refers to whether or not another antigen (the Rh antigen) exists on the red blood cells. If the Rh antigen exists on the plasma membrane of the red blood cells, the blood is called **Rh-positive**. If there is no Rh antigen on the red blood cells, the blood is **Rh-negative**. About 85% of people are Rh-positive. Since the Rh factor is an important description of the blood, whether or not a person is Rh-positive is usually listed with the blood type. Thus, type O positive (O+) blood is type O blood in which the red blood cells have the Rh antigen. Type O negative blood (O-) is type O blood in which the red blood cells do not have the Rh antigen.

How is the Rh factor inherited? There is a separate gene that determines whether or not your red blood cells have the Rh antigen. The Rh-positive allele is dominant, and the Rh-negative allele is recessive. Thus, to be Rh-negative, you must have an Rh-negative allele from each parent. If you have just one allele for Rh-positive, you will be Rh positive.

Since the Rh antigen is an antigen, antibodies can be produced against it. People who are Rh-positive, of course, will not produce the antibody that fights the Rh antigen, because a body will not produce an antibody that attacks its own blood. However, people who are Rh-negative *can* produce the antibody against the Rh antigen.

ON YOUR OWN

8.8 In domestic chicken, a change in one particular allele can affect the development of the feathers, wings, lungs, kidney, and liver. What is this an example of?

8.9 A woman who has type A blood but carries the type O allele marries a man with type B blood who also carries the type O allele. List the possible blood types of the children, including the percentage chance for each.

8.10 Suppose the woman in the previous question was heterozygous and Rh-positive. If the father is Rh-negative, what are the possible Rh-factors for the children and the percentage chance of each?

Genetic Disorders and Diseases

Although many of the health problems that exist are due to infectious agents such as bacteria, protozoa, and viruses, some health problems can be directly linked to a person's genetic code. In general, there are at least five means by which genetic abnormalities occur: **autosomal inheritance, sex-linked inheritance, allele mutation, changes in the chromosome structure**, and **changes in the chromosome number**. We will discuss each individually.

When a genetic disorder is inherited from one of the 22 pairs of autosomes, it is called **autosomal inheritance**.

Autosomal inheritance – Inheritance of a genetic trait not on a sex chromosome

In a genetic disease associated with autosomal inheritance, a gene exists on an autosome that causes a particular problem. Sometimes the genetic disorder is recessive, and thus the offspring need to have both recessive alleles in order to have the disease. In other cases, the genetic disorder comes from the dominant allele, and you must have both recessive alleles in order to not have the disease.

In Module #5, we discussed lactose intolerance, which causes people to get sick when they eat too many dairy products. This is because those who have the disorder cannot properly produce one of the enzymes needed to digest lactose, which is the sugar that gives dairy products their sweetness. This condition is caused by a recessive allele. If a person is homozygous in this allele, she is missing a key piece of information for building the enzyme, and as a result, cannot digest lactose properly.

Using a Punnett square, you can see that it is possible for two parents without this disorder to have a child with it. Suppose we call the allele for lactose intolerance "l." We use a small "l" because it is a recessive allele. The dominant allele, which keeps you from having the disorder, we will call "L." Suppose two heterozygous parents have children. Since they are heterozygous, neither will have the disorder. However, when they have children, the Punnett square will look like this:

	L	l
L	LL	Ll
l	Ll	ll

While there is a 75% percent chance of their children not having the disorder ("LL" and "Ll" will not have the disorder), there is a 25% chance of a child having it ("ll" will have the disorder).

People who are heterozygous in a recessive genetic disorder are called **genetic disease carriers**.

Genetic disease carrier – A person who is heterozygous in a recessive genetic disorder

Since genetic disease carriers have the dominant allele that keeps them from expressing the genetic disorder, they themselves are healthy. However, as the Punnett square demonstrates, their children can end up with the disorder.

Some genetic disorders are actually caused by the dominant form of the allele. Probably the best known case is called **Huntington's disorder**. The vast majority of the people in the world are homozygous recessive in the allele that causes Huntington's disorder. If someone has the dominant allele, however, his nervous system deteriorates rapidly once he hits his forties. He eventually has trouble doing even the simplest of tasks and ends up dying prematurely. The particularly bad thing about this disease is that anyone who has it is very likely to pass it on to his children, since the allele is dominant. Even if someone who is heterozygous in the allele has children with someone that does not have the disorder, the children have a 50% chance of getting the disorder. If the parent has the disorder and is homozygous, all of the children will have the disorder! Unfortunately, most people do not know that they have the disorder until they are forty or so, and by that time, they have had their children.

As you saw in Experiment 8.3, some genetic disorders are sex-linked.

Sex-linked inheritance – Inheritance of a genetic trait located on the sex chromosomes

Hemophilia is probably the best example of a sex-linked genetic disorder. As mentioned in the experiment, hemophilia was particularly well known among the royal families of England and Russia. One reason is that members of the royal family often married their own relatives to keep the blood line "pure." Interestingly enough, this practice enhances the chance of the children having a genetic disease! Since members of a family have similar genetic codes, it is more likely for both parents to be disease carriers or to have the disease if they come from the same family. Thus, having children with the disease becomes much more likely as well. This is one reason that most countries have laws against people marrying others in their own family.

Sometimes, for reasons that we still do not completely understand, an allele will undergo a radical change during meiosis. When this happens, we say that the allele has mutated.

Mutation – A radical chemical change in one or more alleles

Sometimes when this happens, the results are benign. Many times, however, the results are quite bad. **Hutchinson-Gilford progeria syndrome**, for example, results when an allele on one of the human autosomes mutates during meiosis. A child with this allele ages rapidly. By age three or four, the child has many of the characteristics of a very old person: wrinkles, hair loss, arthritis. Because of their rapid aging, people with this disease die extremely young. The oldest living person with this disease died at age 18. Most die by the time they reach age 10.

Another form of genetic disorder comes from a change in the structure of the chromosomes.

Change in chromosome structure – A situation in which a chromosome loses or gains genes during meiosis

When a chromosome has too many or too few genes, the body does not understand all of the genetic information. As a result, disorders occur. For example, **cri-du-chat**, commonly called the "cat's-cry disease" is caused when the fifth chromosome is damaged during meiosis. The chromosome loses some of its genes, and a child produced by the damaged gamete will be mentally retarded. In addition, the child's larynx (the organ used to make sounds) is abnormal, and the infant's cry sounds like the meow of a cat.

The last genetic disorder process that we want to discuss occurs when a person receives too few or too many of a particular chromosome.

<u>Change in chromosome number</u> – A situation in which abnormal cellular events in meiosis lead to either none of a particular chromosome in the gamete or more than one chromosome in the gamete

If a gamete that has none of a particular chromosome goes through fertilization, the resulting offspring will not be homozygous in that chromosome. Alternatively, if it has two of a particular chromosome, then after fertilization, the offspring will have a total of three! The results of this can be quite devastating.

Down's syndrome, for example, occurs when a person has three of chromosome number 21. This happens because either the father's sperm or the mother's egg has two of that chromosome rather than just one. Those who have Down's syndrome (about 1 in 1,000 of the newborns in North America) are mentally retarded. Their skeletons also develop more slowly than they should; thus, they are short and their muscles are slack. Great strides have been made in the treatment of those with this genetic disorder. Even though there is no cure, with the proper physical and mental therapy, most people with Down's syndrome can lead very happy and productive lives.

<u>Summing Up</u>

Some of the most exciting careers in biology are in the field of genetics, and the research going on is truly remarkable. If you liked this subject at all, you might consider doing a little more investigation into genetics and genetic engineering. You can start by visiting the course website that was discussed in the "Student Notes" section of this book. It has some links that will lead you to more in-depth discussions of genetics.

We will be taking another look at genetics in the next module when we discuss the creation/evolution controversy. The science of genetics has probably done more to illustrate the fallacies in the theory of evolution than any other single subject. You will see how in the next module!

Before we leave this module, however, we need to re-emphasize something you learned in Module #7. We told you then that an organism's characteristics are not wholly determined by genetics. There are environmental factors and, for humans, spiritual factors that influence the makeup of an organism. With all of the talk of Punnett squares and the emphasis on phenotypes that are completely determined by genetics, it is important to remind you that genetics is not the only thing at play. Perform the following experiment to remind yourself of this important fact!

EXPERIMENT 8.4
Environmental Factors and Their Effect on Radish Leaf Color

<u>Supplies:</u>

♦ 60 radish seeds (purchase locally)
♦ 2 shallow pans or dishes
♦ Potting soil
♦ Clear plastic wrap
♦ Box to cover one dish

- ♦ Water
- ♦ Lab notebook
- ♦ Magnifying glass (if available)
- ♦ Eyedropper

Object: To observe an effect that the environment has on phenotype

Procedure:

1. Label one dish "light" and the other dish "dark."
2. Place soil in each dish and then add water until the soil is wet.
3. Spread 30 seeds across the top of each dish and cover the dish with clear plastic wrap.
4. Place both dishes in a warm place where there is sunlight, but not direct sunlight.
5. Cover one dish with the box, so that no light will get in.
6. Check the dishes daily.
7. Add water with an eyedropper if the soil begins to dry.
8. Watch for the seeds to sprout. When half of the seeds have sprouted, begin examining them each day. A magnifying glass will be useful in these inspections.
9. The first two leaves on each sprout, which are called **cotyledons** (kot ih lee' duhnz), will be either yellow or green.
10. In your notebook, set up a chart like the one below and fill in the results from your observations. Do this for the next **six** days.

	Light Dish				Dark Dish		
	# Green	# Yellow	% Yellow		# Green	# Yellow	% Yellow
Day 1							
Day 2							
Day 3							
Day 4							
Day 5							
Day 6							

11. Your observations should tell you that the cotyledons of the plants in the light will have some yellow to begin with, but they will turn green as they are exposed to sunlight. In addition, those cotyledons in the dark will continue to stay yellow or turn increasing yellow due to the lack of light.
12. Your results show you that the observable trait of cotyledon color can be influenced by environmental conditions.
13. After your sixth day of recording data, you will be done with the experiment. Clean up all of the materials and return them to their proper places.

ANSWERS TO THE "ON YOUR OWN" PROBLEMS

8.1 a. Since both factors indicate that the peas will be wrinkled, <u>the plant will produce wrinkled peas.</u>

b. This plant has one factor for smooth (S) and one for wrinkled peas (s). We are told, however, that smooth (S) is dominant. Thus, because it has one dominant factor, <u>the plant will produce smooth peas.</u>

c. Since both factors indicate that the peas will be smooth, <u>the plant will produce smooth peas.</u>

8.2 If a pea plant produces yellow peas, one possible set of factors would be "YY." However, since yellow peas are dominant, a plant needs to have only one yellow factor in order to produce yellow peas. Thus, "Yy" is also a possibility. The plant, therefore, can have <u>either "YY" or "Yy"</u> and produce yellow peas.

8.3 Since the green pea factor "y" is not dominant, the only way that a plant can produce green peas is if both of its factors are for green peas. Thus, the only possible way to produce green peas is to have the <u>"yy" factors.</u>

8.4 <u>The "SS" offspring's genotype is, simply, "SS."</u> Since it has the same alleles, <u>it is homozygous.</u> Since "S" means smooth peas, <u>its phenotype is for smooth peas. The "Ss" offspring have the "Ss" genotype; they are heterozygous</u> because their alleles are different; <u>and since the smooth pea allele is dominant, they also have the phenotype for smooth peas. The "ss" offspring are homozygous; they have the genotype "ss"; and they have the phenotype for wrinkled peas.</u>

8.5 We are using "P" to represent the purple-flower allele and "p" to represent the white-flower allele. Since one plant is heterozygous, its genotype is "Pp." The white-flowered plant must have a genotype of "pp." Remember, since the white-flower allele is recessive, a white-flowered plant *must* be homozygous in the recessive allele. The resulting Punnett square, then, is:

	P	p
p	Pp	pp
p	Pp	pp

Note that it doesn't matter whether you put the "Pp" genotype on the top or to the left. As long as one parent is on top of the Punnett square and the other is to the left, you will be able to predict the probabilities of the genotypes produced. <u>50% (two out of four) will have the "Pp" genotype, while 50% (once again, two out of four) will have the "pp" genotype.</u> Since the purple-flower allele is dominant, a "Pp" genotype leads to a purple flower. A "pp" genotype, however, leads to a white flower. Thus, <u>50% will have the purple flower phenotype and 50% will have the white flower phenotype.</u>

8.6 The cross between individuals #1 and #2 does not tell us much, since they have different phenotypes. However, both #3 and #4 have the same phenotype (no tail). When crossed, they have some offspring with tails and some without. Thus, they must each have both alleles. As a result, we know that <u>the no-tail allele is dominant,</u> because these two individuals have both alleles but express the no-tail phenotype. We will therefore call the no-tail allele "N" and the allele for a tail "n." This tells us that #3 and #4 have the "Nn" genotype. Also, since the only way to have a tail is to have both recessive alleles, we know that #2 has the "nn" genotype. Finally, the only way that #1 and #2 can have children with tails is if #1 also has an allele for a tail. Since #1 expresses the no-tail phenotype, however, #1 has the "Nn" genotype.

8.7 If the plant is heterozygous in both traits, its genotype is "Ss" and "Yy." In meiosis, the "S" could go with the "Y" or with the "y." In the same way, the "s" could go with the "Y" or with the "y." Thus the possible gametes are: *SY, Sy, sY, and sy*.

8.8 In this case, one allele is affecting many traits. Thus, this is an example of pleiotrophy.

8.9 Since the woman is type A but carries the O allele, she must have the "AO" genotype. Similarly, the man must have the "BO" genotype. The Punnett square, then, is:

	B	O
A	AB	AO
O	BO	OO

The possible blood types for the children, then, are AB (25%), A (25%), B (25%), and O (25%).

8.10 The Rh-positive allele is dominant and the Rh-negative allele is recessive. If the woman is heterozygous, she has one positive allele (P) and one negative allele (p). If the man is Rh-negative, he must have two Rh-negative alleles (pp), because that's the only way a recessive allele can be expressed. Thus, the Punnet square is:

	P	p
p	Pp	pp
p	Pp	pp

This indicates that 50% of the children will be Rh-positive and 50% will be Rh-negative. Thus, regardless of whether their blood type is AB, A, B, or O, they each have a 50% chance of being Rh-positive.

STUDY GUIDE FOR MODULE #8

1. Define the following terms:

a. True breeding
b. Allele
c. Genotype
d. Phenotype
e. Homozygous genotype
f. Heterozygous genotype
g. Dominant allele
h. Recessive allele
i. Mendel's principles of genetics
   (use updated terminology)
j. Pedigree

k. Monohybrid cross
l. Dihybrid cross
m. Autosomes
n. Sex chromosomes
o. Antigen
p. Autosomal inheritance
q. Genetic disease carrier
r. Sex-linked inheritance
s. Mutation
t. Change in chromosome structure
u. Change in chromosome number

2. Three pea plants have the following alleles for yellow ("Y") and green ("y") peas. What is the genotype and phenotype of each? Note whether they are homozygous or heterozygous.

a. YY   b. Yy   c. yy

3. What process causes gametes to have only one allele, since other human cells have two of each allele?

4. A pea plant which is homozygous in the dominant, axial flower allele ("A") is crossed with a pea plant that is heterozygous in that allele. What are the possible genotypes and phenotypes, along with their percentage chances, for the offspring?

5. A woman is heterozygous in the ability to roll her tongue when extended. If she marries a man who cannot roll his tongue, what percentage of their children will be able to roll their tongues? Remember, the allele for being able to roll your tongue is dominant.

6. Recall that in guinea pig coat color, black (filled circles and squares) is dominant and white (hollow circles and squares) is recessive. What is the genotype of the male parent in the cross below?

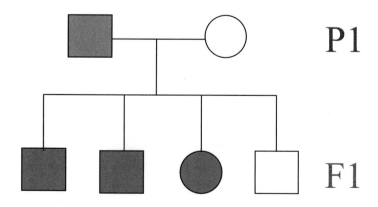

7. The following pedigree is for the presence or absence of wings on a certain insect. The hollow circles and squares represent insects without wings, while the filled circles and squares represent insects with wings. Which is the dominant allele? What are the genotypes of individuals 1-4?

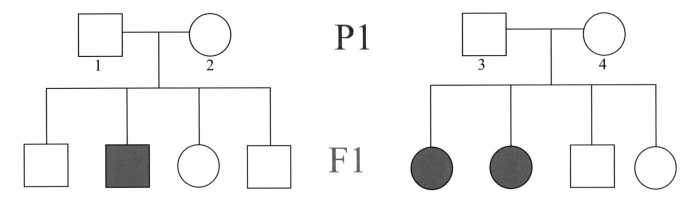

8. Give the possible phenotypes and the percentage chance for each in the dihybrid cross between a pea plant that is homozygous in producing smooth, yellow peas and a pea plant that produces wrinkled, green peas. The smooth and yellow alleles are dominant.

9. Give the possible phenotypes and the percentage chance for each in the dihybrid cross between a pea plant that is heterozygous in producing smooth, yellow peas and another with the same genotype.

10. In fruit flies, the color of the eye is a genetic trait that is sex-linked. What is the percentage of males that will have white eyes when a heterozygous, red-eyed female is crossed with a white-eyed male? What is the percentage of females that will have white eyes from the same cross?

11. In the case of fruit fly eye color again, what must be the genotype of a male fly if, when crossed with a heterozygous female, there is no possibility of having a female with white eyes?

12. If a gamete has two alleles for the same genetic trait, what type of genetic disorder will result in a zygote formed with this gamete?

13. A person carries a genetic disorder but does not have the disorder. How is that possible?

14. Do sex-linked genetic disorders affect men and women the same? If not, which sex is affected more and why?

15. Two individuals have the exact same genotype for a certain trait, but they are not identical when it comes to that trait. How is this possible?

16. A woman with type O blood marries a man with type AB blood. What blood types are possible for their children? What is the percentage chance for each blood type?

17. If a person has type B- blood, what are the possible genotypes for that person? Include the possible genotypes related both to the type of blood as well as the Rh-factor.

18. What term do we use to refer to genetic traits that are governed by more than one set of genes?

MODULE #9: Evolution: Part Scientific Theory, Part Unconfirmed Hypothesis

Introduction

In 1859, Charles R. Darwin published a book entitled *On the Origin of Species by Means of Natural Selection, or the Preservation of Favoured Races in the Struggle for Life.* Typically referred to as *The Origin of Species*, Darwin's book caused a firestorm in the scientific community. In his book, Darwin proposed a theory that attempted to explain the diversity of life that exists on earth. This theory, now known as the theory of evolution, made no reference to God. Instead, it proposed that the same kinds of processes which we see occurring today are, in fact, responsible for all of the species on the planet. In effect, Darwin's book proposed to answer the age-old question, "How did we get here?" without ever referring to a supernatural Creator.

This sent shockwaves throughout the scientific community because, at that time, science was inseparably linked to God. Most scientists were formally trained in the Bible. Many were, in fact, clergy of the church. Back then, it was common for scientific publications to mention God and the Bible often and with reverence. The very idea, then, that Darwin proposed a theory that did not refer to or require God was astonishing! To some, it was horrendous. They could not imagine a scientist who didn't refer to God as a natural part of his scientific inquiries. To others, it was what they had wanted for a long time. They did not believe in God, and finally an idea had come along that told them how we came to be without referring to anything supernatural.

Sadly, Darwin's theory (with ample modifications) has become the standard explanation for the origin of life on this planet. As you will see in this module, however, it is not because of the scientific evidence. Based on the scientific method that we discussed in Module #1, the important part of Darwin's theory never really left the stage of being a hypothesis. Nevertheless, the majority of scientists today consider his theory to be a scientific law! Why is this the case? Hopefully, that is one of the questions this module will answer for you.

Before we begin discussing Darwin and his theory, however, we need to point something out. There are many ways that people of faith have approached the theory of evolution. On one end of the spectrum, there are those who assume that it must be true, since the majority of scientists today believe in it. As a result, they try to make the Scriptures of their faith consistent with it. In the case of Christians and Jews, for example, this means trying to interpret the first few chapters of Genesis in such a way as to allow for the process of evolution. On the other end of the spectrum, there are those who refuse to even consider the theory, because it runs counter to what they believe Scripture teaches. Christians and Jews on this end of the spectrum, for example, will not even consider the theory, because their interpretation of the first few chapters of Genesis flatly contradicts the theory of evolution. Of course, there are many people between these two ends of the spectrum, attempting to either integrate the idea of evolution into their faith on one level or another, or rejecting part or all of the theory because it is considered incompatible with their faith.

Who is right? What is the proper way to view the theory of evolution in the light of faith? We will not answer those questions. Clearly we have very strong ideas on how Christians should approach the theory of evolution, but we do not think that such opinions belong in a science textbook. Questions of Scripture interpretation and fundamental dogma should be left to other, more theological works. Instead, we choose to approach evolution strictly from a scientific perspective. We will examine the theory and see where it stands in the light of current scientific knowledge.

Charles Darwin

Because he is such an important figure in the field of biology, it is necessary to look at the life of Charles Darwin in some detail. This will help us gain insight into how he developed his theory, and, hopefully, you might learn a lesson or two from his story.

Charles Darwin
Portrait from www.clipart.com

Charles Robert Darwin was born in the village of Shrewsbury, England on February 12, 1809. On that same day, far away, Abraham Lincoln was born in Kentucky. Darwin was born into a relatively wealthy family where education and artistic enrichment were stressed. In 1825, he enrolled at the University of Edinburgh, where all the men in his family had been educated. His father wanted him to study medicine, but he was sickened at the first sight of surgery being performed without anesthesia. He also showed little aptitude for the subject, so after two years, he abandoned the study of medicine.

When he left the study of medicine, he transferred to Christ's College in Cambridge, England, to study theology. Contrary to what you might have heard about Darwin, he seemed to be a deeply committed Christian at this point in his life. During this part of his life, he said that he did not "…in the least doubt the strict and literal truth of every word of the Bible" (Julian Huxley and H.B.D. Kettlewell, *Charles Darwin And His World*, [New York, NY: Viking Press, 1965], 15). Since he considered himself a devout Christian, the study of theology came quite naturally to him, and he graduated with a B.A. in theology, Euclid, and the classics.

Although his degree was in theology, Euclid, and the classics, Darwin developed a keen interest in geology while at Cambridge. Thus, when he had the opportunity to accompany a Cambridge professor, Adam Sedgewick, on a geology field trip in the summer of 1831, he jumped at the chance. While on that field trip, he was offered the position of naturalist on the *HMS Beagle*, a ship that planned to circumnavigate the globe. Although you might think it unusual for a ship to employ a naturalist, nearly every ship had such a position available. Darwin eagerly accepted the assignment, and that voyage changed both Darwin's life and the face of science forever.

Before that voyage, Darwin had read a book written by Thomas Malthus entitled *An Essay on the Principles of Population*. In this book, Malthus said that all individuals within a population struggle against other individuals to obtain what is necessary (food, shelter, a mate, etc.) in order to survive and reproduce. While on board the *HMS Beagle*, Darwin also read some of the works of a controversial geologist named Sir Charles Lyell. Lyell was one of the first scientists who rejected the history of the world as told in the Old Testament and tried to show that the same processes we see at work today could, given eons and eons of time, produce all of the geological features in the world. Geologists often summarize Lyell's idea with the catch phrase "The present is the key to the past."

Darwin voyaged on the *HMS Beagle* for five years; during that time, he made many observations. Each time the ship dropped anchor, Darwin collected samples and made observations of the species native to whatever island or land mass he was on. These observations, some of which we will detail later, combined with the ideas of Malthus and Lyell, led Darwin to formulate his theory, which he called "natural selection." Although his theory was completely formulated by the time he left the *HMS Beagle*, he did not publish his book for another 23 years. Part of the delay was due to Darwin trying to perfect his work, but most of it was due to his wife, who recognized the devastating effect that his work could have on the church. She pleaded with him not to publish, and he respected her wishes for some time, but in the end, he felt that he had to communicate his ideas to the scientific world. He therefore published *The Origin of Species* in 1859.

It is important to note that Darwin was a careful, meticulous scientist. He was not the anti-religion crusader that many have made him out to be. If you actually read his work, you will find that it is quite evenhanded. Indeed, Darwin devoted more space to discussing the reasons a scientist might not want to accept his main hypothesis than he did to the discussion of why a scientist should accept it! You will not find that kind of evenhandedness in the majority of scientific writing that occurs today. Indeed, modern scientists (especially evolutionary crusaders) could learn a lot from Darwin's style. Darwin's only real mistake was to allow his faith to erode as a result of the science he pursued on the *HMS Beagle*.

The best illustration of how Darwin's faith eroded while on the *HMS Beagle* and the years after can be found by comparing two statements he made. During the earliest part of his voyage, he wrote in his diary that he often bore the brunt of a good deal of laughter "…from several of the officers for quoting the Bible as final authority on some moral point" (Bern Dibner, *Darwin of the Beagle*, [Cambridge, MA: Burndy Library, 1964], 82). Only a few years after his voyage, however, he stated "…that the Old Testament from its manifest false history of the world, with the Tower of Babel, the rainbow as a sign, etc., etc., and from its attributing to God the feelings of a revengeful tyrant, was no more to be trusted than the sacred books of the Hindoos [sic], or the beliefs of any barbarian" (Dibner, pp. 82-83). Clearly these are the statements of a man whose faith at first seemed strong but then eroded over time to nothing!

Charles Darwin died in 1882 as the result of a long illness. He died a celebrated naturalist whose views were said to usher in a new age of science. He was buried in Westminster Abbey along with such scientific greats as Sir Isaac Newton and Lord William Thomson Kelvin. There is a myth going around the Christian community that Darwin recanted his theory on his deathbed. This is a lie, and historians think that it was started by the widow of Sir James Hope, fleet admiral for the Royal Navy. She claims to have visited Darwin shortly before his death and to have heard him recant his theory and ask to be told how he might be saved. Darwin's own daughter Henrietta, however, said, "…[the admiral's widow] was not present during his last illness, or any illness. I believe he never even saw her…" In addition, she states that "He never recanted any of his scientific views, either then or earlier…The whole story has no foundation whatever" (Paul F. Boller and John George, *They Never Said It: A Book Of Fake Quotes, Misquotes, & Misleading Attributions*, [Oxford: Oxford University Press, 1989], 19-20). Although the story of a deathbed recantation by Darwin is appealing to Christians, it is almost certainly a lie and therefore such a story does not, in any way, honor God.

Although this biography was rather long and involved, it was necessary for four reasons. First, it is important that the phony story of Darwin's deathbed recantation not be spread any further. Second, it is important for you to realize that although Darwin's theory has had devastating effects on the faith of many people, Darwin himself was not an antireligion crusader like many evolutionists are today. Darwin was a careful, dedicated scientist who started his career speaking like a Bible-believing Christian. There is not a shred of evidence that he had any intentions of harming the church. He was merely communicating what he thought were the obvious conclusions of science. Third, it shows that even careful use of the scientific method can result in the wrong conclusion. Despite the fact that Darwin did everything right in terms of the science that he did, we can show that although a portion of his theory is valid, the major conclusion is not. Thus, proper use of the scientific method does not guarantee a correct answer!

Finally, a look at Darwin's life can show you how horrible the results are when you put your faith in science. As we stated in Module #1, science is limited and is constantly changing. What we thought were scientific laws less than a century ago are now known to be wrong. Indeed, as you will see in the rest of this module, we now know that most of Darwin's ideas were very wrong. You simply cannot put your faith in something as limited and subject to change as science. Had Darwin realized that, he would not have allowed his faith in the Bible to be eroded, and he might never have championed this errant theory that has had such a devastating effect on the faith of others!

Darwin's Theory

During his time on the *HMS Beagle,* Darwin had a chance to investigate a small chain of islands called the Galapagos (guh lah' puh gus) archipelago (ark uh pel'uh go). The 13 islands of the archipelago are the result of volcanic activity, and these islands still exist about 600 miles west of the South American nation of Ecuador (on the equator). Although Darwin was pleased to study a wide variety of plant and animal life on these islands, he concentrated on the finches that lived there. Many science historians credit these birds, now known as "Darwin's finches," as inspiring Darwin's theory of evolution through natural selection.

You see, there were (and still are) many different species of finches living in the Galapagos. These species have several common characteristics, but there are specific differences between each species that were of great interest to Darwin. Consider, for example, the figure below, which concentrates on the beaks of three of these finches.

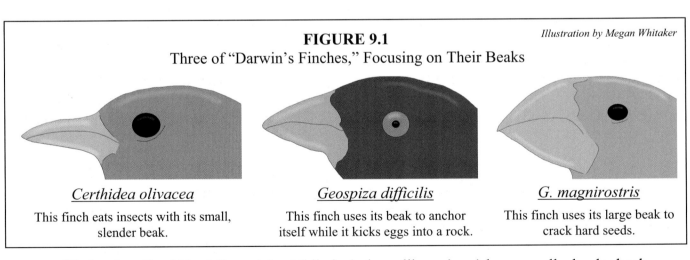

FIGURE 9.1

Illustration by Megan Whitaker

Three of "Darwin's Finches," Focusing on Their Beaks

Certhidea olivacea

This finch eats insects with its small, slender beak.

Geospiza difficilis

This finch uses its beak to anchor itself while it kicks eggs into a rock.

G. magnirostris

This finch uses its large beak to crack hard seeds.

Notice that *Certhidea olivacea* (ser' thih dee' uh aw lih vay' see) has a small, slender beak, while *Geospiza difficilis* (jee' oh spih' zuh dif' uh sil' us) has a larger, stouter beak, and *G. magnirostris* (mag' nih rah' stris) has the largest, stoutest beak of the three. Why do these finches have such different beaks? Well, let's look at what they eat. *C. olivacea* eats tiny insects. *G. difficilis* eats eggs that it steals from nests, but it does not use its beak to break them. Instead, it grabs onto something with its beak and then kicks the eggs into a rock. Finally, *G. magnirostris* eats hard seeds that it must crack with its beak. Each finch seems to have just the right beak for its food source: a small, slender beak for the finch that eats soft insects; a larger, sturdier beak for the one that uses it as an anchor; and the largest, sturdiest beak for the one that uses it to crack hard seeds.

Differences like these fascinated Darwin. You see, the scientists of Darwin's day would have looked at each of these species and assumed that God had designed each one individually and gave each one exactly the beak that it needed to eat the food it was supposed to eat. Darwin, however, imagined something else. He said that other than the beak and a few other differences (size and plumage color, for example), these finches were all remarkably similar. Since they were all so similar, he imagined that they all came from a common ancestor long ago. As the feeding needs of the finches changed, however, this common ancestor began to give rise to many different species of finch, each with a unique beak.

How did Darwin propose that this could happen? Well, he said, look at what happens when any species reproduces. When two people have a baby, for example, the baby has many characteristics in common with his parents. The baby's eye color might be the same as his mother's, and his hair color might be the same as his father's. Nevertheless, the baby usually has some characteristics that do not seem to come from either parent. Some tall professional basketball players, for example, have short parents. Thus, although offspring do tend to resemble their parents, they also have a few characteristics that are quite different from the corresponding characteristics in their parents. It is these differences, Darwin thought, which could be responsible for all of the finches in the Galapagos.

Suppose that long ago, there was only one species of finch living on the islands. If food supplies were to grow scarce, the finches that made up the population of this species would compete with one another for the dwindling food supplies, just as Malthus predicted. When this competition began occurring, any finch that had an advantage would be more likely to win the competition than one who didn't. Thus, suppose a finch was born that had a beak which was stronger than the typical finch. Well, that finch might be able to find a new source of food (hard nuts that other finches couldn't break open, for example). With this new source of food, this strong-beaked finch would most likely win the competition for survival. As it reproduced, then, it would most likely pass on this new, strong beak to at least some of its offspring. Over many, many generations, each time one of these finches was born with an even stronger beak, it would be more likely to survive, because it could continue to find more food than the finches with which it was competing. This competition, combined with the natural differences that arise between parent and offspring, could, over generations, produce a finch whose beak was short and stout, like that of *G. magnirostris*, even if the original species of finch had a small, slender beak, like that of *C. olivacea*.

During that same time (or perhaps later), another finch might have been born whose beak was short and stout, but not short and stout enough to break open new seeds. However, perhaps it was short and stout enough to act as an anchor while the finch broke open eggs with its feet. This would make it easier for that finch to get to the nutrition inside of eggs, giving it an edge in the competition to survive. As time went on, this finch would survive and would pass on its basic beak shape to its offspring. Each time a finch was born whose beak was more ideally suited for the task of anchoring the bird while it broke open eggs, the finch would have a better edge in the competition for survival, making it more likely to live and pass its new characteristic on to more and more offspring. Thus, as time went on, the original finch species that lived in the Galapagos and had a beak ideal for eating insects would eventually give rise to two new species of finch: one with a beak ideal for grasping onto objects and another with a beak that was ideal for breaking open hard seeds.

This is the mechanism by which Darwin imagined that all species of finch he observed could have originated from a single type of finch long ago. He called this mechanism "natural selection," because he said that due to the fierce competition which occurred between members of a species, any individual that had a unique characteristic making it more likely to win the competition would be selected by nature to survive. As time went on, these unique characteristics would continue to "pile up" on one another until eventually, a new species was formed.

Darwin, of course, did not stop there. After all, he imagined, if such a mechanism could be responsible for causing finches to develop new beaks, why couldn't that same mechanism allow them to develop longer, stronger wings, longer, sharper talons, and keener eyesight? If that was so, finches could eventually give rise to eagles! These ideas led Darwin to his overall theory of evolution. At one time, Darwin believed, there was a relatively simple (most likely aquatic) life form that existed on earth. Darwin made no speculations about how that life form developed, but others who followed have constructed wild scenarios that try to explain the formation of this organism without the intervention of a supernatural creator. This life form, Darwin assumed, would begin to reproduce and, as is the case today, variations would occur in the reproduction. These variations, guided by the process of natural selection, would eventually "pile up" so as to form new species. These species would, in the same way, give rise to other species. Thus, over eons of time, Darwin believed that this mechanism could explain the existence of all life forms on the planet.

Hopefully you can see how Malthus and Lyell influenced Darwin's thinking. After all, Malthus gave Darwin the idea that individuals within a species compete with one another in order to survive. This led to Darwin's idea of natural selection. Lyell's concept that the present is the key to the past allowed Darwin to speculate that the same variations which we see in reproduction today could, over vast ages of time, be responsible for all of the variations among all of the species that exist on the planet. In other words, Darwin did not dream up this theory on his own. He was influenced by the works of others.

If you remember our discussion of the scientific method, you will recognize that at this point in the story, Darwin's idea was really no more than a hypothesis. Darwin made a bunch of observations and then proceeded to develop an explanation for those observations. Did the concept of evolution through natural selection ever make it past the stage of being a hypothesis? Well, the answer to that is both yes and no. Hopefully you will see what we mean in the next section.

ON YOUR OWN

9.1 What two concepts promoted by other scientists influenced Darwin in developing his ideas?

9.2 The cheetah is the fastest land-dwelling animal on the planet. It has been observed to reach peak speeds in excess of 70 miles per hour! It uses this speed to catch animals that are often too fast to be killed by most other predators. Use Darwin's reasoning to explain how natural selection could produce such a creature from a slower animal.

9.3 Why is Darwin's hypothesis sometimes called "the survival of the fittest"?

Microevolution and Macroevolution

As we learned in Module #1, once a hypothesis is formed, it is tested against experimental data. If the data continue to support the hypothesis, it eventually becomes a theory. If it does not, it must be altered or discarded. This is the next step in the scientific method, and it is where Darwin's hypothesis ran into some trouble.

After leaving the *HMS Beagle*, Darwin began experimenting with pigeons. He raised and bred them, trying to see if natural selection could result in new species of pigeon. To investigate further, he talked to other animal breeders. He interviewed those who bred dogs, horses, and pigeons, looking to their experience as a guide to whether or not his hypothesis could be correct. Indeed, he found much evidence that confirmed at least part of his hypothesis. He noted several cases of breeders who, over several generations, succeeded in producing pigeons that were so different from the species with which the breeder started that they could reasonably be classified as a new species of pigeon. The same seemed to be the case with dogs and horses as well.

As another piece of evidence for his hypothesis, Darwin compared the domesticated versions of many animals with their wild counterparts. Wild dogs, for example, looked and behaved quite differently from domesticated dogs. In fact, many breeds of domesticated dog cannot reproduce with wild dogs. Thus, by the definition of species as laid out in Module #1, these domestic dogs would be considered a wholly different species from any species of wild dog. Despite these incredible differences, domestic dogs were, many generations ago, simply wild dogs that men began to train and domesticate. Over generations, however, dog breeders would selectively mate those dogs that had what the breeder considered the best traits for domestication. Thus, the "wilder" dogs were not allowed to reproduce, and the tamer dogs were. This "manmade" selection, Darwin realized, mimicked natural selection, allowing the small variations that occurred during reproduction to "pile up," leading to a new species of dog: the domesticated dog.

With these observations, Darwin was able to do two things. First, he established as a valid scientific theory the idea that the natural variations which occur during reproduction could, when guided by natural (or manmade) selection, take one species and pile up so many changes that the result could be something reasonably classified as another species. In other words, he showed that his explanation for the many species of finches in the Galapagos archipelago was scientifically viable. Second, Darwin was able to destroy forever an idea that had been established for generations before him: **the immutability of species**.

<u>The immutability of species</u> – The idea that each individual species on the planet was specially created by God and could never fundamentally change

In other words, scientists of Darwin's day believed that every creature was created during the time of creation and has existed, essentially the same, ever since that time. In the case of dogs, for example, those who held to the idea of the immutability of the species would say that in the Garden of Eden, there were Doberman pinschers, Saint Bernards, dachshunds, and chihuahuas. Each of these breeds of dog continued, essentially unchanged, up to the present. Darwin masterfully showed that this just wasn't true. He showed that all of these breeds of dog came from some original dog ancestor, and the natural variations that occurred in reproduction, guided by natural (or manmade) selection, resulted in the many different breeds of dog that exist to this day.

Although it sounds like Darwin had remarkable success in testing his hypothesis, you need to realize that what Darwin showed to be true was only a small part of his hypothesis of evolution. The idea that one ancestral finch could, over generations, give rise to many different species of finch was revolutionary, but it was not where Darwin's idea stopped. Once he had destroyed the idea of the immutability of the species, he wanted to go much further. He wanted to show that this same process, over millions (or perhaps billions) of years, could, eventually, cause the ancestral finch to give rise to an eagle. This is where Darwin ran into all sorts of trouble when comparing his hypothesis to the data.

Although it was rather easy to show that a species of wild dog could, over time, give rise to several breeds of domestic dog, it was quite another to show that a dog could give rise to a radically different species, such as a horse or a cow! In fact, Darwin found *some* evidence for this idea, but it was inconclusive at best. We will be looking at this data in depth in the next section, but for right now it is enough to say that there was so much data contradicting this part of his hypothesis, that he spent the majority of his book discussing the problems with his hypothesis.

In the end, Darwin found ample evidence that starting with a basic life form (a finch, for example), many other specialized species of this life form (many species of finch) can arise as a result of variation guided by natural (or manmade) selection. However, when it came to showing that a basic life form (once again, a finch) could evolve into a completely different life form (like an eagle) by natural selection, there was precious little evidence for his hypothesis and plenty of evidence against it. This has led scientists to divide Darwin's theory of evolution into two parts: the theory of **microevolution** and the hypothesis of **macroevolution**.

Microevolution – The theory that natural selection can, over time, take an organism and transform it into a more specialized species of that organism

Macroevolution – The hypothesis that processes similar to those at work in microevolution can, over eons of time, transform an organism into a completely different kind of organism

The distinction between macroevolution and microevolution cannot be overemphasized. There is so much evidence to support the idea of microevolution that it is a well-documented scientific theory. There is so little evidence for macroevolution and so much evidence against it that it is, at best, an unconfirmed hypothesis.

Well, if Darwin could not find much evidence in support of macroevolution and found a lot of evidence against it, how did it become so popular among scientists? There are several answers to that question, but one of the most important ones is that at the time, scientists were rather ignorant about a great many things which we take for granted. As a result, Darwin could argue his point rather convincingly.

Darwin basically said that since microevolution is so clearly apparent from a scientific point of view, then macroevolution should also be rather obvious. After all, if finches can change a little over a small amount of time, shouldn't they be able to change a lot over a long period of time? Assuming that the amount of change a given species can experience is essentially limitless, it will just take a little longer for microevolution to slowly lead to macroevolution.

To scientists of Darwin's day, this sounded like a reasonable argument. You see, they didn't know what we know about genetics. They didn't know that the genetic code is responsible for determining the range of characteristics that a species has. Thus, they didn't know that the natural variation we see in reproduction today is simply the result of different alleles being expressed in different individuals. Since we know that the number of alleles in the genetic code of any species is limited, we also know that the natural variation which occurs as a part of reproduction is limited as well. Thus, unlike Darwin argued, the variation that a species can experience is *not* unlimited. It is limited by the number and type of alleles in the species' genetic code. Thus, today we know that macroevolution cannot occur the same way that microevolution occurs.

You see, microevolution is simple to explain. When God created the animals and plants, he built into their genetic code a great amount of variability. As these plants and animals began reproducing, this variability began manifesting itself. This built-in variation was then acted upon by natural selection to create the variations that we see within a particular kind of creature. Thus, God probably created a "typical" dog during the creation week, and then the process of microevolution produced the many variations of dog that we see today. Microevolution, then, is a testament to God's foresight. As the Creator, God knew that the creatures of His world would have to adapt in order to survive. Thus, He built in their genetic codes the ability to change, and microevolution is simply the theory that describes how that change takes place and is directed by the pressures of the environment.

Macroevolution, however, is something quite different. The hypothesis of macroevolution assumes that a given life form has an unlimited ability to change. This means that some process must exist to *add information to the creature's genetic code*. After all, a creature's ability to change is limited by the information in the genetic code. There are only a certain number of genes and alleles of those genes. There is therefore only a certain number of possible variations in genotype and therefore a limited number of possible phenotypes. Thus, in order to get an unlimited amount of change, a creature must somehow find a way to add genes and alleles to its genetic code! This is something altogether different from microevolution and, as we will see in the next few sections, there is precious little data supporting such a hypothesis and quite a lot of data contradicting it.

In the end, then, we can distinguish between microevolution and macroevolution by referring to genetics. If we are talking about a species varying *within* its genetic code, we are talking about microevolution. This is how wild dogs became domestic dogs and how the many varieties of finch formed in the Galapagos archipelago. On the other hand, if we are talking about a species suddenly *adding information to* its genetic code, we are talking about macroevolution. The distinction is quite important, because the former is a well-established scientific theory while the latter is an unconfirmed hypothesis.

ON YOUR OWN

9.4 House sparrows are small, seed-eating birds that are native to parts of Europe, Northern Africa, and the Middle East. In the fall of 1851 and the spring of 1852, one hundred of these birds were brought to Brooklyn, New York. Since then, they have spread throughout the United States. Even though the original 100 birds were all very similar, their descendants are not. In fact, if you study the size of house sparrows in the United States, you will find that they are bigger in the northern parts of the United States and smaller in the southern parts of the United States. Is this an example of microevolution or macroevolution? Can you think of what might have caused this difference in size?

9.5 Some biologists believe that the whale once had a cowlike ancestor that lived on land. This ancestor was very heavy, so it started spending a great deal of time in the water. The water helped buoy it up, making it easier for the animal to walk. As time went on, the animal began adapting to the water, slowly changing its legs into fins and its skin into a substance more ideal for swimming in the water. Eventually, the cowlike creature gave rise to the whale. Is this an example of microevolution or macroevolution?

<u>Inconclusive Evidence: The Geological Column</u>

As we said in the previous section, there is some evidence for the hypothesis of macroevolution. For example, consider some of the facts uncovered by geology. As you should have learned in a physical or general science course, geologists classify rocks as either igneous (the result of lava cooling), metamorphic (the result of rock being transformed under extreme conditions of temperature and/or pressure), or sedimentary (formed from particles of sediment). The sedimentary rock usually forms layers that are called **strata**.

<u>Strata</u> – Distinct layers of rock

In these strata, geologists find the preserved remains of once-living organisms. Those remains are called **fossils**.

<u>Fossils</u> – Preserved remains of once-living organisms

The way that fossils are generally arranged in these strata is quite interesting, as illustrated in the figure below.

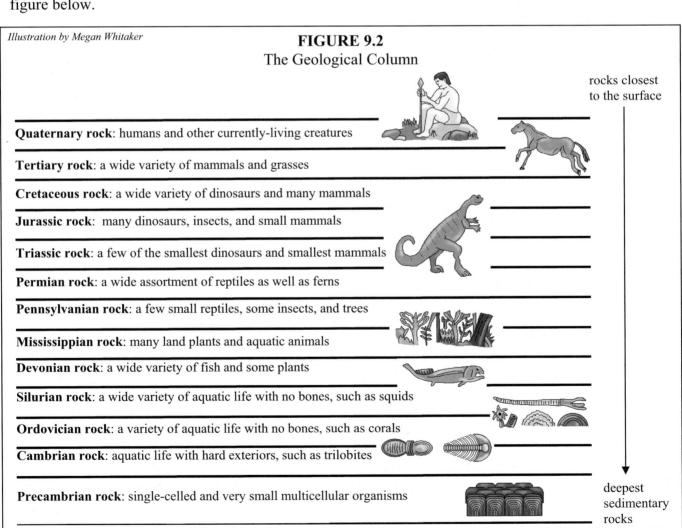

Illustration by Megan Whitaker

FIGURE 9.2
The Geological Column

rocks closest to the surface

Quaternary rock: humans and other currently-living creatures

Tertiary rock: a wide variety of mammals and grasses

Cretaceous rock: a wide variety of dinosaurs and many mammals

Jurassic rock: many dinosaurs, insects, and small mammals

Triassic rock: a few of the smallest dinosaurs and smallest mammals

Permian rock: a wide assortment of reptiles as well as ferns

Pennsylvanian rock: a few small reptiles, some insects, and trees

Mississippian rock: many land plants and aquatic animals

Devonian rock: a wide variety of fish and some plants

Silurian rock: a wide variety of aquatic life with no bones, such as squids

Ordovician rock: a variety of aquatic life with no bones, such as corals

Cambrian rock: aquatic life with hard exteriors, such as trilobites

Precambrian rock: single-celled and very small multicellular organisms

deepest sedimentary rocks

Each row in Figure 9.2 represents a set of layers of sedimentary rock. The rows on top are those sedimentary rocks closest to the surface, and those at the bottom are the deepest sedimentary rocks. Now please realize that the fossils listed in each of the rows are simply the ones that best characterize each layer. There are many other fossils in each set of rocks, but the ones listed in the figures are those that distinguish each set of layers from the others. When the rock layers are illustrated in this way, we say that you are looking at the **geological column**.

Notice the general trend shown in the figure. As you go deeper in the geological column, the fossils seem to get more and more "simple." Now we have already clearly seen that there is no such thing as a simple life form, but remember, we are talking about evidence for Darwin's hypothesis of macroevolution. Darwin and his early supporters did not have the benefit of microscopic analysis and detailed explanations of how organisms work. Thus, they did not really know that there is no such thing as a simple life form. Instead, they saw that the strata which were near the surface of the earth contained fossils of animals like horses, lions, and humans, while the deeper strata gave way to fossils of life forms like reptiles and small mammals. Even deeper, these fossils disappeared and only fossils such as fishes, squids, and trilobites remained. Darwin and his supporters thought that this progression of fossils went from "complex" organisms like human beings to "simple" organisms like squids.

Before we discuss how Darwin and his followers interpreted this geological column, we must make two important points. First, even though Figure 9.2 is a common representation of the geological column, it is not really what the geological column actually looks like! You see, 95% of all fossils that we recover are those of clams and similar organisms. There are, quite literally, fossilized clams in every region of the earth in nearly every layer of rock. So what we are showing in this representation of the geological column is really only about 5% of the fossil record. Thus, it is not a realistic representation of what the geological column actually looks like. Nevertheless, since the geological column is usually discussed in reference to macroevolution, the clams are ignored. Thus, any conclusions you make based on the geological column are, in fact, based on a tiny minority of the available data!

Second, it is very important to realize that the geological column is an idealized representation of the sedimentary rock and fossils that we see on the earth. There is really no place on the planet where you can dig and find every layer of the geological column as well as the fossils in those layers. Instead, you will find one set of fossil-bearing strata in one area of the world, and another set of fossil-bearing strata in another area of the world. These sets often contain some of the same strata, so by comparing many such sets of strata, Figure 9.2 is a representation of what the geological column might look like if you could find all of the strata and their fossils in one place. In other words, it is a theoretical construct that may or may not be accurate. It probably is a reasonable representation of the nature of the fossil record, but nevertheless, it is not pure data. That must be remembered when using it as evidence for or against *any* hypothesis, including that of macroevolution.

Now we can discuss how Darwin and his followers interpreted the geological column. To do that, we must discuss how the geologist Lyell (who influenced Darwin tremendously) interpreted it. Lyell said that the strata shown in Figure 9.2 were laid down sequentially over vast eons of time. Using his idea that the present is the key to the past, Lyell said that the strata seen by geologists were formed when sediments accumulated slowly over time. We see this happening today, he said, and it results in layers of sediment. Eventually, Lyell postulated, various chemical and environmental factors would take a given layer of sediment and harden it into rock. This would result in a single layer of sedimentary rock. As time went on, another layer of sediment would slowly accumulate on top of this layer of rock, eventually forming another layer of sedimentary rock on top of the previous layer. This would happen over and over again, eventually forming the layers of rock seen in the geological column.

Now, of course, all of this is speculation, but it is accepted by many geologists today as the way in which the rock strata in the geological column were formed. Well, if you accept this speculation as fact, then you are left to conclude something rather obvious. The deeper a rock layer is in the geological column, the farther back in earth's past it was formed. After all, since Lyell's process requires the lower rock layers to form before the higher ones can, the rock layers on top should be younger than the rock layers on the bottom. In fact, geologists even try to assign time frames for when these rock strata were supposed to have formed. Cambrian rock, for example, is supposed to have formed between 570 and 500 million years ago. The next set of rock layers (Ordovician rock) is supposed to have formed between 500 and 435 million years ago. Such numbers exist for every layer illustrated in Figure 9.2. They are based on many assumptions, and it is not clear that they are accurate, but they are commonly used in reference to the geological column.

So, from the concrete scientific facts that sedimentary rock generally forms in layers and we find fossils in those layers, we develop a theoretical construct called the geological column. Then, when we add Lyell's speculation to the geological column, we are forced to conclude that the lower the strata in the geological column, the older the rock. If we further add more assumptions, we can even come up with how long ago such rock formed. If this is the case, then fossils found in strata that are low in the geological column are the remains of creatures that lived hundreds of millions of years ago. In the same way, fossils in strata near the top of the geological column must be the remains of creatures that lived in the more recent past.

Based on Lyell's speculation, then, Darwin and his followers argued that the geological column shows us that long ago, only "simple" life forms existed. That is why we see fossils of only "simple" forms in the lowest geological strata. As you look up the geological column, however, the fossils become more and more "complex." This indicated to Darwin that as time went on, life forms got more and more complex. Well, Darwin argued, this is great evidence for macroevolution. After all, macroevolution predicts that life started out simple and, over eons of time and guided by natural selection, more complex forms of life emerged.

Now you have to realize that this conclusion is based on assumptions. You must assume that the geological column is an accurate picture of the earth's sedimentary rocks and the fossils found in them. You also have to believe that those rocks are formed as Lyell and many geologists today speculate: by the slow accumulation of sediment over eons of time. The geological column is evidence for macroevolution only if those two assumptions are right.

So the big question is, are those two assumptions right? The answer, from a scientific standpoint, is that we don't know. The geological column is probably a reasonable representation of the sedimentary rocks on earth. However, the issue of how those rock strata are formed is quite tricky. Since this is not a geology course, we do not want to spend a lot of time on the specifics. Instead, we will simply say that scientists have seen layers of sediment form slowly as a result of a process much like that suggested by Lyell. These sediment layers look a lot like "soft" versions of the sedimentary rocks we see today. Thus, it is reasonable to assume that sedimentary rocks *can* form that way. However, scientists have also seen that natural catastrophes like floods and volcanic eruptions can lay down many layers of rock virtually overnight. Geologists who have spent time studying the results of the eruption of Mount St. Helens in the state of Washington, for example, have documented the formation of a huge wall of sedimentary rock that has many strata in it, all in the span of about five hours.

Science tells us, then, that rock strata can be formed either slowly in a process like that suggested by Lyell or quickly as a result of natural catastrophes. So, if you *want* the geological column to provide evidence for macroevolution, it does. You simply have to assume that it was formed much in the way Lyell suggested that it was formed. If, on the other hand, you *don't want* the geological column to provide evidence for macroevolution, you can assume that it was not formed that way. Instead, you can assume that it was formed quickly as the result of one or more catastrophes. For example, we consider the major parts of the geological column to be the result of the worldwide flood that happened during Noah's time. An excellent book written by Dr. Steven Austin entitled *Grand Canyon: Monument to Catastrophe* uses the specific example of the Grand Canyon to provide convincing evidence that this is, indeed, the case. Of course, those who believe that the geological column was formed according to the speculations of Lyell also have evidence of their own, so the final answer is not clear.

Since, from a scientific point of view, we really don't know whether the geological column was formed according to the speculations of Lyell or by catastrophe, the data from the geological column is inconclusive. IF the geological column was formed according to the speculations of Lyell, then it is excellent evidence for macroevolution. IF, on the other hand, it was formed by one or more catastrophes, then it is excellent evidence against macroevolution. Creation scientists, for example, believe that the worldwide flood can easily explain most of the geological column. Thus, they believe that the geological column is evidence against macroevolution because it indicates that most of the fossilized organisms lived at the same time, in direct contradiction to the evolutionary view. Evolutionists, on the other hand, think that Lyell's speculations are the best way to explain the fossil record, so they believe that the geological column is evidence for macroevolution.

ON YOUR OWN

9.6 What is the big assumption that must be made in order to interpret the geological column as evidence for macroevolution?

9.7 Why is the geological column not conclusive evidence for or against macroevolution?

The Details of the Fossil Record: Evidence Against Macroevolution

Well, if the geological column holds no real evidence for or against macroevolution, where must we look next? We can look the same place Darwin did. After examining the geological column, Darwin looked at the details of the fossil record. The study of the fossil record is called **paleontology** (pay' lee un tah' luh jee), and it provides strong evidence against macroevolution.

Paleontology – The study of fossils

If macroevolution did occur, paleontologists should be able to find series of fossils that demonstrate how one species slowly evolved into another. If wild dogs, for example, did eventually give rise to horses, then there should be fossils of animals that are somewhere between a dog and a horse. Darwin called these life forms (which he assumed must have existed) **intermediate varieties**. Today, we call them **intermediate links** or **transitional forms**, because they represent a link (or transition) between one species and another. Unfortunately for Darwin, there were only a few examples of fossils that might be interpreted to be intermediate links, and even for those fossils, their status as intermediate links was quite questionable.

This lack of intermediate links was the most vexing problem that Darwin had with his hypothesis. In fact, in his book, he stated:

> Geological research, though it has added numerous species to existing and extinct genera, and has made the intervals between some few groups less wide than they otherwise would have been, yet has done scarcely anything in breaking the distinction between species, by connecting them together by numerous, fine, intermediate varieties; and this not having been affected, is probably the gravest and most obvious of all the many objections which can be raised against my views. (Charles Darwin, *The Origin of Species*, 6th ed, [New York, NY: Collier Books, 1962], 462)

Notice what he says here. Darwin's hypothesis says that one species eventually led to another. Thus, there should be "fine, intermediate varieties" of fossils in between species. The fact that there weren't was a problem that he called "grave." Additionally, note that he admits there are "many objections" which can be raised against his views. As we stated before, Darwin was very open about the flaws that he saw in his macroevolutionary hypothesis. In fact, he devoted many pages of his book to detailing the many objections which could be raised against his views!

Although Darwin could not find any good examples of intermediate links in the fossil record, he had a hope. He figured that geology and paleontology were still in their infant stages; therefore, they just hadn't found the intermediate links yet. He was convinced that as time went on, however, geologists and paleontologists would find them. Thus, he assumed that the intermediate links were currently just "missing" from the fossil record, but they would be found in time. Critics of macroevolution quickly coined the phrase "missing link" to emphasize that the fossil record was devoid of any evidence for macroevolution.

Well, what of these missing links? Has paleontology uncovered them? The answer to that is an unequivocal *no*. Read, for example, the words of Dr. David Raup, the curator of the Chicago Field Museum of Natural History and an expert on the fossil record.

> Well, we are now about 120 years after Darwin, and knowledge of the fossil record has been greatly expanded…ironically, we have *even fewer examples* of evolutionary transition than we had in Darwin's time. By this I mean that some of the classic cases of Darwinian change in the fossil record, such as the evolution of the horse in North America, have had to be discarded or modified as the result of more detailed information. (David Raup, *Field Museum of Natural History Bulletin*, **50**:25, 1979 – emphasis added)

So Dr. Raup says that the missing links are still missing. Darwin saw this fact as strong evidence against macroevolution, and Dr. Raup says that the situation is worse now than ever!

Since Dr. Raup's quote is more than 20 years old, you might wonder whether paleontology has discovered anything in the past two decades to make the situation any better for macroevolution. The answer is a clear and convincing *no*. Consider, for example, this summary of the state of paleontology in regard to macroevolution:

> …according to Darwin...the fossil record should be rife with examples of transitional forms leading from the less to more evolved...Instead of filling the gaps in the fossil record with so-called missing links, most paleontologists found themselves facing a situation in which there were only gaps in the fossil record, with no evidence of transformational intermediates between documented fossil species. (Jeffrey H. Schwartz, *Sudden Origins*, [New York, NY: John Wiley & Sons, 1999], 89)

In other words, Dr. Schwartz (a macroevolutionist) is admitting that instead of finding transitional forms, paleontologists find mostly gaps. Now please understand that this flies in the face of what Darwin proposed. In fact, on the back cover of Dr. Schwartz's book, we read, "Darwin may have argued that new species emerge through a slow, gradual accumulation of tiny mutations, but the fossil record reveals a very different scenario – the sudden emergence of whole new species, with no apparent immediate ancestors."

Now think about this for a minute. The hypothesis of macroevolution tries to explain something about earth's past. Since no one was around back then to tell us whether or not macroevolution actually happened, it is necessary to look for data that either support or contradict the hypothesis. Well, if you're looking for data about earth's history, where is the *first* place you would look? You would look in the fossil record! What does the fossil record say? It says that macroevolution never happened! Do you see what we mean when we say that scientists don't believe in macroevolution today because of the evidence? If the fossil record (the main place you look for information about earth's past) shows no evidence for macroevolution, scientists simply should not believe in it.

At this point, you might be thinking, "If the fossil record points so strongly against macroevolution, why do any scientists today believe in it?" Well, there are a couple of reasons. Some scientists are so committed to the idea of macroevolution that they have come up with special "variations" of macroevolution that attempt to "explain around" the fact that the fossil record does not support the idea. In fact, both Dr. Raup and Dr. Schwartz expose the lack of transitional forms in the fossil record specifically to promote one of these variations, which is called **punctuated equilibrium**. In this variation, it is assumed that the transitional forms that link one species to another do not live for very long. As a result, there is not much chance of them fossilizing. We will discuss this idea in more depth later in this module.

The second reason some scientists still cling to the hypothesis of macroevolution is that even though the fossil record has mostly gaps, there are *some* fossils that can be pointed to as *possible* transitional forms. Their status as transitional forms is highly questionable, but if you really need to believe in macroevolution, these fossils can give you at least some hope.

Consider, for example, the very famous fossil of a creature called *Archaeopteryx* (ar kee op' ter iks), which evolutionists want to believe is an intermediate link between reptiles and birds. This fossil is found in Jurassic rock, which (see Figure 9.2) contains remains of dinosaurs as well. In the geological column, Jurassic rock is underneath Cretaceous rock, which is underneath Tertiary rock. Although a few fossilized birds are found in other strata, the vast majority of bird fossils in the fossil record come from either Tertiary rock or the Quaternary rock that lies on top of it. Now remember, according to the assumptions of macroevolutionists, this would mean that birds did not really exist in significant numbers until the times during which Tertiary and Quaternary rock formed. Thus, during the times when Cretaceous and Jurassic rock were forming, macroevolutionists assume that birds had not yet evolved to any significant degree. Since *Archaeopteryx* is found in Jurassic rock, macroevolutionists conclude that it lived prior to most birds and that it could therefore be one of the transitional forms linking birds to their common ancestor, which is assumed to be some form of reptile. The figure below shows you a picture of the fossil of *Archaeopteryx* as well as an artist's rendition of what the creature might have looked like.

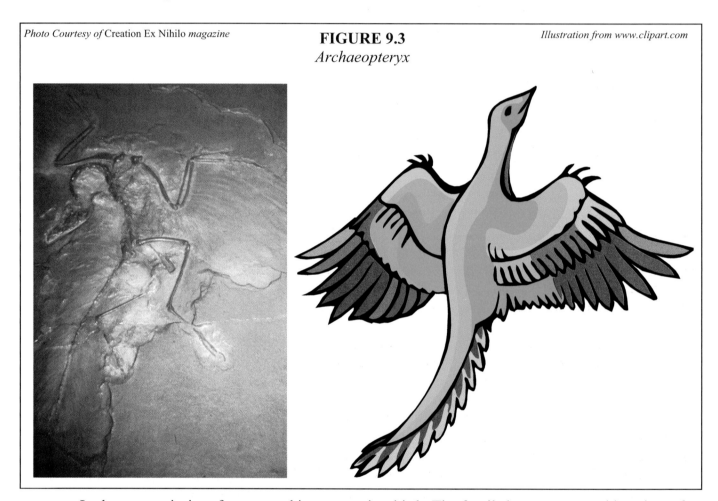

Photo Courtesy of Creation Ex Nihilo *magazine*   **FIGURE 9.3**   *Illustration from www.clipart.com*
Archaeopteryx

In the vast majority of respects, this creature is a bird. The fossil shows very good imprints of feathers, and analysis of these feather imprints indicates that they are the kinds of feathers you see on birds that are living today. In fact, flightless birds that are living today have different feathers from those of birds that fly, and the fossil imprints indicate that *Archaeopteryx* had feathers of a flying bird.

In addition, the bones preserved in the fossil are very similar to the bones of birds that are living today. The skull, for example, shows that *Archaeopteryx* had a brain very similar to flying birds that are living today, and the fossilized inner ear indicates that *Archaeopteryx* had senses of hearing and balance that are comparable to flying birds that are living today. After performing X-ray scans of the skull and working with computer models, Dr. Timothy Rowe of the University of Texas at Austin said, "This animal had huge eyes and a huge vision region in its brain to go along with that and a great sense of balance. Its inner ear also looks very much like the ear of a modern bird." (University of Texas at Austin Press Release, August 4, 2004, http://www.utexas.edu/cons/news/imaging.html, retrieved 01/06/05). In addition, paleontologists have been able to confirm bone structures that indicate *Archaeopteryx* had the same kind of lung design that birds living today have. In the end, then, *Archaeopteryx* seems to be a bird.

Why do many paleontologists consider *Archaeopteryx* a transitional form between reptile and bird? Because *Archaeopteryx* has teeth (which birds living today do not have), and it has claws on its wings, as shown on the illustration in the figure. No adult bird living today has claws on its wings. Some young birds (like the juvenile touraco or the juvenile hoatzin) have claws on their wings when they are young, but they lose them by the time they are adults. Some adult birds, like the ostrich, have structures on their wings that a few texts call "claws," but they are better called "spurs," because they do not have the actual structure of claws.

Because of these minor differences between *Archaeopteryx* and birds living today, macroevolutionists want to believe it is a transitional form between bird and reptile. After all, most (but not all) reptiles have teeth, and most (but not all) reptiles have front and back claws. Thus, the teeth in *Archaeopteryx* are supposed to represent reptilian teeth that had not quite "evolved away," and the claws on its wings are supposed to represent front reptilian claws that had not quite blended in to the wing.

There are at least two problems with this interpretation. First, it assumes that birds living today are the only representations of proper birds. This is a rather myopic view of the natural world. We have many fossils that tell us a lot about the kinds of creatures that lived on this earth but are not living now. Are we to ignore them? If we do not ignore these creatures, we find that there were other birds that had teeth. In fact, there is a whole subclass devoted to such extinct birds, subclass Odontornithes (oh' don tor' nih theez). As a result, it is not clear that the teeth on *Archaeopteryx* are all that special.

The second problem with this interpretation is that it puts a lot of emphasis on rather minor structures in the animal. Based on its feathers, bone structure, lung structure, etc., *Archaeopteryx* seems to be a true bird. These are the main features we look at to determine whether or not something is a bird. To concentrate on two minor features, one of which exists in extinct birds, seems to be ignoring the vast majority of the data. Those who want to believe that *Archaeopteryx* is a transitional form will counter that we only see these structures in extinct birds, and therefore it is reasonable to assume that they were a part of the evolutionary process. However, species go extinct rather regularly, as demonstrated by the fossil record itself. The fact that all of the birds with these characteristics are now extinct is not surprising, since extinction is a major part of the fossil record.

Archaeopteryx at least illustrates how difficult it is for macroevolutionists to come up with transitional forms in the fossil record. If *Archaeopteryx* is a transitional form, it is a very late one. It must have been one of the very last creatures on the hypothetical macroevolutionary line between reptile and bird. Of all the transitional forms necessary to turn a reptile into a bird, it seems odd that the only one that has been found is so incredibly birdlike. Why isn't there a transitional form that is not so ambiguous?

As we said before, there are a few fossils like *Archaeopteryx* that macroevolutionists can present as transitional forms. The problem is, like *Archaeopteryx*, these supposed transitional forms are incredibly similar to one of the two types of creatures they are supposed to be linking. It seems much safer to conclude that these are just specialized versions of the creatures they are similar to, as opposed to transitional forms between two different kinds of creatures.

Another classic example of this kind of fossil is *Australopithecus* (aw stray' low pih' thih kus) *afarensis* (ah fuh' ren sis). This creature is supposed to be a transitional form between apes and humans. Although fossils of several creatures from genus *Australopithecus* have been discovered over the years, the best example is a partial skeleton of the *A. afarensis* species that has been nicknamed "Lucy." This skeleton was discovered by Donald Johanson in 1974 near Hadar, Ethiopia. This was not the first discovery of fossils from this genus, and it was not the last. However, this skeleton is important because it is the most complete skeleton we have from this genus. A sketch of that skeleton is shown below.

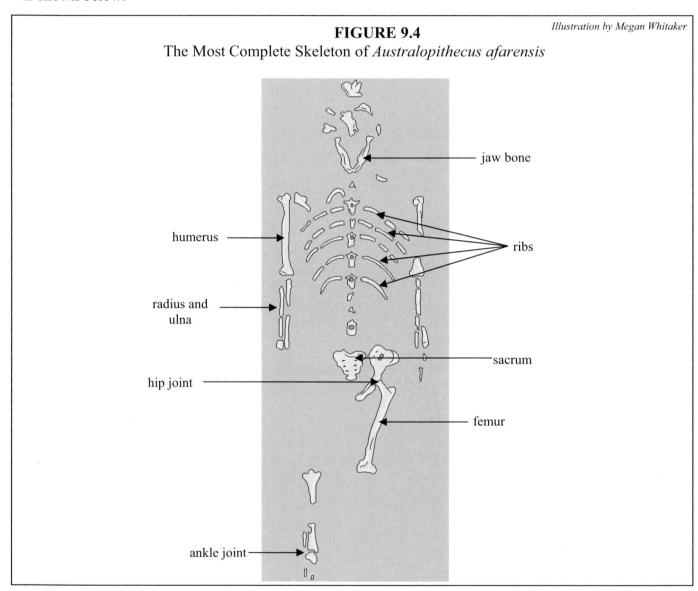

FIGURE 9.4

Illustration by Megan Whitaker

The Most Complete Skeleton of *Australopithecus afarensis*

Although this might not look like a complete skeleton to you, it is actually quite a find. One of the problems with studying fossils is the fact that they tend to be very incomplete. As a result, scientists are left to speculate on the nature of a creature based often on far too little evidence. Many fossils were proclaimed intermediate links based on just a few tiny fragments of bone. Perhaps the greatest example was Nebraska Man. Scientists found a single tooth among the remains of some ancient tools. Based on *that single tooth*, scientists proclaimed that they had found one of the intermediate links between man and ape. They had artists draw full pictures of Nebraska Man, based only on a tooth! The pictures looked very convincing – a hulking, brutish figure that indeed looked like a link between man and ape. Later on, it was conclusively shown that the tooth actually belonged to a peccary, which is a certain type of pig! Thus, you should never believe the drawings that you see in books. Find out what *fossils* exist for the supposed creature. You will most likely be amazed at how few there are!

This specimen of *Australopithecus afarensis*, however, is quite a nice specimen, so we can learn a lot from it. When scientists examine such a specimen, they typically look at a few "key" bones that are characteristic of a certain kind of creature. For example, jaw bones are often very distinctive in most animals. Notice how the jaw bone of *Australopithecus afarensis* is shaped like a "V." This is very distinctive of an ape. Also, the ratio between the size of the humerus and the size of the radius or ulna is also helpful in determining the kind of creature that *Australopithecus afarensis* was. Once again, the ratio of these bones indicates that *Australopithecus afarensis* is an ape. In fact, each bone in this entire skeleton indicates that *Australopithecus afarensis* is an ape.

If the bones indicate that it is an ape, why is *Australopithecus afarensis* considered an intermediate link? Well the hip joint and ankle joint can be constructed in such a way as to make *Australopithecus afarensis* stand upright in a relatively comfortable manner. This is unusual in apes. Most apes tend to be comfortable on all fours, whereas humans tend to be comfortable standing upright. Thus, because it is *possible* that *Australopithecus afarensis* might have stood upright, it is considered an intermediate link between man and ape.

There is a problem with this interpretation, however. Recent studies seem to indicate that Lucy most likely walked by using her knuckles. For example, Dr. Brian Richmond examined the wrist bones of two species in genus *Australopithecus*: *A. afarensis* and *A. anamensis*. They demonstrated that the wrists of these two species were quite similar to those of modern chimpanzees, which walk on both their hind legs and their knuckles. Thus, based on the wrist, you would classify Lucy as an ape that walks on its knuckles.

This is further supported by a study by Dr. Fred Spoor, who did CAT scans on the inner ears of members of genera *Australopithecus* and *Paranthropus*. Genus *Paranthropus* contains fossil creatures that are very similar to those in genus *Australopithecus*. In fact, some paleontologists think that these two genera should be combined, but that is still hotly debated. Spoor's study is important, as the inner ear is where balance is maintained, and if a creature walks upright, it needs a significantly different sense of balance from a creature that does not. What did Spoor find? His paper states, "…the semicircular canal dimensions in crania from southern Africa attributed to Australopithecus and Paranthropus resemble those of the extant great apes." (Fred Spoor *et al.*, "Implications of early hominid labyrinthine morphology for evolution of human bipedal locomotion," *Nature* **369**:645-8, 1994). The term "extant" just means "currently living." Thus, Spoor's conclusion is that the inner ear of the members of genus *Australopithecus* is very similar to that of currently-living apes. This indicates that they probably did not walk upright, at least not habitually.

Once again, then, whether or not *Australopithecus afarensis* is an intermediate link depends on your point of view. If you want to believe in macroevolution, you can look at the *possibility* that it could stand upright and conclude that it is a transitional form. If you do not want to believe in macroevolution, you can look at the major features of the skeleton and the details of its wrists and inner ears and conclude that it is not.

The point, however, is quite clear. If macroevolution happened, the fossil record should be littered with intermediate links, as Darwin predicted. Instead, macroevolutionists can only present a few highly questionable ones. These supposed intermediate links closely resemble one of the two species they are supposed to link together. If intermediate links truly existed in the fossil record, you would think that at least one unambiguous intermediate link could be found somewhere!

ON YOUR OWN

9.8 If macroevolution really occurred, would you expect to find more fossils of individual species or of intermediate links?

9.9 What features on *Archaeopteryx* make macroevolutionists think that it is an intermediate link? What features make creation scientists think that it is not?

9.10 Why do macroevolutionists consider *Australopithecus afarensis* an intermediate link between man and ape? Why do creation scientists think that it is not?

The Cambrian Explosion

Before we leave our discussion of the fossil record, we have to mention probably the greatest problem that faces macroevolutionists today. The problem exists as a result of some fossils that were "rediscovered" in the mid-1980s, and based on those fossils, we know that at least the bottom portion of the geological column as presented in Figure 9.2 (and all introductory biology and geology textbooks to our knowledge) is incorrect. You see, as presented in Figure 9.2, the geological column is used to give the impression that the fossil record indicates that only the "simplest" multicellular life existed in the years represented by the Cambrian rock (570-500 million years ago), and then through the years represented by the Ordovician rock (500-435 million years ago), the life got more complex. This supposedly continued through to the times represented by the Silurian rock (410-435 million years ago) as well as the time represented by Devonian rock (360-410 million years ago). As a result, the diverse life that we see in the oceans did not fully evolve until about 400 million years ago. Now remember, all of these numbers for how many millions of years ago these things happened are based on a lot of assumptions, and we don't have time to go into those assumptions. However, we are discussing how macroevolutionists interpret the geological column, and this is how they do it.

Based on the "rediscovery" we mentioned earlier, we now know that the macroevolutionists' view of the lower layers of the geological column is simply false. What do we mean by "rediscovery?" Well, in the early 1900s, Charles Walcott (a paleontologist) discovered a lot of fossils in a layer of Cambrian rock called the "Burgess Shale." However, these fossils were not what Walcott expected. Remember, according to the geological column, only the "simplest" of multicellular life was supposed to have existed in the times represented by Cambrian rock. However, Walcott found thousands of fossils of very complex life. In fact, by the time the collection was complete, Walcott had found representatives *from every major animal phylum* that exists in our classification scheme (see Figure 9.5 on the next page).

What did this stunning discovery mean? It meant that the bottom of the geological column as presented in textbooks (still to this day) is *wrong*. Walcott found fossils of the "simple" animals that were supposed to be in Cambrian rock, but he also found *thousands* of examples of animals that were too complex to have evolved in the short time represented by Cambrian rock. According to the geological column, some of these animals were supposed to have formed much, much later, in the times represented by Silurian and Devonian rock. Why in the world were these complex fossils found in Cambrian rock? Walcott had no real answer to that question.

In other words, Walcott had discovered something *revolutionary*! This was a find that would radically alter the scientific world's understanding of the geological column. What, then, do you think he did? What would you do if you discovered something so revolutionary? I would think that you would publicize it so that the whole world would see it. That's not what Walcott did, however. He wrote a few modest articles in an extremely obscure scientific journal (*Smithsonian Miscellaneous Collection*), and then he *reburied* the fossils in his laboratory drawers, and he never mentioned them again. It would be *80 years* before the fossils were "rediscovered" and their importance understood.

What is their importance? As we mentioned, they tell us that the first few layers of the geological column as presented in the geological column are simply false. After all, the geological column pictures the animals in the "simple" phyla evolving into the animals of the more complex phyla over a period of about 170 million years (from Cambrian rock to somewhere in the middle of Devonian rock). This, however, is not what the fossils say. The fossils discovered by Walcott (more than 60,000 in all, some of which are reconstructed in the figure below) and those discovered later by others tell us that *all* of the major animal phyla can be found in Cambrian rock.

FIGURE 9.5

Illustration by Megan Whitaker

Reconstructions of Some of the Fossils Found in Cambrian Rock

These are illustrations of what some of the creatures fossilized in Cambrian rock might have looked like. Some of them are alive today; many are extinct. The creatures found in Cambrian rock come from *all* of the major animal phyla.

These fossils lead to a serious problem for macroevolution. All macroevolutionists agree that macroevolution takes a long time. No one presently understands how such a huge amount of evolution could have taken place in the relatively "short" time supposedly represented by Cambrian rock. Even if the currently-assumed ages for Cambrian rock are correct, it took "only" 70 million years for evolution to go from the simplest animal phyla to the most complex animal phyla. At present, no one understands how this could have happened in the context of macroevolution. In fact, the problem is so well-known in the field of paleontology, it has a name. It is called the **Cambrian explosion**, which refers to the fact that there seems to have been an "explosion" of life in Cambrian times.

Not only is the "short" time a problem for macroevolution, but the fossils themselves present a real problem as well. Even though the fact that there are no intermediate links in the fossil record is a well-known problem for macroevolution, the problem is *much more* dramatic in Cambrian rock. After all, a *huge* amount of macroevolution had to have occurred in the time represented by Cambrian rock, but there is just no evidence for it. There aren't even possible transitional forms like *Archaeopteryx* or *Australopithecus*. In fact, the creatures that are fossilized in Cambrian rock just appear there suddenly, exactly as you would expect if each of these creatures was simply made by God.

One of the most honest summaries of the problem that the Cambrian Explosion presents to macroevolution is given by prominent macroevolutionist Richard Dawkins:

> It is as though they [fossils] were just planted there, without any evolutionary history. Needless to say this appearance of sudden planting has delighted creationists...Both schools of thought (Punctuationists and Gradualists) despise so-called scientific creationists equally, and both agree that the major gaps are real, that they are true imperfections in the fossil record. The only alternative explanation of the sudden appearance of so many complex animal types in the Cambrian era is divine creation and (we) both reject this alternative. (Richard Dawkins, *The Blind Watchmaker*, [New York, NY: W.W. Norton & Company, 1996], 229-230)

Notice how honest Dr. Dawkins is in this quote. He tells us that the fossils seem to have just "appeared" there, without any evidence of evolution whatsoever. He says that both schools of macroevolutionists (we will discuss what "Punctuationists" and "Gradualists" are later) agree that even though the fossils seem to indicate that the creatures were created divinely, they must reject that notion, and they must therefore assume that the appearance of "planting" is just due to imperfections in the fossil record. What does he mean by "imperfections?" He means that the intermediate links did, indeed, exist, but they just happened to have not been fossilized. This is what you are forced to believe if you look at the fossil record and reject the most obvious conclusion that it presents: that the organisms preserved there were divinely created.

<u>Structural Homology: Formerly Evidence for Macroevolution, Now Evidence against It</u>

Another piece of evidence that Darwin detailed in his book came from comparing the structures that are found in representatives of vastly different species. The study of similar structures in different species is called **structural homology** (hum awl' uh jee).

<u>Structural homology</u> – The study of similar structures in different species

The best way to understand why Darwin considered structural homology to be evidence for macroevolution is to show you a figure.

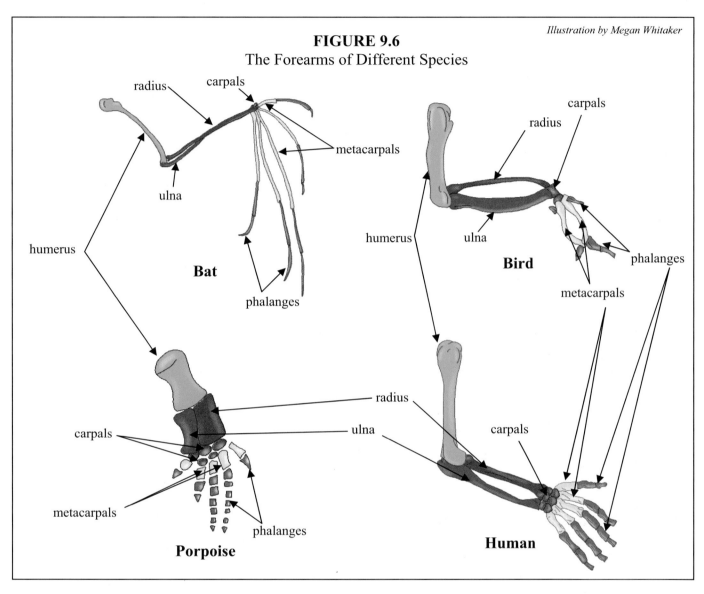

FIGURE 9.6
The Forearms of Different Species

Illustration by Megan Whitaker

Notice in the figure that although these arm bones come from very different species, they have an amazing amount in common. They all are jointed about halfway down. They all have a single, upper arm bone, the humerus (hyou' mer us), which is green in the figure. They then have two lower arm bones, the radius, which is colored purple, and the ulna (uhl' nah), which is colored blue. They all end in a palm consisting of carpals (car' puhls), which are colored red, metacarpals (met uh car' puhls), which are colored yellow, and phalanges (flan' jeez), which are colored gray. Indeed, with the exception of the bird, they all have five sets of phalanges!

Darwin said that these incredible similarities among such vastly different species were excellent evidence that they all had a common ancestor. After all, he said, that common ancestor probably had a forearm which was jointed in the middle, had a humerus, followed by a radius and an ulna, followed by carpals, metacarpals, and phalanges. This general forearm structure, then, over eons of time and guided by natural selection, simply adapted to the needs of each individual species as it arose. Thus, much like you can tell that a man and his brother's grandson are related by a common ancestor because of their striking similarities, you can assume that man and the bat are related by a common ancestor because of the striking similarities in their forearms.

In Darwin's time, structural homology was very strong evidence for macroevolution. How could vastly different species have such similar characteristics unless they were all related by a common ancestor? If they all had a common ancestor, then clearly macroevolution would have to have occurred in order to turn this common ancestor into these vastly different species, right?

That sounded like a great argument in Darwin's time because scientists back then had no idea how traits were passed on from generation to generation. With the advent of Mendelian genetics, however, scientists finally began to understand how this happens. As scientists began to understand genetics and DNA better, they developed technology to actually determine the sequence of nucleotide bases in an organism's DNA. This spelled the end of structural homology as evidence of macroevolution.

You see, if structural homology was the result of common ancestry, it should show up in the genetic codes of the organisms that possess similar structures. Take, for example, the forearms shown in Figure 9.6. If the forearms of the bat, bird, man, and porpoise look so similar because they all inherited their forearms from a common ancestor, then the parts of their DNA that contain the information regarding the forearms should be similar. After all, traits are passed from parent to offspring through DNA. If each one of these creatures inherited its forearm structure from a common ancestor, then the portions of DNA which contain information about the forearm would all have come from that same common ancestor. As a result, those portions of the DNA should be similar from organism to organism.

Is this the case? Is structural homology the result of similar DNA sequences? No, it is not! Read the words of Dr. Michael Denton:

> The evolutionary basis of homology is perhaps even more severely damaged by the discovery that apparently homologous structures are specified by quite different genes in different species…With the demise of any sort of straightforward explanation for homology one of the major pillars of evolution theory has become so weakened that its value as evidence for evolution is greatly diminished. (Michael Denton, *Evolution: A Theory in Crisis*, [Bethesda, MD: Adler and Adler, 1985], 149 and 151)

Notice what Dr. Denton says here. Apparently homologous structures are specified by quite different genes in different species. Well, if they are specified by different genes, then there is *no way* that the homologous structures could have been inherited from a common ancestor. As basic genetics tells us, the only way to inherit something from an ancestor is through the genetic code! This is why Dr. Denton concludes by saying that the value of homology as evidence for evolution is greatly diminished.

To creation scientists, Figure 9.6 actually offers excellent evidence for a Creator. After all, any good engineer, once he finds a design that works, tends to stick with that design and simply adapts it from situation to situation. Thus, structural homology is, to creation scientists, evidence of common design, not common ancestry.

Think about what we have learned so far. Darwin's two greatest evidences for macroevolution were the geological column and structural homology. As we have seen, based on today's understanding of science, the geological column provides inconclusive evidence, and structural homology now provides *evidence against macroevolution*. In addition, the fossil record provides evidence against macroevolution. Is this the end of the story? Actually, no. It turns out that we haven't even studied the data that speak out against macroevolution the strongest. We saved that for the next section!

ON YOUR OWN

9.11 What major scientific breakthrough led to structural homology changing from evidence for macroevolution to evidence against it?

Molecular Biology: The Nail in Macroevolution's Coffin

Of all the scientific data that provide evidence against evolution, perhaps the most important come from the field of molecular biology, which studies the properties and structures of the molecules important to biology. Think back to what you learned in the three previous modules. Aside from DNA, what is the most important type of molecule in the chemistry of life? The protein! As a result, a large amount of the research effort in molecular biology centers on understanding proteins.

Early on, molecular biologists noticed something rather amazing. There are certain proteins that are common to many species. Most animals, for example, have the protein **hemoglobin**. This protein transports oxygen through the bloodstream to the cells. In addition, most organisms have the protein **cytochrome** (sye' tuh krohm) **C**, which takes part in cellular metabolism. Interestingly enough, these proteins are not identical from species to species. In other words, the cytochrome C that you find in a bacterium is a bit different from the cytochrome C that you find in a human.

How do these proteins vary from species to species? Well, what determines the structure and function of a protein? In Module #5, you learned that the sequence of amino acids within a protein determines its structure and function. Thus, if you were to examine the amino acids in the cytochrome C of different species, you would find slightly different sequences. For example, Table 9.1 lists the amino acid sequences on the same portion of cytochrome C for several different types of organisms.

TABLE 9.1
Partial Cytochrome C Amino Acid Sequences

Organism	Amino Acid Sequence (partial)
Horse	Gly-Leu-Phe-Gly-Arg-Lys-Thu-Gly-Glu(NH_2)-Ala-Pro
Kangaroo	Gly-Ile-Phe-Gly-Arg-Lys-Thu-Gly-Glu(NH_2)-Ala-Pro
Human	Gly-Leu-Phe-Gly-Arg-Lys-Thr-Gly-Glu(NH_2)-Ala-Pro
Chicken	Gly-Leu-Phe-Gly-Arg-Lys-Thr-Gly-Glu(NH_2)-Ala-Glu
Tuna	Gly-Leu-Phe-Gly-Arg-Lys-Thr-Gly-Glu(NH_2)-Ala-Glu
Moth	Gly-Phe-Gly-Arg-His-Thr-Gly-Glu(NH_2)-Ala-Pro-Gly
Yeast	Gly-Ile-Phe-Gly-Arg-His-Ser-Gly-Glu(NH_2)-Ala-Glu(NH_2)

Now remember what we are looking at in this table. Each three-letter abbreviation represents a specific amino acid. Thus, "Gly" stands for the amino acid glycine, "Leu" stands for the amino acid leucine, and so on. Don't worry about the NH$_2$. That's an amine functional group that really has no bearing on our discussion. Also, notice that some of the three-letter abbreviations are red. Red amino acids indicate that they are different from the amino acid expected if you use the horse sequence as the standard. Thus, compared to the horse sequence, the kangaroo sequence has one amino acid that is different. However, the yeast sequence has four amino acids different from the horse sequence.

What do we see in studying Table 9.1? Well, first notice that all of these sequences are very similar. That's not at all surprising, because the protein is the same in each case: cytochrome C. It performs the same basic function in each organism, but in order to be able to work with the specific chemistry of each organism, it is slightly different in each specific case. That's where the differences in the sequences come from. The cytochrome C section shown here, for example, is nearly identical between the horse and the kangaroo. The only difference is the second amino acid, which is leucine (Leu) in the horse and isoleucine (Ile) in the kangaroo. Because of this one difference, the cytochrome C of a horse will not work in a kangaroo or vice-versa.

What does all of this tell us? Well, think about how proteins are made. They are made in the cells *according to the instructions of DNA*. Thus, by looking at the amino acid sequences in a protein that is common among many species, you are actually looking at the differences between specific parts of those organisms' genetic code: the part that determines the makeup of that protein. If macroevolution is true, then that portion of the genetic code should reflect how "closely related" the two species are. If two species are closely related, the DNA sequences that code for a common protein should be very similar. If they are only distantly related, however, the DNA sequences that code for that same protein should have more significant differences between them. Looking at the differences between the amino acid sequences of a common protein, then, is a way to determine just how many differences exist between corresponding sections of the DNA of the organisms in question.

For example, the portion of the amino acid sequence for cytochrome C shown in the table is 11 amino acids long. Of those amino acids, there is only one difference between the horse and the kangaroo. We can therefore calculate the percentage difference between the cytochrome C amino acid sequence in a horse and the cytochrome C amino acid sequence in a kangaroo.

$$\text{percent difference} = \frac{1}{11} \times 100 = 9.1\%$$

Comparing the amino acid sequences in cytochrome C for the yeast and the horse, however, there are four differences. Thus, the percent difference is

$$\text{percent difference} = \frac{4}{11} \times 100 = 36.4\%$$

These two comparisons, then, tell us that the portion of a horse's genetic code that determines the makeup of this part of cytochrome C protein is much closer to the kangaroo's than it is to a yeast's. From a macroevolutionary point of view, this would tell us that the horse is more closely related to the kangaroo than to the yeast. In other words, in some macroevolutionary scheme in which one life form evolves into another, the kangaroo would be closer to the horse than would be the yeast.

That makes sense, doesn't it? After all, a yeast is considered, by macroevolutionists, to be a rather "simple" life form, whereas kangaroos and horses are rather complex. As a result, it makes sense that a horse is more closely related to a kangaroo than a yeast. Even though comparing this portion of cytochrome C in the horse, kangaroo, and yeast makes sense in terms of macroevolution, as you look across the data that is out there, you find that it causes serious problems for macroevolution.

Consider, for example, the bacterium *Rhodospirillum* (roh doh spuh ril' um) *rubrum* (roob' rum). When its cytochrome C amino acid sequence is compared to vastly different organisms, nothing makes sense in terms of the macroevolutionary hypothesis. Table 9.2 shows the percentage difference between the amino acid sequence in a *Rhodospirillum rubrum*'s cytochrome C and the amino acid sequence of other organisms' cytochrome C.

TABLE 9.2
Percent Differences between a Bacterium's Cytochrome C and That of Other Organisms

Organism	Percentage Difference from the Bacterium
Horse	64%
Pigeon	64%
Tuna	65%
Silkworm moth	65%
Wheat	66%
Yeast	69%

Now remember what macroevolution says. It says that "complex" life forms evolved from "simple" ones. Well, the "simplest" life form on the planet is a bacterium. Of the organisms listed in the table, the yeast (a single-celled fungus) is probably the next "simplest" life form. Increasing in complexity then come the silkworm moth, followed by the tuna, followed by the pigeon, followed by the horse. Thus, macroevolution would assume that the bacterium is most closely related to the yeast, then to the silkworm moth, etc., etc., all the way up to the horse. As a result, then, the yeast's cytochrome C should be most similar to that of the bacterium, the silkworm moth's cytochrome C should be the next most similar, and so on. According to the data, however, each organism in the table is essentially as closely related to the bacterium as any other organism on the table! If anything, the bacterium is more closely related to the *most complex* organisms, not the least complex ones!

In other words, the data presented in Table 9.2 show none of the evolutionary relationships that *should* exist if macroevolution really occurred. Instead, these data seem to indicate that the bacterium is just as different from the horse as it is from the yeast! As you look at more and more data like this, you will find that this is the pattern of the vast majority of the data. Regardless of the protein studied, the amino acid sequences seem to indicate that each individual type of organism is just as different from one type of organism as it is from another. Just to make it clear that the data is really overwhelming on this point, we present a few more tables like Table 9.2.

TABLE 9.3
More Percentage Differences between Cytochrome C Amino Acid Sequences

The cytochrome C of a lamprey eel compared to other organisms		The cytochrome C of a carp compared to other organisms		The cytochrome C of a pigeon compared to other organisms		The cytochrome C of a horse compared to other organisms	
Organism	*% diff*	*Organism*	*% diff*	*Organism*	*% diff*	*Organism*	*% diff*
Horse	15%	Horse	13%	Horse	11%	Pigeon	11%
Pigeon	18%	Pigeon	14%	Carp	14%	Turtle	11%
Turtle	18%	Turtle	13%	Turtle	8%	Carp	13%
Carp	12%	Lamprey	12%	Lamprey	18%	Lamprey	15%

Now think about what these data tell us. We are comparing the cytochrome C's of a mammal (horse), a bird (pigeon), a reptile (turtle), a fish (carp), and an eel (lamprey). According to macroevolution, eels would have come first, later fishes would have evolved, followed by amphibians, followed by reptiles, followed by birds, and finally followed by horses. Thus, the eel should be most similar to the fish, then the reptile, and then the bird, and it should have the least in common with the horse. Instead, the data tell us that the cytochrome C in the lamprey eel is most like the carp, *but then next most like the horse, and has essentially the same differences between the turtle and the pigeon*!

If you look at the data presented in Tables 9.2 and 9.3, it is clear that you can establish no macroevolutionary trends. This is the case with the vast majority of the data collected from molecular biology. If you map the amino acid sequences of virtually any protein and compare the differences between organisms that have that protein, you will generally find no macroevolutionary trends. Instead, each kind of organism seems to be equally or nearly equally different from every other kind of organism. As is the case with all of science, there are exceptions to this general rule, but those exceptions are quite rare.

Even though the exceptions are rare, some macroevolutionists actually highlight those exceptions as evidence for evolution. You see, if you pick and choose your data very carefully, you *can* find examples of molecular biological data that *seem* to indicate a macroevolutionary trend. For example, look at Table 9.4.

TABLE 9.4
A Highly Selective View of Some Cytochrome C Sequences Compared
to Human Cytochrome C Sequences

Organism	% Difference Compared to Human
Rhesus monkey	1%
Rabbit	9%
Horse	12%
Tuna	20%
Screw Worm	25%
Wheat	38%
Yeast	41%
Bacterium	65%

Now this table does seem to indicate a macroevolutionary trend, doesn't it? After all, as you go down the table, you are getting to "simpler" and "simpler" life forms. As the life forms get "simpler," their cytochrome C sequences seem to get more and more different from that of a human, don't they?

The problem with this table is that in order to construct it, you must *ignore 99% of the data* from molecular biology and choose only the 1% of the data that agrees with the hypothesis of macroevolution. This is clearly not responsible science. In science, we must look at all of the data, and the data from amino acid sequencing provide strong evidence against macroevolution.

ON YOUR OWN

9.12 A molecular biologist details the amino acid sequences in a common protein for the following creatures: a human, a rat, an amoeba, a fish, and a frog. Assuming macroevolution did occur, list these creatures in terms of increasing similarity between their protein and the human protein.

9.13 Why is the comparison of amino acid sequences in common proteins such a useful tool in determining whether or not macroevolution occurred?

Macroevolution Today

Because the data speak so strongly against the hypothesis of macroevolution, it has changed greatly since Darwin's time. Two of the most important changes to his theory are typically called **neo-Darwinism** and **punctuated equilibrium**. They each address different problems that occur when you compare macroevolution to the data, so we must discuss them individually.

Neo-Darwinism came first. Remember, Darwin pretty much established the theory of microevolution, but when it came around to providing evidence for macroevolution, he was stuck. When scientists began to understand genetics, they understood why he got stuck. As we said before, microevolution is simply the variation of a type of organism *within* its genetic code. Thus, microevolution is understandable in terms of our current understanding of genetics. Macroevolution, however, is a completely different story. In order for macroevolution to occur, a species would have to *add* information to its genetic code. How could that happen? Well, neo-Darwinists say that it could happen by mutation. You see, Darwin thought that all of macroevolution could occur by the normal changes that happen during the reproductive process. When scientists came to understand genetics, however, it became clear that those kinds of changes only help us to understand how microevolution works. If macroevolution were to occur, something else would have to *add information* to the genetic code. That something else, according to neo-Darwinism, is mutation.

Remember, mutation is a mistake that alters the genetic code. Now as we discussed previously, that alteration is often bad for the organism, but that doesn't have to be the case *every* time a mutation occurs. Suppose that a few *good* mutations could occur in a species. Those mutations might, by blind chance, produce a new trait in the organism that made it more fit to survive. This, trait, then, would be passed on, adding new information into the genetic code of the species!

What really excites neo-Darwinists about this idea is not only does it provide a means by which an organism can add information into its genetic code, but it also allows them to explain away at least some of the discrepancy between macroevolution and the fossil record. After all, mutations result in dramatic changes between parent and offspring. If macroevolution occurred by these dramatic changes, then perhaps there shouldn't be as many intermediate links as Darwin first suspected. This modification, then, seems to kill two birds with one stone.

The main problem with this view is that when we observe mutations in the lab, or when we study nature and reconstruct cases in which mutations happened, they do not lead to an *increase* in the information of an organism's genetic code. Instead, they end up *destroying* information in the genetic code. Consider, for example, the case of antibiotic-resistant bacteria. As we mentioned in Module #4, antibiotics have been used in the treatment of bacteria-related illness for some time. However, as antibiotics are used, bacteria can develop strains that are immune to them. This, of course, causes problems, as a new antibiotic must be found to stop this resistant strain of bacteria.

Antibiotic-resistant bacteria have often been cited as evidence for macroevolution, but they are not. Some antibiotic-resistant bacteria provide even more evidence for microevolution, and some antibiotic-resistant bacteria provide even further evidence *against* macroevolution. How can we say this? Well, you first have to know how bacteria develop resistance to antibiotics.

As you might recall from Module #2, bacteria can add to their genetic codes using conjugation and transformation. In conjugation, a bacterium that has a certain gene can pass that gene to a bacterium that doesn't. In transformation, a bacterium that does not have a certain gene can absorb it from a dead bacterium that had the gene. Most antibiotic-resistant bacteria get their resistance in one of these two ways. When a population of bacteria is exposed to an antibiotic, it kills most of them. However, one or two of the bacteria might have a set of genes that makes them immune to the antibiotic. If they can pass those genes via conjugation or transformation, several bacteria can survive the antibiotic, and as they asexually reproduce, they will build an entire population that is resistant to the antibiotic.

There is actually a third method by which bacteria can add information to their genetic codes. It is called **transduction**, and it involves the aid of a virus. If a virus infects a population of bacteria, genetic information can be passed from one bacterium to another as a result of the infection spreading. If a bacterium can survive the infection, it might have extra genetic material as a result. It is possible that resistance to an antibiotic could spread this way as well.

Of course, if a bacterium becomes resistant to an antibiotic by any one of these methods, the genes which provide resistance must have *already existed* in some bacteria. As a result, this cannot be cited as evidence for macroevolution. In macroevolution, *new genes* must be created. Old genes don't help the process. One of the most convincing studies that demonstrates that bacteria become resistant to antibiotics through old genes was published in 1988. In that study, individual bacteria from the intestines of explorers who had been frozen *before* the development of antibiotics were shown to *already* be resistant to several antibiotics that had not been developed until *after* the explorers were frozen. Thus, the genes that produced resistance were already there. This kind of immunity to antibiotics is more evidence of microevolution. Genetic characteristics that were *already a part* of some bacteria's genetic codes cause certain individual bacteria to survive the harsh conditions of the antibiotic. These "selected" bacteria then pass on their trait of resistance to their offspring.

Although most antibiotic-resistant bacteria get their resistance from genes that already exist, there are some examples of bacteria that become resistant to antibiotics through mutation. In each case that has been studied, however, the resistance comes as a result of *loss of information* in the genetic code. For example, bacteria have complex proteins in their plasma membranes that allow nutrients into the cell. Normally, these proteins work very efficiently. Many antibiotics seem to get into the bacterium through these proteins. Thus, the proteins not only bring in food, but they also bring in the antibiotic that kills the bacterium. Well, a mutation that destroys these proteins (or causes them to work less efficiently) will also stop the antibiotic from getting into the cell (or at least cause the cell to take it in at a much lower rate). This, of course, makes the bacterium immune to (or at least able to survive) the antibiotic, but it also makes the bacterium weak, as the bacterium cannot get as much food as before.

Now think about it. What caused the bacterium's resistance in this case? *A loss of information caused it.* If the bacterium had not lost information about how to make an efficient means of getting food into the cell, it would have died. The loss of information, which would in general make it less fit to survive, helped it to survive the specific case of being exposed to the antibiotic. Did the mutation provide a survival advantage? Yes! Did it add information to the genetic code? No! If mutations are to provide the means by which macroevolution occurs, they must add information to the genetic code. So far, there has been no indication that they do. Mutations either leave the information content of the genetic code the same, or they reduce it. As a result, they cannot be used to "power" macroevolution.

Now the fact that mutations which add information to the genetic code cannot be found has never bothered neo-Darwinists much. After all, they can always say that sometime, way back when, there were mutations that could do so. The problem for neo-Darwinists, however, is still in the fossil record. Even though neo-Darwinism reduced the number of intermediate links needed for evolution, it did not reduce the number to near zero, which is essentially what exists in the fossil record. This is the point that Dr. Raup, Dr. Schwartz, and Dr. Dawkins were making when they each wrote what we quoted in the past sections regarding the fossil record. Even with the revision of neo-Darwinism, the fossil record just doesn't provide nearly enough transitional forms. As a result, yet another revision needed to be made to the hypothesis of macroevolution.

The latest revision is called **punctuated equilibrium**. Those who hold to this revision are called "Punctuationists," while those who hold to neo-Darwinism are generally called "Gradualists." This is because neo-Darwinism views evolution as a fairly gradual process, while punctuated equilibrium views it as a process that occurs in quick punctuations followed by long times in which no evolution occurs. These are the two macroevolutionary "camps" that Dawkins wrote about in the quote that we presented in the section on the Cambrian Explosion. In his book, he discusses the differences between these two camps. There are a lot of differences between the camps, but as the quote we presented earlier suggests, they are united by their dislike for creationists.

In punctuated equilibrium, mutations still add genetic information to the fossil record, but they add it in steps that occur over very short time intervals. In between these very short time intervals, no macroevolution occurs. The idea here is that a group of organisms might be suddenly exposed to high levels of toxic chemicals or radiation. This would speed up the mutation rates among the individuals. Most of the resulting offspring would die, but a few lucky ones would get several good mutations all at once. These many mutations would add a lot of information to the genetic code, and the offspring would be more fit to survive. As the high levels of toxic chemicals or radiation continued, these few lucky offspring would produce more offspring with more mutations. Once again, most would die, but a few lucky ones might get several more mutations that added even more information, making them even more fit to survive. After just a few generations, the radiation or chemicals would be gone, and a new species would have emerged. That species would exist, essentially unchanged, for millions of years until another episode of high mutation rates occurred.

Notice what punctuated equilibrium does. It still counts on mutations to add information to the genetic code of the species that is evolving, but it expects those mutations to occur over only a few generations. Thus, *the intermediate links would not exist for very long*. What does that accomplish? Well, if the intermediate links didn't live for very long, their *chance of being fossilized is very small* compared to that of the final product, which would live for millions of years essentially unchanged. Thus, by altering the hypothesis, we can now account for the fact that no intermediate links appear in the fossil record.

Now of course, this still doesn't explain why structural homology is not echoed in the genetic code, and it still doesn't tell us why the vast majority of molecular biology data indicates no macroevolutionary trends, but at least the problem with the fossil record is fixed! This is pretty much the current state of macroevolutionary thinking today. Its proponents must cook up such scenarios to explain away the lack of data in support of it.

Punctuated equilibrium actually presents a new problem for macroevolutionists. If evolutionary change happens as quickly as punctuated equilibrium exists, it becomes hard to understand how it could happen in populations that sexually reproduce. After all, if an individual gets several large mutations at once, it is significantly different from the other members of its species. However, when an individual is significantly different from the other members of its species, it is hard for that individual to sexually reproduce with them. If punctuated equilibrium really did occur, it would have to occur *slowly* enough to allow the mutant organism to sexually reproduce with others of its own species (so that its mutation could be passed on to the next generation), but *quickly* enough so that the mutants would not appear in the fossil record. Either that, or there would have to be *several* mutants in one generation, all of them having essentially the same mutations so that they could still sexually reproduce with one another. Both these scenarios seem to be rather far-fetched.

ON YOUR OWN

9.14 What is the main difference between Darwin's hypothesis of macroevolution and the neo-Darwinist hypothesis?

9.15 What does punctuated equilibrium explain that neo-Darwinism and Darwin's original hypothesis cannot?

9.16 The three graphs below are hypothetical graphs that plot macroevolutionary change (on the y-axis) versus time (on the x-axis). Which graph represents Darwin's original hypothesis, which represents neo-Darwinism, and which represents punctuated equilibrium?

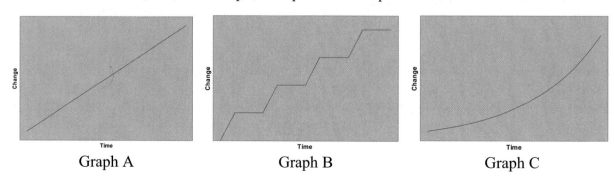

Graph A   Graph B   Graph C

Why Do So Many Scientists Believe in Macroevolution?

In this module, we have laid down some of the basic data relevant to evolution. We have shown that while microevolution is a well-established scientific theory, macroevolution is, at best, an unconfirmed hypothesis. There is no convincing data in favor of the hypothesis and plenty of data contradicting it. Given these facts, why do so many scientists believe in macroevolution? There isn't a simple answer to this question. Instead, there are a host of reasons that tend to lead to this situation.

The first reason is that scientists are simply indoctrinated at a very early age to become macroevolutionists. Since alternatives to macroevolution are not allowed in most classrooms (high school *or* college), students are left with no other explanation to one of life's most critical questions. If you think about it for a moment, you will realize how absurd such a situation is. The whole idea of intellectual pursuit is based on considering alternative ideas. Nevertheless, most schools will not even allow the consideration of any idea for life's origin other than macroevolution. One is forced to wonder why macroevolutionists are so afraid to allow their idea to compete with others! After all, if the evidence supports macroevolution, it should "win" any honest debate with any other competing hypothesis. Nevertheless, when educators attempt to bring an alternative to macroevolution to the classroom, macroevolutionists mount legal battles to keep it from happening. That seems odd for scientists who believe that their theory can be supported by the data.

The next reason so many scientists believe in macroevolution is a consequence of the first. Those of us who do not believe in macroevolution are regularly ridiculed on college campuses. The best way to illustrate this is to look at the personal experiences of one of these authors, Dr. Jay L. Wile.

Personally, I began experiencing the ridicule heaped on creationists when I attended graduate school at the University of Rochester. At that time, I had to choose a professor who would be my research advisor. His role was to direct my research and provide me with the training necessary to become a scientist in my chosen field. I picked my research advisor based on the kind of work that he did. The man I chose did what I considered to be the most interesting chemical research that went on at the University of Rochester.

While he was, in fact, an incredibly gifted scientist, my chosen research advisor was a committed macroevolutionist. As my creationist views became apparent to him, he ridiculed them. Of course, he would not discuss the data relevant to macroevolution with me. Instead, he simply made fun of me. In addition, one of the other members of our research team, a man I still consider to be the greatest scientist with whom I have ever worked, also constantly made fun of my creationist stand. Such a situation was very difficult for me. After all, I admired these two men. I aspired to become the caliber of scientist that they both were. Nevertheless, they made fun of my most strongly held scientific beliefs! Throughout my graduate school career, I concluded that if I was not as strongly convinced as I was by the data, I would have left my creationist stand behind to escape the ridicule of these scientific role models.

Sadly, Dr. Wile's story is very common among creation scientists. When macroevolutionists encounter students who do not believe as they do, the response is rarely an intellectual discussion of the facts related to macroevolution. Instead, the response is often public ridicule.

The third reason that most scientists still believe in macroevolution stems from the fact that most of the scientists have never investigated the data. There is, of course, a very good reason for this. Any time that a scientist spends researching something not directly related to his field, the less likely he is to become a great scientist. Today's science is so intricate and so specialized that a person must devote himself completely to his field, or he will be left behind. Thus, since scientists have been indoctrinated over the years to believe that macroevolution is the only scientifically sound belief system, they are not about to waste time and energy researching the facts surrounding the issue. The wasted time could damage their careers.

In addition to the time constraints that inhibit scientists in examining questions outside of their field, there is a definite pressure in the university setting to either believe in macroevolution or simply stay quiet about it! For example, a professor of biology at San Francisco State University was forbidden to teach the introductory biology course that he had been teaching for more than a decade because he began stressing the design elements that are prevalent in the world around us. The biology department felt that this would "confuse" students when they later reached a course on evolution, so they forbade him to teach that course! Another example comes from Oregon Community College. A professor of biology at that school was *fired* because he refused to state, in class, that macroevolution was a scientific fact! His former department chair said that he was an excellent teacher, and his student evaluations indicate that he was a challenging and well-loved teacher. Nevertheless, he was fired because he actually paid attention to the data and realized that macroevolution is, at best, an unconfirmed hypothesis. Creationist professors regularly feel great pressure from their universities to keep quiet about their creationist views, or they might very well lose their jobs.

In the end, then, scientists are indoctrinated at a very early age to be macroevolutionists; they are ridiculed throughout their careers if they believe otherwise; they usually cannot spare the time necessary to look into the facts regarding macroevolution; and they are often punished for believing anything else. These issues keep scientists from learning the data necessary to help them understand the fact that macroevolution is no more than an unconfirmed hypothesis.

Although this assessment is rather grim, we must point out that times are changing. More and more books questioning macroevolution are being written by respected university professors. These books have obviously had an impact. In 2004, a Gallup poll showed that only about one-third of Americans believe that Darwin's theory of evolution is supported by the evidence. Forty-five percent of Americans believe that God created human beings in their present form about 10,000 years ago. In addition, about 15% of high school teachers teach both creation and evolution side by side, and close to 20% of them reject macroevolution. Despite the fact that macroevolutionists have been trying to monopolize the education systems for decades, macroevolution has not gained much ground with students and teachers.

Before we end this module we must make one cautionary note. We have only scratched the surface of the hypothesis of macroevolution and the various data that relate to it. If you are interested in learning more about this interesting topic, you should go to the course website that we discussed in the "Student Notes" section of the book. There are many links to several places where you can learn more details about the subjects we have discussed here as well as other topics that we simply did not have the time to address.

ANSWERS TO THE "ON YOUR OWN" PROBLEMS

9.1 Darwin was influenced by <u>the idea of a struggle for survival</u> as proposed by Thomas Malthus. In addition, Sir Charles Lyell's idea that <u>the present is the key to the past</u> influenced him.

9.2 <u>Darwin would say that in the beginning, only slow predators existed. As food got scarce, however, any predator born that was slightly faster than the others would be able to get to more food; thus, this predator would be naturally selected to survive and would pass on its speed to its offspring. As generation after generation passed, each time a predator was born that was slightly faster than its peers, it would be naturally selected to survive. Thus, the extra speed would "pile up" generation after generation until, eventually, a cheetah was formed.</u>

9.3 Darwin believed that any variation which gives an organism an advantage in the struggle for life makes that organism more likely to live. <u>If the organism has an advantage, we could say that it is more "fit" for survival than those without that advantage. Thus, the "fittest" organisms will tend to survive.</u>

9.4 <u>This is an example of microevolution.</u> The sparrows remained sparrows; some just got bigger as time went on, and others got smaller. This is variation within a genetic code. In general, <u>a larger bird is better suited for colder climates</u>, so most likely, any larger sparrows that were born in the north were more likely to survive. As generation after generation passed, then, the larger birds that were more suited to the cold climates end up being the dominant sparrows in the colder, northern parts of the U.S.

9.5 <u>This scenario is an example of macroevolution.</u> In order for something like a cow to turn into a whale, radical changes must be made to the genetic code. This is what must happen in macroevolution.

9.6 In order for the geological column to be interpreted as evidence for macroevolution, <u>you must assume that each strata of rock was laid down individually over long periods of time, according to the speculations of Lyell</u>. That way, the lower rock strata can be interpreted as older than the upper rock strata.

9.7 <u>It is not conclusive because whether the geological column supports or contradicts macroevolution depends on assumptions that cannot be confirmed.</u> Since scientists have seen rock strata form both as the result of a Lyell-like process and as the result of natural catastrophes, we have no idea which process is responsible for forming the geological column. Without knowing how the geological column was formed, we have no idea whether it gives evidence for or against macroevolution.

9.8 <u>You would expect more intermediate links than individual species.</u> After all, to get from one species to another there must have been several steps along the way. This would result in many intermediate links for each new species.

9.9 <u>It has teeth and claws, which are not common in birds. These features lead macroevolutionists to conclude that it is part reptile. However, it has feathers that are designed for flight, the lung design of a bird, and the vision and balance of a bird. This makes creation scientists think it is just a bird.</u>

9.10 <u>Evolutionists think that since it is possible that this creature stood upright, it represents a link between man and ape. Creation scientists disagree because every bone in its body is characteristic of an ape, and wrist and inner ear studies indicate that it did not walk upright.</u>

9.11 Structural homology turned from evidence for macroevolution to evidence against it <u>when scientists learned about genetics and how to map out nucleotide sequences</u>.

9.12 According to a macroevolutionist, the order of creatures in terms of *increasing* protein similarity to human protein would be:

<u>amoeba, fish, frog, rat</u>

Since macroevolutionists believe that life evolved from the simple to the complex, the more complex the organism is, the more similar its protein will be to the corresponding human protein. As we pointed out in the text, this is precisely what we *don't* see in the data!

9.13 Since DNA codes for the amino acid sequences in a protein, <u>tracking the amino acid sequences is like comparing corresponding strands of DNA in different creatures</u>. We assume the strands are corresponding because they code for the same protein.

9.14 <u>The real difference is mutations</u>. Neo-Darwinists use mutations to add information to the genetic code. Darwin thought that the normal variations that are the result of reproduction could drive macroevolution.

9.15 <u>Punctuated equilibrium explains the lack of intermediate links in the fossil record</u>. By constructing the scenario, punctuated equilibrium says that intermediate links only live during a brief period in time. This makes them very unlikely to be fossilized.

9.16 Darwin's original hypothesis assumed that the natural variation as a result of reproduction was responsible for macroevolution. This would result in very slow change. Thus, the graph that changes the slowest <u>(A) represents Darwin's original hypothesis</u>. Neo-Darwinism allowed for faster change using mutations, so <u>(C) represents neo-Darwinism</u>. In punctuated equilibrium, there are brief periods of change followed by long periods of no change. Thus, <u>(B) represents punctuated equilibrium</u>.

STUDY GUIDE FOR MODULE #9

1. Define the following terms:

a   The immutability of species
b. Microevolution
c. Macroevolution
d. Strata
e. Fossils
f. Paleontology
g. Structural homology

2. Where did Darwin do most of the work which led to his hypothesis of evolution?

3. Did Darwin ever recant his scientific beliefs?

4. What was the main idea that Thomas Malthus's work gave to Darwin?

5. What was the main idea that Sir Charles Lyell's work gave to Darwin?

6. What age-old concept was Darwin able to dispel with his research?

7. Suppose a herd of horses were living in an area where food near the ground was scarce but there was plenty of food in the trees. If, after several generations, the horses gave rise to giraffes that could easily reach the food in the trees, would this be an example of microevolution or macroevolution?

8. Consider a fish population that is trying to survive under conditions of extremely cold water. If, over several generations, the fish develop thicker fat layers under their skin for better insulation, is this an example of microevolution or macroevolution?

9. From a genetic point of view, what is the main difference between microevolution and macroevolution?

10. In this module, we studied four main sets of data: the geological column, the fossil record, structural homology, and molecular biology. For each set of data, indicate whether it is evidence for or against macroevolution or if it is inconclusive. Briefly explain why.

11. Name two creatures that macroevolutionists claim are intermediate links and why they are not really intermediate links.

12. What is the Cambrian Explosion? Why is it a problem for macroevolution?

13. What are the four ways a bacterium can become resistant to an antibiotic?

14. If a bacterium has a mutation that makes it resistant to an antibiotic, does information get added to its genetic code?

15. Consider the following amino acid sequences that make up a small portion of a protein:

a. Gly-Ile-Gly-Gly-Arg-His-Gly-Gly-Glu(NH$_2$)-Glu-Glu(NH$_2$)-Lys-Lys-Lys

b. Gly-Leu-Phe-Gly-Arg-Lys-Ser-Gly-Glu(NH$_2$)-Gly-Glu(NH$_2$)-Ala-Arg-Lys

c. Leu-Ile-Gly-Gly-Arg-His-Ser-Gly-Glu(NH$_2$)-Ala-Glu(NH$_2$)-Arg-Arg-Arg

Which protein would you expect to be the most similar to a protein with the following subset of amino acids?

Gly-Ile-Phe-Gly-Arg-His-Ser-Gly-Glu(NH$_2$)-Ala-Glu(NH$_2$)-Arg-Arg-Lys

16. Based on macroevolutionary assumptions, which organism's cytochrome C should most resemble that of a yeast: a kangaroo or a bacterium?

17. What main problem with Darwin's hypothesis did neo-Darwinism hope to solve?

18. What problem with Darwin's hypothesis did punctuated equilibrium attempt to solve?

19. How would an adherent to punctuated equilibrium explain the lack of intermediate links in the fossil record?

20. What problems mentioned in this module still exist for those who believe in punctuated equilibrium?

Cartoon by Speartoons

A pivotal moment in the hypothesis of macroevolution

MODULE #10: Ecology

Introduction

Have you ever wondered why certain creatures can only be found in certain places on the earth? For example, where can you find polar bears? You can either find them in zoos or in the Arctic. Why? If polar bears can live in zoos, they must be able to live in climates other than that found in the Arctic. Why, then, do they only live in the Arctic? The answer to that question lies in a discussion of **ecology**.

<u>Ecology</u> – The study of the interactions between living and nonliving things

In the science of ecology, we study how organisms are specifically designed to interact with other organisms as well as their physical surroundings.

When we study ecology, we often break up what we are studying into different subgroups, much like the science of taxonomy breaks organisms into many subgroups. In the "taxonomy" of ecology, we typically classify things according to these groups:

<u>Population</u> – A group of interbreeding organisms coexisting together

<u>Community</u> – A group of populations living and interacting in the same area

<u>Ecosystem</u> – An association of living organisms and their physical environment

<u>Biome</u> – A group of ecosystems classified by climate and plant life

If you look at the definitions, you will see that the groups go from narrowest to broadest. **Population** is the narrowest, as it includes only one species (remember, we defined species as a set of organisms that can interbreed). A **community** is made up of one or more populations, however, so it is a broader category. An **ecosystem** includes not only several populations, but also their physical surroundings, so it is even more broad; and a **biome** (bye' ohm) is a very broad classification, encompassing several ecosystems.

In order to assure that life can continue on any part of the earth, there must be a critical balance between the organisms living in an ecosystem and the nonliving chemical and physical processes that play critical roles in the support of life. If any one of these creatures or factors were to disappear or to cease functioning properly, the ecosystem would fall out of balance, potentially creating havoc. In addition, if a creature or process that isn't supposed to be there is suddenly added, similar havoc could ensue.

One of the best examples of what happens when an ecosystem is thrown out of balance comes from the country of Australia. In 1859, an Australian sheep rancher named Thomas Austin decided that he wanted to hunt rabbits on his ranch. There was only one problem: rabbits didn't live in Australia back then. So he decided to import a few (24 to be exact) from England, where he had grown up. Well, in only six years, his land was home to more than 10,000 rabbits! He had hunted to his heart's content, claiming to have killed more than 20,000 rabbits, but they continued to overrun his ranch. They destroyed his grasslands, making it impossible for his sheep to graze. In essence, his sheep were starving, all because he had imported 24 rabbits six years earlier.

What happened? Well, Australia has its own unique ecosystems. Several specific species of animals and plants live in that region of the earth, and they all have particular roles to play in order to make sure that life can exist there. When the rabbits were introduced into Australia, it upset those ecosystems. You see, in England, rabbits have several natural predators. Rabbits are, in essence, the main food supply for many predators in England. Thus, God has designed them to reproduce rapidly, knowing that many, many rabbits are needed each year to meet the food needs of several predators in England. None of the creatures in Australia, however, eat rabbits. With no predators, the rabbits were able to reproduce rapidly, with nothing to hold their numbers in check. In just a few years, Australia was overrun by rabbits!

By 1883, the Australian government decided this was a major problem. The rabbits had spread out more than 1,000 miles from Austin's ranch and were destroying some of the major grazing lands in the country. As a result, the government decided that something had to be done. What would you do to fix this problem? Would you import a bunch of natural predators (such as hawks or foxes) to kill the rabbits? Well, that might take care of the rabbits, but then what would happen when the rabbits were gone? You would still have a creature (the fox or hawk) that didn't belong in the ecosystem. No one knows what damage that would cause!

The first thing that the government tried was to hire exterminators to kill the rabbits. They spent an enormous amount of money on poisons, rabbit traps, and hunters. When that didn't work, they even built a huge fence (42 inches high and almost 1,200 miles across) in an attempt to contain the rabbits. None of this worked. Rabbits were still overrunning Australia, destroying thousands of acres of prime grazing land.

By 1937, scientists had discovered and tested a virus that was harmful only to the English strain of rabbit that was imported to Australia. After extensive research, they decided that the virus would harm no other creature in Australia except the foreign rabbit. Thus, they infected many rabbits with the virus and released them into Australia. For a while, nothing happened, and scientists thought something had gone wrong. In 1950, however, a huge number of dead rabbits were found in a certain part of Australia. They had died of the viral disease. Soon, many more were dying. It turns out that mosquitoes which had bitten the diseased rabbits were slowly infecting nondiseased rabbits. Soon, the rabbit population was under control.

Does that mean rabbits are gone from Australia? No, they are not. You see, as the rabbit population dwindled, so did the rate of infection. After all, the fewer rabbits there are, the fewer mosquitoes can get the virus by biting diseased rabbits. With fewer mosquitoes that carry the virus, then, even fewer rabbits get newly infected. Thus, the rabbit population will probably never die in Australia. However, by finding a way to kill large numbers of them, scientists were able to control an immense threat to the grazing lands of Australia.

This story helps to illustrate the way in which ecosystems across the world are finely balanced. If one creature is removed or added to an ecosystem, the results could be disastrous. As a result, ecology is an important part of biology. Generally, we hear about ecology in the context of pollution. When the media needs to hear scientific opinions regarding pollution and its effects, they usually interview ecologists. Ecology, however, is a much broader science. There is a lot more to be learned about ecosystems than what pollution does to them. This module will give you a glimpse of this fascinating subfield of biology.

Energy and Ecosystems

In order to survive, all organisms need energy, and lots of it. Where does that energy come from? Ultimately, almost all of it comes from the sun. As you learned in Module #1, autotrophic organisms convert the energy from the sun into food for themselves. This food gives them the energy to sustain their life functions. We say that "almost all" of the energy used in ecosystems comes from the sun because, as we learned in Modules #2 and #3, not all autotrophs use the sun's energy to make their food. Some autotrophs (chemosynthetic ones, for example) use alternative forms of energy (chemosynthetic autotrophs use the energy stored in chemicals). Since these types of autotrophs are rare, only a small fraction of the energy used in ecosystems comes from a source other than sunlight, so it is safe to say that "almost all" ecosystem energy comes from the sun.

No matter what autotrophs use as their energy source, we have already learned that they are called **producers**, because they produce food for themselves and for other creatures. The creatures that use this food (one way or another) we called **consumers**. In this module, we will be a little more specific. **Herbivores**, as you should recall, eat only plants. Since they get their food directly from the producers, we call them **primary consumers**.

Primary consumer – An organism that eats producers

Of course, God has designed His creatures to get food not only directly from the producers, but also from other consumers. **Carnivores**, for example, eat other organisms that are not producers. Since carnivores do not get their food directly from the producers, we call some of them **secondary consumers**.

Secondary consumer – An organism that eats primary consumers

Finally, carnivores that eat other carnivores are called **tertiary** (ter' she air ee) **consumers**.

Tertiary consumer – An organism that eats secondary consumers

These relationships, called **trophic** (troh' fik) **levels**, are shown in the figure below.

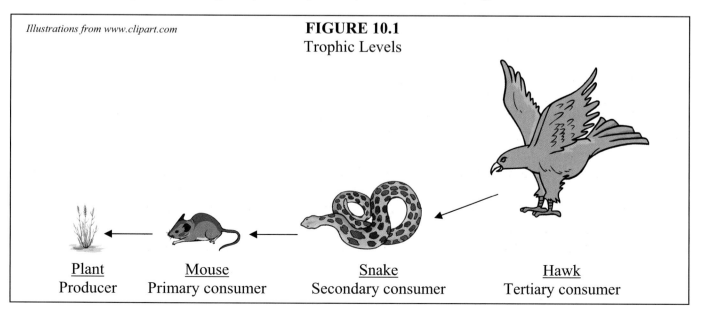

Illustrations from www.clipart.com

FIGURE 10.1
Trophic Levels

Plant	Mouse	Snake	Hawk
Producer	Primary consumer	Secondary consumer	Tertiary consumer

A diagram like the one shown in Figure 10.1 is often called a **food chain**. Since the plant is first, we say that it is lowest on the food chain. Likewise, the hawk is last, so we say that it is highest on the food chain. Although such diagrams are useful, they tend to be a little simplistic. For example, a hawk can change its trophic level. In the food chain diagrammed above, the trophic level of the hawk is tertiary consumer, because it is eating the snake. Sometimes, however, hawks will eat mice. When they eat mice, they are secondary consumers. Thus, the trophic levels of some organisms can change. As a result, a more complex diagram, called a **food web**, is often used to provide a more accurate description of the feeding relationships among organisms in an ecosystem.

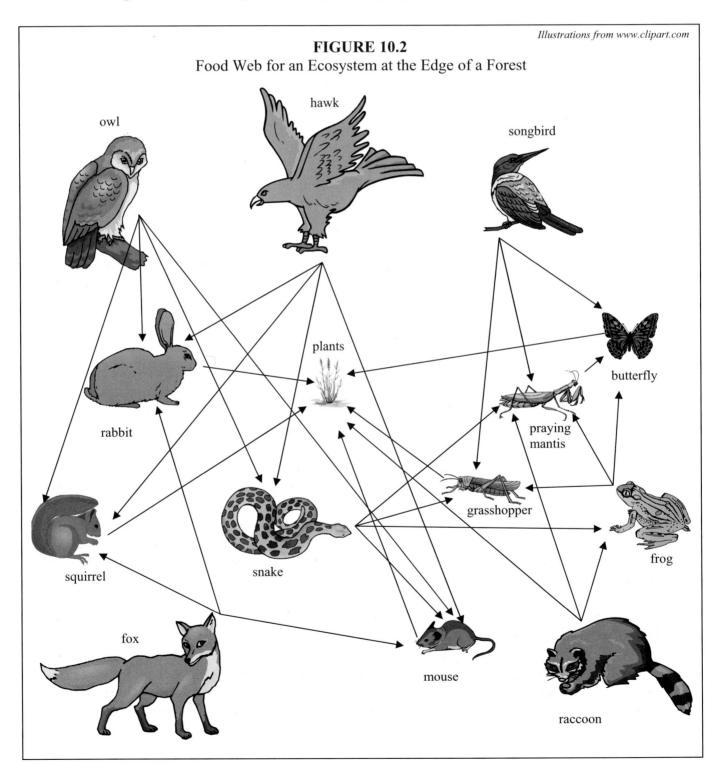

Illustrations from www.clipart.com

FIGURE 10.2
Food Web for an Ecosystem at the Edge of a Forest

Using the food web, you can easily see how some creatures remain at the same trophic level while others vary theirs. The fox, for example, is shown as always being a secondary consumer, because it is shown eating only the mouse, rabbit, or squirrel. All three of these creatures are primary consumers, so the fox is always a secondary consumer. Look at the raccoon, however. It is shown eating plants, a praying mantis, and a frog. When the raccoon eats plants, it is a primary consumer. When it eats a praying mantis or a frog, it is a tertiary consumer. This is why a food web is a much more realistic representation of trophic levels in an ecosystem than is a food chain.

So why are trophic levels so important in an ecosystem? First of all, as we learned from the story of the rabbits that were imported into Australia, every living organism must have some sort of predator, or the organism will overrun its ecosystem. Food webs show you the predator/prey relationships in an ecosystem. What about the creatures at the edge of the food web in Figure 10.2? Do the fox, owl, hawk, songbird, and raccoon have predators? Of course they do! If they did not, they would overrun the ecosystem in no time at all. Why aren't their predators pictured in the figure? Well, at some point, you have to limit your diagram of a food web because it is just too complex when you try to include all predator/prey relationships in it. Also, some predator/prey relationships are rather difficult to diagram. For example, some species of songbird actually prey on the hawk. They don't eat adult hawks, but they eat hawk eggs before the eggs have had a chance to hatch. A food web showing a songbird eating a hawk would look a little silly, but nevertheless, songbirds do prey on hawk eggs. So Figure 10.2, as is the case with most attempts to diagram creation, is a highly simplified picture of what really goes on in an ecosystem.

The second reason that trophic levels are important is that they actually track energy as it moves through the ecosystem. When energy comes into the ecosystem, it is usually through the sun. The energy of the sunlight is used by producers to make food for themselves. When the producers are eaten by primary consumers, energy moves to the next trophic level. As secondary consumers eat primary consumers, energy moves up to the next trophic level.

It turns out that each time energy moves up a trophic level, *a lot of that energy is lost*. This is a very important point to remember:

Energy is lost each time it moves up a trophic level in an ecosystem.

How is that energy lost? There are two ways. As you have already learned, when cellular respiration occurs, some energy is lost in the form of heat. In addition, when a fox eats a rabbit, it doesn't eat the whole thing; it eats only the meat and fat. The fur, skeleton, etc. of the rabbit are not used by the fox. Nevertheless, it took energy to form and maintain these structures. All of that energy is lost because it does not end up being used by other consumers! Of course, God designed His creation very carefully, so the energy used to make those structures is not really lost. We will discuss what actually happens to it in a moment. For right now, however, let's just assume it is lost.

What does this loss of energy between trophic levels mean? Well, it means that the lower an organism's trophic level, the less energy the ecosystem must invest in it. Plants, for example, require the least amount of energy. When a primary consumer eats a plant, however, up to 90% of that energy is lost. As a result, it takes a *lot* of plants to satisfy the feeding needs of only one primary consumer. Once again, when a secondary consumer eats a primary consumer, much of the energy contained in the primary consumer is lost. Thus, it takes a *lot* of primary consumers to meet the feeding needs of just one secondary consumer. Likewise, it takes a lot of secondary consumers to meet the feeding needs of one tertiary consumer.

One of the best ways to illustrate the consequence of energy loss in an ecosystem is via the **ecological pyramid**.

Ecological pyramid – A diagram that shows the biomass of organisms at each trophic level

Of course, this definition does us little good if we are not familiar with the term **biomass**!

Biomass – A measure of the total dry mass of organisms within a particular region

For example, suppose you went to a pond and caught all of the fish in that pond. You could calculate the biomass of the fish by drying them so that there is no water left in their tissues and measuring the mass of the resulting dry organic matter.

Biomass, then, tells us something about how much biological matter exists in a certain region. The ecological pyramid separates biomass out according to trophic level so that you can get a feel for how much energy must exist at each level. For example, ecologist Howard T. Odum extensively studied the ecosystem in Silver Springs, Florida. The ecological pyramid for this ecosystem based on his study is displayed in Figure 10.3.

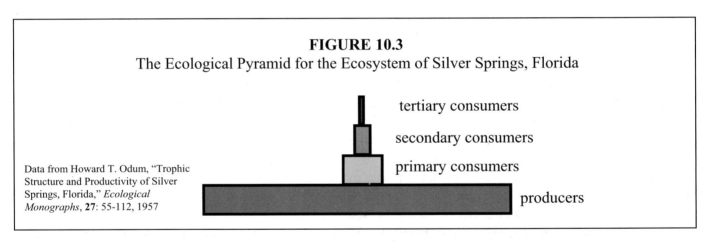

FIGURE 10.3
The Ecological Pyramid for the Ecosystem of Silver Springs, Florida

Data from Howard T. Odum, "Trophic Structure and Productivity of Silver Springs, Florida," *Ecological Monographs*, **27**: 55-112, 1957

tertiary consumers

secondary consumers

primary consumers

producers

In this figure, the length of the rectangle is an indication of the amount of biomass at each trophic level. As you can see, the biomass of producers is huge compared to the biomass of primary consumers, which is large compared to the biomass of secondary consumers, which is large compared to the biomass of tertiary consumers. If no energy were lost from trophic level to trophic level, the biomass of each trophic level would be the same. However, since energy is lost at each trophic level, there must be more and more creatures as you go to lower and lower trophic levels. It is important to note that you should look at the differences in trophic levels in terms of the *percentage* change, not the absolute change in the length of the rectangles. In other words, it is irrelevant what the actual size of each rectangle is. Instead, the only thing that is important is the percentage change between each rectangle.

Now before we leave this section, we hope that you have noticed an entire class of organism that we have simply ignored in this discussion. We have ignored the decomposers. We have done this on purpose, because it is hard to fit them into this kind of framework. Decomposers are rather indiscriminate; they tend to feed at all trophic levels. They feed on dead plants, dead primary consumers, dead secondary consumers, and dead tertiary consumers. As a result, they have their own unique trophic level. It is hard, therefore, to include them in a food chain or a food web. In addition, the biomass of decomposers is remarkably variable depending on the ecosystem studied. As a result, they do not fit nicely into ecological pyramids either. Therefore, we simply ignored them as a part of our discussion. As mentioned in previous modules, however, decomposers are vitally important to an ecosystem. In fact, the decomposers take care of the energy that is "lost" between trophic levels. They eat the parts of the organisms that the consumers do not eat and, as a result, the energy eventually gets back into creation. Thus, the decomposers are God's way of making sure that the energy "lost" from trophic level to trophic level is not lost to creation.

ON YOUR OWN

10.1 Looking at Figure 10.2, identify all possible trophic levels of each organism in the figure.

10.2 Based on Figure 10.2 and the idea of an ecological pyramid, would you expect more owls or more mice in the ecosystem that is illustrated there?

10.3 Based on the ecological pyramid in Figure 10.3, between which two trophic levels is the greatest amount of energy lost?

Mutualism

Although the previous section concentrated on the predator/prey relationship, there are other relationships that exist between organisms in an ecosystem. For example, we have already discussed the phenomenon of mutualism, a specific form of symbiosis in which two or more organisms live in a mutually beneficial relationship. In Module #3, for example, we discussed the termite and the protozoa from genus *Trichonympha*. The termite eats wood but cannot digest one of the principal components of wood, cellulose. Since the termite cannot digest cellulose, it is not able to eat wood without some help. The help comes from protozoa of the genus *Trichonympha*, which live in the gut of termites and eat the cellulose that the termites cannot digest. This keeps the termites alive while providing an ample food supply for the protozoa. In Module #4, we also learned about the mutualistic relationships between algae and fungi, which we call lichens, as well as the mutualistic relationships between plants and fungi, which we call mycorrhizae.

It turns out that there are many, many examples of mutualism in God's creation. The three that you have already learned about are examples from the land, but there are also incredible mutualistic relationships that exist between organisms underwater. Consider, for example, the amazing life of the **clownfish** (see Figure 10.4). These fish get their name from their bright orange and white markings. What makes them unique, however, is that they live among the stinging tentacles of **sea anemones** (uh nem' uh neez).

Sea anemones are normally deadly to fish. Their tentacles are full of stingers, each of which delivers a paralyzing poison to any fish that it touches. Once the fish is paralyzed, the anemone eats it. This doesn't happen to clownfish, however. When a clownfish touches the tentacles of a sea anemone, it does not get stung. Although biologists do not know all of the details of why the clownfish does not get stung, they are relatively certain it has to do with the chemical makeup of a layer of mucus that covers the clownfish. This mucus is very similar to sea anemone mucus, and biologists speculate that the mucus fools the sea anemone into thinking that the clownfish is actually a part of the sea anemone. As a result, the sea anemone does not sting the clownfish!

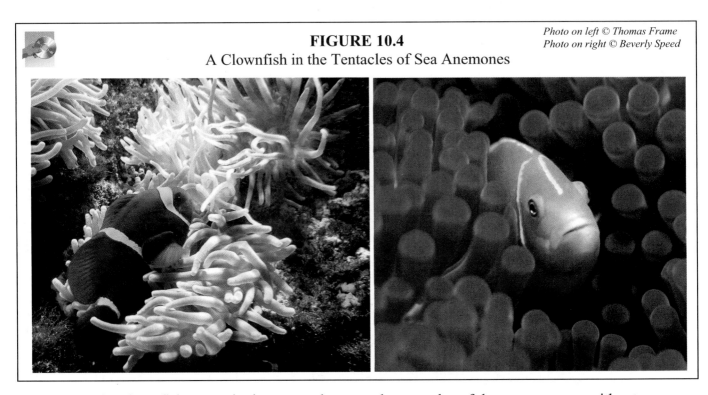

Photo on left © Thomas Frame
Photo on right © Beverly Speed

FIGURE 10.4
A Clownfish in the Tentacles of Sea Anemones

So the clownfish can swim in, out, and among the tentacles of the sea anemone without any adverse effects. As a result, the clownfish is protected from predators. After all, any predator that goes into the tentacles of the sea anemone will be killed. Thus, the clownfish benefits from the relationship by being protected from predators. The sea anemone benefits because the motion of the clownfish swimming near or in the anemone tantalizes predator fish. When the predator fish tries to attack the clownfish, the clownfish darts deep into the tentacles of the sea anemone. Often, the predator fish will be overcome with the desire to eat the clownfish and will not realize that it is being led into the tentacles of death. Thus, the sea anemone benefits because the clownfish tends to attract food to it. Recent studies also indicate that the activity of clownfish in a sea anemone frightens butterfly fish, which eat sea anemones. Thus, not only do the sea anemones protect the clownfish, but the clownfish also protect the sea anemones!

Another example of mutualistic symbiosis can be found in the relationship between the **blind shrimp** and the **goby** (goh' bee), pictured in Figure 10.5. These two organisms exist in most tropical seas, and they are usually found together. They both live in a hole on the ocean floor that has been dug by the blind shrimp. If you've ever been to the beach and tried to dig a hole in the sand under the water, you will have some idea about how hard it is for the blind shrimp to maintain its home. Ocean currents are constantly throwing debris into the hole, so the shrimp has an almost never-ending job of clearing the debris away from the hole. This puts the blind shrimp in danger, because (as its name implies) it cannot see very well. It therefore has no way of knowing whether or not it is in danger of predators. This is where the goby comes in.

The goby, a fish, has good vision. When the shrimp needs to clear the hole of debris, the goby goes out with it. While the shrimp digs, the goby keeps watch. The entire time that the shrimp is digging, it keeps one of its feelers on the goby. If the goby sees a predator, it signals the blind shrimp by flicking its tail, and they both head down the hole at lightning speed. Since the hole is too small for predators to follow, they are protected from danger.

FIGURE 10.5

Photo © Gary Bell/oceanwideimages.com

The Blind Shrimp and the Goby

Once again, this is a mutually beneficial relationship. The blind shrimp benefits because it is warned about the presence of predators. The goby benefits because it has access to a hole in which it can hide from predators. Since the goby has no means by which to dig a hole, it would not have such protection were it not for the blind shrimp. Even if it were able to find a hole that was unoccupied, the hole would fill up quickly from the debris carried by ocean currents. With the help of the blind shrimp, however, the goby gets a permanent home in which it can hide from predators.

The last example of symbiosis that we want to cover is rather striking. It is the relationship that exists between two fish: the Oriental sweetlips (the big fish in Figure 10.6 on the next page) and the blue-streak wrasse (the small fish in Figure 10.6). It is an interesting relationship, because the blue-streak wrasse acts as a *toothbrush* for the Oriental sweetlips. You see, the Oriental sweetlips, like many fish, have teeth. If they didn't, there would be some underwater creatures without predators. As we have already seen, this would have disastrous effects on the ecosystem. Now although teeth are useful, they actually are hard to maintain. If you do not clean them, they rot and fall out. Thus, the Oriental sweetlips must have some means by which it cleans its teeth. How does a fish clean its teeth? A person cleans his teeth by brushing them. A dog doesn't brush its teeth to clean them. Instead, it chews on hard, edible substances (like bones). This tends to flake the plaque and other cavity-causing materials off its teeth. A fish could, in principle, do that, but there are very few hard, edible substances available to many fish. Thus, the Oriental sweetlips must find some other way of getting its teeth clean. The figure below shows you how this happens.

FIGURE 10.6

Photo © Gary Bell/oceanwideimages.com

The Oriental Sweetlips and the Blue-Streak Wrasse

After spending its day feeding on little fish, the Oriental sweetlips decides that it is time to get its teeth cleaned. To do this, it looks for a particular spot on the coral reef, which is often called a "cleaning station." When it finds the cleaning station, it swims up to it with its mouth wide open, and little blue-streak wrasses dart out of the coral and *swim right into the mouth or up the gills of the Oriental sweetlips*! These little fish then proceed to eat all of the plaque and other problem-causing materials off the Oriental sweetlips' teeth. This, of course, cleans the sweetlips' teeth, and it also provides an excellent meal for the blue-streak wrasses. If all of this isn't amazing enough, the Oriental sweetlips also knows not to chomp down on the blue-streak wrasses when they finish their work. It allows the little fish to swim back out of its mouth so that it can have its teeth cleaned again tomorrow. Once again, this is a mutually beneficial relationship. Without the blue-streak wrasses, the Oriental sweetlips would lose its teeth. Without the Oriental sweetlips, the blue-streak wrasses would not have such fine meals.

Studying these incredible relationships should give you a great appreciation for the power and ingenuity of God. It should also tell you something about the hypothesis of macroevolution, which we studied in the previous module.   Remember, macroevolution hopes to explain the origin of all species on this planet through competition and natural selection. Think about this hypothesis in light of the mutualism that we have just studied. According to macroevolution, species should compete with one another for survival. The many, many examples of mutualism in creation tell us that species do not necessarily compete. They often help each other out. This flies in the face of a fundamental assumption in macroevolution.

Mutualism presents a much greater challenge to the hypothesis of macroevolution, however. You see, macroevolution can never hope to explain many of the incredible examples of mutualism in creation. For example, think about how macroevolution would have to explain the relationship between the Oriental sweetlips and the blue-streak wrasse. At some point in time, macroevolutionists would say, the sweetlips' ancestors probably had no teeth. In a number of generations, however, teeth began to form in a few of the ancestor's offspring. Now, in order for these teeth to avoid rotting and falling out, this new fish would have to develop the instinct for seeking out the wrasse, allowing the little fish to swim into its mouth, and not eating the little fish when it was done with its work. This instinct, of course, would have to evolve *at exactly the same time* that the Oriental sweetlips' teeth evolved. That's not enough, however. *At the exact same time that the teeth and instincts evolved in the Oriental sweetlips*, the blue-streak wrasse would have to *independently* develop the instinct to swim right into the Oriental sweetlips' mouth without fear of being eaten! Remember, if all of these things didn't happen in the same exact generation, the system would not work.

Obviously it is ridiculous to believe in all of these chance coincidences occurring at the same time. Over and over again, however, this is what the scientist who believes in macroevolution *must* believe. When a scientist begins to look at the world around him, he sees far too many of these "happy coincidences." Among all the precisely balanced properties of atoms and subatomic particles, the incredibly complex web of interactions that makes life on earth possible, the complicated nature of life itself, and the amazing design features of the animals we see in the world around us, it becomes obvious that this fantastic world around us could never have appeared by chance. It must have been designed by a powerful and intelligent designer. As Sir Frederick Hoyle (England's foremost astrophysicist, who died in 2001) says,

> ...A common sense interpretation of the facts suggests that a superintellect has monkeyed with physics, as well as with chemistry and biology, and that there are no blind forces worth speaking about in nature. The numbers one calculates from the facts seem to me so overwhelming as to put this conclusion almost beyond question. (Frederick Hoyle, *Engineering and Science* November, 1981, p. 12)

Mutualism is just one more example of creation providing witness for its Creator.

ON YOUR OWN

10.4 Although macroevolution can never hope to explain most of the examples of mutualism in creation, it can explain some. Which of the three examples of mutualism that you read about in this section would be easiest to explain in terms of macroevolution? How would macroevolution attempt to explain the relationship?

The Physical Environment

So far, our discussion of ecosystems has revolved around the associations between organisms. Specifically, we have discussed the feeding relationships that exist between organisms and the phenomenon of mutualistic symbiosis. However, the definition of ecosystem includes something besides the associations that exist between organisms; it also includes the physical environment. The physical environment is composed of all nonliving things in the ecosystem. This includes, but is not limited to, such things as weather, temperature, chemicals in the ecosystem, the gases that the organisms breathe, the availability of water, and so on. These are the things we want to begin discussing in this section.

Now if you think about it, in order to begin to study the physical environment of an ecosystem, you first have to determine what its limits are. For example, although you could say that the whole earth is a single ecosystem, it would be awfully hard to study. After all, the physical environment changes dramatically from place to place around the earth. In deserts, water is scarce, whereas in the ocean, water is obviously quite plentiful. In the Arctic, the temperature is always rather low, whereas near the equator, the temperature is usually high. Thus, in order to make an ecosystem understandable, we must establish limits. The more you restrict the region that you are considering, the easier it is to understand the ecosystem.

In Figures 10.1 and 10.2, for example, we highlighted the trophic levels in an ecosystem at the edge of a forest. By restricting ourselves to the creatures in a region like that, our food web was manageable. We could, however, define an ecosystem that is even easier to understand. If you were to walk into your backyard (or a park) and measure a region of grass that is two feet wide by two feet long, you would also have an ecosystem. This ecosystem would have the benefit of being very easy to study. Mostly, there would be producers (the grasses and other plants) and primary consumers (insects and an occasional squirrel or rabbit). If your cat came into that region and ate a mouse, you would have a secondary consumer. Of course, the drawback of such a narrowly defined ecosystem is that you would not learn an enormous amount by studying it, because there is not much diversity of life in it. In the end, then, an ecosystem really consists of a region that we choose to define. If we define it narrowly, it is an easy-to-understand ecosystem, but we will not learn an enormous amount from it. On the other hand, if it is large, it will be quite complex, but we can learn a great deal from it.

Once we have defined an ecosystem, we can begin to look at its physical environment. Now, as we have already discussed, there are a whole host of things that make up the physical environment of an ecosystem, and it is simply far too complex to study even the majority of these things. As a result, we will concentrate on just a few of the more important substances that make up the physical environment of an ecosystem: **water**, **oxygen**, **carbon**, and **nitrogen**. By looking at how these substances are distributed in an ecosystem, you will get some idea of how incredibly complex and wonderful God's creation is!

ON YOUR OWN

10.5 A biologist is studying two different ecosystems. One is defined as a 30-yard stretch of ocean shoreline. The other is a forest that runs one mile across and half of a mile wide. Which ecosystem will be the easiest to study? Which will most likely reveal more information?

10.6 The following is a list of things found near a pond. Which are part of the ecosystem's physical environment?

a. water   b. rocks   c. fish   d. algae   e. grass   f. mud

The Water Cycle

One of the most important substances for life on earth is water. All organisms on earth need water to live. Some need more water than others, but every living thing needs at least some water in order to survive. Thus, water is vital to all ecosystems. It is important, then, to understand where water comes from and where it goes to in an ecosystem. Consider, for example, what happens to water in an ocean shore ecosystem.

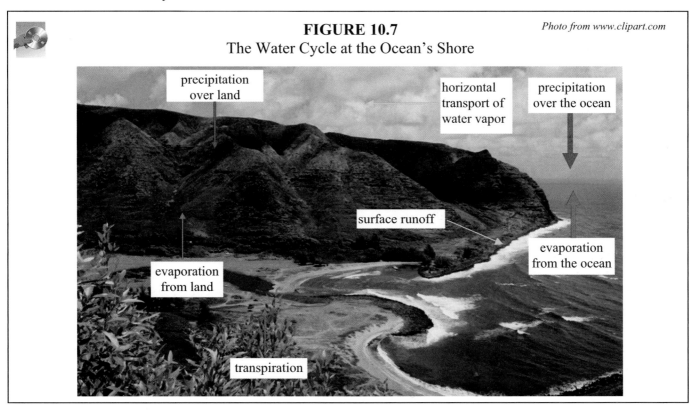

FIGURE 10.7
The Water Cycle at the Ocean's Shore

Photo from www.clipart.com

This diagram represents what we call the **water cycle** for this ecosystem. It attempts to detail where water comes from and where it goes.

Looking at Figure 10.7, you will see that water seems to cycle from the ocean to the shore and back again. Water evaporates from the ocean and travels into the air, where it eventually forms clouds. Clouds are simply water droplets (or tiny ice crystals) suspended on fine dust particles in the air. The clouds eventually cause precipitation (rain, snow, sleet, etc.), which brings water back into the ocean. Interestingly enough, however, more water evaporates from the ocean than what falls back into the ocean as precipitation. If that were the end of the story, the oceans would slowly run dry. Fortunately for us, however, that's not what happens.

The water vapor that evaporates from the ocean forms clouds that travel through the air. If they travel far enough, those clouds can cause precipitation that falls on the land. When that precipitation falls on land, there are several things that can happen. First, it can evaporate from the land, going back into the air again. Plants can also absorb the water from their roots. Interestingly enough, however, a large amount of that water ends up back in the air, because water actually evaporates from the leaves of plants. This is called **transpiration** (tran spuh ray' shun).

Transpiration – Evaporation of water from the leaves of a plant

As you will learn in a later module, transpiration is an important part of how water travels through a plant. For right now, however, the important aspect of transpiration is that it allows water that was on land to end up back in the air.

Well, it turns out that if you add up the amount of water that evaporates from land and the water that goes back into the air as a result of transpiration, you will find that it is *less* than the amount of water that falls on land in the form of precipitation. If that were the end of the story, the land would get more and more waterlogged. However, the excess water trickles back into the sea as surface runoff (or as part of a river). So you see that while most of the evaporation in the ecosystem comes from the ocean, excess water from land replenishes the water in the ocean. That way, the ocean does not run out of water. Thus, there is a fine balance between the horizontal transport of water vapor, which takes water from the ocean, and the surface runoff, which brings water back into the ocean.

It turns out that the water cycle does more than make sure that water stays where it should be in an ecosystem. It is also responsible for transporting nutrients from one part of an ecosystem to another, or even *between one ecosystem and another*. For example, in the ocean shore ecosystem depicted in Figure 10.7, the surface runoff water carries with it nutrients that were contained in the soil and sand near the shore. These nutrients include chemicals that are the result of decomposers feeding off dead creatures and minerals that were held in the soil and sand. Organisms in the ocean use these nutrients, so it is important that they get brought in from land via the water cycle.

The role of the water cycle in transporting nutrients is best seen through a study of a **watershed**.

Watershed – An ecosystem where all water runoff drains into a single body of water

The figure below shows an example of a watershed, as well as the water cycle that occurs there.

FIGURE 10.8
The Water Cycle in a Watershed

Photo from www.clipart.com

Looking at Figure 10.8, you can see that water comes into the watershed in one way: through precipitation. It leaves in one of three ways: through transpiration, evaporation, or stream flow. Along the way, a lot of other things happen to the water. Water can flow over the land and into the stream as surface runoff. It can also be absorbed by the soil. Once in the soil, it can become part of the groundwater, which is water that flows beneath the surface of the land. That groundwater eventually flows along the base of the soil until it reaches the stream. Finally, it can be absorbed by the roots of plants, which means it will eventually leave the watershed as a result of transpiration.

Watershed ecosystems play a vital role in the distribution of nutrients throughout the earth. The soil of a watershed is rich with nutrients, and if the watershed's body of water is a river (such as the one pictured in Figure 10.8), that river can carry water out of the watershed and to other ecosystems that need the nutrients. The plants in the watershed moderate the amount of nutrients carried away by the river, as they limit the amount of soil that can be pulled into the river by surface runoff or groundwater flow. Without these plants, too many nutrients would flow into the water, causing algal blooms that could destroy the water ecosystem or the ecosystem to which the nutrients are being carried. This would also deprive the soil of nutrients that promote plant growth. This is why ecologists are worried about deforestation. If too many forests are cut down, watersheds will not be able to control nutrient flow in the water cycle. This could result in disastrous effects to local ecosystems.

Despite the obvious need for strong forest and plant life in watershed ecosystems, many third-world countries find that the economic benefit derived from cutting down the trees is too great to worry about long-term environmental effects. As a result, forests are being leveled in places like India, Mexico, and Brazil, destroying the local watersheds. In the United States, environmental regulations have forced logging companies to replant trees at a *faster* rate than they cut them down. As a result, there is more forest coverage in the United States today than there has been in the past 100 years. If other countries can force their people to do the same, watersheds across the world would be better protected.

Although this gives you an idea of how the water cycle works, it should be obvious to you that the water cycle is different in each ecosystem. Just as the water cycle in an ocean shore ecosystem is different from the water cycle in a watershed, it is different in just about every other ecosystem as well. Despite these differences, however, there are similarities, so once you have learned the basics of the water cycle, it is easier to interpret the specific water cycle for a specific ecosystem.

ON YOUR OWN

10.7 There are at least three different means by which water can potentially leave an ecosystem. Which of these means exists in a watershed but not an ocean shore environment?

10.8 Suppose you studied a watershed ecosystem in depth and measured the amount of water that evaporates, the amount of water that goes through transpiration, and the amount of water that the watershed gains through precipitation. If you were to add the total amount of water that leaves the watershed by both evaporation and transpiration, would you expect that to be larger or smaller than the amount of water that the watershed gains through precipitation?

The Oxygen Cycle

Although some organisms exist without oxygen, most organisms are aerobic, which means they require oxygen in order to survive. Thus, oxygen is obviously an important part of the physical environment in any ecosystem. The air that we breathe is about 21% oxygen. This often surprises students, because they think that since organisms breathe air in order to get oxygen, the air must be mostly oxygen. In fact, less than one-fourth of the air is oxygen. Almost all of the rest of it is nitrogen, and a small part of it is made up of several other gases including argon, carbon dioxide, and ozone. It turns out that this mixture of gases is simply ideal for supporting life. When you take chemistry, you will find out why this is the case.

Well, if the mixture of gases that makes up the air we breathe is ideal for the purpose of supporting life, that mixture should never be allowed to change. This is a problem, since organisms are constantly breathing oxygen from the air, using up the oxygen. If this used oxygen were not continually being replaced by new oxygen, the amount of oxygen in the air would drop to a percentage much lower than 21%. Eventually, there wouldn't be enough oxygen to sustain life. This is not the case, of course, because the Creator has designed ecosystems to continually replenish the oxygen used by organisms. The physical processes that cause oxygen to be used up and replenished are collectively called **the oxygen cycle**, which is depicted in the figure below.

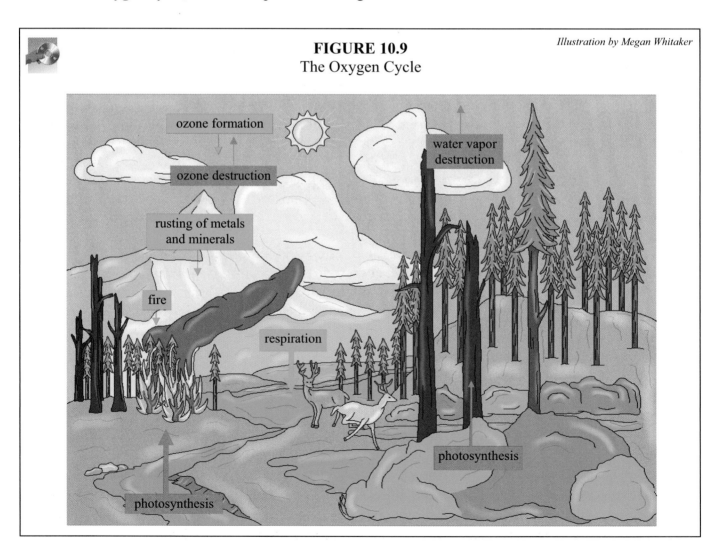

FIGURE 10.9
The Oxygen Cycle

Illustration by Megan Whitaker

In the oxygen cycle, oxygen is used up in many different ways, as illustrated by the blue arrows pointing downwards. The thicker the arrow, the more oxygen that is used up in the process. Aerobic organisms (both aquatic and land-based) use up oxygen as a part of their respiration. Natural and human-made fires use up oxygen. The rusting of metals is actually a chemical process that uses up oxygen. Finally, some oxygen is converted into ozone by the energy of the sun. This actually turns out to be a good thing, because ozone forms a shield (the ozone layer) that protects us from certain harmful rays which come from the sun. When you take chemistry, you will learn more about this fascinating substance.

Although oxygen is used up by all of these processes, the amount of oxygen in the air does not diminish, because there are also processes by which oxygen is restored to the air, as illustrated by the lavender arrows in the figure. The major process by which this happens is photosynthesis. Green plants on land perform photosynthesis, as do phytoplankton in the water. As we learned in Module #3, phytoplankton are responsible for the majority of all photosynthesis on earth. In order to perform photosynthesis, the organisms require carbon dioxide (CO_2) and water, as well as energy from the sun. The amount of water in the air, of course, is regulated by the water cycle. As we will learn in the next section, the amount of carbon dioxide in the air is regulated by the carbon cycle. Thus, the cycles that we are discussing here are not isolated from one another. Instead, they all work together to regulate the physical environment of the many ecosystems on earth. We simply separate them in order to make them each a little more understandable.

One thing that students often fail to understand about the oxygen cycle is that although phytoplankton and plants produce oxygen via photosynthesis, *they also use oxygen through respiration.* After all, why do producers perform photosynthesis? They do so to make food for themselves. What do they do with that food? They perform cellular respiration, as we discussed in Module #6. What happens in cellular respiration? Glucose molecules *and oxygen molecules* are used up. Thus, when the plants actually use the food that they make via photosynthesis, they end up taking oxygen out of the air. Of course, when they make more food, they put it back into the air. Overall, plants make *a lot more* food than they ever use. Thus, they do more photosynthesis than respiration, and as a result, they make more oxygen than they use. Nevertheless, they do use oxygen, and it is important to understand that.

Even though photosynthesis is the main process by which oxygen is replenished, it is not the only one. When certain rays from the sun hit ozone molecules, the ozone is destroyed and oxygen is formed. Also, when other rays hit water vapor in the air, the water can decompose into hydrogen and oxygen. This, then, is another way that oxygen is replenished in the air around us. The amount of water vapor in the air is controlled by the water cycle, once again demonstrating that these cycles are not independent of one another.

ON YOUR OWN

10.9 Suppose producers made only the food that they needed and never made any extra. What would happen to the oxygen level in the atmosphere? What would happen to the consumers of creation?

10.10 As you learned previously, Biosphere II was an attempt at making an artificial, enclosed ecosystem that was self-sustaining. It did not work. Suppose you were working on Biosphere III and were having problems keeping the right amount of oxygen in the air: the oxygen level continually rose, resulting in too much oxygen. What could you do in an attempt to fix the problem?

The Carbon Cycle

As you learned in Module #5, the chemistry of life is built on the chemical element carbon. Indeed, in order for a molecule to be considered organic, it must have carbon atoms in it. The **carbon cycle** regulates the amount of carbon in ecosystems, principally by keeping careful track of carbon dioxide (CO_2). The carbon cycle is illustrated in the figure below.

FIGURE 10.10
The Carbon Cycle

Illustration by Megan Whitaker

The amount of carbon in an ecosystem can be tracked using carbon dioxide because, at one point or another, most of the carbon from organic molecules passes through a stage in which it is part of a CO_2 molecule. Carbon that makes up a living organism's body, for example, will be converted to carbon dioxide by decomposers after the organism dies. When organisms eat food (either from producers or consumers), the carbon in that food is eventually converted to carbon dioxide by respiration. Thus, by seeing where carbon dioxide goes to and comes from in an ecosystem, we are actually tracking most of the carbon in that ecosystem.

We see, then, that carbon dioxide enters the air through a variety of different processes. When decomposers do their job, the dead organic matter is converted to carbon dioxide. The respiration of all aerobic organisms results in carbon dioxide. When something burns (either by natural or human-made means), carbon dioxide is produced. All of these processes tend to increase the amount of carbon dioxide in the air.

Although carbon dioxide is constantly being added to the air, it is also constantly being taken away. Since photosynthesis requires carbon dioxide, an enormous amount of carbon dioxide is taken out of the air by this process. In addition, carbon dioxide also dissolves into the ocean, where it is used in several chemical processes, such as the means by which some marine organisms make their shells. Some of the dissolved carbon dioxide also reacts with other chemicals in the ocean to form minerals, which get stored in the sediments at the bottom of the sea.

It turns out that the amount of carbon in the atmosphere is important to track for reasons other than trying to see where the carbon is in an ecosystem. Carbon dioxide itself has a very important task in keeping the earth hospitable for life. It helps keep the planet warm. How does it do that? Well, carbon dioxide is one of the principal gases involved in a process known as the **greenhouse effect**.

Greenhouse effect – The process by which certain gases (principally water vapor, carbon dioxide, and methane) trap heat that would otherwise escape the earth and radiate into space

You see, the light shining on the earth from the sun warms the planet; however, the earth tends to radiate a lot of that light back out into space. This cools the planet. If that were the end of the story, the earth would be a VERY cold place. It would be so cold, in fact, that life could not exist anywhere on its surface.

Fortunately, however, that is not the end of the story. It turns out that certain gases (principally water vapor, methane, and carbon dioxide) actually trap much of the light that is radiated by the earth. As a result, the energy from that light is not lost. Instead, it is held in the air, and the net effect is to warm the planet. Because of the greenhouse effect, the earth is warm enough to support life. To become more familiar with the concept of the greenhouse effect, perform the following experiment.

EXPERIMENT 10.1
Carbon Dioxide and the Greenhouse Effect

Supplies:

♦ Thermometer (It must be able to read temperatures from room temperature to at least 100 degrees Fahrenheit. The smaller the thermometer, the better.)
♦ A large, clear Ziploc® freezer bag (It must be large enough for the thermometer to fit inside once it is zipped.)
♦ Sunny windowsill (If it's not sunny today, just wait until it is.)
♦ Plastic, two-liter soda pop bottle
♦ Vinegar
♦ Baking soda
♦ Teaspoon

Object: To observe the ability of carbon dioxide to absorb energy from sunlight

Procedure:

A. Running the experiment with air
1. Open the Ziploc bag.
2. You want to fill the bag with air. Do this by holding the bag wide open, with the open side facing down. Then, raise the bag as high as you can and quickly lower it. This should fill the bag with air.
3. Put the thermometer in the bag and zip it closed. You should now have a Ziploc bag that is partially inflated with air and has a thermometer inside.
4. Place the bag on a sunny windowsill. Arrange the bag so that you can look through it and read the thermometer.
5. Allow the bag and thermometer to sit for 15 minutes.
6. Read the temperature from the thermometer.

B. Running the experiment with carbon dioxide
1. Take the bag off the windowsill.
2. Open the bag and remove the thermometer.
3. Place the thermometer in a safe place while you prepare the next part of the experiment.
4. Fill the two-liter bottle about one-third of the way with vinegar.
5. Measure out one teaspoon of baking soda and add it to the vinegar. The contents of the bottle will begin to bubble. Those bubbles tell you that a gas is being formed. The gas is carbon dioxide. Wait for a while. This will allow the carbon dioxide to push the air out of the bottle.
6. Repeat step #5. By doing that step twice, you are ensuring that there is only carbon dioxide in the bottle above the vinegar and baking soda.
7. Measure out another teaspoon of baking soda, but this time, keep it in the spoon.
8. Open the Ziploc bag and press it flat to remove any air in it. Keep it handy.
9. This has to be done quickly. Add the baking soda to the vinegar and then quickly hold the Ziploc bag over the opening of the bottle. Don't worry about making sure all of the baking soda lands in the bottle. In this step, speed is important. Immediately close the bag around the bottle opening, so that the carbon dioxide coming from the bottle goes into the bag. Allow this to continue until the bubbling slows down significantly.
10. Although the bag will not be significantly inflated, it will contain a lot of carbon dioxide. Carefully lift the bag off the bottle and put the thermometer in the bag.
11. Quickly zip the bag closed. You should now have a bag with carbon dioxide and a thermometer in it.
12. Go to the windowsill and place the bag in the same position that you did in part A of the experiment.
13. Allow it to sit for 15 minutes.
14. After the 15 minutes, read the temperature.

C. Analyzing the data

Look at the two temperatures. They should be different. The temperature that you got in part B should be higher than the one you got in part A. Why? Well, carbon dioxide absorbs energy from sunlight. This is what causes the greenhouse effect. In your experiment, because carbon dioxide absorbed the energy from the sunlight, the bag got hotter when it contained more carbon dioxide.

Since carbon dioxide is one of the principal gases that participates in the greenhouse effect, the amount of carbon dioxide in the atmosphere must be regulated very carefully by the carbon cycle. After all, if the amount of carbon dioxide in the air decreased significantly, the earth would get colder and colder, perhaps turning into a frigid wasteland. If, instead, the amount of carbon dioxide in the atmosphere increased significantly, the greenhouse effect would get stronger and stronger, leading to a warmer and warmer planet. If the planet got too warm, disastrous things could happen to certain ecosystems. The north and south poles, for example, are filled with tons and tons of ice. If those regions of the earth were to get too warm, that ice could melt, causing floods around the globe. Also, most ecosystems are very sensitive to temperature. A significant change in the temperature of some ecosystems could destroy much of the life contained in them.

Some people are worried that this very thing is occurring today. Because we burn a large amount of coal, oil, and wood as a part of our modern lifestyle, humanity has been adding carbon dioxide to the atmosphere. As a result, the amount of carbon dioxide in the atmosphere has been rising steadily for the past 100 years or so. Some people are worried that the air is getting so rich in carbon dioxide that the greenhouse effect is already becoming enhanced and the earth is already getting abnormally warm. They call this phenomenon **global warming**, and they are convinced that the earth will simply get too hot if something isn't done to stop the buildup of carbon dioxide in the air.

Although the fear that too much carbon dioxide in the air could lead to global warming is based on sound scientific reasoning, reality is just a bit more complex than that. To see what we mean, take a look at the data presented in Figure 10.11.

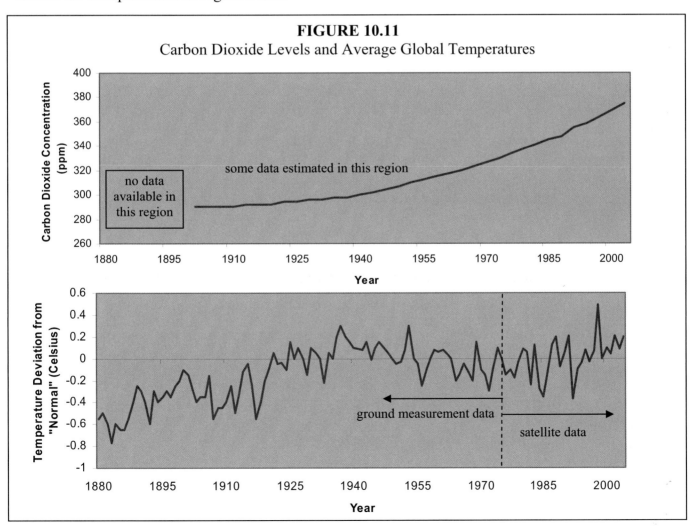

FIGURE 10.11
Carbon Dioxide Levels and Average Global Temperatures

Before we figure out what the data tell us, let's make sure you understand the graphs themselves. In the top half of the figure, the concentration of carbon dioxide in the atmosphere (in parts per million) is graphed versus the year the measurement was taken. These data are from the Mauna Loa observatory in Hawaii, and they make up the largest set of data regarding carbon dioxide concentration in the atmosphere. Notice that the data do not start until 1900, and we do not have a complete data set until the late 1950s. Thus, some of the data between 1900 and the late 1950s are estimated.

In the bottom half of the figure, the *change in global temperature* is plotted versus the year in which the change was measured. This means that the line representing zero in the graph is the "normal" average global temperature (about 50° F). Data that lie below that line represent years in which the average global temperature was lower than normal, and data that lie above that line represent years in which the average global temperature was warmer than normal. The data from 1880 to 1978 are from temperature measurements that were taken on the ground (K. H. Bergman,: *International Journal of Environmental Studies*, Vol.20, 1983 p.93). Although these data are illustrative, they are not the most accurate data, as nearly 70% of the earth is covered by water. Thus, these data ignore the vast majority of the earth. The data from 1979 to the present are taken by NASA satellites (http://www.ghcc.msfc.nasa.gov/MSU/msusci.html, retrieved 01-12-05). These data are ideal, as the satellites measure the temperature of the *entire* earth.

Now what do all of these data tell us? Well, the top graph tells us that there has been a steady increase in carbon dioxide levels in the air since about 1920. If you look at the bottom graph, however, you do not see a corresponding increase in the temperature of the earth. Instead, from about 1880 to 1920, the average global temperature increased in a very shaky pattern by about 0.5 degrees Celsius (0.9 degrees Fahrenheit). After that, however, the temperature change varies up and down quite a bit, but continues to hover around zero. In other words, over the time that the amount of carbon dioxide in the air increased steadily, the average temperature of the earth, on average, did not change significantly.

Does this mean that the amount of carbon dioxide in the air does not affect the temperature of the earth? No. We know that the greenhouse effect is real, or we wouldn't be here. These data tell us that reality is more complex than theory. Remember, in order to see a change in the greenhouse effect, there must be a *significant* change in the amount of carbon dioxide in the air. Unfortunately, scientists do not know what a significant change would be. If the amount of carbon dioxide in the air doubled, would that be significant? From the standpoint of the greenhouse effect, we really do not know. What we know for sure, however, is that the sum total of all carbon dioxide produced by human activity is approximately 3% of the carbon dioxide produced by the other processes that make up the carbon cycle. One could argue, then, that this amount of added carbon dioxide is simply not significant compared to all of the other processes that add carbon dioxide to the air.

Also, the way in which carbon dioxide is added to the air is very important in the greenhouse effect. When people burn fuels, carbon dioxide is not the only gas that is released. Many other gases are released as well. Some of these gases tend to reflect light rather than absorb it. This actually *reduces* the amount of energy absorbed by the earth, causing a net cooling effect. It could be that any increase in the greenhouse effect due to human-produced carbon dioxide is offset by the cooling caused by the other chemicals associated with human activity.

Although there are many things we do not know about the greenhouse effect, right now the reliable data indicate that the earth is not warming, at least not in any significant way. In addition, we know that the earth has been much warmer in the past. A team from Harvard University reviewed over 240 scientific studies on global temperatures in the past, and their research indicates that the earth was significantly warmer between the ninth and fourteenth centuries than it is today. As a result, it is awfully hard to believe that modern burning of fuels is leading to any kind of global warming, as people did not burn nearly as much fuel in the Middle Ages as they do now!

If the data seem to indicate that global warming is not happening, and if scientists agree that it has been much warmer in the past than it is now, why does the media seem to say that global warming is happening? There are a few possible reasons. First, most of the people who talk about the earth's ecosystems really do not understand them very well. Thus, they tend to use their own personal experiences and attempt to relate those experiences to the earth as a whole.

For example, many people who talk about global warming discuss the fact that in their area of the world, the winters have been unusually warm over the past few years. Thus, they say, global warming must be happening. This, of course, neglects the fact that the entire world is often quite different from one person's little corner of it. While one area of the world might be experiencing unusually warm winters, another part of the world might be experiencing unusually cool summers. As a result, the *global* temperature does not change very much, despite what might be happening in one region of it.

This can also be applied to other issues as well. For example, it is common to hear that certain large ice sheets which have been the same size for a long, long time are now starting to melt. This is then used to conclude that the earth as a whole is getting warmer. However, this is not sound scientific reasoning. The fact that an ice sheet is melting is only evidence that *the part of the world that holds the ice sheet* is warming up. For example, a large ice sheet in the Antarctic Peninsula, the LarsenB ice shelf, has experienced a remarkable decline in size over the past few years. Some say that this is evidence for global warming. The problem is that the west Antarctic ice sheet has been *thickening* over the past few years. Thus, while the Antarctic Peninsula is getting warmer, the western part of the Antarctic is getting cooler. It turns out that temperature measurements of Antarctica indicate that overall, the continent of Antarctica has actually been cooling since the 1960s. Does that mean that global cooling is happening? Of course not! It just means that when you talk about global warming, you must consider the entire earth, not just one part of it.

Another problem is that people who have a political agenda to push can often distort the facts in order to make their point more persuasive. For example, it is not unusual for people to say that the earth has experienced measurable warming over the past 100 years and that is due to the carbon dioxide produced by people. While the first part of that statement is most certainly true, these people simply neglect to tell their audience that this warming occurred in the early 1900s, *before* human beings were burning much fuel. Thus, the earth did warm up a bit in the past 100 years, but it was not because of rising carbon dioxide levels. That did not happen until after the warming ended.

Finally, if you dig hard enough, you can find *some* evidence that the earth is warming up as a whole. For example, land-based temperature measurements do show a slight warming trend, even over the past 20 years. The problem with these data, as we mentioned before, is that they ignore 70% of the earth's surface, because they ignore what is happening over the oceans. When the entire globe is measured by satellites, no warming trend is seen. Thus, to believe that global warming is occurring, you are forced to rely on the *least reliable* data and you are force to ignore the *most reliable* data. That is not good scientific practice.

In the end, then, if you want to make reliable statements about how the earth's temperature is changing, you need to look at global indicators. The best global indicator is the satellite temperature data, and they indicate no global warming. In addition, radiosonde balloons (balloons that carry disposable weather-measuring equipment into the air) have been used to measure global temperatures since 1979. These balloon measurements also indicate no significant warming trend over the entire earth. Thus, the two most reliable data sources indicate that there has been no significant warming trend since 1979. The best scientific conclusion to make, then, is that global warming is not happening.

ON YOUR OWN

10.11 If a tree dies and slowly rots away, does this add carbon dioxide to or remove it from the air?

10.12 If you were able to see carbon dioxide as it interacted with the surface of the ocean, you would see the ocean absorb an enormous amount of carbon dioxide. Only a portion of that carbon dioxide actually dissolves in the ocean, however. A large amount of the carbon dioxide goes into the ocean but is not dissolved. If it isn't dissolved, what is it used for?

The Nitrogen Cycle

Most people know that carbon, oxygen, and water are all important for life on earth. However, many people do not know that nitrogen is critically important to living organisms as well. In fact, all organisms need nitrogen, as nitrogen is in many of the chemicals that make life possible, such as DNA, RNA, and proteins. Thus, it is important for us to understand how organisms get the nitrogen that they need to survive and what happens to that nitrogen after those organisms die.

The majority of the air that we breathe (78%) is made up of nitrogen gas. As a result, you might think that organisms can get the nitrogen right out of the air that they breathe. You would be wrong, however, since the chemical form of nitrogen gas (N_2) is not very reactive. This is a problem, because organisms need to use nitrogen in their biosynthesis, and for a substance to be used in biosynthesis, it must chemically react with other substances. Thus, organisms must use nitrogen in more chemically active molecules such as nitrates (chemicals containing NO_3^-), nitrites (chemicals containing NO_2^-), or ammonia (NH_3).

How do these organisms get such chemically active molecules? Well, one way is through the process of **nitrogen fixation**, which converts nitrogen gas into chemically active molecules that many organisms can use. Sometimes, the nitrogen fixation is done by the physical environment. The intense heat formed in lightning strikes, for example, causes nitrogen gas to react with water and oxygen in the air, making nitrates and nitrites. These nitrates and nitrites dissolve in the water that is in the atmosphere, eventually falling to the ground in the form of rain. However, most of the nitrogen fixation is not done by the physical environment; it is done by **nitrogen-fixing bacteria**, which have been designed to take nitrogen gas from the air and convert it to ammonia. Some of that ammonia is used by plants, and the rest is converted into nitrites and nitrates by **nitrifying bacteria**.

Now we know that all of this might be a bit confusing, so we want to illustrate the nitrogen cycle for you.

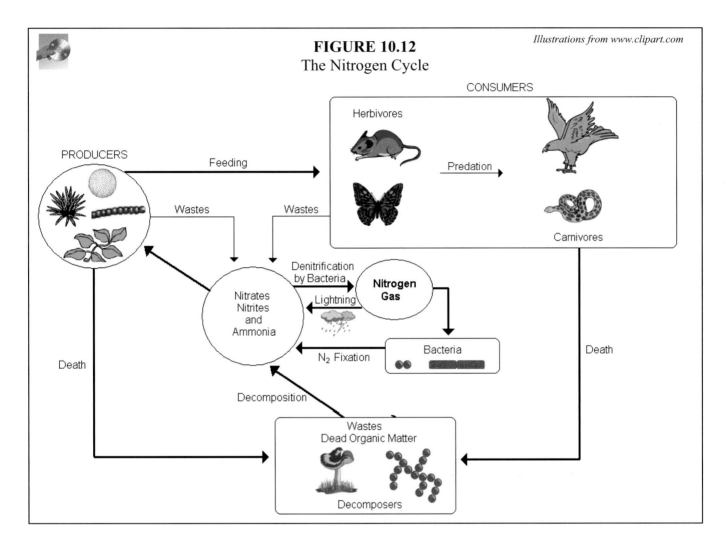

FIGURE 10.12
The Nitrogen Cycle

Illustrations from www.clipart.com

    To get your orientation, look at the oval with the boldfaced "Nitrogen Gas" in it. That's nitrogen in the air. The nitrogen gas can be taken in by nitrogen-fixing bacteria (such as the cyanobacteria you learned about in Module #2) and through nitrogen fixation can be turned into ammonia, which can be further turned into nitrates and nitrites. Lightning can also form nitrates and nitrites. The ammonia, nitrates, and nitrites are then used by the producers (plants and algae, for example). Primary consumers then eat the producers, getting their nitrogen that way. Carnivores then feed on primary consumers, bringing nitrogen up the food chain. So that's how nitrogen travels up the food chain. It goes from the air, through the nitrogen-fixing bacteria (or through lightning) so that more chemically active nitrogen-containing molecules are made. Those more active molecules are then used by the producers, and the nitrogen is then transferred to the consumers when they eat the producers.

    What happens to all of this nitrogen? Well, some of it is excreted by the producers and the consumers as waste. Those wastes often contain the chemically active molecules that can immediately be used by the producers. The rest of the nitrogen stays with the organism until it dies. At that point, decomposers such as fungi and saprophytic bacteria begin decomposing the dead organic matter, turning the nitrogen-containing molecules back into the nitrates, nitrites, and ammonia that the producers can use. In the end, then, the nitrogen that has gone up the food chain eventually gets recycled so that it can travel up the food chain once again.

There is one part of the figure we have not discussed yet. Sometimes, the more active nitrogen-containing chemicals are used by certain bacteria that end up taking the nitrogen and remaking nitrogen gas. Thus, the nitrogen cycle doesn't just pull nitrogen gas out of the air; it also puts it back in. This process is called **denitrification**, and it effectively takes nitrogen out of the food chain until nitrogen-fixing bacteria can once again turn it into the more active nitrogen-containing compounds.

ON YOUR OWN

10.13 Suppose you are once again designing Biosphere III. In your initial tests, you find that the amount of nitrogen gas in the air of your biosphere is decreasing as time goes on, but the amount of nitrates, nitrites, and ammonia are fairly constant and cannot be decreased much without affecting the organisms that need them. What should you look at in order to try to fix this problem so that nitrogen does not have to be pumped into your biosphere?

Summing Up

In this module, we have only scratched the surface of the science of ecology. In studying ecosystems, we have only concentrated on two types of relationships between organisms (trophic and symbiotic relationships) and four processes related to the physical environment (the water, oxygen, carbon, and nitrogen cycles). However, even this cursory glance at ecosystems should tell you something rather profound: ecosystems are intricately balanced, complex systems. If even one of the many, many processes that regulate the physical environment were to change significantly, life would cease to exist. If even one organism is introduced into an ecosystem in which it does not belong, a disaster can occur!

This intricacy and complexity, of course, is a testament to the power and knowledge of God. Just think about it! God has placed all of the organisms in just the right ecosystems so that every organism has enough predators to keep its population in check but not so many that the population dies out. He has designed some organisms to work together in order to survive. In addition, He has set up literally millions and millions of chemical and physical processes that regulate the physical environments of all ecosystems on the earth. Is there any way that all of this could just have happened by chance? Of course not! Even the most basic ecosystem in the world is already more intricate and complex than the most sophisticated machine that human science has ever developed. Clearly, systems of that complexity require an incredibly knowledgeable and powerful Designer. That Designer has left His fingerprints all over His creation. It is up to us to recognize that and give Him the praise that He deserves.

ANSWERS TO THE "ON YOUR OWN" PROBLEMS

10.1 The trophic levels of the organisms in Figure 10.2 are summarized in the table below:

Organism	Possible Trophic Levels
Hawk	secondary consumer, tertiary consumer
Songbird	secondary consumer, tertiary consumer
Squirrel	primary consumer
Plants	producers
Butterfly	primary consumer
Owl	secondary consumer, tertiary consumer
Rabbit	primary consumer
Praying mantis	secondary consumer
Snake	secondary consumer, tertiary consumer
Grasshopper	primary consumer
Raccoon	primary consumer, tertiary consumer
Fox	secondary consumer
Mouse	primary consumer
Frog	secondary consumer, tertiary consumer

10.2 There would be more mice. Since energy is lost at each trophic level, there should be more primary consumers than secondary consumers.

10.3 Remember, we do not look at the absolute change in the length of each rectangle. Instead, we look at the percentage change. For example, the rectangle for primary consumers is 1/8 (12.5%) the size of the one for producers. Thus, there was a 7/8 (87.5%) change in the step from producers to primary consumers. The secondary consumer rectangle, however, is about 1/3 (33%) the size of the primary consumer rectangle. Thus, there was only a 67% change going from primary consumers to secondary consumers. Likewise, it looks like there is about a 67% change going from secondary consumers to tertiary consumers.   This indicates that a huge number of producers are needed to support the feeding needs of the primary consumers. Thus, the largest amount of energy is wasted between the producers and the primary consumers.

10.4 The clownfish/anemone mutualistic symbiosis could be explained by macroevolution. According to the hypothesis, a clownfish might have been born with the protective mucus layer as a result of mutation. When the clownfish inadvertently swam into the anemone, it did not die and was protected from its predators. That strain of clownfish was naturally selected to survive because it was so well protected from predators. Thus, it passed on its trait to its offspring. This is easiest to explain because only the clownfish has to evolve, and it needs to evolve only one characteristic. In the case of the blind shrimp and the goby, both of the creatures would have to independently evolve several instincts that allow them to work together. In the same way, the relationship between the Oriental sweetlips and the blue-streak wrasse would require instincts to evolve independently in two separate creatures.

10.5 The 30-yard stretch of ocean shore is the simplest ecosystem, because it is smaller. The biologist will probably learn more from the other ecosystem, however.

10.6 The physical environment includes all nonliving surroundings. Thus, <u>a, b, and f</u> are part of the physical environment. The others are living organisms.

10.7 <u>Stream flow does not carry nutrients out of an ocean shore ecosystem</u>. If there is a stream leading to the ocean, it is putting nutrients into the ocean. We call this an estuary.

10.8 Think about it. Since water continually leaves the watershed by way of the stream, in order to keep the stream from running dry, a lot more water has to enter the ecosystem than that which evaporates and transpires. As a result, <u>the amount of precipitation will be larger</u>.

10.9 <u>The level of oxygen in the air would plummet</u>, because all of the aerobic organisms would be using oxygen, but the producers would be using up just as much oxygen in respiration as they produced in photosynthesis. As a result, the producers would not make any extra oxygen, and the oxygen used by the consumers and decomposers would never be replenished. <u>The consumers would also eventually starve</u>. After all, the primary consumers eat the producers to get the extra food that the producers made but did not use. Thus, they would starve, and when they were gone, the other consumers would have nothing to eat, so they would starve as well.

10.10 <u>You would need to get rid of some producers in the biosphere</u>. This would decrease the level of photosynthesis being done, decreasing the amount of oxygen produced.

10.11 When the tree rots, it is decomposed by decomposers. This <u>adds carbon dioxide to the air</u>.

10.12 <u>The majority of the carbon dioxide is going into the ocean so that it can be a part of the photosynthesis of phytoplankton</u>.

10.13 Think about it. If nitrogen is constantly being pulled out of the atmosphere, but the amount of nitrates, nitrites, and ammonia is not increasing, the nitrogen is probably being "trapped" somewhere in the nitrogen cycle. Looking at the figure, one way the nitrogen could get trapped is in dead organic matter. If that matter is not recycled back into the biosphere, nitrogen will have to be continually pulled from the air to get the organisms the nitrogen they need. Thus, <u>you should check the levels of decomposers you have in your biosphere</u>. Perhaps you do not have enough decomposers to handle all of the dead organic matter.

STUDY GUIDE FOR MODULE #10

1. Define the following terms:

a. Ecology
b. Population
c. Community
d. Ecosystem
e. Biome
f. Primary consumer
g. Secondary consumer
h. Tertiary consumer
i. Ecological pyramid
j. Biomass
k. Transpiration
l. Watershed
m. Greenhouse effect

2. When fruits or vegetables are imported into the U.S. from a foreign country, they are always very closely inspected for insects, even though the vast majority of insects are not really harmful. Why is the inspection done?

3. For the following abbreviated food web, list all possible trophic levels for each organism.

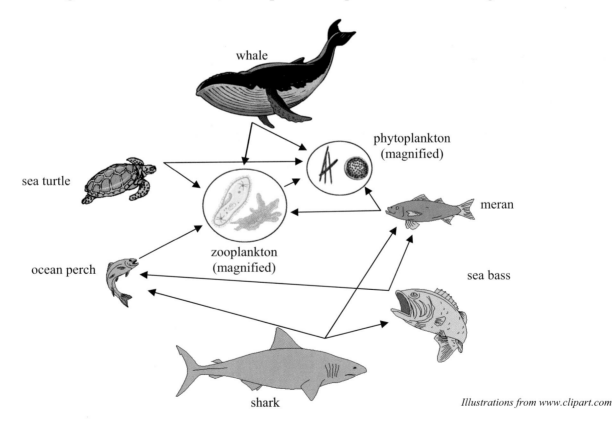

Illustrations from www.clipart.com

4. Consider the following ecological pyramid:

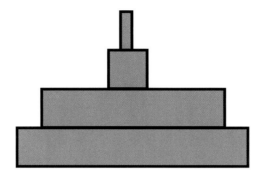

a. Which two trophic levels have the greatest disparity in biomass?

b. Between which two trophic levels is the smallest amount of energy wasted?

5. Name the participants in the three new symbiotic relationships that we learned in this module. Briefly describe the roles of each participant.

6. What fundamental assumption of macroevolution does mutualism seem to contradict?

7. In the water cycle of an ocean shore ecosystem, more water evaporates from the ocean than falls back into the ocean in the form of rain. Why doesn't the ocean lose water?

8. What does the water cycle accomplish besides balancing the water in an ecosystem?

9. What is the possible consequence if deforestation occurs in a watershed?

10. What is the principal means by which oxygen is taken from the air? What is the principal means by which it is restored to the air?

11. Name the other ways that oxygen is removed from the air.

12. Name the other ways that oxygen is replenished in the air.

13. Name the ways in which carbon dioxide is removed from the air.

14. Name the ways in which carbon dioxide is replenished in the air.

15. What human activity worries those who think that global warming is a problem?

16. Is human-produced global warming occurring now?

17. What is nitrogen fixation? What type of organisms perform it?

18. What two ways does the nitrogen in organisms get put back into the environment?

MODULE #11: The Invertebrates of Kingdom Animalia

Introduction

We've taken a break from looking at individual groups within biology's classification scheme and have, instead, looked at more global issues such as chemistry, genetics, and ecosystems. A major part of biology, however, is studying the amazing creatures that God has placed in His marvelous creation, so it is time to start doing that again. We have two kingdoms left to discuss in this course: kingdom Plantae and kingdom Animalia. We will actually jump back and forth between these kingdoms over the rest of the course, but for right now, we will start our discussion with kingdom Animalia.

There are well over 20 phyla in kingdom Animalia, depending on the taxonomy scheme that you use. Now before you get too worried, we assure you that we will not cover every one of them in this course! Instead, we will pick a few representative phyla and learn some basic characteristics of each. That way, you will get a good overview of the creatures that make up this kingdom. When we get to phylum **Chordata** (kor dah' tah), however, we will spend a great deal of time discussing it. Why? Well, that's because most of the creatures that students consider to be "animals" (birds, fish, and mammals) belong in that phylum, since they are the animals with which students are most familiar. However, kingdom Animalia is much more diverse than what is represented by the members of phylum Chordata. In fact, you might be surprised at some of the creatures that biologists call animals!

Kingdom Animalia is often split into two groups: **invertebrates** (in vur' tuh brates) and **vertebrates** (vur' tuh brates). Although they are not official taxonomy groups, biologists use the terms often, so it is good to know them.

Invertebrates – Animals that lack a backbone

Vertebrates – Animals that possess a backbone

You, of course, are a vertebrate, because you have a backbone. Worms and insects do not have backbones, so they are invertebrates.

It turns out that invertebrates make up all of the phyla in kingdom Animalia except for phylum Chordata. As you might imagine, then, there are far more invertebrates in kingdom Animalia than there are vertebrates. However, vertebrates are the organisms with which we are most familiar. As a result, we will spend an undue amount of time on them in a later module. For now, we will begin with invertebrates. We will start off with some of the lesser-known invertebrates and then, in the next module, we will focus on a single phylum of invertebrates, phylum Arthropoda (are thruh' pah duh). Before we do that, however, we need to discuss something that is used in our classification scheme: **symmetry** (sim' uh tree).

Symmetry

When looking at an organism, it is often possible to split that organism into two identical halves. For example, a person has a left side and a right side. Each side seems to be identical in every way; this kind of symmetry is illustrated in Figure 11.1 on the next page.

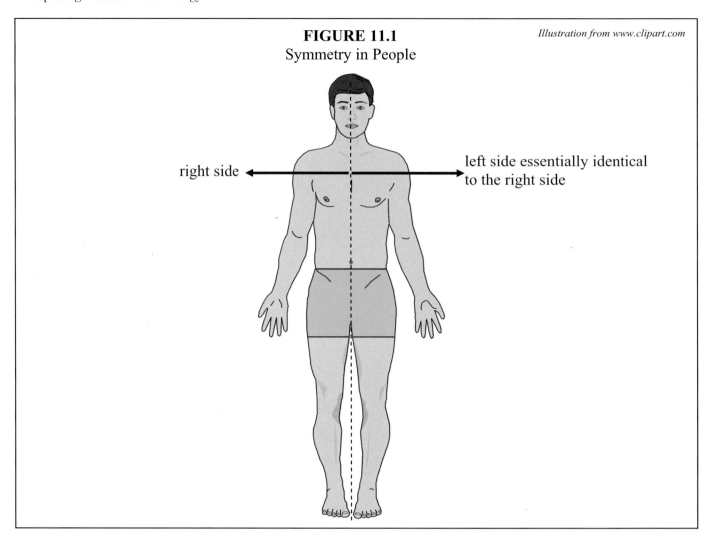

FIGURE 11.1
Symmetry in People

Illustration from www.clipart.com

right side ← → left side essentially identical to the right side

Although this is the most familiar kind of symmetry in the animal kingdom, there are actually a total of three different kinds of symmetry that we can find in creation: **spherical symmetry, radial symmetry, and bilateral** (bye lat' uh ruhl) **symmetry.**

Spherical symmetry – An organism possesses spherical symmetry if it can be cut into two identical halves by any cut that runs through the organism's center.

Radial symmetry – An organism possesses radial symmetry if it can be cut into two identical halves by any longitudinal cut through its center.

Bilateral symmetry – An organism possesses bilateral symmetry if it can only be cut into two identical halves by a single longitudinal cut along its center which divides it into right and left halves.

Hopefully, you can tell from these definitions that humans have bilateral symmetry. If these definitions are not clear, however, take a look at Figure 11.2 on the next page, which illustrates them all.

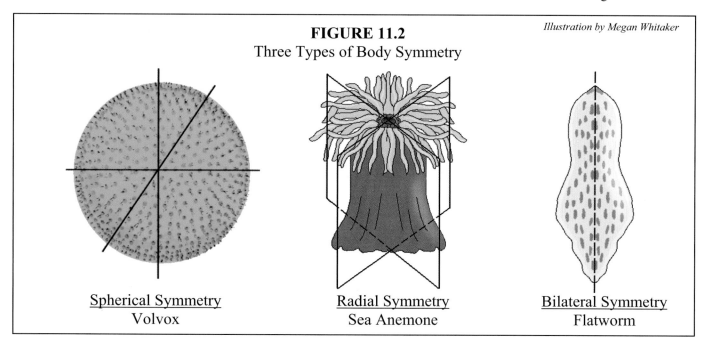

FIGURE 11.2
Three Types of Body Symmetry

Illustration by Megan Whitaker

Spherical Symmetry
Volvox

Radial Symmetry
Sea Anemone

Bilateral Symmetry
Flatworm

Generally, you find radial and bilateral symmetry in the animal kingdom. However, some animals don't have any kind of symmetry.

Please note that these symmetry distinctions are not perfect. For example, you have bilateral symmetry, but there are some differences between your right and left sides when it comes to your internal organs. For example, your right lung is composed of three lobes, while your left lung has only two. This difference, however, does not keep you from having bilateral symmetry. In the same way, if you develop a wart on your right hand, you do not "lose" your bilateral symmetry. Symmetry refers to the overall characteristics of a creature. As a result, minor differences between two halves of a creature do not discount that creature's overall symmetry. Be sure that you understand how to recognize symmetry by performing the following "On Your Own" problem.

ON YOUR OWN

a-c. Illustrations from www.clipart.com
d. Illustration by Megan Whitaker

11.1 Identify the symmetry (if any) possessed by these animals :

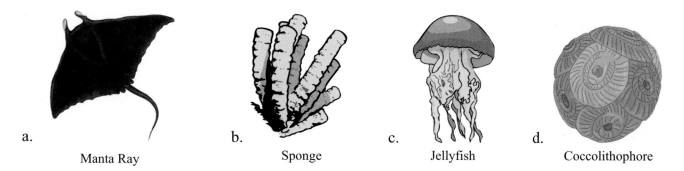

a.
Manta Ray

b.
Sponge

c.
Jellyfish

d.
Coccolithophore

Now that we've looked at the symmetries that exist in creation, it is time to study some of the major phyla of invertebrates. As we said earlier, we will use a creature's symmetry (or lack thereof) as one criterion for placing invertebrates into their respective phyla.

Phylum Porifera: The Sponges

Have you ever washed dishes or cleaned up a mess with a sponge? Most likely, the sponge that you used was synthetic, but its structure is based on a whole group of creatures commonly called "sponges," the members of phylum **Porifera** (poor if' uh ruh).

Although there are a handful of freshwater species, most sponges live in marine environments. They are often mistaken for plants because they are anchored to an immobile object and are therefore unable to move. Nevertheless, they are classified in the animal kingdom because of their cellular structure and the way they feed.

Sponges are amazing creatures. They possess no symmetry and take on a variety of shapes. These shapes include flat, tubular, branched, cuplike, and vaselike. Some sponges are no larger than the size of a fingernail, whereas others are large enough for you to sit in! A few examples of these incredible creatures are shown in the figure below.

FIGURE 11.3
Some Members of Phylum Porifera

Photos © Corel, Inc.

Barrel Sponge

Basket Sponge

Tube Sponge

Diver Behind a Strawberry Vase Sponge

These creatures have an interesting anatomy. Sponges have two layers of cells that are separated by a thin, jellylike substance. The outer layer of cells is called the **epidermis** (ep uh dur' miss), and the jellylike substance is called the **mesenchyme** (mes' uhn kime).

Epidermis – An outer layer of cells designed to provide protection

.Mesenchyme – The jellylike substance that separates the epidermis from the inner cells in a sponge

Although the mesenchyme is not made up of cellular material, it is necessary to the life of the sponge, as you will learn in a minute. Please note that the term "mesenchyme" is used to refer to other things as well. Unfortunately, there are some terms in biology that have many different definitions. Simply treat this definition as one of a few possible definitions for the word mesenchyme.

A sponge is supported by a network of **spicules** that are found mostly in the mesenchyme. They weave throughout the mesenchyme, providing a framework that supports the sponge. The spicules are made of lime (calcium carbonate) or silica, depending on the species of the sponge. In some sponges, these spicules actually extend through the epidermis, giving the sponge a spiny or velvety look. Some sponges do not have spicules, however. These sponges are supported by a tough web of a protein called **spongin**. Typically, sponges that have spicules feel hard and spiny, whereas sponges that have spongin are soft. Whether or not a sponge has spicules and what those spicules (if they exist) are made of help biologists further classify sponges.

Although sponges have interesting anatomy, their feeding habits really make them amazing! Since they cannot move, they cannot seek out prey. Instead, they must force their prey to come to them. How do they do this? Well, it all starts with the canals and cavities that exist in the body of the sponge. These canals and cavities are lined with **collar cells**, which are also called **choanocytes** (koh an' uh sytes).

Collar cells – Flagellated cells that push water through a sponge

These cells have flagella that beat constantly, pulling water through the sponge. As the water passes through, the algae, bacteria, and organic debris contained in the water are extracted and eaten by the sponge.

Sponges have no organs, so they must have specialized cells that take care of digestion. These cells are called **amoebocytes** (uh mee' buh sytes).

Amoebocytes – Cells that move using pseudopods and perform different functions in different animals

The amoebocytes in sponges travel freely in the mesenchyme. They digest the food that the sponge has extracted from the water and transport the digested food to the parts of the sponge that need it. In addition, they take in waste products from the inner cells and travel to the epidermis, where the waste products are released. Amoebocytes also exchange gases with the surroundings, bringing needed gases (such as oxygen) to the inner cell layers. On top of all this, these useful little cells also produce the lime or silica that makes up the spicules.

Once food has been extracted and digested, the water must be expelled to allow fresh water to be drawn in. This is accomplished by more collar cells. Thus, sponges are continually pulling water in and pushing it back out. During that process, the water is "cleaned" of algae, bacteria, and organic debris. Now you see why sponges have so many holes in them. Water must be continually pumped into and out of the sponge in order to provide it with food. Thus, it must be filled with many intertwined canals to allow for the passage of so much water.

Because of these canals, as well as the properties of spongin, sponges that contain spongin are very useful for people. As we mentioned before, the sponges you use for cleaning are usually synthetic, but for a long time, people used natural sponges for cleaning because the web of spongin that supports the sponge makes a soft, absorbent material. They did not use living sponges, because living sponges usually have a strong, foul odor. Instead, people used the "skeletons" of long-dead sponges. You see, spongin does not decay very quickly. Thus, much like the bones of a vertebrate, the spongin of a sponge stays around long after the sponge has died.

Sponges are used for many purposes other than cleaning. During Roman times, they were used to pad the armor of soldiers. They are still used today as painting tools, and certain surgical swabs are made from sponges. Because sponges are so useful, there is a very profitable "fishing" industry built around them. Most of this "sponge fishing" occurs in either the Mediterranean Sea or the Gulf of Mexico, where the spongin-containing species are plentiful.

Sponges have several modes of reproduction at their disposal. They can reproduce asexually by budding. They can also regenerate. In other words, if a portion of a sponge is cut away, it can grow into a new sponge. Also, during periods of freezing temperatures, sponges can produce a **gemmule** (jem' yool).

Gemmule – A cluster of cells encased in a hard, spicule-reinforced shell

A gemmule, much like a cyst, can survive through a long period of inclement weather. Once conditions become favorable again, the gemmule will break open and a sponge will grow from the encased cells.

Sponges also have a sexual mode of reproduction. Under certain conditions, collar cells can produce either eggs or sperm. These gametes are then released into the flow of water that the collar cells are maintaining. If a sperm cell and an egg cell meet, fertilization occurs, and a zygote is formed.

If you have a microscope, you can learn more about sponges by performing the following experiment.

EXPERIMENT 11.1
Observation of the Spicules of a Sponge

Supplies:

♦ Microscope
♦ Prepared slide: sponge
♦ Lab notebook
♦ Colored pencils
♦ Natural sponges (optional)

Object: To observe a specimen from phylum Porifera and note the complexity of this animal's support structure

Note: You can observe many sponges at an aquarium near you. Sponges can be iridescent, encrusting, round, or tubular. Crabs, shrimp, and brittle stars sometimes seek shelter within the maze of pores in a sponge, converting the sponge into a living apartment house. Natural (but dead) sea sponges can also be purchased at art supply stores and hobby shops.

Procedure:

1. Set up the microscope as instructed in previous experiments.
2. Place the prepared slide under the microscope.
3. Observe under low power and draw what you see in your notebook. This slide shows you the spicules, which make up the support system of the sponge. They are produced by amoebocytes and come in a variety of shapes: needle, multipronged "jack," hooked, or barbed. The shape of the spicules is used to classify sponges.
4. Look for different shapes of spicules. See if you can relate any of the shapes you see to the ones listed above.
5. Observe under high power and draw one microscope field.
6. (Optional) If you have purchased sponges from an art store, slice off a small section and make a wet mount by wetting the slice and covering it with a coverslip. The thinner the slice, the more you will see.
7. Observe under low power and high power and sketch a section of the sponge in your notebook, one for each power.
8. Clean up and return the equipment to its proper place.

ON YOUR OWN

11.2 One biology book calls sponges "tireless, natural pumps." Why is this a good description of sponges?

11.3 If a sponge has plenty of food but cannot distribute it to all of its cells, what is the sponge missing?

11.4 A sponge feels hard and prickly. Does it contain spicules or spongin?

Phylum Cnidaria

The next set of invertebrates we want to discuss are the members of phylum **Cnidaria** (nih dahr' ee uh). Members of this phylum, which include jellyfish, sea anemones, and hydra, have two basic forms: the **polyp** and the **medusa**.

Polyp – The sessile, tubular form of a cnidarian with a mouth and tentacles at one end and a basal disk at the other

Medusa – A free-swimming cnidarian with a bell-shaped body and tentacles

Even though many cnidarians have either one or the other of these forms, there are those with both. Various cnidarians are pictured in the figure below, illustrating the difference between the medusa form and the polyp form.

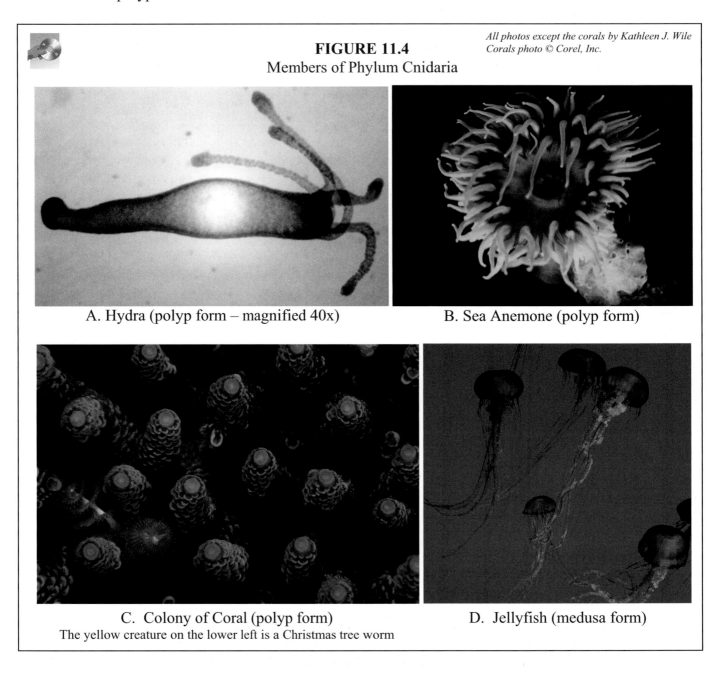

All photos except the corals by Kathleen J. Wile
Corals photo © Corel, Inc.

FIGURE 11.4
Members of Phylum Cnidaria

A. Hydra (polyp form – magnified 40x)

B. Sea Anemone (polyp form)

C. Colony of Coral (polyp form)
The yellow creature on the lower left is a Christmas tree worm

D. Jellyfish (medusa form)

Before we look at individual organisms in phylum Cnidaria, we want to discuss the characteristics that are common to all members of this phylum. Cnidarians, like sponges, have an outer layer of cells and an inner layer of cells separated by a jellylike layer. Each of these layers of cells is an **epithelium** (ep ih thee' lee uhm), and they are separated by a **mesoglea** (mez uh glee' uh).

Epithelium – Animal tissue consisting of one or more layers of cells that have only one free surface, because the other surface adheres to a membrane or other substance

Mesoglea – The jellylike substance that separates the epithelial cells in a cnidarian

The epithelial layers of cells in a cnidarian are home to **nerve cells** and **contractile cells**. The nerve cells sense outside stimuli and coordinate the organism's response to it, while the contractile cells bend the organism's body to produce movement.

All members of phylum Cnidaria have radial symmetry. Their bodies consist of tentacles, a mouth, and a saclike gut. The tentacles catch prey by releasing **nematocysts** (nih mat' uh sists).

<u>Nematocysts</u> – Small capsules that contain a toxin which is injected into prey or predators

When most creatures are touched by the tentacles of a cnidarian, nematocysts are injected into the creature. Depending on the chemistry of the creature, the toxin released by the nematocyst either paralyzes the creature or causes an irritating sting. If it paralyzes the creature, the cnidarian can then eat it. Although the main function of the nematocysts is capturing prey, they are also quite effective at warding off predators.

Once a cnidarian has captured its prey, the creature is pulled into the cnidarian's mouth and extracellular digestion begins in the saclike gut. After the extracellular digestion reaches a certain stage, the cells of the inner epithelial layer engulf the partly digested food, and the digestion continues inside the cells.

Interestingly enough, cnidarians do not have structures for either respiration (breathing) or excretion of waste products. As you should recall, these duties are done by the amoebocytes in sponges. Cnidarians, however, have such thin epithelial layers that gases and waste products can be directly exchanged with the surroundings through the body itself, getting rid of the need for respiration and excretion systems.

Reproduction in phylum Cnidaria occurs both sexually and asexually. The most common form of asexual reproduction is budding, just as it is for sponges. Cnidarians also reproduce sexually at least once during their life cycle. Now that we've discussed the similarities of the organisms in this phylum, let's look more closely at the organisms pictured in Figure 11.4.

<u>Specific Members of Phylum Cnidaria</u>

The organism pictured in Figure 11.4a is a tiny creature known as the **hydra**. A sketch is shown in Figure 11.5 on the next page, highlighting its major structures. As you can see in Figure 11.4a, one end of the hydra is covered in tentacles. They surround the mouth so that prey can be pulled into the mouth. These tentacles are covered with nematocysts, as is the hydra's outer epithelial layer. The nematocysts actually lie within special structures that contain a pressure-sensitive trigger. When something brushes up against the trigger, the nematocyst is unleashed, stinging whatever set off the trigger. Isn't that neat? This little creature has a natural "booby trap" that is designed to be released when something touches it. The triggers on these cells are very sensitive and very reliable. Can you imagine such a thing forming by chance? Of course not. The kind of engineering needed to design such a system is far too complex to have arisen by chance. This is just one more tangible evidence of God's handiwork in His creation!

FIGURE 11.5
A Sketch of the Hydra

Illustration by Megan Whitaker

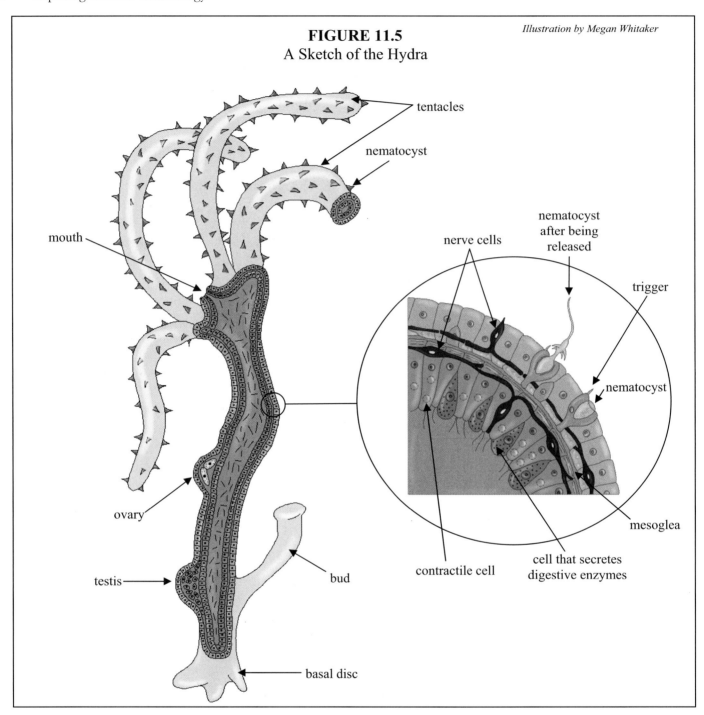

By studying the drawing, you can get an idea how the extracellular/intracellular digestion of the hydra works. When prey is subdued by the nematocysts, the tentacles pull it into the mouth so that it enters the hydra's gut. There, the prey is broken down by the digestive enzymes secreted by cells lining the gut. Once it is in small enough pieces, other cells take in the pieces and finish off the digestion. Waste products are then released into the surroundings right through the body of the hydra.

Hydra reproduce both sexually and asexually. The asexual method is budding. The bud starts as a small bump on the side of the hydra. It grows tentacles and elongates until it actually looks like a miniature hydra. It then separates from the adult, anchors itself to something, and is a new hydra that is genetically identical to the parent.

For sexual reproduction, egg or sperm gametes are formed in the hydra and pack together in little bumps in the animal's body wall. The bumps in which sperm cells form are called **testes** (test' ez), while the bumps that contain the egg cells are called **ovaries** (oh' vuh reez).

<u>Testes</u> – Organs that produce sperm

<u>Ovaries</u> – Organs that produce eggs

Although most hydra produce either eggs or sperm, a few hydra have been observed that actually form both! Eggs and sperm are released in the water, and when a sperm meets an egg, fertilization occurs.

When the hydra was first discovered, many biologists thought that it was one of the "missing links" that Charles Darwin had discussed. Remember from Module #9 that Darwin acknowledged the appalling lack of transitional forms in the fossil record, but he thought that maybe they had not been discovered yet. Thus, they were termed "missing links" by his critics. When the hydra was first discovered, it was hailed as a macroevolutionary link between plants and animals. After all, here was a creature that stayed in one place (like a plant) but ate food (like an animal). As with all such "missing links," however, the more biologists learned about the creature, the more they realized that it was not a link between plants and animals. Instead, the hydra is 100% animal. To learn more about this interesting little creature, perform the following experiment. This is a microscope experiment, so if you do not have a microscope, please read through the experiment to learn as much as you can.

EXPERIMENT 11.2
Observation of a Hydra

<u>Supplies:</u>

♦ Microscope
♦ Prepared slide: *Hydra*
♦ Lab notebook
♦ Colored pencils

Object: To observe the hydra as a typical member of phylum Cnidaria

Note: Other members of phylum Cnidaria might be observed at your local zoo or aquarium. For those who wish to study this phylum more intensely, live or preserved specimens may be purchased from a biological supply house. Check the course website that we mentioned in the "Student Notes" section for links to such supply houses.

<u>Procedure:</u>

1. Set up the microscope as instructed in previous experiments.
2. Place the prepared hydra slide under the microscope and observe under low power.
3. Draw the entire hydra in your notebook. Using Figure 11.5 as a guide, label the following structures: tentacles, mouth, bud (if one exists), basal disc, nematocyst, and testis or ovary (if one is present).
4. Move the slide so that a tentacle is in the center of the microscope field.
5. Turn to medium power, focus, and recenter the tentacle.

6. Turn to high power and focus. Note the shape of the cells in the tentacles. Can you find the nematocysts? They should look like "bumps" on the tentacle. You might even see a hairlike projection coming from one of the nematocysts. That's a stinger. Draw what you see and label what you can.
7. Turn back to low power and move the slide so that one side of the body is in the center of the field.
8. Turn to medium power, focus, and recenter.
9. Turn to high power and focus.
10. Draw a portion of the wall of the body.
11. Clean up and put your equipment away.

Although we spent a great deal of time on the hydra, there are many other interesting organisms that make up phylum Cnidaria as well. A sea anemone is pictured in Figure 11.4b. Remember from Module #10 that the sea anemone has a symbiotic relationship with the clownfish. The clownfish can brush up against the tentacles of the sea anemone without getting stung by its nematocysts. How is that possible? Well, if you thought that the nematocysts in the hydra were well-engineered, get a load of this.

The nematocysts of a sea anemone are not triggered by a pressure-sensitive device like the hydra's. Instead, the sea anemone has a *chemical recognition system* that looks for a specific chemical or set of chemicals. Biologists are not completely sure of the details, but studies indicate that the mucus of the clownfish does not contain the chemical or chemicals that the nematocysts use as a trigger. Thus, the clownfish never sets off the sea anemone's nematocysts, and it can nestle comfortably among its tentacles. Now think about that for a moment. A chemical recognition system for releasing nematocysts is much more complicated than a pressure-sensitive booby trap. Both such systems are present in the members of phylum Cnidaria. Furthermore, the organisms in this phylum are *the only ones* that possess nematocysts. The design apparent in this phylum should make you once again marvel at the power and wisdom of the Creator!

Another set of cnidarians are collectively called **corals**, an example of which is illustrated in Figure 11.4c. These organisms are very tiny polyps that live in self-made stonelike structures. Much like the amoebocytes in sponges, these creatures have cells that produce the stonelike substance that is used to form a cuplike "house" in which the polyp lives. Although the substance that this residence is made of is much like stone, when the coral is alive, it stays flexible enough to open and close. When the coral feeds, the cup is open, and the coral extends tiny tentacles that catch plankton in the water. When the coral is not feeding, the cup stays closed, protecting the animal inside. Corals are important because they gather together and form huge colonies, actually attaching their cuplike structures to one another. As corals begin to die, more corals attach themselves to the top of the dead ones, enlarging the size of the colony. These colonies get so large that they form great reefs (coral reefs) that can stretch for miles. These coral reefs form important ecosystems in warm, shallow marine waters and are home to thousands of other organisms.

Jellyfish are shown in Figure 11.4d. Although a jellyfish is usually recognized in the medusa form, some jellyfish (the ones in genus *Aurelia*, for example) actually go through the polyp form before they turn into the medusa form! A sketch of the life cycle of an *Aurelia* (aw reel' yuh) is shown in Figure 11.6 on the next page.

FIGURE 11.6

Illustration by Megan Whitaker

The Life Cycle of a Common Jellyfish (from genus *Aurelia*)

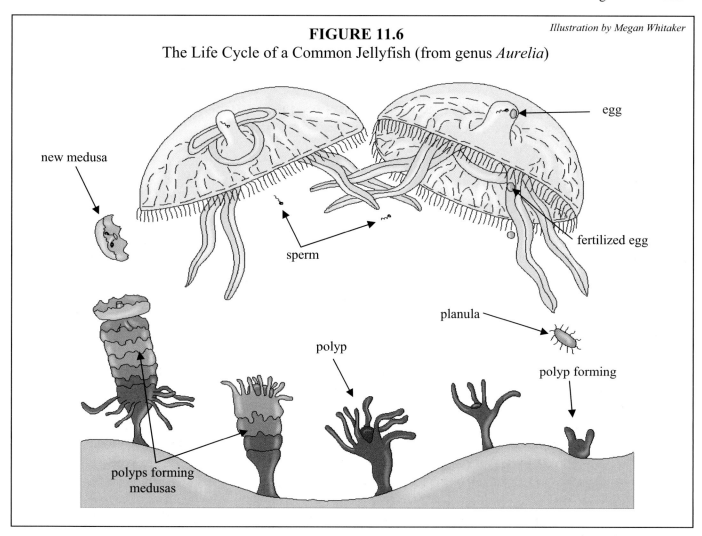

Unlike many cnidarians, the common jellyfish exists as either male or female during the medusa stage. The male releases sperm that swim in the water. Some of these sperm will enter the gut of the female, where an egg is waiting to be fertilized. When fertilization takes place, the zygote leaves the gut through the mouth and clings to the female's tentacles for a while. After some development, the zygote becomes a **planula** (plan' yuh luh), which swims away and attaches itself to an underwater base. It then grows tentacles and a mouth, forming a polyp. During its time as a polyp, the jellyfish might asexually reproduce by budding. Eventually, however, the polyp undergoes a dramatic change. It becomes almost cylindrical, with stacks of rings. These rings eventually separate, each forming a medusa.

If you are ever in the ocean swimming and you see a jellyfish, just remember that this recognizable form only represents one part of this fascinating creature's life cycle. If you were to examine the bottom of the ocean, you would probably see the polyp form as well. Of course, if you see a jellyfish, you should immediately leave the water! Why? Well, like all cnidarians, the jellyfish has nematocysts in its tentacles. If you are unfortunate enough to brush up against those tentacles, the jellyfish will release its nematocysts into you. Now, you are far too big for the jellyfish to paralyze, but the poison released into your system will cause great swelling and irritation. Some people also have an allergic reaction to the poison, and that allergic reaction has been known to result in death. So even though jellyfish are quite beautiful and very interesting, it is best to stay away from them!

ON YOUR OWN

11.5 An organism has a mouth, tentacles, and a gut. It also has bilateral symmetry. Can it be placed in phylum Cnidaria?

11.6 If a clownfish were to brush up against a hydra, would the hydra release its nematocysts into the clownfish?

11.7 A lot of animals need some sort of respiratory and excretion systems. Why don't cnidarians need these?

11.8 If a jellyfish reproduces asexually, is it in polyp form or medusa form?

<u>Phylum Annelida</u>

It is now time to discuss some of the more familiar invertebrates, the members of phylum **Annelida** (ann uh lee' duh). Most likely, you have fished with members of this phylum, found them under rocks, or seen them crawling along the sidewalk after a heavy rain. Most likely, you have even squished them underfoot. You see, phylum Annelida is made up of worms. It turns out, however, that there are several different types of worms in the animal kingdom, far too many for just one phylum. As a result, phylum Annelida contains only one particular type of worm, the **segmented worm**.

The most familiar type of segmented worm is the common earthworm, a member of genus *Lumbricus* (loom brih' cus). The basic anatomy of the common earthworm is illustrated in the figure below.

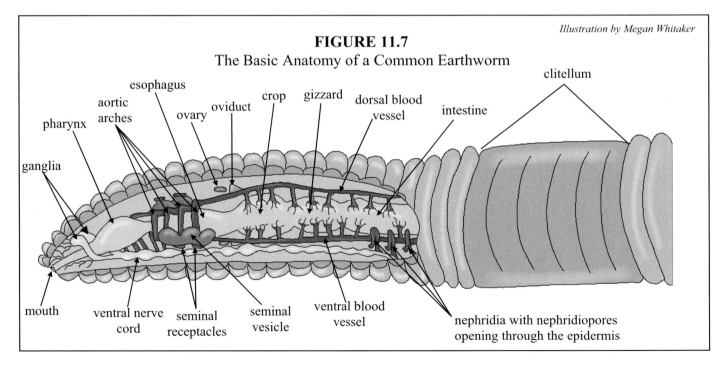

Illustration by Megan Whitaker

FIGURE 11.7
The Basic Anatomy of a Common Earthworm

These worms inhabit almost every bit of moist soil in the world. They possess bilateral symmetry and are made up of segments that look like little rings stacked next to one another. An earthworm also has a **clitellum** (klye tel' uhm), which is a barrel-shaped swelling that usually starts at the thirty-second segment and covers all segments up to thirty-seven. This structure aids in reproduction (as we will discuss later), and it helps us to distinguish the head of the worm from its tail.

That last statement might surprise you a bit. When you look at an earthworm, it certainly doesn't seem to have a head and a tail. Nevertheless, it does. In the animal kingdom, we usually distinguish the region of the organism that contains the head from the region that contains the tail with the terms **anterior end** and **posterior end**.

Anterior end – The end of an animal that contains its head

Posterior end – The end of an animal that contains its tail

The anterior end of the earthworm is usually a little more pointed than the posterior, and it is also usually darker than the rest of the body. The easiest way to tell the anterior end from the posterior end is to look for the clitellum. The anterior end is closest to the clitellum. The earthworm's mouth is at the tip of the anterior end, and it has another opening at the tip of its posterior end called the **anus**.

The earthworm moves with a combination of two muscle layers and small bristles called **setae**. The worm has a **circular layer** of muscles that, when contracted, makes the earthworm long and thin. When the other layer (the **longitudinal layer**) contracts, the earthworm gets short and thick. In order to move, the earthworm uses its posterior setae to anchor its posterior end while it contracts its circular muscles. This stretches the earthworm out. Since the posterior end is anchored, however, the net effect is to simply push the anterior end forward. Once the circular muscles have contracted completely, the earthworm then uses it anterior setae to anchor its anterior end and at the same time releases its posterior setae. After that, the longitudinal muscles contract. Since the anterior is anchored and the posterior is not, this causes the posterior to move forward as the longitudinal muscles cause the earthworm's body to become shorter. This is how the earthworm crawls along a surface and tunnels through dirt.

Feeding Habits of the Earthworm

The earthworm is the first animal we have studied that has a complete digestive system. It ingests soil into its mouth by using its powerful **pharynx** (fa' rinks) to create a sucking action. The soil is then passed into the **crop**, where it is stored for a while. Eventually, the soil makes it to the **gizzard**, where muscular contractions grind the soil, breaking into small pieces any vegetation, refuse, or decaying organic matter that happens to be in the soil. These pieces are what the worm actually digests as food. The rest of the soil is passed down through the **intestine** and out through the anus. Digestible materials are digested by enzymes in the intestine. The digested food is then absorbed by blood that circulates through the walls of the intestine. The blood then transports the digested food to cells throughout the worm's body.

Although a lot of waste is excreted through the anus, some of the waste products of metabolism are gathered in small organs called **nephridia** (nuh frid' ee uh). These organs, which function like your kidneys, are in every segment of the earthworm's body except the first three and the very last one. They push the waste products out of the earthworm through tiny holes called **nephridiopores** (nuh frid' ee uh poors).

The feeding habits of the earthworm actually produce two beneficial results for the earthworm and one for the ecosystem in which it resides. First, of course, the earthworm gets the food it needs to survive. Second, it also loosens the soil, allowing the earthworm to tunnel though it. In other words, in order to tunnel through soil, the earthworm actually eats it! Finally, the earthworm's feeding habits make the soil in which it resides very fertile. By loosening the soil, it helps oxygen and water filter into the soil, which makes it easier for plants to absorb them. Also, because the earthworm is constantly tunneling up and down in the soil, it takes the nutrient-rich soil at the surface and mixes it with the mineral-rich soil below the surface, producing a fertile layer of topsoil. Finally, the excretions of earthworms are full of nutrients that plants need for biosynthesis. Soil cannot stay fertile year after year without the help of earthworms!

This is, yet again, a great example of the design that is apparent in creation. Since plants constantly extract minerals from the topsoil in which they grow, there would be no minerals left in only a few short years. In addition, when plants die, they decay on the surface. Thus, without the aid of earthworms, there would be a nutrient-rich layer of soil at the surface, but those nutrients could not be absorbed by the plants' roots. At the same time, there would be no minerals for the roots to absorb, because they would be too low for the plants' root systems to reach. The worm, however, was created to continually mix the lower soil with the upper soil, recycling the nutrients and bringing up more minerals. Now you know why all moist soil contains earthworms. They are the caretakers that God has made for the plants, allowing them to continue to grow!

The Respiratory and Circulatory Systems in an Earthworm

Looking back at Figure 11.7, you can see that there are still some parts of the earthworm's anatomy that we have not yet discussed. For example, the earthworm is the first animal we have discussed that has a full-fledged **circulatory system**.

Circulatory system – A system designed to transport food and other necessary substances throughout a
        creature's body

Composed of a complex series of blood vessels, this circulatory system allows vital substances such as food and oxygen to travel to all cells within the earthworm's body. Although the earthworm has a full-fledged circulatory system, it does not have a heart. Instead, the two main blood vessels, the **dorsal blood vessel** and the **ventral blood vessel,** are linked together by a series of strongly muscled vessels known as **aortic** (ay or' tik) **arches**. These aortic arches pump the blood, causing it to move through the ventral blood vessel. Along the way, smaller blood vessels branch off of the ventral blood vessel, allowing blood to flow to all regions of the earthworm's body. The blood then collects in the dorsal blood vessel, which takes it back to the aortic arches so that the process can start all over again.

The blood picks up nutrients as it flows through the vessels that are a part of the intestine. However, the blood must also take oxygen to the cells, and it must pick up carbon dioxide from the cells so that it can be released as waste. This happens at the surface of the earthworm's body, through its thin epidermis. As the blood flows through the epidermis, it absorbs oxygen that has diffused in from the surroundings and releases the carbon dioxide that it has picked up from the cells, which diffuses out into the surroundings. In other words, the earthworm breathes through its skin (epidermis). In order to be able to absorb oxygen and release carbon dioxide in this way, the earthworm's epidermis is covered in a moist layer called a **cuticle**.

If you have ever seen a shriveled earthworm lying on the sidewalk, you now know why it died. If the earthworm is out in the sun too long, the cuticle begins to dry. This slows the exchange of gases with the surroundings, suffocating the worm. Under normal circumstances, the earthworm can tunnel underground and allow the moist soil to wet its cuticle. If, however, the worm is unlucky enough to be caught on a sidewalk, it cannot tunnel underground, and it therefore suffocates. An earthworm can also suffocate during a rainstorm. If the soil in which it is tunneling gets too wet, the water takes the place of oxygen in the dirt. As a result, there is little oxygen to absorb. Thus, the earthworm quickly tunnels to the surface. While on the surface, of course, it is exposed to birds and other predators, so it will only stay there while the soil remains too wet to inhabit. That's why the best time to look for worms is right after a heavy rain.

How does the earthworm know when to leave the protective soil and expose itself to predators? The earthworm has a complex **nervous system** that allows it to respond to outside stimuli.

Nervous system – A system of sensitive cells that respond to stimuli such as sound, touch, and taste

The nervous system is controlled by a small "brain" composed of two masses of nerve cell bodies that biologists call **ganglia** (gan' glee uh).

Ganglia (singular: ganglion) – Masses of nerve cell bodies

Although not as sophisticated as a real brain, these two ganglia do control the nervous system and coordinate the responses of the earthworm. A large **ventral nerve cord** travels down the bottom of the earthworm, and the cord has a small ganglion at each segment of the worm. Nerves stretch out from these ganglia, sensing touch, light, and certain chemicals. If these nerves sense too much water, for example, they send a message to the ganglion from which they extend, which then sends a message to the main ganglia. The main ganglia then tell the earthworm to head for the surface.

The Earthworm's Reproductive System

Perhaps the most interesting aspect of the earthworm is its method of reproduction. Earthworms are **hermaphroditic** (hur maf ruh dit' ik)

Hermaphroditic – Possessing both the male and the female reproductive organs

In other words, a single worm produces both eggs and sperm. Unlike a hydra which can, on occasion, do the same thing, the earthworm cannot mate with itself. Instead, it must find a partner with which to reproduce.

Here's how earthworm reproduction works. Sperm produced by the testes of an earthworm are stored in **seminal vesicles** until the earthworm mates. Eggs produced in the ovaries are likewise stored in **oviducts**. When an earthworm mates, it finds another earthworm pointing in the opposite direction, and the two worms attach themselves together with a **slime tube**, exchanging their sperm. They each empty their seminal vesicles into small pouches, called **seminal receptacles**, in the other worm. In this way, the sperm of one worm is exchanged for the sperm of another. Once the sperm is exchanged, the worms separate, but their anterior ends each remain wrapped in their own slime tubes.

A few days later, a **cocoon** forms around the clitellum of each worm. Each worm then backs out of the cocoon, leaving the slime tube behind as well. As the cocoon passes over the oviducts, eggs are released into it. Then, as the cocoon passes over the seminal receptacles, sperm are released as well. When the worm is completely free of the cocoon, the cocoon seals, and fertilization takes place. In a few weeks, a young earthworm will break out of the cocoon.

Now we realize that we have thrown a lot of information at you. This is the most complex organism that we have studied so far, so it is only natural for your head to be swimming right now with all of these facts. They are important facts, however, and you need to remember them. Specifically, you need to be able to label all of the organs shown in Figure 11.7 as well as be able to explain the use of each. This is a lot of information, so you will have to study these sections a few more times in order to retain it all. The earthworm, however, is the only organism in this module that you will have to know in such a detailed fashion.

Other Segmented Worms

Earthworms, of course, are not the only kind of segmented worms. The leech is another kind of segmented worm that you might have heard about. Leeches live in marine, freshwater, and moist soil environments and move about by using two suckers, one on the posterior end and the other on the anterior end. These same suckers are used to latch onto their prey and suck out its blood. Although most leeches actually feed on small prey, killing them before they begin ingesting their blood, some leeches simply attach to their prey and suck out its blood while the prey stays alive. This particular type of leech was used in medicine at one time. Before we knew much about medicine, it was believed that many diseases were the result of "bad blood." To get rid of the "bad blood," leeches were applied to the sick person. We know now, of course, that such a practice can be harmful to the patient, but it was common medical practice for quite some time! Interestingly enough, however, recent studies indicate that in certain specific cases, the use of leeches can be beneficial to a patient.

Other, more exotic worms also exist in this phylum. For example, the figure below shows two animals that you might not, at first glance, think of as worms. Nevertheless, they both belong to class Polychaeta (pol ee kay' tuh), which is a class in this phylum.

Feather-duster worm photo from the MasterClips collection

FIGURE 11.8

Christmas tree worm photo from www.clipart.com

Two Members of Phylum Annelida, Class Polychaeta

A. Feather-Duster Worm   B. Christmas Tree Worm

The worm pictured in Figure 11.8a is a marine worm called the "feather-duster worm." It lives in a tube which it constructs, and has tentacles that stick out the end of the tube. The tentacles are covered with cilia and mucus, and when algae or organic matter in the ocean water stick to the mucus, the cilia move them down into the mouth of the worm. Its name comes from the fact that the tube and cilia-covered tentacles combine to look like a feather duster. When frightened, the worm retracts into its tube for protection.

The worm pictured in Figure 11.8b is another marine-dweller, the "Christmas tree worm." Like the feather-duster worm, it makes a tube in which it lives. The tube is trumpet-shaped, and its tentacles are arranged in whorls called "radioles" (ray' dee ohlz). When the radioles are extended in order to catch food, the worm looks very much like a Christmas tree, which is where it gets its name. These worms are quite beautiful, and the radioles come in many colors, depending on the species and the chemicals in the environment. Like the feather-duster worm, the Christmas tree worm can retreat back into its tube when threatened.

ON YOUR OWN

11.9 If an earthworm's mouth is fully functional but it cannot ingest soil, what organ is malfunctioning?

11.10 What process will stop if the earthworm's cuticle dries up?

11.11 In a dissected earthworm, you see a nerve chord with ganglia at each segment right next to a long blood vessel. Is blood flowing towards the anterior or the posterior in this blood vessel?

11.12 An earthworm's seminal vesicle is empty but its oviducts are full. Has the earthworm mated yet?

If you purchased the dissection kit described in the beginning of the book, you should perform the following earthworm dissection so that you will be more familiar with the earthworm's anatomy. If you did not, please read through the experiment to learn as much as you can.

EXPERIMENT 11.3
Earthworm Dissection

┌─────────────────────────┐
│ **Be careful!** Dissection │
│ tools are SHARP! │
└─────────────────────────┘

Supplies:

NOTE: The course website discussed in the "Student Notes" section at the beginning of this book has several pictures of an earthworm dissection.

♦ Dissecting tools and tray that came with your dissection kit
♦ Earthworm specimen
♦ Magnifying glass
♦ Laboratory notebook

Object: To become more familiar with the earthworm's anatomy through dissection

<u>Procedure</u>:

All photos by John Skipper

1. Examine your earthworm specimen carefully. Rub your fingers lightly across the surface until you feel bristles. Those bristles are the **setae**. Write in your laboratory notebook how many setae you find on each of the worm's segments.
2. Using a magnifying glass, try to find the **nephridiopores**, which are tiny holes near the bottom of many segments anterior to and posterior to the clitellum.
3. Examine the **clitellum**. In your laboratory notebook, write down how many segments there are in it.
4. Now you are ready to begin the dissection. Place the specimen ventral side (the side with the setae) down on the tray. Pin the anterior and posterior end to the pad, as shown below.

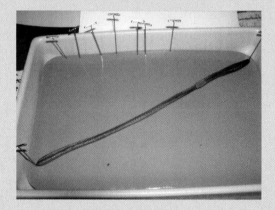

5. Use your scissors to cut through the body wall. Start about an inch posterior to the clitellum and just to the left of center. Being careful not to cut anything but the body wall, extend your cut all the way to the anterior end, as shown below:

6. Pull apart the edges of the cut and peer in. You can probably see the **intestine**. The space between the body wall and the intestine is called the **body cavity**. Notice that the body cavity is separated by partitions that run from the body wall to the intestine. These are called **septa**.

septa

intestine

7. Using forceps (tweezers) and your probe (the tool with the metal pointer), break the septa so that you can peel back the body wall.
8. Peel back the body wall on both sides of the cut and pin it down, as shown in the picture below:

dorsal blood vessel

intestine

gizzard

crop

seminal vesicles

seminal receptacle

esophagus

pharynx

ganglia (the two little spots)

seminal receptacle

9. Now the internal structures should be visible. Make a drawing of your dissected earthworm in your laboratory notebook. As you identify the structures listed below, label them in your drawing. Note any structures you could not see, as well as any organs that you saw but could not identify.
10. Using Figure 11.7 and the picture above as guides, identify the following digestive structures:

Pharynx - A thick-walled structure in the area of segments 4-7
Esophagus - The structure that extends from the pharynx to about segment 14
Crop - A bulge just posterior of the esophagus
Gizzard - The structure posterior to the crop
Intestine - The structure that extends from the gizzard to the anus
Seminal vesicles - The larger structures on either side of the esophagus
Seminal receptacles - The smaller structures on either side of the esophagus

11. Review the functions of these structures as described in the text.
12. Once again, using Figure 11.7 and the picture above as guides, identify the following circulatory system structures:

Dorsal blood vessel - A dark vessel running along the dorsal side of the intestine. It might actually lie on the intestine.
Aortic arches - You will have to remove the seminal vesicles (see picture above) and septa to see the arches clearly. Remove them only from the left side of the earthworm and examine the aortic arches that are revealed. They will look like tubes surrounding the esophagus.
Ventral blood vessel - Use your probe to move aside (*do not remove*) the intestine near the posterior end of your cut. This should reveal the ventral blood vessel, which looks very similar to the dorsal blood vessel.

13. Review the functions of these structures as described in the text.
14. Locate the **nephridia**. The best way to do this is to extend your cut another two inches to the posterior. Without tearing the septa, use your probe to lift up the intestine from this region and

then use your magnifying glass to look under the intestine and find the nephridia. They will be in all segments except the first three and the last one, so there should be plenty to see. If you cannot find them, don't worry. They are the most difficult of the earthworm's internal structures to find.

15. If your dissection has been a bit sloppy, these next steps might not turn out too well. Using Figure 11.7 and the picture on the previous page as guides, try to find the **ganglia** that form the earthworm's "brain." They should be just anterior to the pharynx. Follow the **ventral nerve cord** from the ganglia. Note the small ganglion (which looks like a bulge) that appears at each segment.

16. To get a better look at the reproductive structures as well as the ventral nerve cord, remove a portion of the digestive track. Do so by cutting across the intestine near the clitellum. Make a similar cut just posterior to the pharynx. You have now cut out a section of the digestive tract. Gently pull that section (that portion of the intestine, gizzard, crop, and esophagus), out of the earthworm.

17. Look for the ventral nerve cord. It might have been damaged when you removed the portion of the digestive tract. Can you see the ganglia on the nerve cord better now?

18. You should notice that the seminal vesicles and seminal receptacles are also below this portion of the digestive tract, not just to either side of it.

19. Dispose of your specimen.

20. Clean and dry your dissection tools, tray, and pins. Put everything back in its proper place.

Phylum Platyhelminthes: The Planarian

Although the members of phylum Annelida are the worms with which you are most familiar, there are other kinds of worms that are just far too different to be contained in the same phylum. The members of phylum **Platyhelminthes** (plat ee hel min' theez) are more commonly called "flatworms," because compared to the cylindrical segmented worms, members of this phylum are rather flat. A common example of the organisms in this phylum is the planarian, pictured in the figure below.

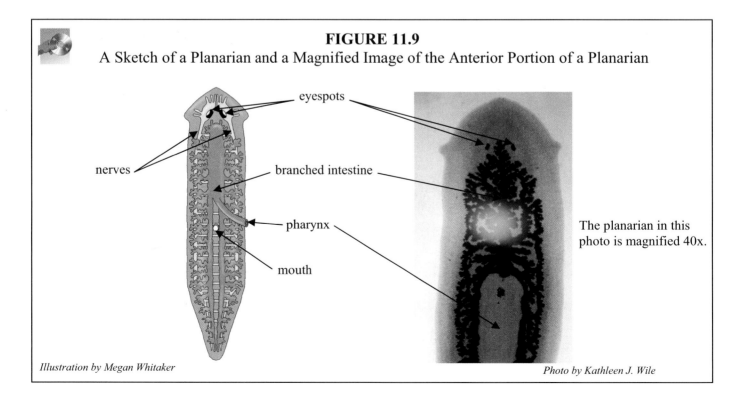

FIGURE 11.9

A Sketch of a Planarian and a Magnified Image of the Anterior Portion of a Planarian

eyespots

nerves

branched intestine

pharynx

mouth

The planarian in this photo is magnified 40x.

Illustration by Megan Whitaker

Photo by Kathleen J. Wile

Planarians live in fresh water and swim there freely. Although planarians live in fresh water, there are many other flatworms that live in marine environments as well.

A planarian usually eats small organisms that it grabs with its mouth. Digestive enzymes are then secreted through the mouth, and, after the organism is broken down into small particles, a tube-like pharynx is extended out of the mouth, sucking up the small particles. These particles then go into the intestine, where more digestive enzymes continue to break them down. The intestine itself is highly branched and runs throughout most of the body. Unlike the members of phylum Annelida, planarians do not need a circulatory system, because the intestine branches extend throughout most of the body. This allows all cells to be close enough to the intestine that digested food can get to them simply by diffusion.

The nervous system of planarians is a little more complex than the one you studied in earthworms. At the head of the organism lies a mass of nerve tissue that is best described as a "brain." This "brain" has nerve branches that go throughout the head and also attach to the worm's eyespots. Although planarians do not really see with these eyespots, they can sense light and move towards it. Planarians move towards light in order to seek out prey such as photosynthetic organisms. Two longitudinal nerves extend from the brain to the posterior of the body, and those nerves are connected to each other by a series of transverse nerves that go across the body. This complex nervous system allows the planarian to have senses of taste, smell, and touch, as well as the ability to sense light.

Planarians sexually reproduce much like earthworms. They are hermaphroditic, but can only reproduce upon the exchange of sperm with a mate. One interesting way that planarians can asexually reproduce is by **regeneration**.

Regeneration – The ability to regrow a missing part of the body

In order to asexually reproduce, a planarian simply tears itself in half. Then, the two halves each regenerate their missing half, producing two planarians where there was only one before. Although some earthworms have the ability to regenerate, they never use it for reproductive purposes. They only use it in case of an accident. While planarians can use regeneration to heal from an accident, they also use it deliberately as a means of asexual reproduction!

If you have a microscope, you can learn more about planarians by performing the following experiment.

EXPERIMENT 11.4
Observation of a Planarian

Supplies:

♦ Microscope
♦ Prepared slide: planarian
♦ Lab notebook
♦ Colored pencils

Object: To observe a planarian as an example of organisms from phylum Platyhelminthes

Note: For those wishing to study the flatworms more thoroughly, a local veterinarian or medical doctor might have preserved specimens of tapeworms and/or flukes. Also, live planarians may be purchased from a biological house. Live planarians are extremely interesting. They react to stimuli like air (use an eyedropper), movement, touch, and chemicals, including salt water. If split in half longitudinally, they will regenerate.

Procedure:

1. Set up the microscope as instructed in previous experiments.
2. Place the prepared slide under the microscope.
3. Observe under low power. You should be able to see the two eyespots, the intestine, the pharynx, and the mouth. You may be able to see the longitudinal and transverse nerves.
4. Draw what you see, and using Figure 11.9 as a guide, label everything that you can identify. Note that the mouth is more than halfway down the body, at the end of the pharynx. Looking closely, you might see the tubules and flame cells that are part of the excretory system.
5. Increase to medium power and refocus. Can you see any details that you were not able to see before?
6. Increase to high power and refocus. Once again, look for any details you might have missed at lower settings.
7. Clean up and put your equipment away.

Other Members of Phylum Platyhelminthes

Although planarians make a good case study in phylum Platyhelminthes, there are many other types of flatworms as well. For example, there are a host of parasitic flatworms called **tapeworms** and **flukes**. These two classes of flatworm are quite different from each other, but they have much in common. Mostly, they do not have as elaborate nervous or digestive systems as do the planarians. After all, as parasites, they need not look for prey, and their host typically does a great deal of digestion for them. Instead, the main features of parasitic flatworms are mechanisms that protect them from the digestive juices and infection-fighting mechanisms of their host. They also have suckers, hooks, or both to help them hold their position within the host.

ON YOUR OWN

11.13 Compare the digestion of a planarian with a fungus. What are the similarities?

11.14 A flatworm has very complex digestive and nervous systems. Is this flatworm likely to be parasitic? Why or why not?

Phylum Nematoda

Members of phylum **Nematoda** (nem uh toh' duh), often called roundworms, are tiny (often microscopic), cylindrical worms that possess bilateral symmetry. Like their common name implies, members of this phylum are round, rather than flat. Thus, they look more like members of phylum

Annelida than members of phylum Platyhelminthes. Typically, however, they are much smaller than the members of phylum Annelida, and they do not have segmented bodies.

Roundworms are particularly hardy. They live in virtually every environment in creation. You can find them in dirt, in the snow of the arctic tundra, in the heat of hot springs, and in the terribly hot, high-pressure environment of hydrothermal vents at the bottom of the ocean. Their bodies are essentially composed of a tube within a tube. The outer tube, which is tapered on both ends, is made up of the epidermis and a cuticle that the epidermis secretes. The inner tube is a digestive canal that is open on both ends. Three of the creatures in this phylum are shown below.

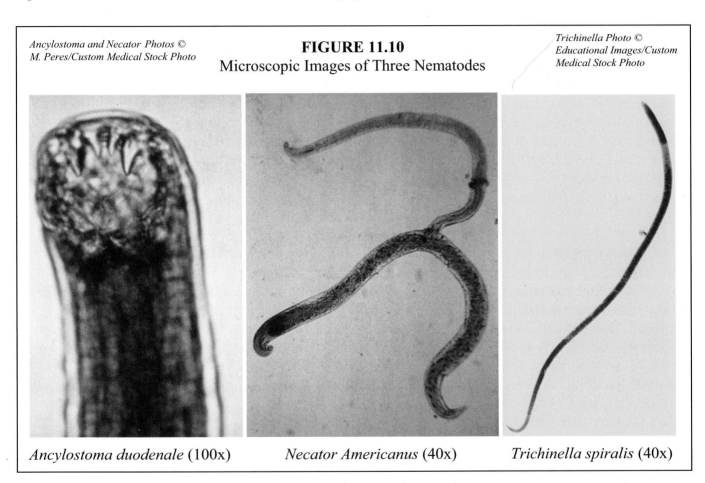

Ancylostoma and Necator Photos ©
M. Peres/Custom Medical Stock Photo

FIGURE 11.10
Microscopic Images of Three Nematodes

Trichinella Photo ©
Educational Images/Custom Medical Stock Photo

Ancylostoma duodenale (100x)   *Necator Americanus* (40x)   *Trichinella spiralis* (40x)

Many members of this phylum are parasitic. People can be infected by about 30 species of roundworms, including the three pictured above. A roundworm's cuticle protects it from the digestive juices of its host, allowing the worm to live in the host's intestine. Notice the hooks on the mouth of the *Ancylostoma* (an' sih luh stoh' muh) *duodenale* (due ah' duh nal) specimen. Those hooks are used to attach to the intestinal wall so that it can feed on the host's blood. *Necator* (nuh kay' tor) *Americanus* (uh mehr' ih kahn' us) does essentially the same thing. Both of these nematodes can cause anemia, abdominal pain, diarrhea, and weight loss.

We want to concentrate a bit on *Trichinella* (trik ih nel' uh) *spiralis* (spuh ral' us), which is shown on the far right side of the figure. Their life cycle is typical of the life cycle of many parasitic roundworms. Adult *Trichinella* worms live in the intestines of pigs and certain game animals. When the female reproduces, the young make their way to the host's blood vessels, and from there, they go to the host's muscle tissue. Once in the muscle tissue, the young develop for about 14 days and then form cysts, which allow them to survive there for a long time.

When a person eats pork or other game animals that are infected with the worms, the digestive juices in the person's small intestine free the juvenile worms from their cysts, and they begin to develop and mature, reaching adulthood in about two days. As they mature, they cause severe irritation of the intestinal tract, resulting in abdominal pain, nausea, vomiting, and watery stool. When the worms mature, they mate. The young produced by this mating migrate to the muscle tissue and cause muscle damage which often results in muscle pain, swelling, and joint pain. Such infection is called **trichinosis** (trik in oh' sis). Although most people who contract trichinosis recover in about six months, the disease often results in permanent heart or eye damage. Nearly 5% of trichinosis cases are fatal.

Trichinella spiralis infection in people is usually caused by eating poorly cooked pork. Despite rigid standards of meat inspection in the U.S., infected pork can sometimes pass inspection because the worm cysts are very difficult to detect. If infected pork is not cooked thoroughly, the cysts can survive, resulting in the infection discussed above. Many Old Testament scholars consider some of the dietary laws of the Israelites (see Leviticus 11:7, for example) God's way of protecting His people from trichinosis. To completely kill the worm cysts in pork, the meat must be frozen at 5 °F for 21 days or -22 °F for 25 hours. Clearly, such processing was not available to the people of Old Testament times; thus, God decided to protect His people from trichinosis by simply forbidding them to eat pork.

This is an important point. The Israelites in Old Testament times didn't know anything about *Trichinella spiralis*. In order to protect His people, then, God simply forbade them to eat pork. Now to someone who was alive back then, such a rule might have seemed arbitrary and unfair. After all, they might have asked themselves, why can't we eat pork? Is God simply trying to keep us from enjoying ourselves? No, of course not. God was simply taking care of His people by forbidding them to eat something that was unhealthy for them! If you ever think that one of God's rules is designed to make life more difficult for you, just remember trichinosis. God's rules protect us from dangers that we do not understand, just like God's law forbidding pork protected the children of Israel from a disease that they did not understand.

Phylum Mollusca

Phylum **Mollusca** (muh lus' kuh) contains many organisms, including clams, oysters, snails, and squid. Although these organisms might seem rather different to you, they actually share several features. Most members of phylum Mollusca have a **mantle**, a **shell**, a **visceral** (vis' er uhl) **hump**, a **foot**, and a **radula**.

Mantle – A sheath of tissue that encloses the vital organs of a mollusk, makes the mollusk's shell, and performs respiration

Shell – A tough, multilayered structure secreted by the mantle, generally used for protection, but sometimes for body support

Visceral hump – A hump that contains a mollusk's heart, digestive, and excretory organs

<u>Foot</u> – A muscular organ that is used for locomotion and takes a variety of forms depending on the animal

<u>Radula</u> – An organ covered with teeth that mollusks use to scrape food into their mouths

The common snail (shown in the figure below) is a good example of the members of this phylum.

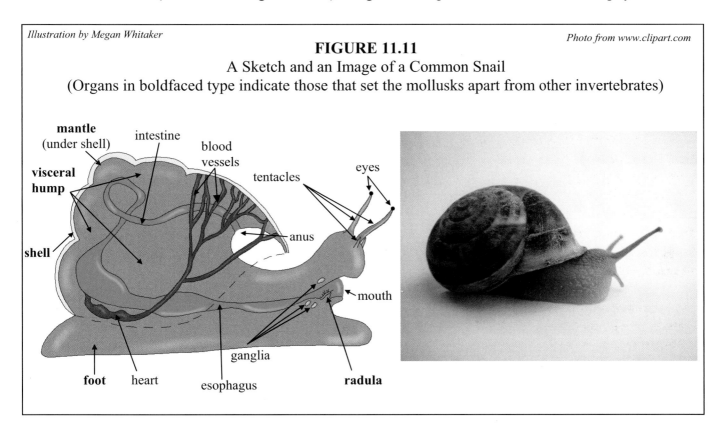

Illustration by Megan Whitaker   *Photo from www.clipart.com*

FIGURE 11.11
A Sketch and an Image of a Common Snail
(Organs in boldfaced type indicate those that set the mollusks apart from other invertebrates)

The snail is a **univalve**, meaning it has only one shell. Organisms like clams are **bivalves**, because they have two shells, one on the top and one on the bottom.

<u>Univalve</u> – An organism with a single shell

<u>Bivalve</u> – An organism with two shells

Univalves are sometimes called **gastropods** (gas' truh podz), while bivalves are sometimes called **pelecypods** (puh less' uh podz). In a snail, the shell exists for protection. When threatened, the snail can retreat into its shell, and, in the case of most snails, a "door" can be shut, sealing the shell.

The snail moves by laying down a thin layer of slime upon which it glides by rhythmically contracting its foot. This method of locomotion allows the snail to move at the "incredible" rate of 3 meters (about 9 feet) per hour! While moving at this speed, the snail grazes on plant material by extending its radula and grating the plant material. The food is digested in the intestine, and indigestible substances are expelled through the anus.

The foot is directly below the visceral hump, which is contained in the shell. Surrounding the visceral hump, under the shell, is the mantle, which is rich in blood vessels. The snail's head contains its main sensory organs: two pairs of tentacles. The longer pair ends in light-sensitive spots that the snail uses for a rudimentary sense of sight. The smaller pair of tentacles has nerves that provide senses of touch and smell.

ON YOUR OWN

11.15 If you were to observe a sidewalk that a snail had just traveled across, what would you expect to find?

11.16 Clams often burrow into the sand. What organ do you think they use when they do this?

Summing Up the Invertebrates

Since there are over 20 phyla of invertebrates in kingdom Animalia, it is simply impossible to discuss them all in a first-year biology course. We have tried to discuss the major ones, but please realize that there are several phyla that we have left out completely. Phylum Echinodermata, for example, contains sea urchins and starfish. There are several phyla of worms that we did not even mention, and even in the phyla that we did discuss, we did not talk about all of the major classes. Nevertheless, the organisms that we did discuss should give you a good idea about the general characteristics of the invertebrate animals. In order to try to give you a more complete view of at least one phylum of the invertebrates, we will spend all of next module on one phylum: Arthropoda.

Before you begin the study guide, we need to remind you to use the study guide and "On Your Own" problems as a guide for what you need to know on the test. Clearly, there is no way for you to remember all of the information that we provided in this module. As a result, we will only require you to know the intricate details of the earthworm's anatomy. Although we discussed the anatomy of other organisms in some detail, you will only need to know the general principles that we discussed. Thus, the only detailed anatomy you will find in the study guide is that of the earthworm. This tells you that for the test, you will not be required to know the detailed anatomies of the other organisms we studied.

ANSWERS TO THE "ON YOUR OWN" PROBLEMS

11.1 a. The manta ray can only be cut in identical pieces if the cut separates left from right. This means it possesses bilateral symmetry.

b. There is really no way to cut the sponge to get identical halves. It therefore has no symmetry.

c. The jellyfish can be cut in identical halves with any cut so long as it is longitudinal (up and down). This gives it radial symmetry.

d. The coccolithophore, a microscopic creature, is pretty much the same everywhere, so it has spherical symmetry.

11.2 Sponges can be described as pumps because they continually pull water into themselves in order to extract their food and then push it back out again.

11.3 The sponge is missing its amoebocytes. Sponges use amoebocytes to distribute food to their cells. Without them, there is no way for the food to get there.

11.4 Only spongin has the soft, absorbent feel we are used to in a sponge. Spicules, on the other hand, are hard and prickly because they are made of lime or silica. Thus, this sponge contains spicules.

11.5 Although the mouth, tentacles, and gut are common to all cnidarians, the bilateral symmetry is not. Cnidarians must have radial symmetry, so it is not in phylum Cnidaria.

11.6 Yes, the hydra would release nematocysts into the clownfish. Unlike the sea anemone, the hydra's mechanism for releasing nematocysts is pressure-sensitive. Thus, *anything* that brushes up against the hydra will be stung.

11.7 Cnidarians do not need respiratory and excretion systems because their bodies are so thin that gases can easily diffuse in and out directly through the body. As a result, there is no need for an elaborate system to perform the function of taking in or releasing gases.

11.8 When in medusa form, jellyfish reproduce sexually. Thus, the jellyfish must be in polyp form.

11.9 The pharynx sucks soil into the mouth. Thus, the pharynx must be malfunctioning.

11.10 When an earthworm's cuticle dries up, its respiration stops.

11.11 If the nerve cord has ganglia, we are looking at the ventral nerve cord. This means that the blood vessel next to it is the ventral blood vessel. Since the ventral blood vessel carries blood away from the aortic arches (which are on the anterior side) to the rest of the worm, it must be moving to the posterior of the earthworm.

11.12 If the oviducts are full, then it is at or past the stage where it is ready to mate. If its seminal vesicles are empty, however, then it must have already exchanged sperm. Thus, the earthworm must have already mated, but it must still be waiting for the cocoon to finish forming.

11.13 <u>The digestion of a planarian and fungus are somewhat similar because the fungus digests its food completely before taking it in, and the planarian digests its food partially before taking it in.</u> Thus, although the process is not identical, there is at least some digestion before ingestion in both creatures.

11.14 <u>The flatworm is most likely not parasitic.</u> Remember, parasitic flatworms don't need much of a digestive system because their host digests most of their food for them. Also, they do not need much of a nervous system because they do not need to seek out prey.

11.15 <u>You would expect to find a trail of slime.</u> This is because the snail excretes a layer of slime upon which it glides. You would probably also find excrement.

11.16 Even though we did not study clams explicitly, they are members of phylum Mollusca, and a mollusk uses its foot for locomotion. Since burrowing in sand is a form of locomotion, the clam uses its <u>foot</u>.

STUDY GUIDE FOR MODULE #11

1. Define the following terms:

a. Invertebrates
b. Vertebrates
c. Spherical symmetry
d. Radial symmetry
e. Bilateral symmetry
f. Epidermis
g. Mesenchyme
h. Collar cells
i. Amoebocytes
j. Gemmule
k. Polyp

l. Medusa
m. Epithelium
n. Mesoglea
o. Nematocysts
p. Testes
q. Ovaries
r. Anterior end
s. Posterior end
t. Circulatory system
u. Nervous system
v. Ganglia

w. Hermaphroditic
x. Regeneration
y. Mantle
z. Shell
aa. Visceral hump
bb. Foot
cc. Radula
dd. Univalve
ee. Bivalve

2. Do the vast majority of animals have backbones?

3. Determine the symmetry of the following organisms: *Illustrations from www.clipart.com*

a.   b.   c.   d.

4. How do sponges get their prey?

5. If a sponge is soft, does it contain spicules or spongin? What purpose do these substances serve in a sponge?

6. What is the predominant mode of asexual reproduction in a sponge?

7. What roles do amoebocytes play in the anatomy of a sponge?

8. When does a sponge produce gemmules?

9. What is the difference between the nematocysts of a hydra and those of a sea anemone?

10. Why do cnidarians not need respiratory or excretory systems?

11. Some biology books say that jellyfish live "dual lives." Why?

12. If a jellyfish reproduces sexually, what form is it in?

13. What is another name for a large coral colony?

14. Name all of the structures in the diagram below:

Illustration by Megan Whitaker

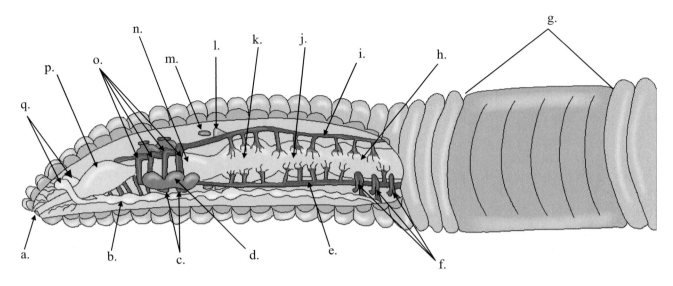

15. What benefits do earthworms give the plants in the soil that they inhabit?

16. If you pick up two earthworms and the first feels very slimy near the clitellum and the second does not, what can you conclude about the first earthworm?

17. What similarities exist between the hydra's sexual reproduction and the earthworm's? What differences exist?

18. What will happen to an earthworm if its cuticle gets dry?

19. Why don't planarians need circulatory systems?

20. If a flatworm has no complex nervous or digestive systems, is it most likely free-living or parasitic?

21. What is the main mode of asexual reproduction in a planarian?

22. Place each organism in one of the following phyla: Porifera, Cnidaria, Annelida, Mollusca, Platyhelminthes

        a. sea anemone   b. clam.   c. sponge   d. flatworm   e. segmented worm

MODULE #12: Phylum Arthropoda

Introduction

As you have already learned, invertebrates make up the vast majority of the animal kingdom. In this module, we will study the most populous phylum of invertebrates, phylum **Arthropoda** (ar thrah' poh duh). This phylum contains crayfish, lobsters, spiders, scorpions, and insects. In fact, it contains more species than *all of the other phyla in kingdom Animalia combined!* Obviously, then, this important phylum deserves an in-depth look.

Arthropods are all around us. They crawl on the ground, fly in the air, and skim across the water. They have an amazing effect on the environment. Arthropods, for example, help plants reproduce by carrying pollen from one plant to another. They also produce many useful items such as silk, wax, honey, and drugs. Although arthropods are very necessary for the earth's ecosystem, they can also be quite dangerous. Some arthropods transmit deadly diseases, while others have been responsible for the destruction of millions of acres of crops. A study of these important animals promises to be quite interesting!

General Characteristics of Arthropods

Although this phylum is vast and diverse, there are many common characteristics that unite arthropods. These characteristics are important to know and understand.

Common Characteristic #1: An Exoskeleton

All arthropods have an **exoskeleton**.

<u>Exoskeleton</u> – A body covering, typically made of chitin, that provides support and protection

As the definition states, the exoskeleton is generally made of **chitin** (kye' tin). This chemical has the useful property of being both tough and flexible. In addition to chitin, there is usually a mineral substance in the exoskeleton of an arthropod that makes it hard. The hard, tough exoskeleton can be thought of as a suit of armor that an arthropod wears. It is flexible enough to move with the creature, but it is tough enough to provide a good measure of protection.

Unlike suits of armor, however, exoskeletons also serve another purpose. The invertebrates that we have studied so far have not needed any support. They either float in the water (like medusae), attach themselves to an object (like sponges), or build themselves a container (like clams). The arthropods, however, must be able to move about or fly. As a result, their fleshy bodies must have support. People (and all vertebrates) get their support from their skeleton, a network of bones that runs inside the body. Arthropods get their support from their exoskeleton. Thus, you could say that while people (and all vertebrates) have their skeletons on the inside, arthropods have their skeletons on the outside. That's where the term "exoskeleton" comes from.

Although the exoskeleton is necessary for the existence of arthropods, it comes at a cost. You see, the exoskeleton is heavy. It is so heavy, in fact, that it limits the growth potential of an arthropod. As an arthropod increases in size, the amount of exoskeleton must increase as well. This causes the arthropod to get quite heavy. For each arthropod, there comes a point at which the creature's muscles just aren't strong enough to carry around the weight of the exoskeleton. Thus, each arthropod is limited as to how big it can get. Class **Crustacea** (kruh stay' shuh) contains the largest arthropods, some of which can grow to 12 feet. The other classes of arthropods contain species that rarely get much larger than 11 inches.

Not only does the exoskeleton limit the growth *potential* of an arthropod it also makes it hard for the arthropod to grow during its life cycle. The exoskeleton is secreted by the arthropod's epidermis and forms around the body, but it cannot grow. Thus, as the body gets bigger, the exoskeleton gets more and more constricting. As a result, an arthropod must **molt** several times throughout the course of its lifetime.

Molt – To shed an old outer covering so that it can be replaced with a new one

Most arthropods molt by secreting enzymes that eat away at the exoskeleton, weakening it. They then take in water, swelling the body until the exoskeleton breaks away. This is done while a new exoskeleton is being produced under the old one. Once the old one is gone, the arthropod's new exoskeleton will be larger than the body of the arthropod because of the swelling that was caused by the excess water that was taken in. As a result, once the arthropod gets rid of the excess water, the body has room to grow again. As the arthropod continues to grow, however, it will once again get constricted, at which time it will molt again.

Common Characteristic #2: Body Segmentation

Like the organisms in phylum Annelida, arthropods are segmented. This segmentation is quite different from the annelids' segmentation, however. In arthropods, the body is divided into three major divisions: the **head**, the **thorax** (thor' aks), and the **abdomen**. These divisions can sometimes be further segmented. In addition, some arthropods have the thorax and head united in a single segment called the **cephalothorax** (sef uh loh thor' aks).

Thorax – The body region between the head and the abdomen

Abdomen – The body region posterior to the thorax

Cephalothorax – A body region composed of the head and thorax fused together

This segmentation is necessary in order to allow the exoskeleton to shift with the movements of the body. The segments can move back and forth, like "joints" in the "armor."

Common Characteristic #3: Jointed Appendages

The term "arthropoda" actually means "joint-footed." Needless to say, then, one of the features common to all arthropods is that their appendages are jointed. This, of course, is not unusual. Without joints, there would be no way to bend the appendages, which would make walking and grasping things much more difficult! Vertebrates, therefore, have jointed appendages as well. However, the setup is quite different from that of an arthropod. In a vertebrate (such as a person), the muscles form over the joint and move the joint from above. Because arthropods have exoskeletons, however, the muscles form *under* the exoskeleton, moving the joints from underneath.

Common Characteristic #4: A Ventral Nervous System

In order to react to stimuli, seek out prey, and seek protection from predators, arthropods have a nervous system. Two ganglia form a brain, much like that of an earthworm, but more complex. Again, like an earthworm, a ventral nerve cord runs from the ganglia to the posterior. The fact that it is placed at the bottom of the body (that's why we call it "ventral") is no accident. This placement provides maximum protection. It is protected not only by the exoskeleton, but also by the bulk of the body. Instinctively, arthropods do everything they can to avoid exposing their undersides, because instinct tells them that this negates the body's ability to protect the ventral nerve cord.

The nervous system is fed with information through various sensory organs. **Antennae** in the head region provide touch, taste, and smell sensations to the nervous system. In addition, all arthropods have some sort of eyes. There are two different types of eyes in phylum Arthropoda: **compound eyes** and **simple eyes**.

Compound eye – An eye made of many lenses, each with a very limited scope

Simple eye – An eye with only one lens

Scanning electron microscope image of spider by Ric Felton, www.semguy.com

FIGURE 12.1
Simple Eyes and a Compound Eye

Scanning electron microscope image of fly eye © B. Plowman/Custom Medical Stock Photo

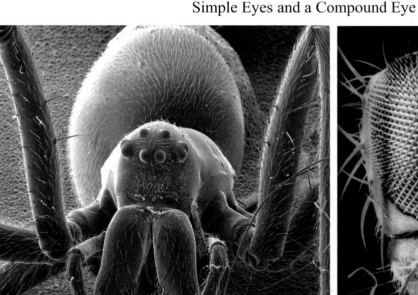

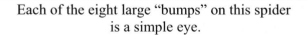
Each of the eight large "bumps" on this spider is a simple eye.

In this fruit fly's compound eye, each little bump is an individual lens.

Now don't be fooled by their names. First of all, no eye is simple. It takes an enormous amount of engineering to come up with a system that can detect light, turn that light energy into electrical signals, send the electrical signals to the brain (or ganglia), and have the brain convert those signals into an image! Only God can create such a marvel. Also, a "simple" eye is not necessarily less desirable than a compound eye. For example, the human eye has only one lens; thus it is a simple eye. Nevertheless, it is a marvelously-engineered organ and provides better overall sight than the eye of any other species. Some animals can see farther than people can, and some can see in less light, but when you consider all factors such as range, sharpness, sensitivity, and color depth that the human eye provides, there is simply nothing else like it in creation!

Spiders have simple eyes. Because there is only one lens and because that lens is small, the eye does not cover much area. If you want to get an idea of what a spider sees, take a look through a thin straw. That's the kind of area that a spider's eye covers. Flies, on the other hand, have compound eyes. As a result, their sight covers a greater area, but, since the lenses are individual, the image is rather strange. A fly sees many versions of the same image, each slightly tilted with respect to the other, because the lenses are slightly tilted relative to one another. Thus, the fly gets a "mosaic" view of the world, whereas a spider gets a "tunnel" view.

Common Characteristic #5: An Open Circulatory System

Arthropods have quite an unusual circulatory system. In order to bring vital substances to every cell in the body, a heart in the dorsal (upper) region of the body pumps blood into short vessels that *empty out into different cavities of the body*! This allows blood to flow right over all of the cells in that cavity. In a sense, then, arthropods are always bleeding internally.

Open circulatory system – A circulatory system that allows the blood to flow out of the blood vessels and into various body cavities so that the cells are in direct contact with the blood

Of course, once released to flow throughout the body, the blood has to be collected again and then recycled back into the heart. You will learn more about how this circulatory system works when you study crayfish anatomy in depth.

ON YOUR OWN

12.1 Suppose an arthropod lives in an ecosystem which contains no predators. The arthropod would never, ever be threatened. Would it need an exoskeleton in this situation?

12.2 If you count the cephalothorax as one body segment, how many major segments does an arthropod with a cephalothorax have?

12.3 An organism moves its joints with muscles that lie on top of the joint. Is the organism an arthropod?

12.4 If an arthropod cannot taste something, but the rest of the nervous system is operational, what sensory organ is malfunctioning?

12.5 Why do we say that arthropods are constantly bleeding internally?

<u>Class Crustacea: The Crayfish</u>

The first arthropods that we will study come from class **Crustacea**. Most crustaceans live in either fresh water or marine environments (like the crayfish or the lobster), although some species (like the pill bug, which is often called the "roly-poly") are land dwellers. Since the crayfish is the "standard" crustacean to study, we will start there.

The crayfish, often called the "crawdad," lives in fresh water. It looks much like a lobster, which is the marine "equivalent" of the crayfish. Lobsters, however, grow to be much larger than crayfish. If you go to virtually any river in the United States and look at the shoreline, you will probably see crayfish busily moving up and down the shallow waters of the shore, looking for food. Many fishermen will tell you that crawdad meat is one of the best baits that you can use to catch fish. In the southern United States (and many other places), crayfish are used in a variety of popular dishes. Of course, you have to exercise a bit of caution when you try to catch them, because their claws (chelipeds) can give you a pretty good pinch!

The crayfish has a detailed external anatomy that deserves some attention. A photo of the crayfish, showing most of its major exterior features, is given below.

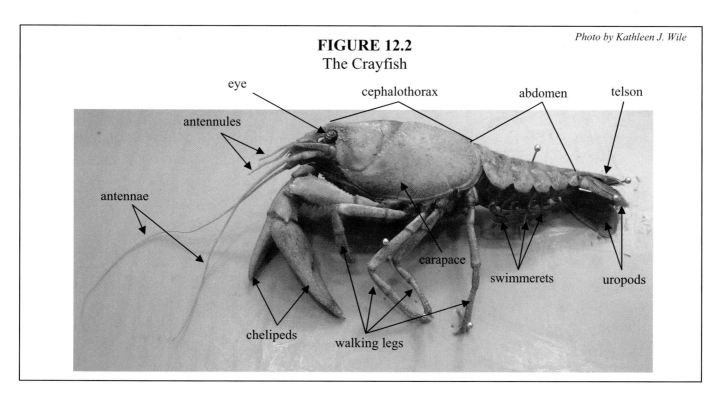

FIGURE 12.2
The Crayfish

Photo by Kathleen J. Wile

The first thing that you should notice is that the crayfish is one of those arthropods with a cephalothorax. The cephalothorax is covered in a single plate called the **carapace** (kehr' uh pace). The abdomen is separated into six segments, each with its own protective plate of exoskeleton.

The six sets of appendages shown in the crayfish above all perform different functions:

> **Walking legs** - These appendages are used for locomotion when the crayfish is on land or moving on the bottom of the lake or river in which it lives.

> **Swimmerets** - These aid in swimming as well as reproduction. In male crayfish, the first and second swimmerets transfer sperm to the female during mating. In females, the swimmerets carry both the eggs and the developing offspring.

> **Uropods and telson** - These appendages form the flipper-shaped tail that the crayfish uses for swimming.

> **Chelipeds** (kee' luh pedz) - The chelipeds (usually called "claws") are used for defense as well as to grab onto prey.

> **Antennules** - These small antennae aid the creature in balance and provide taste and touch sensations.

> **Antennae** - These longer appendages are much more sensitive than the antennules, providing the crayfish with strong senses of taste and touch.

These appendages are all controlled by the nervous system, giving you some idea of how complex it is.

The Crayfish's Respiratory System

The crayfish gets its oxygen from the water through two sets of **gills** that are located in the cephalothorax. Gills are amazing organs that are present in most of the animals that live underwater. As you know, all aerobic organisms need oxygen to survive. For land-dwelling creatures, the oxygen is readily available from the surrounding air. But how do organisms that live underwater get their oxygen?

Well, oxygen gas is dissolved in the water. The problem is how to get the oxygen *out* of the water and *into* the creature. In the case of hydra, jellyfish, and other invertebrates that we studied in the previous module, the organism is designed to allow oxygen to diffuse into the body, while waste gases diffuse out. Although this system works well for these creatures, it will not work for many water-dwelling animals. Crayfish, for example, are protected and supported by an exoskeleton. No oxygen is going to diffuse through that! Even though other water dwellers do not have exoskeletons (the vertebrate fish, for example), they often have scales or some other covering that would prevent such an exchange of gases with the environment.

How then, do these creatures get the oxygen that they need? They use gills, which take oxygen that has been dissolved in the water and transfer it to the bloodstream. At the same time, carbon dioxide in the blood is released into the gills, which transfer it to the water. The gills, therefore, act as a transfer station, transferring dissolved oxygen from the water and into the blood while at the same time transferring carbon dioxide from the blood and into the water.

The two sets of gills in a crayfish reside in two gill chambers that are slightly posterior to the head on each side of the crayfish. One of these sets of gills is shown in Figure 12.3 on the next page.

FIGURE 12.3
The Gills on a Crayfish

Photo by John Skipper

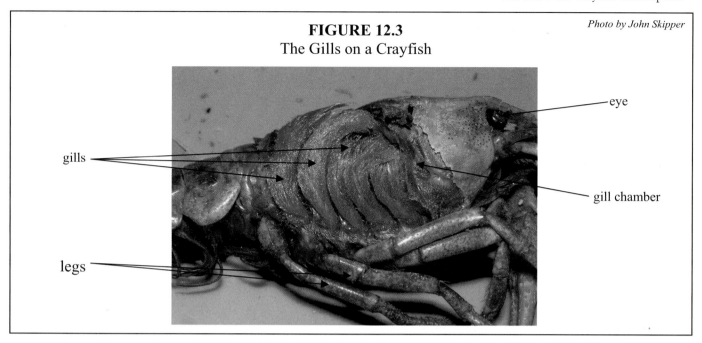

eye

gills

gill chamber

legs

The crayfish has small openings on the ventral side of the **gill chambers** that allow water from the surroundings to flow inside to the gills. Blood traveling through the gills can then release carbon dioxide into the water and absorb oxygen. Do you see what's happening in the gills? Since the exoskeleton cannot allow gases to be exchanged with the surroundings, God has designed gills that can do what the entire body of a hydra or jellyfish does. Of course, to make this system work, there has to be continuous circulation of water through the gill chambers. This is accomplished by the motion of the crayfish's swimmerets as well as tiny appendages near the mouth called **maxillae** (mak sil' ay – singular is **maxilla**). Although the maxillae are involved in keeping water flowing through the gills, they also help the crayfish handle food.

Interestingly enough, many water-dwelling crustaceans (including the crayfish) can actually store up a lot of water in their gill chambers so that they can make brief excursions out of the water. The gills continue to exchange oxygen and carbon dioxide with the stored water until all of the oxygen is used up. After that, the crayfish must return to the water or it will suffocate. This situation is very similar to a land-dwelling animal that holds its breath while underwater. The lungs can continue to absorb oxygen from the stored air, but eventually, the animal must return to the surface of the water and breathe again. Crustaceans that have this ability typically use it to seek out new food sources near the shore of the lake, river, stream, or ocean in which they live. Some use it to move from one body of water to another.

Before we leave this section, it is important to dispel a myth that seems to be very popular in schools today. As you might already know, water is composed of two hydrogen atoms connected to an oxygen atom (H_2O). Thus, in a way, there is oxygen in every water molecule. For some reason, this makes some students (and a few teachers, unfortunately) think that gills somehow decompose water, separating the hydrogen and oxygen atoms. Then, the creature uses the oxygen atoms to perform respiration. *This is completely untrue.* The only way that gills can get oxygen is to extract the *oxygen gas that is dissolved* in the water. If a crayfish (or even a fish) were put in oxygen-poor water, it would suffocate immediately. In other words, if the water does not have adequate oxygen dissolved in it, a crayfish (or fish) can actually drown! The only way these creatures can get oxygen is if that oxygen has been dissolved in water. The oxygen atoms that are in a water molecule are not available for respiration!

Think about the last time you saw an aquarium with lots of fishes and other water-dwelling animals in it. If you paid careful attention, you probably saw bubbles coming up from the bottom of the tank. Those bubbles were the result of air being pumped into the aquarium so that oxygen from the air can dissolve in the water, replenishing the oxygen used by the creatures in the aquarium. This keeps the aquarium's inhabitants from drowning. If an aquarium (or fish bowl) does not have air being pumped into it, it either has lots of plants and phytoplankton performing photosynthesis to replenish the oxygen being used up by the animals, or it has only a few animals in it. If there are only a few animals in the tank, the oxygen that dissolves into the water from the air above the surface is enough to replenish the oxygen used by the few animals in the tank.

The Crayfish's Circulatory System

Circulation is closely tied to respiration. After all, if the gills get oxygen from the water but the animal cannot send it anywhere, the oxygen wouldn't do much good, would it? Transporting oxygen to the places where it is needed is the job of the circulatory system. As we mentioned in the previous section, arthropods have open circulatory systems. These circulatory systems are very interesting, because they allow for direct contact between the tissues and the blood. This is quite different from the circulatory system that you find in vertebrates. In the vertebrate circulatory system, the blood never leaves the blood vessels. Instead, the exchange of substances between the blood and the tissues occurs through thin-walled vessels called "capillaries." In arthropod circulatory systems, however, the blood stays in blood vessels for only part of its trip through the animal. It is then dumped into the tissues, bathing them with the substances they need. The figure below illustrates many different systems in the crayfish. For right now, we will focus on the circulatory system.

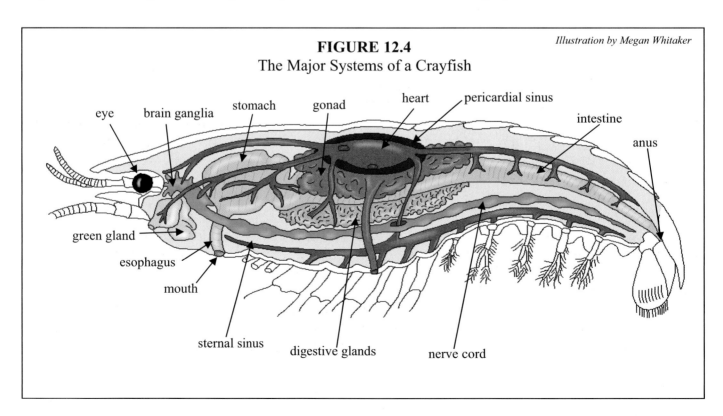

FIGURE 12.4

Illustration by Megan Whitaker

The Major Systems of a Crayfish

A crayfish has a **heart** in the dorsal part of its body. It rests in a cavity called the **pericardial** (pehr uh kar' dee uhl) **sinus** (sye' nus). Blood collects in this cavity, and it enters the heart through one of three openings in the heart's surface. Each opening has a valve that closes when the heart is ready to pump. Once it absorbs the blood and closes these valves, the heart pumps blood through a series of blood vessels that are open at the other end. These vessels dump the blood directly into various body cavities. This allows blood to bathe every cell in the area, giving up the oxygen it is carrying and absorbing the carbon dioxide that the cells need to give up.

Gravity causes the blood to fall into the **sternal** (stir' nuhl) **sinus**, where it is collected by blood vessels that are open at one end. Unlike the blood vessels that dump the blood into the body cavities, these vessels carry the blood back towards the pericardial sinus. On its way there, the blood is passed through the gills, where it can release the carbon dioxide it has collected from the tissues and pick up a fresh supply of oxygen. The blood also passes through a **green gland**, which cleans it of impurities and dumps those impurities back into the surroundings. Once the blood has passed through the gills and the green gland, it makes its way back to the pericardial sinus to begin the trip all over again.

Why do arthropods have this unusual form of circulatory system? Well, it turns out to be a very efficient means by which gases and other vital substances can be sent to the cells. After all, the blood is in direct contact with the cells that it needs to supply. This obviously makes exchanging gases and other substances quite easy. If the open circulatory system is so efficient, why don't other organisms have it? Well, some do. Mollusks, for example, have an open circulatory system. Such systems come at a cost, however. Animals with open circulatory systems rely on gravity to make their blood flow properly. After all, once the blood is released from the blood vessels, gravity must pull it down through the tissues and into the sternal sinus. If the animal happens to be on its back, blood doesn't flow properly, and if the animal stays on its back for an extended amount of time, tissues can die. Thus, although an open circulatory system provides for an efficient exchange between the blood and the tissues, it does restrict the ways in which the animal can position itself. Animals with closed circulatory systems can stay on their backs for as long as they want, because they do not rely on gravity for proper blood flow.

Now before we go any further, we want you to stop a moment and think about what you have been reading. The crayfish has a means by which it can extract dissolved oxygen from the water and transport it throughout its body. The system that does the transporting allows direct contact between the blood and the cells. In addition, the chemical nature of the blood allows it to pick up oxygen where it is supposed to (in the gills) and release it where it is supposed to (in the cells). At the same time, the blood can pick up carbon dioxide where it is supposed to (in the cells) and release it where it is supposed to (in the gills). Isn't that amazing?

As amazing as all of this is, we are not done exploring the wonders of the crayfish's circulatory system. Remember, the exoskeleton protects and supports the crayfish, but this exoskeleton can be damaged. Crayfish (and many other crustaceans) can lead pretty violent lives. When attacked by a predator, a crustacean can usually defend itself fiercely. As we already mentioned, the claws of a crustacean can be used to battle off enemies. Well, in the midst of such battles, it is not uncommon for a crustacean to lose one of its appendages. Its antennae could be broken off, it could lose a leg, or it could even lose a claw! You would think that such a disaster would cause all of the crayfish's blood (which is bathing the tissues) to flow out into the surroundings, killing the creature.

That's what you would think, but it's not what happens! You see, the crayfish and most other crustaceans have a double membrane in each appendage. When the crayfish loses an appendage, the membrane seals the resulting hole, keeping the blood in the body! If this isn't incredible enough, once the membrane seals the hole, the crayfish can *actually regenerate the missing appendage*. This isn't a means of asexual reproduction, as is the case with the planarian. It is a repair mechanism built into most crustaceans! Crustaceans seem to know all about this marvelous mechanism, because if a crayfish is caught by one of its appendages, it will willingly break it off in order to escape!

Clearly there is a lot of detailed engineering going on here. Not only is there an enormous amount of design evident in just the circulatory system of crustaceans, but there are fail-safe mechanisms as well! In the case of disaster, built-in systems react to minimize the damage and then rebuild whatever appendage was lost. That engineering is just one more example of the fact that God's creation is a continuing testament to His majesty!

The Crayfish's Digestive System

Crayfish are scavengers, eating virtually anything that can be digested. In order to eat something, the crayfish first uses its **mandibles** to break the food into small chunks. The food then enters a short **esophagus** and goes into a **stomach** that essentially has two regions. The first region, which is on the anterior side of the stomach, grinds the food into fine particles. These fine particles are then sent to the other region of the stomach, which is on the posterior side. This region sorts the particles. If they are small enough, they are sent directly to **digestive glands** which secrete enzymes, completing the digestion process. If the particles are too large to be digested immediately, they are sent to the **intestine**. As they travel through the intestine, they are exposed to digestive enzymes, which digest what they can. Anything that remains at the other end of the intestine is considered indigestible and is expelled out the anus.

The Crayfish's Nervous System

A crayfish's brain is comprised of two **ganglia**, each of which has a **nerve cord**. These nerve cords join together posterior to the stomach and run along the ventral side of the crayfish. At regular intervals, there are ganglia that continually process the signals running down the nerve cord. This system is fed information from various appendages throughout the body. The crayfish has compound eyes that send sight information to the brain. The antennules and antennae send taste and touch information to the nervous system. Also, tiny bristles are found all over the crayfish's body, providing a sense of touch. These bristles are necessary because the exoskeleton is so hard that it is not very sensitive to touch. The bristles make up for that, providing touch sensations to parts of the body that otherwise would not have them.

One rather interesting feature of the crayfish's nervous system can be found at the base of the antennules. As we mentioned before, antennules help the crayfish keep its balance. Here's how: at the base of each antennule is a **statocyst** (stat' uh sist).

Statocyst – The organ of balance in a crustacean

These statocysts are little containers that are lined with tiny hairs, each providing a sense of touch. Inside each hair-lined container, you will find a grain of sand, called a **statolith** (stat' oh lith). This statolith shifts when the crayfish is knocked off balance. The hairs detect that shift and send a message to the brain. The brain then sends signals to the muscles, which work the abdomen, swimmerets, telson, and uropods until the crayfish has righted itself.

The Crayfish's Reproductive System

Reproduction in the crayfish begins in the **gonad**, which produces gametes.

Gonad – A general term for the organ that produces gametes

In males the gonad is called the **testis** (plural is **testes**), and in females the gonad is called the **ovary** (plural is **ovaries**). In male crayfish, sperm are formed in the gonad and transferred to the first and second pairs of swimmerets. Crayfish usually mate in the fall. Male crayfish deposit their sperm into special containers that the female has. The female then stores the sperm until spring.

In the spring, the female produces eggs. When the eggs travel through the oviduct, they are fertilized by the sperm and go to the swimmerets. The fertilized eggs attach themselves to the swimmerets and develop for approximately six weeks, at which point the eggs hatch. The newborns look like miniature versions of the parents, and they tend to cling to the mother for many weeks after they hatch. During the first year of life, the average crayfish molts seven times. This is because the body experiences rapid growth, and the exoskeleton gets too restrictive. After a year is passed, crayfish tend to molt only twice per year. The average crayfish lives four to eight years.

Other Crustaceans

We have concentrated on the crayfish in this module because it is a good representative of the organisms from class Crustacea. Of course, there are many, many different kinds of crustaceans, a few of which are presented in the figure below.

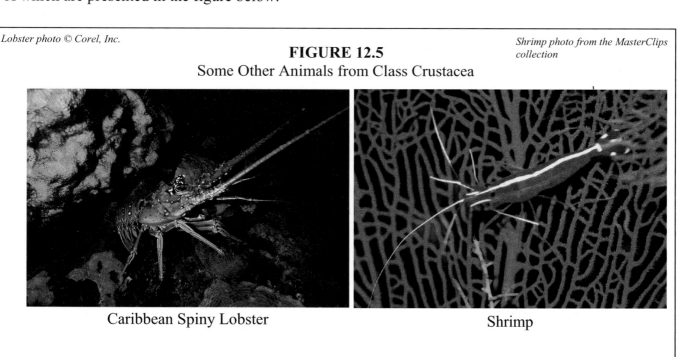

Lobster photo © Corel, Inc.

Shrimp photo from the MasterClips collection

FIGURE 12.5
Some Other Animals from Class Crustacea

Caribbean Spiny Lobster

Shrimp

Exploring Creation With Biology

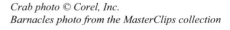

| Sally Lightfoot Crab | Barnacles |

Lobsters, shrimps, and crabs live in virtually every marine environment and are popular dishes for many seafood lovers. Another marine crustacean that you see in nearly every marine environment is the barnacle (bar' nuh kuhl). Many students mistake barnacles for mollusks, because they appear to have a shell. Well, that's not really a shell. It's an exoskeleton formed out of calcium rather than chitin. Constantly moving appendages bring water into the exoskeleton, allowing the barnacle to breathe and take in food. Barnacles will stick to virtually anything. Although they usually attach themselves to rocks, barnacles have been found on boats, lobsters, large clams, and even whale teeth!

<u>An Important Note</u>

As we did with the earthworm, we have thrown a lot of material about the crayfish at you. Since this creature is such a good example of an arthropod, we want you to remember all of it. In other words, like the earthworm, you need to be able to point out all of the major organs and systems in a crayfish and know how they work. Thus, you will need to commit Figures 12.2 and 12.4 to memory. In addition, you will need to have a solid working knowledge of all the systems discussed in this section. This may seem like a lot, but it is really necessary. The good news is that although we will study several more examples of arthropods, you will not need to have such a detailed knowledge of them. Instead, you will just need to know the basics.

ON YOUR OWN

12.6 Although the gills of a certain crayfish seem to be working fine, the crayfish suffocates because it cannot get fresh water into the gill chambers. What organs are not working properly?

12.7 Trace the flow of blood from the pericardial sinus and back again. Mention the following organs in your discussion: pericardial sinus, sternal sinus, green glands, heart, and gills.

12.8 A crayfish loses its claw in a fight. What happens?

12.9 If a crayfish cannot stay upright in the water, what organ is most likely not working?

If you purchased the dissection kit, perform the following dissection lab. If not, go on to the next section.

EXPERIMENT 12.1
Crayfish Dissection

Be careful! Dissection tools are SHARP!

Supplies :

♦ Dissecting tools and tray that came with your dissection kit
♦ Crayfish specimen
♦ Magnifying glass
♦ Laboratory notebook

NOTE: The course website discussed in the "Student Notes" section at the beginning of this book has several pictures of a crayfish dissection.

Object: To become more familiar with the crayfish's anatomy through dissection

Procedure:

1. Examine your crayfish specimen. Identify all of the exterior features labeled in Figure 12.2.
2. Turn the crayfish ventral side up on the dissection tray and examine the mouthparts.

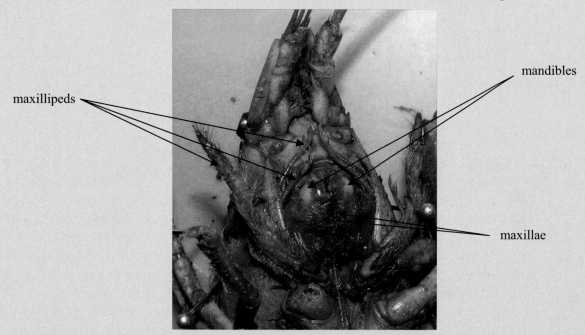

Mandibles - One pair of mandibles can be found posterior to the antennae. These are the jaws of the crayfish. Notice that, unlike human jaws, these mandibles open and close horizontally.
Maxillae - These can be found just posterior to the mandibles. You should find two pairs.
Maxillipeds - These are often called "jaw feet," because they appear to be tiny legs or feet. There are three pairs of them, and they are used by the crayfish to hold food in place. One pair is posterior to the maxillae, and the other two pairs are anterior to the maxillae.

3. Once you have located each of these mouthparts, use your forceps to remove them all from one side of the crayfish. You can remove them by simply grasping them with your forceps and plucking them out. Lay them on the dissecting tray side by side to examine their relative sizes. Note their relative sizes in your notebook.

4. Determine the sex of your crayfish. You can do this by closely examining the swimmerets. In males, the two most anterior pairs of swimmerets are modified for sperm transfer. As a result, they are larger than the others and prong-shaped. In females, the two most anterior pairs of swimmerets are small in size relative to the others. Record the sex of your crayfish in your notebook.

5. Now you are ready to look inside the crayfish. Place your specimen dorsal side up and use your scissors to cut the crayfish's carapace from the posterior end of the carapace to just behind the eyes. Then make a transverse cut just behind the eyes. These two cuts are illustrated by the white lines in the photo below:

6. Carefully remove the carapace in the two pieces determined by your cuts. This should expose the **gills** (see Figure 12.3). Examine the structure of the gills. Count the gills and record the number in your notebook.

7. To make things easier, pull off the walking legs. Carefully remove the internal tissue on the dorsal side of the crayfish. Using Figure 12.4 and the pictures below as guides, locate the following structures:

> **Heart** - The heart should be the structure closest to the top. Notice the main blood vessels attached to it.
> **Gonad** - Depending on the sex, you will see ovaries (female) or testes (male). The ovaries are darkly-colored and should look like a mass of eggs. The testes are small and white with coiled ducts attached.
> **Digestive glands** - These are two lightly colored masses on both sides of the body cavity.

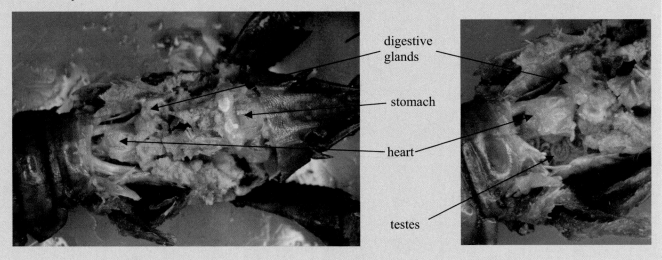

8. Now you need to cut open the abdomen to reveal the intestine. Cut the dorsal side of the abdomen exoskeleton from the anterior end to the telson, and then open the abdomen. Look for a tube that runs its length. That's the **intestine**. You actually should see two tubes. The smaller, darker one is the **dorsal blood vessel**. Notice also the tissue beneath. These are the **abdominal muscles**, which is the part of the shrimp, lobster, or crayfish that we eat.

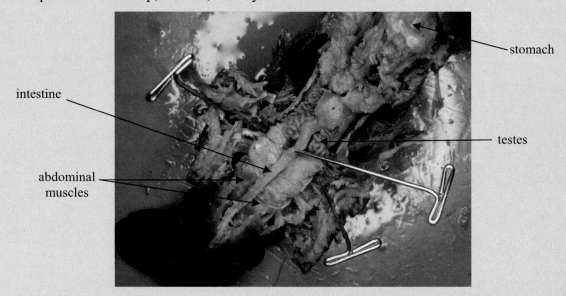

intestine

stomach

abdominal muscles

testes

9. Follow the intestine forward to find the **stomach**.
10. Make a drawing of your crayfish as it looks now. Label all of the structures.
11. Next, you need to remove some of the internal organs. Do this by cutting the muscles just behind the eyes that lead to the stomach. Pull the stomach so that you can reach under it with your knife or a probe and cut the **esophagus**. Once you have done that, you should be able to pull the stomach and intestine out. This will bring a lot of other organs with it!
12. Now that you have cleared out the digestive organs, you should be able to see the **green glands** just posterior to and below the antennules. You may have to remove some tissue in the head in order to see them.

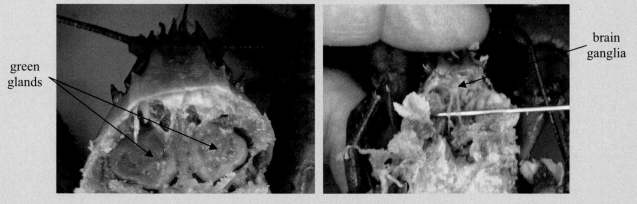

green glands

brain ganglia

13. Looking between the eyes, you should see a mass of white tissue. This is the crayfish's **brain**. Once again, you might need to clean out some tissue to see it.
14. Try to trace the nerve cord from the brain to the abdomen. There is some hard tissue near the bottom that you will need to cut through in order to do this.
15. Make a drawing of the crayfish as it appears now. Label the structures you were able to find, and note in your laboratory notebook any structures you were unable to find.
16. Clean up everything. Wash and dry your tray and tools. Put everything away.

Class Arachnida

Are you ready for chills to run up and down your spine? Well, you had better be, because in this section, we are going to discuss class **Arachnida** (uh rak' nih duh), the class that contains spiders. For some reason, most people have a revulsion against spiders. This is probably because spiders typically live in the dark and attack their prey quickly with no warning. This seems to strike to the very core of some people's greatest fears.

Although spiders are greatly feared, there are only a few that are actually harmful. Many spider bites can hurt, but only a few spiders, like the black widow or the brown recluse, are poisonous to human beings. In fact, many spiders are actually quite beneficial to humans and to the ecosystem in general. Spiders help keep the population of insects in check. If it were not for these fearsome little arthropods, insects would overrun most ecosystems!

Organisms in class Arachnida have five common characteristics:

➢ **Four pairs of walking legs**
➢ **A cephalothorax instead of separate head and thorax**
➢ **Usually have four pairs of simple eyes**
➢ **No antennae**
➢ **Respiration done through organs known as "book lungs"**

You probably know the first characteristic. It is usually the first thing taught to students as a means by which to separate arachnids from insects. Arachnids have eight legs (four pairs), while insects have six legs (three pairs).

The second characteristic listed tells you that arachnids have only two body segments: a cephalothorax and an abdomen. Remember that some arthropods have three segments (head, thorax, and abdomen). If you look back at Figure 12.1, you can count the "bumps" on the spider to see that there are four pairs of simple eyes, just as the third characteristic indicates. Unlike crayfish, spiders do not have antennae. Finally, spiders breathe through **book lungs**, which we will discuss in a later section.

Since spiders are representative of class Arachnida, we will take an in-depth look at them. As we mentioned at the close of the previous section, we will present a lot of information in this section for which you will not be held responsible on the test. Instead of requiring you to memorize a lot of this material, we just want you to get a good appreciation for the organisms in this class. How will you know what information you will be responsible for on the test? Read through this section, and then use the "On Your Own" questions and the study guide to help you determine this. For example, we will present the anatomy of the spider in a little while, pointing out many of the key organs in the spider's body. However, if you look at the study guide, you will see no diagram that asks you to point out the major organs in a spider. Thus, even though we point out all of the major features of spider anatomy, you will not be held responsible for it on the test. However, the study guide does ask you the characteristics that set arachnids apart from other arthropods, so that is something you will need to know for the test.

The Spider

The figure below shows two different spiders. They probably illustrate the most striking difference between the various species of spiders: some spiders spin webs; others do not.

FIGURE 12.6
The Tarantula and the Garden Spider

Photos from www.clipart.com

Tarantula   Garden Spider

Spiders are cunning predators that use many different means to catch their prey. Although their most common means of catching prey involves weaving a web (we will discuss this in a moment), many spiders do not use webs to hunt. Tarantulas, for example, stalk their prey, sneaking up and pouncing on unaware creatures. Although the tarantula strikes fear into the heart of most people who see it, the common species of this spider is not poisonous to humans. Of those that are, the poison usually results only in a stinging rash accompanied by swelling. Nevertheless, partly because of their size and partly because of their reputation, tarantulas are some of the most feared organisms on the planet. Some people use this to their advantage. For example, a jewelry store owner had been robbed several times. To curb the robberies, he placed several large tarantulas in his showcase window after business hours each night. He also posted a sign that said, "Danger! This area is patrolled by tarantulas." Not surprisingly, the robberies stopped.

Of course, spiders are best known for weaving webs because this is the way that most of them capture prey. They weave these webs out of spider silk, one of the most incredible substances in all of creation. Spider silk is a very flexible substance, but at the same time it is very strong. Most man-made materials are either strong or flexible. It is very difficult to engineer a substance that is both. You've probably played with an abandoned web at one time, plucking the silk and watching it vibrate. This is a great illustration of spider silk's flexibility. What you probably don't know, however, is the strength of spider silk. It is rather easy to break the strands of a web, because the silk strands are so thin. However, if you were to weave a rope out of spider silk, *it would be stronger than a steel pipe of the same size*, yet it would be almost as flexible as a rope. A spider silk rope that is just a bit thicker than a garden hose, for example, can support the weight of two full Boeing 737 airplanes!

To illustrate just how incredible spider silk is, we want you to read about one application that scientists use it for today. In physics research, scientists are trying to develop a process called "laser-induced nuclear fusion." If it could ever be developed, it would result in a limitless, safe, and completely clean source of energy. This process involves taking a laser and inducing a nuclear reaction that turns hydrogen into helium. This is a violent reaction but, when done on a tiny sample, it is controllable. To hold the sample in place, scientists need a strong substance that is very lightweight. Do you know what scientists use to hold the sample? They use *spider silk*. Spider silk gives them the strength that they need but does not weigh much at all. As a result, it is an ideal substance for use in laser-induced nuclear fusion systems! The "lowly" spider may one day help us develop clean and inexpensive energy.

There is a host of different ways that spiders use their silk to catch their prey. Most weave webs. Some species of spider build a **sheet web**, which is a single, flat sheet of sticky silk. The spider hangs on the underside of the sheet and, when an insect gets caught, it paralyzes the creature and pulls it through the web. Some spiders spin tangles of webs that have no real discernible pattern. These webs are typically called **tangle webs**. The most geometrically stunning web is the **orb web**, like the one pictured in Figure 12.6. This web consists of concentric circles of sticky silk that are supported by "spokes" of nonsticky silk. These webs can span great distances, making them very efficient traps. The spider sits at the center of the web with its legs on the spokes. When an insect gets caught in the web, it begins to struggle. The spider can tell which part of the web that the insect is in because the spoke nearest the insect will shake more than the others. When the spider senses which spoke is shaking the most, it follows the spoke until it finds the insect.

Of course, there are spiders that produce silk but do not weave webs. The **trap door spider** digs a shallow hole in the ground and then weaves a "trap door" out of silk. The trap door is attached to the ground at one end but is free to move everywhere else. The spider holds on to tiny "handles" that it weaves into the trap door and keeps it slightly ajar. When an insect or crustacean walks by, the spider jumps out of the trap door, startling and killing its prey. It then drags it back into the hole and closes the trap door, feasting on its catch. Other spiders spin single strands of silk with very sticky ends. They launch the silk at their prey, catching them and reeling them in.

Now think about this for a moment. Consider the fact that spider silk is a marvelous material which is both strong and flexible. Then realize that spiders use this to create such engineering marvels as orb webs, trap doors, or projectiles that can be reeled back in! Is there anything more marvelous in all of creation? Well, believe it or not, there is! As you study more and more science, you will see that these kinds of engineering marvels exist all over creation, and there are some even more awesome than the wonder of spider silk and how it is used. Through examples such as this one, we know that life is no accident. Such incredible systems cannot develop by chance. Creation clearly tells us what an awesome God we have!

The Major Points of Interest in Spider Anatomy

Figure 12.7 is a sketch of the major systems in a spider. We will not discuss all of them, because many organs in the spider are pretty much the same as those in the crayfish. Of course, the most striking difference between the organs of a spider and those of a crayfish is the presence of **silk glands** and **spinnerets**. When a spider spins silk, it produces that marvelous substance in its silk glands and uses its spinnerets to spin the silk.

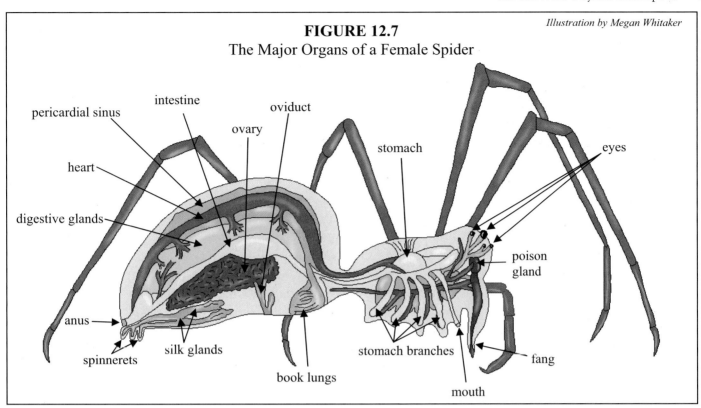

FIGURE 12.7
The Major Organs of a Female Spider

Regardless of how a spider catches its prey, once the prey is caught, most spiders behave similarly. Spiders sink their **fangs** into the prey and use their **poison glands** to inject a paralyzing poison into the prey. This immobilizes the creature, and the spider then secretes digestive enzymes into its body. The tissues that these enzymes partially digest are then sucked through the mouth and into the stomach. The poison of a spider is rarely harmful to humans. It is usually a very weak poison, designed to paralyze tiny prey. There are, however, a few species (the black widow and brown recluse spiders, for example) that have a bite which can be deadly to humans.

The **book lungs** of a spider do the same job that the gills do in a crayfish, except of course for the fact that gills extract dissolved oxygen gas from the water, while the spider's book lungs extract oxygen from the air. Air enters the exoskeleton through a slit in the abdomen, which is called a **spiracle** (spear' uh kuhl). There, it encounters an organ that has several thin layers, almost like the pages of a book. As the air mingles with these "pages," oxygen is absorbed by the blood and carbon dioxide is released.

If you are interested in learning more about spiders, please visit the course website that we discussed in the "Student Notes" section at the beginning of this book. You will find links to information on many aspects of spider biology, including more information on the wonders of spider silk.

ON YOUR OWN

12.10 Most spiders have poison glands and produce poison that they inject into their prey. Does that mean that we must fear most spiders because they are poisonous?

12.11 Where are the abdomen and cephalothorax in Figure 12.7?

Classes Chilopoda and Diplopoda

It seems natural to discuss classes **Chilopoda** (kye lah' puh duh) and **Diplopoda** (duh plah' puh duh) right after class Arachnida, because the members of these classes are likely to make your skin crawl as well. Class Chilopoda is home to the arthropods normally called "centipedes," and class Diplopoda contains those arthropods usually called "millipedes." Although often used interchangeably, these two terms refer to two completely different creatures.

The term "centipede," which in Latin means "hundred legs," refers to the members of class Chilopoda. These arthropods have flat bodies that are divided into several segments, each of which contains a pair of legs. Their common name is misleading, however, because centipedes do not have anywhere near 100 legs. The head of a centipede contains antennae for sensory perception and several mouth parts. The body segment directly behind the head contains the first pair of legs, which have poisonous claws. These claws immobilize the insects and small animals that the centipede eats. The common centipedes with which you are familiar are rather small, but certain tropical species can reach lengths of up to one foot long. Although the bite of a centipede is painful to humans, it is rarely dangerous.

The term "millipede," which in Latin means "thousand legs," refers to the members of class Diplopoda. Once again, unlike their common name implies, these arthropods do not have anywhere near 1,000 legs. They do have many more legs than centipedes, however, because each of their body segments contains two pairs of legs instead of just one. This is not the only difference between centipedes and millipedes, however. Millipedes have bodies that are rounded, rather than flat. Typically, their antennae are shorter than those of centipedes, and they do not have the poisonous claws that centipedes have. In fact, while centipedes are fierce predators, millipedes are typically docile. They move along the ground slowly, eating vegetation and organic debris. Often, when threatened, millipedes will simply roll into a ball, hoping that their strong exoskeletons will protect them.

Centipede photo © Corel, Inc.

FIGURE 12.8
Centipede and Millipede

Millipede photos by Dr. Jay L. Wile

legs

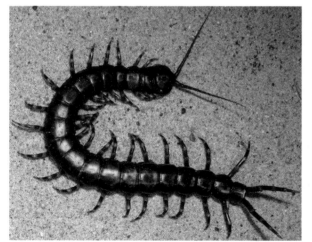

A centipede is a fierce predator with a pair of legs on each body segment.

A millipede is a docile arthropod that curls up in a ball when threatened. It has two pairs of legs per segment.

Class Insecta

We now come to the largest class in creation, class Insecta. There are over a million known species of insects, and the number grows each year as new species are found and classified. In fact, insect species make up more than three-fourths of the species in the animal kingdom. They have the following characteristics that separate them from other members of phylum Arthropoda:

> ➢ **Three pairs of walking (or jumping) legs**
> ➢ **Usually have wings at some stage of their life**
> ➢ **One pair of antennae**
> ➢ **Three segments: head, thorax, and abdomen**

Although these characteristics are common to all insects, there is amazing variability in these and other characteristics within this huge class.

Insect Legs

Most insects use their legs for walking. As you would expect for arthropods, insect legs are jointed. They typically connect to the thorax of the insect, and the variety throughout class Insecta can be illustrated by the many different kinds of insect legs. For example, the front pair of legs on a praying mantis isn't just for walking. Each leg has a powerful claw that the mantis uses to capture prey. Flies, on the other hand, use all of their legs for walking, but the legs come equipped with sticky pads that allow them to walk up walls and even upside down on ceilings! The grasshopper uses its two most powerful legs for making great leaps, while the water strider has legs that are bristled at the end, allowing it to actually walk on water. Other insects use their legs to make sounds, while still others use them as a means to store food. This should give you a glimpse of the variety that exists in this class.

Insect Wings

Although most insects have two pairs of wings, some have only one pair, and a very few have none at all. There are four basic types of wings in class Insecta:

1. **Membranous wings**
2. **Scaled wings**
3. **Leather-like wings**
4. **Horny wings**

Membranous wings are thin, transparent, and have a detailed network of veins that are visible. A fly's wings, for example, are membranous. **Scaled wings** have delicate scales that cover the wings. As a result, they are not transparent. The scales are easy to rub off, as if they were made of powder. Moths and butterflies have scaled wings. **Leather-like wings** are wings that appear to be a part of the exoskeleton. They are typically laid over a second, membranous pair of wings so as to protect them. Grasshoppers have one pair of membranous wings under one pair of leather-like wings. **Horny wings** are often hard to distinguish from leather-like wings. They also are used to cover and protect membranous wings. They are tougher than leather-like wings, however, and they typically cover almost the entire insect, rather than just the membranous wings as in the case of leather-like wings. Beetles, such as ladybugs, tend to have horny wings.

The Basic Anatomy of an Insect

Since there is so much variety in class Insecta, it is hard to pull out a "representative" insect to study. Nevertheless, we will go ahead and use the grasshopper as our representative insect. The figure below contains sketches of the outside and inside of a grasshopper.

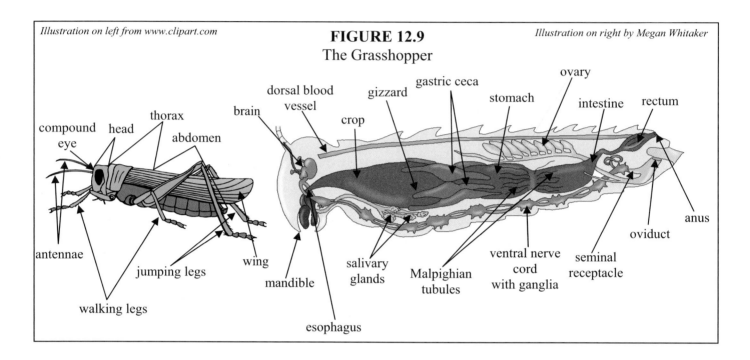

Illustration on left from www.clipart.com   **FIGURE 12.9**   *Illustration on right by Megan Whitaker*
The Grasshopper

Respiration and Circulation in Insects

One of the most interesting aspects of insects is the fact that even though they have *no respiratory system*, they get plenty of oxygen. How is this possible? Well, insects have an elaborate system of interconnecting tubes called **tracheas** (tray key' uhs). These tubes are connected to the outside through a series of small holes in the exoskeleton called spiracles. The network of tracheas is so complex and thorough that air runs throughout the body, providing oxygen to all tissues! That's why you see no lungs in Figure 12.9. Air goes directly to the tissues, where oxygen and carbon dioxide are directly exchanged with the cells.

If insects have no respiratory system, why do they have a circulatory system, as is evidenced by a dorsal blood vessel? Well, cells need more than just oxygen to survive, and they need to expel more than just carbon dioxide. Thus, as the blood flows over the tissue (remember, this is an open circulatory system), it picks up other cellular waste products and delivers vital substances other than oxygen (such as food). Where do the waste products go? Well, insects don't have green glands, but they do have **Malpighian** (mal pig' hee ahn) **tubules**, which reside in the vicinity of the intersection between the stomach and the intestine. When blood flows over the Malpighian tubules, it is cleaned of the waste products, and they are put into the intestine for elimination through the anus.

Another interesting aspect of insects is that many of them *have both simple and compound eyes* as a part of their nervous system. Typically, the compound eyes actually provide vision, while the simple eyes just look for the presence of light. The signals from both sets of eyes are fed into the brain, which is attached to the ventral nerve cord. This makes up the "backbone" of the nervous system. Besides the eyes, the nervous system also gets information from tactile hairs that provide touch, taste, and smell information. Some insects even have a sense of hearing, but the mechanism for detecting sound is not where you would expect. In a grasshopper, for example, there is a vibrating membrane (called the tympanic membrane) attached to the ventral nerve cord. The vibrations of this membrane provide a sense of hearing. Rather than being on the head, however, it is actually in the first segment of the abdomen!

The Feeding Habits of Insects

The grasshopper has a **mandible** which is designed for chewing. Although this is the case for the majority of insects, there are many different mouth structures in the insect world. This is because there are many different feeding habits apparent in class Insecta. Mosquitoes, for example, do not have mandibles. Instead, they have mouthparts designed to puncture and then suck. Flies, on the other hand, have an almost spongelike mouth that is used to absorb food. Finally, because butterflies primarily eat nectar from flowers, they have a mouth designed to siphon.

Once food has entered the mouth, however, most of the differences between insects disappear. The food is mixed with secretions from the salivary glands. Enzymes and water in these secretions begin breaking down the starches that the insect has ingested. The food then passes through the esophagus and goes on to the crop, where it might be stored for later use. Once the food is needed, it is sent to the gizzard, where it is ground into fine particles. The gizzard empties into the stomach, where it is mixed with digestive enzymes which come from the **gastric ceca** (see' kuh). Undigested food then passes through the intestine, to the rectum, and out the anus.

Reproduction and Development in Insects

Perhaps the most extraordinary thing about insects is the means by which they develop from a fertilized egg into an adult. As is the case for almost all arthropods, the female and male sexes are completely separate in class Insecta. Females receive sperm from males during mating but, like the crayfish, store it in seminal receptacles for a time. When the female lays her eggs, they are fertilized by the stored sperm. We call this stage of the new insect's development the **egg stage**. Once the egg hatches, things get really interesting.

When they hatch, most insect young are in a stage called the **larva stage**. In this stage, the young insect, regardless of its species, tends to resemble a segmented worm. It eats and molts over and over again, eventually entering the **pupa stage**. In this stage, the insect forms some sort of case around itself. The case might be formed of exoskeleton, or it might be woven from filaments. During this stage, *everything* changes. The organs are rearranged and reshaped, body structures are dismantled and reformed, and an amazing transformation occurs. When the transformation is complete, the insect breaks out of its case and enters the **adult stage** of life. In the adult stage, the insect has all of the features and organs that are normally associated with its species.

Now, of course, you should have heard about this process before. It is called **metamorphosis** (met uh mor' fuh sis) and is usually discussed in terms of the butterfly. When a butterfly's eggs hatch, the young are called caterpillars. They resemble segmented worms at this point. This, then, is the larva stage for the butterfly. Eventually, the caterpillar weaves a **chrysalis** (krih' suh lis). This is the casing that forms in the pupa stage. When the insect emerges from the chrysalis, it is an adult butterfly. Moths also go through a very similar form of metamorphosis, but the larva of a moth typically forms a cocoon rather than a chrysalis. What you might not have been aware of until now is that the *vast majority* of insects go through this transformation process. Flies, for example, go through the same process. When they are in their larva stage, they are called maggots, and when they are in their pupa stage they do not use a cocoon or chrysalis but rather a shell of exoskeleton.

In fact, if an insect does not go through the metamorphosis process described above, it goes through a similar one that involves only three stages. Thus, biologists classify the development of an insect as either **complete metamorphosis** or **incomplete metamorphosis**.

Complete metamorphosis - Insect development consisting of four stages: egg, larva, pupa, and adult

Incomplete metamorphosis - Insect development consisting of three stages: egg, nymph, and adult

When an insect develops through incomplete metamorphosis, it hatches from its egg stage into its **nymph stage**. In this stage, the insect looks like a miniature version of its adult form, but the proportions seem wrong. It lacks wings and reproductive organs. It molts several times during the nymph stage and, when it finally develops wings and reproductive organs, it is considered an adult. Examples of both types of metamorphosis are given in the figure below.

FIGURE 12.10
Complete and Incomplete Metamorphosis

A praying mantis exhibits incomplete metamorphosis. The egg hatches a nymph (left), which becomes an adult (right).

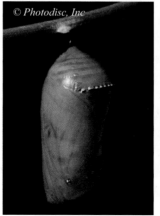

A monarch butterfly exhibits complete metamorphosis. The egg hatches a larva (left), which forms a chrysalis to become a pupa (middle). The adult (right) emerges from the chrysalis.

Now please realize that there is nothing "incomplete" about incomplete metamorphosis. It is a full life cycle in which each stage of life is exquisitely designed. We simply use the term "incomplete" to signify that it has fewer stages than the other type of metamorphosis that is seen in class Insecta.

ON YOUR OWN

12.12 An insect's outer wings are incredibly tough. Most likely, what kind of wings are they?

12.13 You can suffocate an insect by wrapping up its body, except for the head, in plastic wrap. Why, if the mouth is exposed to air, does the insect still suffocate?

12.14 An insect cannot digest food in its stomach due to a lack of digestive enzymes. Which organ is most likely not working?

12.15 An insect goes through a nymph stage in its development. Does it undergo complete or incomplete metamorphosis?

A Few Orders in Class Insecta

Class Insecta contains many different orders. Obviously, then, there is simply no way to go through them all. Nevertheless, to give you some feel for the diversity of life in this class, we need to at least touch on some of the major orders.

Order Lepidoptera: The Butterflies and Moths

Since we were just discussing the developmental process in insects, and since that process is usually associated with butterflies, we might as well start our insect order discussion with order **Lepidoptera** (lep uh dahp' tur uh). As we mentioned before, butterflies and moths (in their adult stage) have scaled wings. This is the major characteristic that sets them apart from most other orders in class Insecta.

Do you know the difference between a butterfly and a moth? Well, there are differences, but they are rather subtle. In its adult stage, the antennae of a moth usually look feathery, while the antennae of a butterfly look like straight stalks with knobs on the end. Also, the body of the adult is much slimmer in butterflies than it is in moths. Finally, when butterflies rest on the ground or sit on top of a plant, they typically hold their wings up vertically. Moths, on the other hand, tend to hold their wings out horizontally when resting.

Although butterflies and moths can be quite beautiful, they are not always the most desirable things to have around. Many butterflies eat ravenously while in their larva stage. Some have been known to cause millions of dollars' worth of crop damage every year. Also, if certain types of moth lay their eggs in your closet, the larvae that hatch can destroy the clothes stored there. That's why many people put mothballs in their closets. Mothballs are small samples of a chemical called napthalene. They emit an odor that repels moths so that when they are placed in a closet, moths will not enter the area and lay eggs.

Order Hymenoptera: Ants, Bees, and Wasps

Scaled wings, of course, are not the only kinds of wings in class Insecta. Members of order **Hymenoptera** (hi muh nahp' tur uh) have membranous wings. In fact, "hymen" means "membrane" and "ptera" means "wings" in Greek. Now that statement might confuse you in light of the fact that the title of this subsection includes ants as being a part of order Hymenoptera. Ants don't have wings, do they? Well, it turns out that they do, but you rarely see them. You'll learn why in a moment. Members of this order usually have stingers, too.

All species within order Hymenoptera are what biologists call **social insects**, which sets them apart from all other orders except Isoptera, which contains the termites. They exist within a society in which they have very particular functions. The best example of this is bees. In a hive, bees belong to one of three groups: queen, drone, or worker. The vast majority of bees in the hive are worker bees. They are actually female bees that do not have reproductive capabilities. Instead, their egg-laying organ, the ovipositor, is a barbed stinger. When they sting an enemy, they release a poison with the stinger. The barb in the stinger, however, lodges in the victim, making it impossible for a bee to remove its stinger. Thus, the bee will rip off its last abdominal segment, leaving the stinger behind. This increases the effect of the poison, but it kills the bee. As their name implies, worker bees do all of the work. They build and maintain the hive, collect food for the hive, care for the eggs, care for the larvae once the eggs hatch, and protect the queen.

The queen bee is the only female with reproductive capabilities. Because she can lay eggs, however, her stinger is not barbed. Thus, she can sting as many times as she wants. A given hive has only one queen, so workers protect her at all times. If the queen dies, the hive is out of luck, right? Wrong. You see, while the queen is alive, she produces a chemical that biologists call "queen substance." This substance is transferred to all workers, and it attracts workers to her. When the queen dies, the workers notice that they are no longer being given that substance, so they feed some of the developing larvae a special high-protein food that they secrete. This causes those larvae to develop reproductive organs. The first one to enter adult stage is the new queen, and the rest are killed.

The drone bees are the only males in the hive. They are useless for all situations except mating. Their mouths are not designed to gather nectar from flowers, so they cannot make honey. They also do not help maintain the hive. They simply wait their turn to mate with the queen. Since drones are good only for reproductive purposes, worker bees have been observed killing them or pushing them out of the hive when food becomes scarce!

As you no doubt are aware, bees make honey. What is honey and why do they make it? Honey is partially digested nectar that the bees have sucked out of flowers. It serves as food for the queen and the drones, and it is also a backup supply of food for the workers when they cannot find nectar. The honey is stored in honeycombs, hexagonal storage chambers made out of wax that is secreted by (you guessed it) the workers. The workers then fan the nectar with their wings in order to evaporate excess water, concentrating the partially digested nectar until it becomes honey. It turns out that no engineer could have designed a better system for the storage of honey. The hexagonal shape of the storage chambers allows them to store a large amount of honey with a minimal amount of wax. The wax that the bees use to make these storage chambers is so strong that two pounds of wax can hold 50 pounds of honey! Bees are so industrious that they make far more honey than they need under normal conditions, so people (and some animals) take advantage of this by using the honey for themselves.

Now stop and think about this for a moment. The queen controls her "subjects" with a special chemical. When that chemical is no longer supplied, the workers "know" that they need a new queen, so they produce another substance that, when fed to the developing larvae, forms a new queen. At the same time, these workers store honey in a structure that science tells us is *perfectly* engineered for the task. If honey runs short, the workers also somehow "know" who among them (the drones) are expendable. Isn't that amazing? This is just one more of the incredibly designed systems in creation that simply shout out the glory of God!

We still haven't told you why you don't see wings on ants, have we? Well, you needed to understand the concept of social groups in insects first. The ants that you see most of the time are worker ants, which do not have reproductive capabilities. These ants do not have wings. Only ants that can reproduce have wings. They never venture out of the anthill, so you never see them. Even if you were to see them, you might still not see wings. The male ant dies shortly after mating, and the female ant loses her wings once she mates. Thus, only a small portion of any given ant species has wings, and those individuals either don't live long or don't keep their wings long. No wonder you never see them!

Order Coleoptera: The Beetles

No, we're not going to discuss the "Fab Four" here. (If you don't know who the "Fab Four" are, ask your parents.) Instead, we are going to discuss the largest order within class Insecta (indeed, the largest order in the animal kingdom), order **Coleoptera** (ko lee ahp' tur uh). The name of this order literally means "sheath wing" in Greek. That is a very adequate description of what sets it apart from most other orders. All members of this order have horny wings. These wings are thick, sometimes colorful, and they typically cover the creature's entire body, protecting the membranous wings underneath. The horny wings make beetles look like little war machines, and that look is often enhanced by horns on their head or extended mandibles that can viciously cut through their enemies.

Beetles have a voracious appetite. Sometimes that is a good thing, and sometimes it is a bad thing. Many beetles feed on insects that damage crops. The ladybug, for example, feeds exclusively upon aphids and other crop-destroying insects. The fact that the ladybug has a huge appetite works out all the better for farmers. There are some beetles, however, (the Japanese beetle and the rice weevil, for example) that destroy crops themselves. The fact that they have a huge appetite is not good for farmers!

One particularly interesting beetle that we want to discuss is called the bombardier beetle. This ugly little beetle has one of the most beautiful defense mechanisms in all of creation: a fully equipped chemical weapon. This weapon begins with storage vessels that contain a mixture of two chemicals: hydroquinone and hydrogen peroxide. Under normal conditions, these chemicals would react, but while stored in the bombardier beetle's vessels, they are kept from reacting by the presence of a third chemical which inhibits the reaction. When the bombardier beetle feels threatened, however, it fills an empty reaction chamber in its body with the chemicals. Two other chemicals, catalase and peroxidase, are then added. These chemicals cause the hydrogen peroxide and hydroquinone to react violently. The violent reaction produces a great amount of heat and pressure in the reaction vessel. The beetle then points its tail in the direction of trouble and opens a valve between the reaction vessel and the tail. A jet of steam that has a temperature of roughly 200° F shoots out the tail in the direction of danger. Any potential predator is immediately burned and frightened away! The bombardier beetle can perform this feat up to 20 times per day! The bombardier beetle is just another in the long list of organisms on this planet that tell us over and over again that this world was *designed*.

Order Diptera: Flies, Gnats, and Mosquitoes

We now come to the order with the most annoying creatures, order **Diptera** (dip' tur uh). All organisms in this order have a pair of membranous wings that they use to fly. Their second pair of wings is much smaller and is used as stabilizers during flight. Some research indicates that they might be used to determine the insect's air speed as well. Members of this order have no other defense mechanism except the ability to get away quickly. Their exoskeleton is rather weak; they have no stingers; and their mouths are designed to either pierce or suck, not to bite. Thus the only way that these creatures stay alive is to fly away. Given the fact that they are so prevalent on warm, sticky days, they obviously are good at it!

Flies are pests, but they can actually be quite dangerous. The average housefly carries millions of bacteria and viruses in its intestinal tract and on its body. When a fly lands on a food source, it eats by secreting digestive juices onto the food and then sponging it back up. Many of the microorganisms and viruses it carries are transferred to the food in that process. Thus, if a fly lands on your food and begins to eat it, you run the risk of being infected by those microorganisms and viruses, some of which are pathogenic to humans.

Mosquitoes also carry potentially pathogenic microorganisms and viruses. In Module #3 we studied how the *Plasmodium* sporozoan, which causes malaria, is carried by the mosquito. Mosquitoes also carry some types of parasitic worms and the yellow fever virus. These creatures are quite effective at infecting you, because they feed on your blood. Their mouthparts are perfectly designed to taper into four needles that easily pierce the skin. Since your blood tries to clot in order to prevent bleeding, the mosquito then injects a chemical that counteracts the blood's clotting mechanism. This keeps the blood flowing freely, allowing the mosquito to get all that it wants. The problem is, of course, that any pathogens which the mosquito carries get mixed into your bloodstream along with the chemical that keeps the blood from clotting!

Order Orthoptera: Grasshoppers and Crickets

The last order that we want to discuss is order **Orthoptera** (or thahp' tur uh). We will not spend much time on it, though, because we have already made an in-depth study of the grasshopper, one of its members. Members of this order have one pair of leather-like wings that cover and protect their membranous wings. They typically have a pair of legs that are quite a bit larger than the other two pairs of legs, and they use them for jumping.

Grasshoppers and crickets are well known for their chirping, which is technically called "stridulation" (strid yoo lay' shun). In order to stridulate, a grasshopper has a row of small pegs on the largest joint of each of its hind legs. These pegs are rubbed over the veins of the wings, causing the sound that we call "chirping." Usually, only the male stridulates. Each species has its own unique song, so stridulation serves to attract females to males for the purpose of mating.

To review what you have learned in this section, perform the following experiment.

EXPERIMENT 12.2
Insect Classification

Supplies:

♦ Laboratory notebook
♦ Specimens in the pictures below
♦ Specimens collected from another source (In order to get more practice at recognizing the orders discussed in this book, you should try to find more examples of insects. Depending on the time of year and the area of the country that you live in, you may be able to collect some on your own. If you cannot, a library or museum might have an insect collection that you can study. If you know anyone in 4-H, they most likely have access to an insect collection. If an insect doesn't fit into one of the five orders that we have discussed here, do some research and see what order it does belong in!)

Object: To become familiar with classifying insects

Procedure: In your notebook, write down the letter of each specimen and the order to which it belongs. The orders for the specimens given below are listed in the solutions and tests guide, after the answers to the study guide for this module.

ANSWERS TO THE "ON YOUR OWN" PROBLEMS

12.1 <u>Yes, it would still need an exoskeleton.</u> Exoskeletons provide support for the body. Without it, the arthropod would be nothing more than a mass of tissue and organs.

12.2 Since the cephalothorax is a combination of the head and the thorax, only the abdomen is left. Thus, it would have <u>two major segments</u>.

12.3 <u>The organism is not an arthropod.</u> The muscles in an arthropod are inside the exoskeleton. Thus, the joints are controlled from the bottom.

12.4 Since antennae provide the taste sensation, <u>the antennae must not be working</u>.

12.5 <u>Since arthropods have an open circulatory system, blood is constantly leaving the arteries and flowing directly on the tissue.</u>

12.6 In order to move fresh, oxygen-rich water into the gill chambers, the swimmerets and maxillae move back and forth. Thus, <u>the swimmerets and maxillae must not be working</u>.

12.7 <u>The blood collects in the pericardial sinus and then proceeds into the heart. The heart pumps the blood through open-ended blood vessels that dump the blood onto the tissues. It then collects in the sternal sinus and is picked up by vessels that take it to the gills to absorb oxygen and get rid of carbon dioxide. The blood also gets cleaned in the green glands and then goes back to the pericardial sinus to start the process all over again.</u>

12.8 <u>When the claw falls off, a membrane closes to prevent bleeding. Then, the claw grows back over time.</u>

12.9 Since the statocyst keeps the crayfish's balance and tells it which way is up, <u>the statocyst must be malfunctioning</u>.

12.10 Most spiders have poison glands, but the poison they make is rarely harmful to humans because it is designed to subdue small prey. <u>Thus, we need not fear most spiders as poisonous.</u>

12.11 Since arachnids have a cephalothorax, the head and thorax are one segment. Notice in the figure that there is a "pinched" region of the body near the center. That marks the change from the cephalothorax to the abdomen. Thus, <u>the abdomen is posterior to the pinched region, and the cephalothorax is anterior to the pinched region</u>.

12.12 Horny wings are the toughest. Thus, <u>they are probably horny wings</u>.

12.13 Insects get air through tiny pores (spiracles) in their exoskeleton, not through their mouths. Thus, it is irrelevant that the mouth is free. <u>The insect suffocated because it could not take in enough air through the spiracles, which were blocked by the plastic.</u>

12.14 It is the gastric ceca's job to secrete digestive enzymes. Thus, <u>the gastric ceca are not working</u>.

12.15 The nymph stage is only present in <u>incomplete metamorphosis</u>

STUDY GUIDE TO MODULE #12

1. Define the following terms:

a. Exoskeleton
b. Molt
c. Thorax
d. Abdomen
e. Cephalothorax
f. Compound eye

g. Simple eye
h. Open circulatory system
i. Statocyst
j. Gonad
k. Complete metamorphosis
l. Incomplete metamorphosis

2. Name the five common characteristics among the arthropods.

3. Identify the structures in the following diagram:

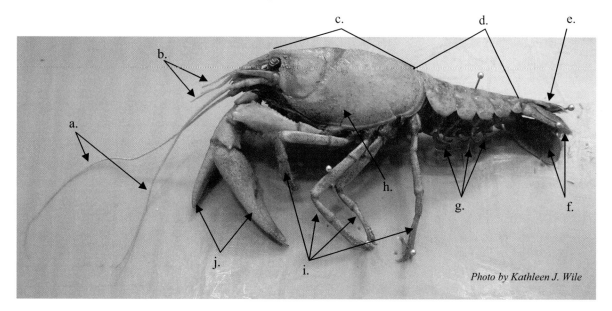

Photo by Kathleen J. Wile

4. Identify the organs in the following diagram:

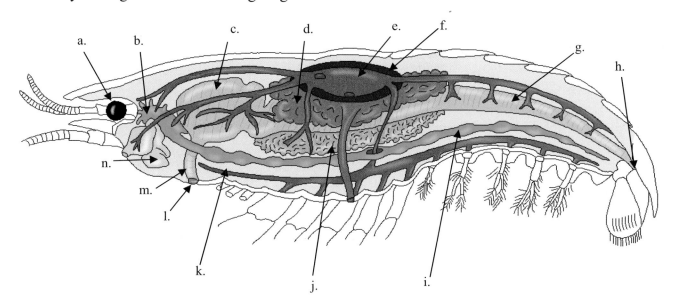

5. Explain the flow of blood in a crayfish, starting from the pericardial sinus.

6. What purpose does the green gland serve?

7. What structures (besides the gills and gill chamber) are vitally important for respiration in a crayfish?

8. What happens when a crayfish loses a limb?

9. Where do the fertilized eggs of a crayfish go?

10. Why do arthropods molt?

11. What two appendages are responsible for taste and touch in a crayfish?

12. What five characteristics set arachnids apart from the other arthropods?

13. What are the three basic types of webs that spiders spin?

14. Do all spiders use their silk to spin webs?

15. Why are the spider's lungs called book lungs?

16. What four characteristics set insects apart from the other arthropods?

17. Why don't insects have respiratory systems?

18. If an insect goes through a pupa stage, does it perform complete metamorphosis or incomplete metamorphosis?

19. What four types of wings exist among insects?

20. For each letter below, indicate the order of insects being described:

    a. Insects with two leather-like wings and two membranous wings
    b. Social insects with membranous wings
    c. Insects with two membranous wings and two membranous balancers
    d. Insects with two horny wings and two membranous wings
    e. Insects with scaled wings

MODULE #13: Phylum Chordata

Introduction

After spending two modules concentrating on the invertebrates of kingdom Animalia, it is now time to start talking about the vertebrate animals. Although there are many, many more invertebrates than vertebrates, most of the animals with which you are familiar are vertebrates. As a result, we should spend some time discussing them.

The vertebrates of kingdom Animalia are all found in one phylum: phylum **Chordata** (kor dah' tuh). From fish to reptiles to mammals, all animals that have some kind of backbone are a part of this phylum. What do we mean by "some kind of backbone?" Well, most members of phylum Chordata have **vertebrae** (ver' tuh bray – singular is vertebra).

Vertebrae – Segments of bone or some other hard substance that are arranged into a backbone

When you think of a backbone, this is typically what you imagine. As a result, the term "vertebrate" is generally used to refer only to creatures that have vertebrae. Not all members of phylum Chordata have such an obvious backbone, however. Some creatures in this phylum have a **notochord** (noh' tuh kord) instead.

Notochord – A rod of tough, flexible material that runs the length of a creature's body, providing the majority of its support

In fact, all members of phylum Chordata have a notochord at some point in their development. While some have a notochord throughout their entire life, others have it only in their larval stage. Many chordates have notochords only in the earliest stages of their development. For such an animal, the notochord turns into vertebrae before it is born or hatched.

The fate of the notochord in an organism actually provides the first level of classification in phylum Chordata. You see, this phylum contains so much diversity that it must be split into three subphlya. Subphylum **Urochordata** (yoor' uh kor dah' tuh) contains those creatures that have a notochord through the larva stage but then the notochord actually disappears in the adult stage of their life. Subphylum **Cephalochordata** (sef' uh loh kor dah' tuh) holds those organisms that have notochords throughout their entire life. Finally, subphylum **Vertebrata** (vurt uh braht' uh) is made up of those vertebrates that have a notochord during their early development, but it turns into a true backbone before the animals are born or hatched. This last subphylum is the group usually referred to as "vertebrates."

Some kind of backbone is not the only common characteristic among organisms within phylum Chordata. All members of this phylum also have a **dorsal nerve cord**. Remember, the arthropods we studied in the previous module had a ventral nerve cord. Thus, chordates have their nerve cord in precisely the opposite location – on top of the body. In arthropods, the nerve cord is placed on the lower portion of the body so that, in addition to the exoskeleton, the entire upper part of the body protects it. As a result, arthropods are always reluctant to expose their undersides, because the nerve cord has the least protection from that side of the body. Chordates, on the other hand, have their nerve cord on the dorsal side of the body. This works fine for most chordates because, in the case of vertebrates, the vertebrae actually encase the nerve chord, providing optimum protection.

Subphylum Urochordata

Subphylum Urochordata is composed of those chordates that have a notochord only during the larval stage of their development. The best example of such a creature is the strange little **sea squirt**. This interesting marine-dwelling creature gets its name from the fact that in its adult stage, it squirts water through a siphon when it is disturbed. Some biologists refer to sea squirts as "tunicates," because in their adult stage, they cover themselves in a leathery "tunic" that they secrete. The rather interesting life cycle of a sea squirt is illustrated in the figure below.

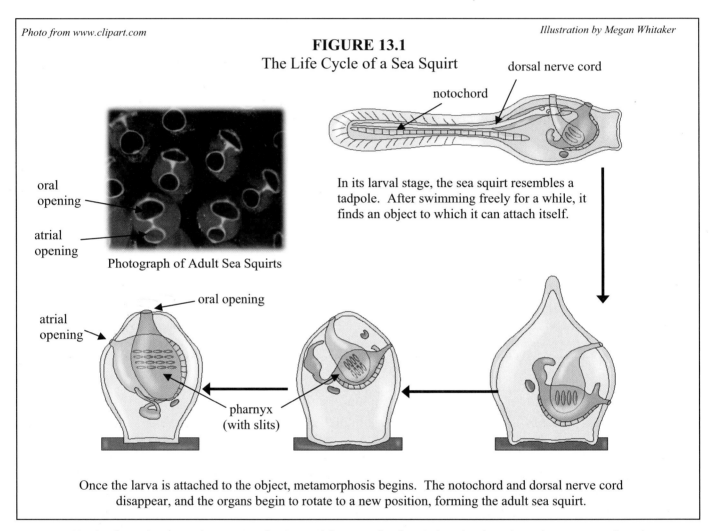

Photo from www.clipart.com   Illustration by Megan Whitaker

FIGURE 13.1
The Life Cycle of a Sea Squirt

dorsal nerve cord

notochord

In its larval stage, the sea squirt resembles a tadpole. After swimming freely for a while, it finds an object to which it can attach itself.

oral opening

atrial opening

Photograph of Adult Sea Squirts

oral opening

atrial opening

pharnyx (with slits)

Once the larva is attached to the object, metamorphosis begins. The notochord and dorsal nerve cord disappear, and the organs begin to rotate to a new position, forming the adult sea squirt.

In its larval stage, the sea squirt resembles a tadpole. It has both a dorsal nerve cord and a notochord. It swims around for a brief period, and then it attaches itself to something on the bottom of the sea. Once attached, the sea squirt begins to develop into its adult form. The nerve cord and notochord are recycled to form new tissues, the organs rotate so that the oral and atrial openings point upwards, and a leathery "tunic" is formed around the animal's exterior for protection and support.

Much like a sponge, the adult sea squirt obtains food by siphoning water into its body through the oral opening and expelling it through the atrial opening. Phytoplankton and zooplankton that come in with this water are filtered out by the slits in the pharynx and then digested as food. Once again, like a sponge, the adult sea squirt does not move. It is firmly attached to whatever structure the larva chose.

ON YOUR OWN

13.1 If the sea squirt feeds like a sponge and stays fixed to a surface like a sponge, why isn't it classified in phylum Porifera?

13.2 What structure in the adult sea squirt performs the function that the notochord performs in the larva?

<u>Subphylum Cephalochordata</u>

Unlike the members of subphylum Urochordata, the organisms in subphylum Cephalochordata retain their notochord throughout their entire life. A representative organism of this subphylum is the **lancelet**. This marine-living creature, also called "amphioxus" (am fee ox' us), is tapered at both ends, looking something like an eel. A sketch of a lancelet is shown in the figure below.

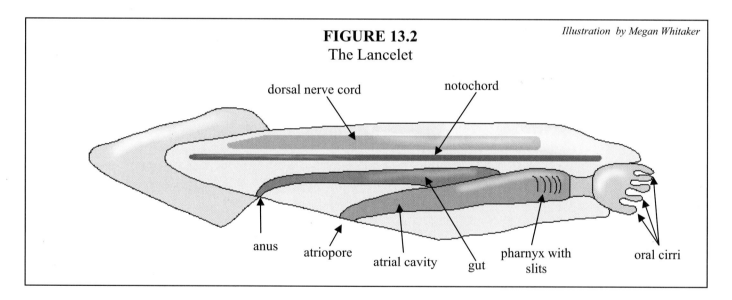

FIGURE 13.2
The Lancelet

Illustration by Megan Whitaker

This interesting little creature (usually less than three inches long) possesses a dorsal nerve cord along with the notochord that makes it a part of this subphylum. Its sensory organs are not as sophisticated as those found in most other creatures in this phylum. As a result, it tends to spend the majority of its time buried in the sand. This allows it to hide from predators without hampering its ability to obtain food. Like the sea squirt, the lancelet takes in food by filtering out the phytoplankton and zooplankton in the water.

Water is drawn into the creature by the beating of cilia located just inside the mouth. This water is first filtered through the **oral cirri** (sear' eye), which clean it of large, indigestible debris (such as the sand in which the lancelet is buried). As the water passes into the pharynx, it travels through slits like those of the sea squirt. The food particles in the water are trapped in the slits, and the water flows out into the **atrial** (ay' tree uhl) **cavity**, eventually leaving the body through the **atriopore** (ay' tree oh por). The food particles that have been trapped in the slits of the pharynx are sent into the gut, where they are digested. Any undigested remains leave through the anus.

Although a relatively unknown organism in the United States, the lancelet is very common in tropical, marine environments. At Discovery Bay, Jamaica, for example, biologists have reported populations of up to 5,000 lancelets per square yard of sand! In many parts of the world, particularly in certain regions of Asia, the lancelet is actually a very important food item. There are some fishermen who make their living solely by harvesting lancelets.

ON YOUR OWN

13.3 A dead lancelet is found to have large particles of sand lodged in the slits of its pharynx. What, most likely, was not functioning properly prior to the lancelet's death?

Subphylum Vertebrata

We now turn our attention to the subphylum that contains the true vertebrates. These organisms have a backbone formed by vertebrae from the day they are born or hatched. There are many, many different creatures in this subphylum. There are so many, in fact, that we cannot discuss them all in this module. As a result, we will revisit this subphylum again in Module #16, where we will discuss reptiles, birds, and mammals. In what remains of this module, we will talk about fish and amphibians.

Before we talk about the individual kinds of organisms in subphylum Vertebrata, however, it is important to discuss the common features that you find among the many classes of organisms that compose this diverse subphylum. Although creatures like fish, amphibians, reptiles, birds, and mammals may seem quite different, they actually have many similarities. It is these similarities that allow us to classify them all in one subphylum.

The Endoskeleton

Unlike the arthropods, which support their bodies with an exoskeleton, vertebrates get their bodily support from an **endoskeleton** (en' doh skel' uh tuhn).

Endoskeleton – A skeleton on the inside of a creature's body, typically composed of bone or cartilage

Now this definition does us little good if we aren't clear on what **bone** and **cartilage** (kar' tuh lij) are.

When you think about bone, what comes to mind? You probably think about a hard, rocklike substance. Believe it or not, however, bone is composed of *living cells* surrounded by a hard substance, which is often called **bone matrix**. Indeed, there are three types of bone cells in the typical vertebrate: **osteoblasts** (ah' stee oh blasts), **osteocytes** (ah' stee oh sytes), and **osteoclasts** (ah' stee oh klasts). Osteoblasts are cells that promote the formation of bone tissue by producing the bone matrix. Osteocytes are mature bone cells that are surrounded by the bone matrix. Bones actually grow and reshape as the vertebrate grows and as its support needs change. Thus, osteoblasts continue to produce new bone tissue. However, sometimes bone tissue must be destroyed as well. Often, this is a part of reshaping the bones to meet new demands placed on the endoskeleton. Other times, a lack of minerals such as calcium will result in bone tissue being destroyed in order to liberate minerals that are stored there. Breaking down bone tissue for these purposes is the job of the osteoclasts. A typical vertebrate bone is illustrated in the figure below.

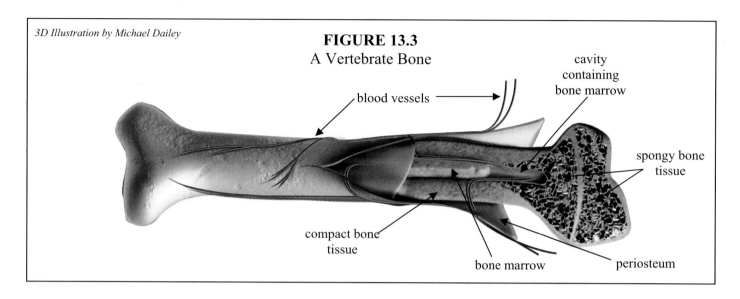

3D Illustration by Michael Dailey

FIGURE 13.3
A Vertebrate Bone

cavity containing bone marrow

blood vessels

spongy bone tissue

compact bone tissue

bone marrow

periosteum

Bones in vertebrates come in all shapes and sizes. They are mostly composed of two different kinds of tissue: **compact bone tissue** and **spongy bone tissue**. Both of these tissues are composed of a protein called **collagen** mixed with calcium-containing salts that harden the tissue. This gives bone its hard, rocklike feel. Within the woven, hardened, collagen fibers, the osteocytes are housed. The main difference between compact bone tissue and spongy bone tissue is how these calcium-hardened fibers are packed together. In compact bone tissue, they are packed together tightly, forming a hard, tough structure that can withstand strong shocks. In spongy bone tissue, on the other hand, the fibers are packed loosely, with a lot of space in between. This gives the tissue a spongy look. Although the tissue looks spongy, it is quite hard. It provides support to the bone without adding a lot of weight.

The compact and spongy bone tissue are both surrounded by a dense membrane called the **periosteum** (pehr' ee ah' stee uhm). This layer contains blood vessels that supply oxygen and nutrients to the cells in the bone. The blood vessels travel through a series of interconnecting canals that are woven throughout the bone, ensuring that all of the bone cells are adequately supplied with oxygen and nutrients. The periosteum also contains nerves. When you break a bone, for instance, these nerves send pain signals to your brain.

In the very center of the bone is a cavity that holds **bone marrow**.

Bone marrow – A soft tissue inside the bone that produces blood cells

So we see that the bones of a vertebrate do more than just support the body. Without bone marrow, vertebrates would have no blood cells! In addition, bone tissue is a repository for excess minerals that are vital to the chemistry of vertebrates. If a vertebrate runs low on these minerals, they can be extracted from the bone tissue through the activity of the osteoclasts.

Now that you know what bone is, what is cartilage? It is composed of collagen fibers like bone tissue, but the collagen in cartilage is slightly different from that of bone. In addition, cartilage is not reinforced with calcium salts. This makes it more flexible than bone, but it also is weaker. When most vertebrates develop prior to birth or hatching, the beginnings of their endoskeleton are formed as cartilage. Later on, this cartilage is replaced by bone tissue. Some vertebrates, however, never replace the cartilage in their endoskeleton with bone. It simply stays as cartilage. This gives them more flexibility in their endoskeleton, but it also provides less protection and support. Even vertebrates that replace cartilage with bone still have some cartilage. It functions as a connective tissue, helping keep the various bones of the skeleton connected. It also cushions bones in certain joints.

Now that we know a little bit about bone and cartilage structure, we need to look at the endoskeleton a little more globally. The skeleton of a typical human being is shown in the figure below.

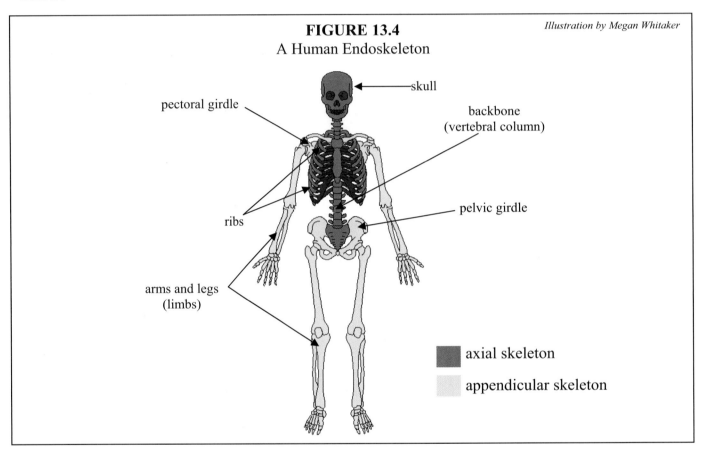

FIGURE 13.4
A Human Endoskeleton

Illustration by Megan Whitaker

Like most vertebrates, humans have endoskeletons that can be split into two major sections: an **axial skeleton** and an **appendicular** (ah pen dihk' you ler) **skeleton**.

Axial skeleton – The portion of the skeleton that supports and protects the head, neck, and trunk

Appendicular skeleton – The portion of the skeleton that attaches to the axial skeleton and has the limbs attached to it

As you can see from the figure, the backbone (often called the **vertebral column**), ribs, and skull make up the axial skeleton (shaded blue in the figure). On the other hand, the **pectoral girdle** and the arms which attach to it, as well as the **pelvic girdle** with the legs which attach to it, make up the appendicular skeleton (shaded beige in the figure). For creatures that have four legs rather than two arms and two legs, the pectoral girdle is the portion of the skeleton to which the anterior limbs attach, while the pelvic girdle is the portion of the skeleton to which the posterior limbs attach.

The Circulatory System

Unlike the open circulatory system that is found in arthropods, vertebrates have a **closed circulatory system**.

Closed circulatory system – A circulatory system in which the oxygen-carrying blood cells never leave
                the blood vessels

A vertebrate's circulatory system begins with a heart that is composed of two, three, or four chambers. When we study some individual organisms later on in this module, we will learn more about heart chambers. The heart pumps blood into **arteries**, which carry it away from the heart. The arteries branch to all parts of the body, sending blood to the **capillaries** (kap' uh lehr eez). The capillaries are tiny, thin-walled blood vessels that allow oxygen to leave the blood and flow out into the tissue. At the same time, the capillaries allow carbon dioxide to enter the blood so that it can be carried away. During this time, however, the blood cells *never* leave the capillaries. Once the blood has given up oxygen and taken in carbon dioxide, it is sent back to the heart in **veins**.

Arteries – Blood vessels that carry blood away from the heart

Capillaries – Tiny, thin-walled blood vessels that allow the exchange of gases and nutrients between
           the blood and the cells of the body

Veins – Blood vessels that carry blood back to the heart

Vertebrate blood is red. This is because the cells that carry the oxygen in the blood are full of a protein called **hemoglobin** (hee' muh gloh' bun) that appears red. The hemoglobin in these cells holds onto the oxygen and allows it to be transported by the blood. These hemoglobin-containing cells are called **red blood cells**, and they give blood its color. When red blood cells have a lot of oxygen, they have a bright red color. When they are low on oxygen, they have a darker red color.

When we look at our skin, we typically see veins, because the veins are usually closer to the skin's surface than are the arteries. Well, since veins carry blood back to the heart, most (but not all) veins carry red blood cells that are low in oxygen. Thus, the blood in most veins is dark red. However, when we look at the veins in our skin, they appear to be blue. Even though they appear to be blue, they are not. They are actually dark red. The blue color is the result of the fact that dark red blood does not reflect light through tissue as well as bright red blood does. As a result, the blood in our veins, when viewed through the skin, appears to be a different color than it really is.

The Nervous System

Vertebrates have the most complex nervous systems of all animals in creation. Figure 13.5 on the next page illustrates the nervous system of a "typical" vertebrate.

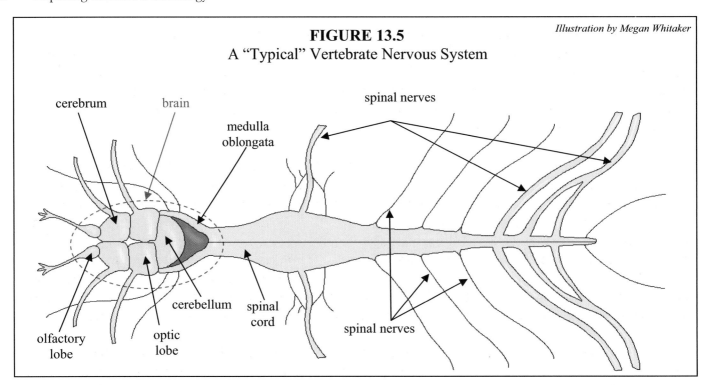

FIGURE 13.5
A "Typical" Vertebrate Nervous System

Illustration by Megan Whitaker

Although this is considered the layout of a "typical" vertebrate nervous system, you must realize that there is a great amount of diversity in subphylum Vertebrata. Thus, many vertebrates have nervous systems that are far more complicated than that shown here. Nevertheless, this is a good starting point for understanding how vertebrates sense and respond to stimuli.

The nervous system is controlled by the **brain**. Most vertebrate brains are segmented into five different types of lobes: **olfactory** (ohl fak' tuh ree) **lobes**, the lobes of the **cerebrum** (suh ree' bruhm), **optic lobes**, the **cerebellum** (sehr' uh bel uhm), and the **medulla oblongata** (muh dul' uh ahb lawn gah' tuh).

<u>Olfactory lobes</u> – The lobes of the brain that receive signals from the receptors in the nose

<u>Cerebrum</u> – The lobes of the brain that integrate sensory information and coordinate the creature's response to that information

<u>Optic lobes</u> – The lobes of the brain that receive signals from the receptors in the eyes

<u>Cerebellum</u> – The lobe that controls involuntary actions and refines muscle movement

<u>Medulla oblongata</u> – The lobes that coordinate vital functions, such as those of the circulatory and respiratory systems, and transport signals from the brain to the **spinal cord**

Notice from the figure that with the exception of one lobe, the **cerebellum**, the lobes come in pairs, each with a right lobe and a left lobe. When this was first discovered, biologists thought that each lobe simply processed information for that side of the body. They thought that the right cerebrum lobe coordinated responses to stimuli for the right side of the creature, and the left lobe did so for the left side. As our knowledge of brain function has increased, however, we have come to realize that it is much more complicated than that. In many vertebrates, the left lobes tend to control the *right* side of the body, while the right lobes tend to control the *left* side of the body. In addition, the lobes tend to do their jobs in slightly different ways. In humans, for example, mathematical and analytical reasoning as well as speech are influenced more by the left lobe of the cerebrum than the right lobe. The right lobe, on the other hand, has more influence over spatial perceptions and many of the functions that tend to affect creativity.

When the brain sends signals to different parts of the body in order to control bodily functions or coordinate responses to stimuli, the signals travel down the medulla oblongata and into the **spinal cord**. The spinal cord runs inside the vertebral column and sends signals to and from the brain. The spinal cord has many **spinal nerves** running out of it so that the signals can be sent to the place that the brain intended for them to go. The nerves also serve a second function. One set of nerves (called sensory nerves) transmits signals to the brain from the many receptors that are located throughout the body. This allows the brain to sense any changes that occur in the environment (either inside the body or outside the body). A second set of nerves (called motor nerves) takes the signals from the brain and sends them to the places that they need to go so the body will do what the brain wants it to do.

Think about what we have here. Receptors sense various stimuli from the environment. The receptors in the eyes sense light; the receptors in the nose sense gaseous chemicals; the receptors in the skin sense motion and touch, etc. These receptors then translate those stimuli into electrical signals that are sent to the brain. The brain processes those signals, figures out what they mean, and decides on a course of action. That course of action is then translated back into electrical signals, which are sent to a *specific set* of nerves in the body. If the brain wants the right arm to move, it sends those signals to nerves in the right arm. The signals go only to the nerves for which the instructions are meant. Those nerves then react to the signal and make the body perform according to the dictates of the brain.

Isn't that amazing? Vertebrates have a nervous system that analyzes and responds to stimuli better, faster, and more efficiently than any computer that human science can devise. This kind of processing occurs in *every vertebrate in creation*! Add to that the fact that humans actually have the ability to think originally, reason deductively, feel emotion, and pass knowledge from generation to generation, and you will soon appreciate how incredibly complex this all is. Clearly something more structured and efficient than the most complicated computer in the world would never have occurred by chance. The marvelous nervous system of the vertebrates is striking evidence that creation is the result of design!

Reproduction

In all vertebrates, the sexes are separate. Males have testes that produce sperm, and females have ovaries that produce eggs. The ways in which fertilization and development occur are many and varied, however. First, there are two types of fertilization: **internal fertilization** and **external fertilization**.

Internal fertilization – The process by which the male places sperm inside the female's body, where the eggs are fertilized

External fertilization – The process by which the female lays eggs and the male fertilizes them once they are outside of the female

While external fertilization occurs in most aquatic vertebrates, internal fertilization occurs in most land-dwelling vertebrates. Now please realize that just because a creature lays eggs, that does not mean it reproduces by external fertilization. Many vertebrates reproduce by internal fertilization and then, after the zygote has developed, lay an egg in which the rest of the pre-birth development takes place. You can always tell an externally-fertilized egg from an internally-fertilized egg, because the internally-fertilized egg has a shell. Externally-fertilized eggs do not.

Reproduction in subphylum Vertebrata is even more diverse because once fertilization occurs, development takes place in one of three ways: **oviparous** (oh vip' ur us), **ovoviviparous** (oh voh vye vip' ur us), or **viviparous** (vye vip' ur us).

Oviparous development – Development that occurs in an egg that is hatched outside the female's body

Ovoviviparous development – Development that occurs in an egg that is hatched inside the female's body

Viviparous development – Development that occurs inside the female, allowing the offspring to gain nutrients and vital substances from the mother through a **placenta** (pluh sent' uh)

Most birds, for example, are oviparous. They lay eggs that hatch outside of the body. Guppies, on the other hand, are ovoviviparous. They form eggs in which the offspring develop. The eggs, however, stay inside the mother until the young hatch. Most mammals are viviparous. There is an attachment between the mother and the offspring called the placenta. It takes nutrients and other vital substances (like oxygen) from the mother and gives them to the developing offspring. The young are then born live.

ON YOUR OWN

13.4 One way that we fight cancer today is through bone marrow transplants. In this procedure, the bone marrow of the sick person is replaced with bone marrow from a healthy person. What cells in the body are most affected by this procedure?

13.5 A shark's skeleton is much more flexible than a human's skeleton. What is a shark's skeleton made of?

13.6 If you see a sealed tube full of bright red blood, did it most likely come from a vein or an artery?

13.7 Blood is traveling back to the heart. Is it in a vein or an artery?

13.8 Although an animal's eyes seem to function properly, it still cannot see. What part of the brain is most likely not working?

13.9 If a person's spinal cord is cut, he loses the ability to move the limbs below the place in which the cut occurred. Why?

13.10 A viviparous mother eats plenty of food but the developing offspring is not nourished properly. What structure is not performing its job?

13.11 Fish typically reproduce when the female lays her eggs and then the male fertilizes them. What kind of fertilization is this? What kind of development is it?

Class Agnatha

Now that we've discussed the characteristics that vertebrates have in common, it is time to get a feel for the wonderful diversity that exists in subphylum Vertebrata. We start our tour of this interesting subphylum with class **Agnatha** (ag na' thuh). Commonly called the "jawless fish," the creatures in this class live in both freshwater and marine environments. Interestingly enough, creatures in this class are often **anadromous** (an uh drohm' us).

Anadromous – A life cycle in which creatures are hatched in fresh water, migrate to salt water as adults, and then go back to fresh water in order to reproduce

The most common example of a creature from class Agnatha is the **lamprey** (lamp' ray) **eel**. In the figure below, three views of adult lampreys are shown.

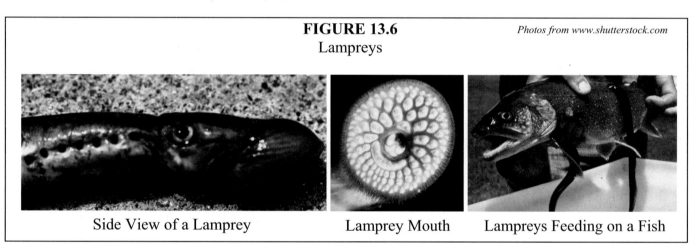

FIGURE 13.6
Lampreys

Photos from www.shutterstock.com

Side View of a Lamprey   Lamprey Mouth   Lampreys Feeding on a Fish

Some lampreys live in fresh water, and the rest are anadromous. When lamprey eggs hatch, the young are in larva form and are called **ammocoetes** (am' uh seetz). While in larva stage, the young behave and look very different from their adult form. They feed by producing strands of mucus that trap food particles floating in the water. They live as larvae for as many as seven years until metamorphosis takes place and they develop into an adult.

As an adult, the lamprey has a round, suckerlike, jawless mouth filled with horny teeth and a raspy tongue. A ring of cartilage supports the mouth, in place of a jaw. It feeds as a parasite, mostly preying on other vertebrates. It does so by attaching to the prey with its mouth. A very complex pumping mechanism in the mouth keeps the lamprey attached by suction, while it scrapes the skin of the prey with its tongue. It then sucks the blood that comes out of the skin as the result of this scraping. The lamprey will continue to feed like this until it is not getting enough blood, and it will then move on to another prey. The unfortunate creature to which it was attached is severely weakened by the loss of blood and usually dies.

The skeleton of the lamprey is made of cartilage. It never forms true bones. The skull is made of overlapping plates of cartilage that encase and protect the brain. The brain has a very small cerebellum but large optic lobes. This illustrates a common phenomenon among vertebrates. Many vertebrates tend to have one set of brain lobes larger than that of the "typical" vertebrate because it is designed to specialize in a particular function. Lampreys, for example, need good eyesight to find prey. As a result, the optic lobes are large, so that the lamprey can process visual information very effectively. Humans, for example, have a large cerebrum, because God has designed us to think. Since the cerebrum is where the processing of information occurs, we need large cerebral lobes.

Although lampreys are considered a delicacy in some parts of Europe, they are generally considered a nuisance. When allowed to breed in a region, they can devastate the local fish population. Since 1835, for example, lampreys have been making their way into the Great Lakes through man-made canals. By the 1940s, their population had grown to the point that they were devastating the population of fish that are the lifeblood of the commercial fisherman. As a result, the U.S. and Canadian governments invested a lot of effort in controlling the lamprey population. They used nets to catch adult lamprey and "poisoned" the water with a chemical that killed lamprey larva but was harmless to the fish population.

By the late 1980s, the lamprey population was under control and the fish populations began to grow again in the Great Lakes. Although the U.S. and Canadian governments claim credit for the lowering of the lamprey population in the Great Lakes, there are many biologists who think, ironically enough, that the pollution problems in the Great Lakes are what actually brought the lamprey population under control. It turns out that lamprey larvae are much more sensitive to pollution than other types of fish, due to their feeding method. As a result, there is a lot of evidence to suggest that the man-made pollution counteracted the effect of the man-made canals and actually saved the commercial fish populations in the Great Lakes!

ON YOUR OWN

13.12 Are there any lampreys that reproduce in marine environments?

13.13 A student says, "lampreys are parasites." How should the student's incorrect statement be modified to make it correct?

13.14 If a vertebrate relies on its sense of smell for survival, what brain lobes will most likely be larger as compared to those of other vertebrates?

Class Chondrichthyes

We now turn our attention to one of the classes in subphylum Vertebrata that some find frightening, class **Chondrichthyes** (kahn drik' theez). In Greek, the word "khondros" means cartilage and "ichthyes" means fish. Thus, the creatures in this class are often referred to as "cartilaginous fish." This, of course, gives away their most important feature: their endoskeleton is made of cartilage, like that of the lamprey. Unlike lampreys, however, members of this class have a very elaborate endoskeleton that includes a jaw. This elaborate skeleton stays flexible because it is not reinforced with the calcium salts that bone tissue is reinforced with. Thus, we sometimes say that the skeleton of these creatures is "not calcified." Some important representatives of class Chondrichthyes are pictured in the figure below.

FIGURE 13.7

Members of Class Chondrichthyes

Blue shark photo ©Cyber Sea Photography
All other photos ©Corel, Inc.

a. Blue Shark

b. Caribbean Reef Shark

c. Blue-spotted Ray

d. Thornback Skate

Now you see why we say this is one of the more frightening classes of vertebrates. Class Chondrichthyes is home to sharks, rays, and skates.

Sharks

Most sharks have been specifically designed to be deadly predators. They have a sleek, torpedo-like shape that allows them to move easily and quickly through the water. Although their skin looks smooth, it is actually covered with tiny scales that make it rather rough.

The shark has six basic types of fins. The **pectoral fins,** near the head, and the **caudal fin**, at the end of the tail, are the fins most used for swimming. By swishing the caudal fin back and forth, the shark can propel itself through the water at great speeds. The **anterior dorsal fin** and the **posterior dorsal fin** are used mostly to stabilize the shark and keep it upright in the water. The **anal fin** performs essentially the same task. The **pelvic fins** are used for stabilization as well, but in males, they also play a role in reproduction. These fins, as well as other major exterior features of the shark are shown in the figure below.

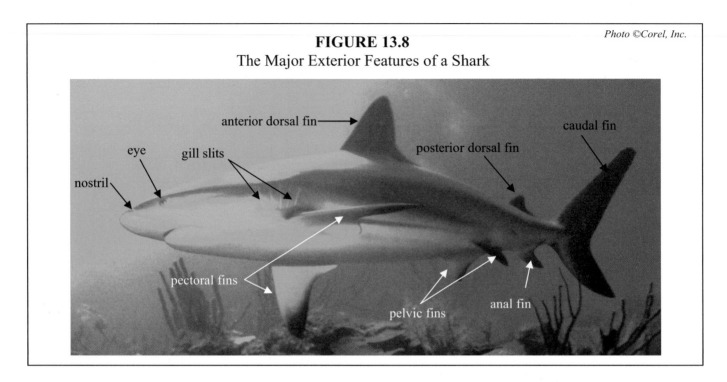

FIGURE 13.8
The Major Exterior Features of a Shark

Photo ©Corel, Inc.

The shark's mouth is typically lined with rows of razor-sharp teeth. Although their endoskeletons are not calcified, their teeth usually are. This makes the teeth strong enough to bite through the skin, flesh, and bones of their prey. If a shark loses a tooth, one of the teeth from a row further back in the mouth advances to replace it, and another grows in place of that one. Thus, the back rows of shark teeth are backup teeth. This allows the shark to feed regardless of minor injuries.

The shark breathes through gills that are rich with capillaries. As the shark swims, water passes over its gills. The blood in the capillaries rids itself of carbon dioxide, which diffuses through the capillary walls and into the water. At the same time, oxygen that has been dissolved in the water diffuses through the capillary walls and is absorbed by the hemoglobin in the blood. It was once thought that all sharks had to swim continuously in order to supply their gills with fresh water. We now know that this idea is wrong. Most sharks have an independent means of pumping water through their gills. This is more of an effort than allowing the water to pass over their gills while swimming, however, so sharks usually swim in order to breathe.

Although sharks do have eyes with receptors that are connected to the optic lobes of the brain, their eyesight is not very keen. The shark also has nostrils that lead to receptors which communicate with the olfactory lobes of the brain. This allows the shark to have a sense of smell which is keener than its eyesight. This sense of smell helps lead the shark to prey. When a shark smells blood, it often goes into a feeding frenzy, madly eating anything it can sink its teeth into.

Even though the shark uses both its eyesight and its sense of smell to hunt prey, neither one of those senses alone would lead it to enough prey to sustain its voracious appetite. The shark's eyesight is not very keen, and even though its sense of smell is good, smells do not travel well under water. As a result, these senses do not provide the shark with enough range to effectively hunt for prey. How, then, does the shark find enough prey to satisfy its appetite? The shark has two highly developed means of searching for prey. A shark has a very sensitive vibration detector and an even more sensitive *electrical field sensor.*

Most sharks have a **lateral line**, a canal that runs the length of the shark's body. This canal is full of receptors that can detect very small vibrations that occur underwater. These receptors indicate to the shark's brain where the vibrations come from, and the shark will investigate them as a possible source of food.

The most sensitive means that the shark has for detecting prey, however, is its three-dimensional electrical field sensor. You see, all animals emit weak electrical signals. The shark has the ability to sense these electrical signals under water and filter out the ones in which it is not interested. It can then wait to sense the electrical signals that come from prey it is interested in eating. When it detects such a signal, its electrical field sensor determines the *precise location* of the prey, and the shark can catch the unfortunate creature *without either seeing or smelling it.*

This incredible phenomenon was first discovered when marine biologists observed sharks feasting on flounders. Flounders are flat fish that lie at the bottom of the ocean, often covered in sand. Because of their coloring, they blend in with the sand, and it is very hard to see them, even if you are looking right at them. Well, biologists would observe sharks that were just swimming along, and then suddenly they would dive down and bite the sand, each time unerringly coming up with a flounder! Even when the scientists could not see the flounder resting on the bottom of the ocean, the shark knew it was there and would deftly grab it.

After extensive research, biologists determined that the shark was homing in on the electrical signals that the flounder sent out! Although the ability to detect electrical signals and use them to home in on prey is amazing enough, the marine biologists were really amazed at the electrical field sensitivity of the shark. It turns out that the shark's ability to detect electrical fields is so sensitive that it is comparable to someone detecting the *precise location of a transistor radio battery more than 1,000 miles away*!

Think about that for a moment. The shark has a *natural* electrical field sensor that is more sensitive (by far) than *anything* that human science can develop. Think what the navy could do with such an electrical field sensor! Our best scientists, however, cannot produce anything close to what the shark has *always had.* The shark's method of finding prey is one of the great testaments to the fact that life was designed by an awesome Creator!

Although sharks are best known as terrifying predators, there are some species of shark that are quite gentle. The whale shark, for example, is the largest fish in creation. Unlike other sharks, it swims slowly and quietly in the ocean, filtering out the plankton in the water as its only food source.

When sharks reproduce, fertilization is internal. The male transfers sperm to the female with a **clasper**, which is a specially-designed portion of the male's pelvic fins. Most sharks are oviparous, releasing the fertilized egg in a capsule that attaches to stationary objects under water. A few sharks, however, are viviparous. This is unusual for vertebrates that are not in class Mammalia, a class that we will study in a later module.

An interesting symbiotic relationship exists between a fish called the **remora** (ruh mor' uh) and many species of shark. Although most fish are wary of sharks, the remora comes right up to a shark and attaches to its body with a sucker organ that is on the top of its head. The remora then begins to eat whatever scraps the shark happens to lose as it is feeding. Especially when it is young, a remora will also eat parasites off the shark's scales. This, of course, provides a health benefit to the shark and, at the same time, provides a food source for the remora. When you see a group of sharks, you will almost always see remoras attached to them.

Rays and Skates

Rays and skates (Figures 13.7c and 13.7d) are flat and thin, not torpedo-shaped like a shark. They are often called "birds of the sea," because their pectoral fins look like huge wings that propel them through the water. They use their tails like rudders, helping to direct them as they swim. They are designed to lie at the bottom of the ocean. Their eyes are on top of their bodies in order to look for danger. Their teeth are blunt, ideal for crushing the shells of mollusks and the exoskeletons of crustaceans that inhabit the ocean floor. Some protect themselves by changing color and blending in with the sand that they inhabit. Others, such as the stingray, have sharp spines that run along their tails. When the stingray feels threatened, it can lash out at the potential predator with its tail, inflicting painful and slow-healing wounds.

One difference between rays and skates can be seen in Figure 13.7. Rays tend to have slender, whiplike tails, while skates tend to have thicker, fleshy tails. Another important difference, not seen in the figure, is the way in which they reproduce. Most rays are ovoviviparous, while some are viviparous. Skates, on the other hand, are oviparous.

ON YOUR OWN

13.15 If you cut the anterior and posterior dorsal fins off a shark, what will happen when it tries to swim?

13.16 When a creature dies, under the right conditions, its remains can be preserved as fossils. Typically, the harder the remains, the more likely they are to fossilize. Thus, if a clam dies, the inside body parts are rarely fossilized, but the shells often are. What parts of the shark would be most likely to fossilize?

13.17 Underwater photographers have batteries in their cameras. When the batteries go dead, photographers sometimes throw them into the water, which is not a good thing to do, because those batteries pollute the water with their chemicals. Some shark photographers have noticed that the sharks sometimes go after those batteries, as if they are prey. Why?

13.18 You see something flat on the ocean floor. It has large pectoral fins that look like wings and a very thin tail. Is it most likely a ray or a skate?

Class Osteichthyes

What most people think of as fish can be found in class **Osteichthyes** (ahs tee ik' theez). The Greek word "osteon" means "bone" while, as you have already learned, "ichthyes" means "fish." Thus, the members of this class are often called the "bony fish." Unlike the creatures that we studied in the previous two sections, members of this class have endoskeletons that are calcified. Thus, they have true bones. Now remember, just because a creature has bones, this does not mean that the endoskeleton is void of cartilage. Vertebrates with truly bony skeletons use cartilage as a connective tissue to keep the bones together.

In fact, most members of this class have only certain regions of their endoskeleton which are calcified.   All bony fish have calcified vertebral columns and skulls, but most of them have ribs that are partly cartilage. The pectoral and pelvic girdles, in the fish that have them, are also usually composed of noncalcified cartilage. Thus, even a "bony fish" is not as bony as you might think! The major exterior features of a typical bony fish are shown in the figure below.

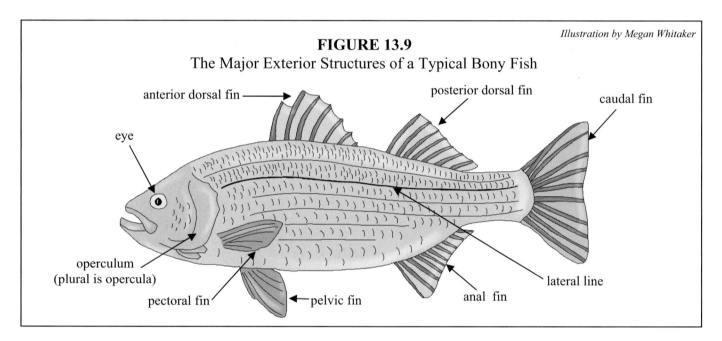

FIGURE 13.9
The Major Exterior Structures of a Typical Bony Fish

Illustration by Megan Whitaker

anterior dorsal fin

posterior dorsal fin

caudal fin

eye

operculum
(plural is opercula)

pectoral fin

pelvic fin

anal fin

lateral line

Notice that, like the shark, a bony fish has anterior and posterior dorsal fins, a caudal fin, pelvic fins, and pectoral fins. Notice also that the lateral line, which senses vibrations in the water, is also present in the typical member of class Osteichthyes. Whereas the shark uses its lateral line to hunt, many bony fishes use their lateral line to warn them of possible predators. Sharks tend to swim towards the vibrations they sense in the water; many bony fishes tend to swim away from them.

Unlike the shark, the typical bony fish has **opercula** (oh per' kyoo lah), which cover the gills. These structures get their name from the Latin word "opercula," which means "cover." In order to breathe, the fish opens its mouth and draws water into it. It then closes its mouth and opens its opercula, forcing the water to run out of its body that way. As the water travels from the mouth to leave through the opercula, it passes over the gills, where the oxygen dissolved in the water diffuses into the capillaries of the gills. At the same time, of course, carbon dioxide diffuses out of the capillaries and into the water. This process is so efficient in bony fish that more than 80% of the oxygen dissolved in the water can be absorbed by the blood in just one pass through the gills!

Since the typical bony fish is flatter than the shark, it needs more help in balancing. This is why the dorsal fins of a bony fish are usually larger than those of the shark, at least in proportion to the fish's body. In addition to being larger than the shark's, the anterior dorsal fins of many bony fish have sharp spines. These spines act as a defense against predators, because they can cause deep jagged wounds. If you have ever grabbed a fish the wrong way while fishing, you have probably experienced the defensive capabilities of a bony fish's anterior dorsal fin!

The body covering of most fish consists of overlapping scales. Special glands beneath these scales secrete **mucus**, a substance that gives fish the slimy feel they have when you touch them. The mucus waterproofs the scales, protects the fish from parasites, and makes the fish more mobile in the water. Studies indicate that this slimy mucus reduces water drag, allowing the fish to swim faster than it would be able to swim without the mucus.

The members of class Osteichthyes have an interesting internal structure that is ideal for an introductory lesson in anatomy. Their internal structure is complex enough for us to point out some features common to all of the other members of subphylum Vertebrata which we will study. At the same time, however, their internal structure is not so complex that it becomes overwhelming, as is the case with many vertebrates. The major internal features of a bony fish are illustrated in the figure below.

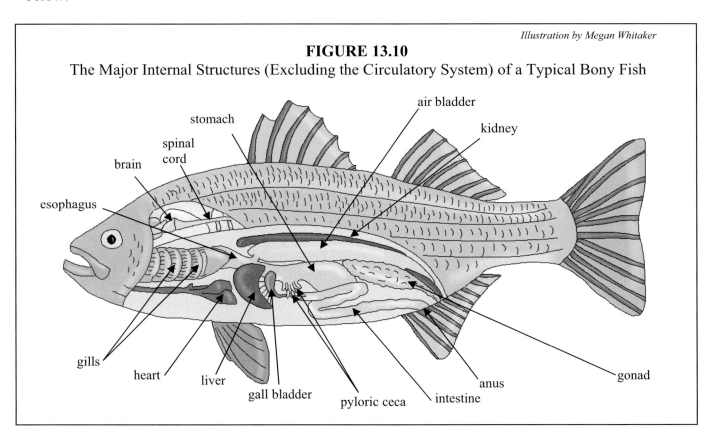

Illustration by Megan Whitaker

FIGURE 13.10

The Major Internal Structures (Excluding the Circulatory System) of a Typical Bony Fish

When a fish feeds, the food travels through its mouth and across its **tongue**, which is lined with **taste buds**. These taste buds are receptors that send signals to the brain, providing the fish with a sense of taste. The food then passes through a muscular pharynx (throat) and into the esophagus, which leads to the stomach. The food is broken down and stored in the stomach until it needs to be digested. Then it is sent to the intestine, where it is digested. Any undigested remains leave the fish through the anus.

The fish has two main organs that aid the intestine in digestion. The **pyloric ceca** secrete some digestive enzymes into the intestine. At the same time, they also secrete into the stomach chemicals that aid in the breakdown of food.   The **liver**, which is connected to the small intestine, secretes **bile**.

<u>Bile</u> – A mixture of salts and phospholipids that aids in the breakdown of fat

Bile tends to speed up the digestion of fats in the intestine. It does not actually contain any digestive enzymes, but it takes large drops of fat and breaks them into smaller drops of fat. These smaller drops digest much more rapidly than the larger drops. Without bile, the fish would release plenty of digestible fats through the anus, because there would not be enough time to digest the large drops of fat before they traveled all the way through the intestine. When food is not flowing through the intestine, a valve shuts the liver off from the intestine, and the bile is concentrated in the **gall bladder**. This concentration makes the bile more effective at its job.

Producing bile is not the only function of the liver. It also aids in storage. If too much glucose (a monosaccharide) is running through the bloodstream, for example, it can convert the glucose to glycogen, which is a polysaccharide. This polysaccharide provides a more efficient means of storing the excess food.

The liver is also important in converting nutrients from one form to another. For example, it changes monosaccharides such as fructose into glucose, which is what the cells would prefer to use for energy. This is why blood sugar is essentially all glucose, even though many different sugars might be ingested. Additionally, the liver can combine nutrients into more useful chemicals. Phospholipids (the major component of cell membranes), for example, are produced in the liver from fats and other chemicals absorbed in the digestion process.

The liver also cleans the blood. For example, some cells in the liver engage in phagocytosis in order to remove dead blood cells, bacteria, and other debris from the blood. In addition, many of the body's natural processes result in byproducts which would be toxic to the body if they were allowed to accumulate. For example, ammonia is a byproduct of the breakdown of amino acids. If ammonia were to build up in the blood or the tissues, it would become toxic. The liver, therefore, converts ammonia to urea, which is not nearly as toxic to bodily tissues. This urea ends up being taken out of the blood by the kidneys, and it leaves the body in the urine. Since God has designed animals so elegantly, their internal processes are efficient, recycling as much as possible. As a result, many of the toxic substances removed by the liver are integrated into the bile that the liver produces.

Most bony fish have an **air bladder**. This organ helps the fish to stay afloat in the water. You see, bony fish are heavier than water. Without an air bladder, most of them would simply sink to the bottom. However, the fish can direct gases from the blood and digestive system to diffuse into the air bladder. This increases the buoyancy of the fish, allowing it to float. If the fish wants to rise to a shallower depth in the water, it simply increases the amount of gases in its air bladder. This increases its buoyancy, allowing it to rise. Alternatively, if the fish wants to move to a lower depth, it simply releases gases from its air bladder, causing the fish to sink.

The fish's brain is composed of the lobes that are shown in Figure 13.5. The olfactory lobes are generally large compared to the overall size of the brain, because the fish's sense of smell is keen. Many fish are able to detect incredibly small amounts of substances by their smell. This is the main way that predatory bony fish seek out prey. Fish also have good eyesight and, as a result, have large optic lobes in their brains. The sense of sight is not as useful underwater as the sense of smell, however, mostly due to the fact that the amount of light underwater decreases with increasing depth.

As with all vertebrates, the sexes are separate in fish. The gonad of a male fish is its testes, while the gonad of a female fish is its ovaries. Most bony fish are oviparous. The female lays eggs, and then the male covers them with **milt**, a milky-white substance that contains its sperm. Fertilization, therefore, is external in most bony fish. The eggs are then left to develop and hatch. When the female lays her eggs, we usually say that she is **spawning**.

The circulatory system of a fish is also instructive. The figure below is a sketch of the basic anatomy of a fish's circulatory system.

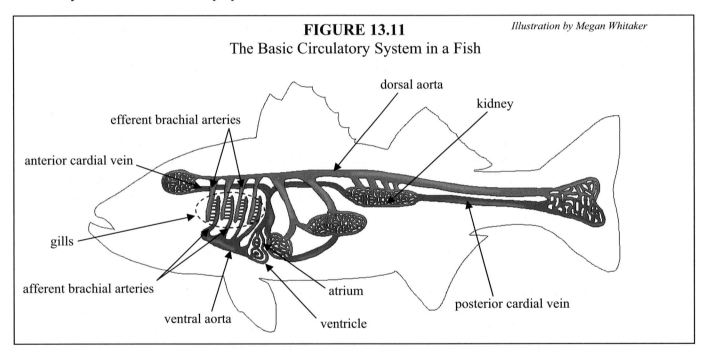

FIGURE 13.11

Illustration by Megan Whitaker

The Basic Circulatory System in a Fish

dorsal aorta

kidney

efferent brachial arteries

anterior cardial vein

gills

afferent brachial arteries

ventral aorta

atrium

ventricle

posterior cardial vein

Notice that we have colored some of the blood vessels blue and the rest red. This is the typical convention in biology. Red represents oxygen-rich blood, while blue represents oxygen-poor blood. To be accurate, the colors should really be bright red for oxygen-rich blood and dark red for oxygen-poor blood. It would be hard to tell those two colors apart, however, so we use red and blue instead. Also realize that the blood vessels drawn in Figure 13.11 are not all of the blood vessels in a fish! These are just the major blood vessels along with the capillaries that service the major structures in the body. In order to deliver nutrients and necessary gases to every cell in the body, there are obviously many capillaries that we just do not show.

The blood flow begins with a **two-chambered heart**. In most of the organisms that we studied in previous modules, the heart was a single muscle with a single cavity that holds blood before it is pumped. As the circulatory system of an animal gets more complex, however, its heart must be designed to accommodate that complexity. The next step in complexity is the two-chambered heart of the fish. The two chambers are called the **atrium** (ay' tree uhm) and the **ventricle** (ven' trih kuhl).

Atrium – A heart chamber that receives blood

Ventricle – A heart chamber from which blood is pumped out

As its definition implies, the atrium receives blood from veins that carry blood back to the heart. The blood is stored there until the ventricle is ready to receive it. At that time, the atrium empties into the ventricle, and the ventricle pumps the blood back out again. While the ventricle is pumping, the atrium fills again, waiting for the next chance to empty into the ventricle.

The blood leaves the heart through the large blood vessel called the **ventral aorta** (ay or' tuh), which branches into a series of smaller vessels called **afferent** (af' uh rent) **brachial** (bray' kee uhl) **arteries**. These arteries supply blood to the capillaries in the gills, where the blood releases the carbon dioxide that it carries and accepts oxygen from the water that pours over the gills. Now the blood is oxygen-rich, and it leaves the gills through **efferent** (ef' uh rent) **brachial arteries**. These arteries dump blood into the **dorsal aorta**. Some of the blood goes to the brain and other cells in the head, supplying them with nutrients and oxygen. At that point, this blood is oxygen-poor and is carried back to the heart in the **anterior cardial** (kar' dee uhl) **vein**.

If the blood goes the other way in the dorsal aorta, it is sent to various organs and cells in the rest of the body. Blood that passes through the capillaries of the intestine picks up nutrients as well as waste while it drops off oxygen to the intestine's cells. The blood will carry those nutrients to the rest of the cells in the body. The blood that passes through the **kidney** is filtered of its waste materials. Once the blood has done its job, the **posterior cardial vein** carries it back to the heart, where the whole process starts again.

Fish are **ectothermic** (ek toh thur' mik), which means they are "cold-blooded."

Ectothermic – Lacking an internal mechanism for regulating body heat

Since they are cold-blooded, fish are very sensitive to the temperature of the water in which they live. In general, their internal temperature decreases when the temperature of the water around them decreases, and it increases when the temperature around them increases.

It is important for you to know the organs and circulatory system of the fish because they are similar to those of humans. Although the human organs and systems are more complex, those of the fish do essentially the same job. The intestine, liver, kidneys, mouth, tongue, and esophagus perform basically the same functions in humans as they do in fish. Although humans have lungs while fish have gills, and although the human heart has four chambers instead of two, the fish is still a reasonable model of what goes on in the circulatory system of a human. At the same time, however, the fish's systems and organs are not as complex as those of a human being. Thus, they are a reasonably easy set of systems and organs to learn and, at the same time, they provide some insight into the systems and organs of your own body. Thus, you will be expected to know the ins and outs of Figures 13.10 and 13.11. You will need to be able to identify and list the functions of the organs shown in Figure 13.10. In addition, you will need to be able to identify the structures shown in Figure 13.11, explain how the blood flows through them, and identify whether the blood is oxygen-rich or oxygen-poor in each.

The Diversity of Class Osteichthyes

The diversity observed in this class is dizzying. There are an incredible number of different species of fish in both freshwater environments and marine ecosystems. We have studied some of them already in other modules. The goby, clownfish, Oriental sweetlips, and blue-streak wrasse all participate in amazing symbiotic relationships. There are, of course, thousands of other fish to study – far too many to discuss in an entire course, not to mention a single module. Nevertheless, the figure below shows a few of these diverse species.

FIGURE 13.12
A Few Members of Class Osteichthyes

a. Carp   b. Salmon   c. Betta

d. Lionfish   e. Spiny Puffer   f. Butterfly Fish

The **carp** (Figure 13.12a) is a freshwater fish typically referred to as a "trash fish." This is because it feeds off the bottom of the lake or river in which it lives, giving the illusion that it is eating trash and decaying matter. While carp do, in fact, eat the remains of decaying plants, they also eat other fish's eggs, algae, living plants, worms, snails, mussels, and many other animals. The carp is also called a trash fish because it can live in water that is too polluted to support most other species of fish. Carp are so bony that it is very hard to eat them, but many ornamental fish ponds are filled with them because they often have interesting colors. Ornamental carp are often called "koi." Carp are prolific and breed rapidly. They also tend to "take over" the waters in which they live, driving other fish away. Typically, they do this by stirring up mud and uprooting vegetation.

The **salmon** (Figure 13.12b) has a lifestyle much like the lamprey eel. There are many different species of salmon, but the Atlantic salmon is very common, especially in restaurants. It lives its adult life in the cold waters of the Atlantic Ocean. In late spring or early summer, it travels back to the cold, swiftly flowing freshwater stream in which it was born in order to reproduce. It goes back to that stream no matter what obstacles lie in its way. The salmon pictured in the figure, for example, are jumping a seven-foot waterfall to get back to their stream. As the female salmon spawns, she lays as many as 20,000 eggs. The males fertilize them, and then the salmon float back down the stream into salt water. Some salmon (the Pacific salmon, for example) spawn only once and then die afterwards. When the eggs hatch, the young stay in fresh water for a few years and then head to the ocean.

In Figure 13.12c, we see the **betta**. This small fish lives in the warm fresh waters of Southeast Asia. Because they are so beautiful, bettas are cultivated as aquarium fish. They are sometimes called Siamese fighting fish because the males are so territorial that they will attack any other betta males that they see. They will even attack their own reflection in a mirror! As a result, when you buy a betta in a pet store, it will generally be in its own little bowl. Surprisingly enough, bettas are tolerant of many other species of fish. Aquarium owners who have spent some time researching bettas and their behavior often have one (and only one) male betta in their aquarium, peacefully coexisting with other species of fish.

Moving to marine environments, we see the lionfish in Figure 13.12d. Its bright colors come from specialized cells called **chromatophores** (kroh mat' uh forz). These chromatophores produce pigments that give the fish its colors. Sometimes the pigment is of one color. Other times, several overlapping cells produce different colors of pigments, and the coloring of the fish is the result of those colors mixing. The lionfish's colors are meant as a warning. Their enlarged pectoral fins are tipped with poisonous glands. Fish seem to associate the colors of the lionfish with danger and steer clear. Humans should steer clear as well, because the poison of the lionfish can be deadly.

The **spiny puffer** (Figure 13.12e) is, most of the time, a small, harmless-looking fish. It uses its beaklike mouth for breaking open hard corals or mollusks and eating the body inside. It only appears as it does in the figure when it feels threatened. When this happens, it takes in large quantities of water and "puffs" itself up to many times its normal size. The spines then stand out, giving the spiny puffer the appearance of an underwater porcupine. It is therefore sometimes called the "porcupine fish." Although considered very tasty, especially in Oriental cultures, the flesh of many spiny puffer species can be deadly if not prepared properly by an expert chef.

Finally, the **butterfly fish** (Figure 13.12f) is a general name given to several brightly colored, tropical marine fish. These fish have an interesting defense mechanism. The chromatophores in their posterior region produce a very dark pigment. They are arranged in a large disk, giving the visual impression of a large eye. Some fish think that this large "eye" means the butterfly fish is a large predator fish, and as a result, they stay away. Other fish that are not afraid of the "eye" still mistake it for a real eye and attack the back of the butterfly fish. This gives the fish a chance to swim away from the predator.

ON YOUR OWN

13.19 A student sees a portion of a rib from a fish and notices that it is made of cartilage. Based on this, the student says that the fish cannot be a member of class Osteichthyes. Is the student correct? Why or why not?

13.20 A fish expels a lot of digestible fat from its anus. What organ is probably not functioning properly?

13.21 Most arteries carry oxygen-rich blood, as the blood is traveling away from the heart in order to supply oxygen to the tissues. Name an artery in the fish that carries oxygen-poor blood.

If you purchased the dissection kit, you will want to perform the following dissection experiment so that you can learn more about the anatomy of bony fishes.

All photos by John Skipper

EXPERIMENT 13.1
Perch Dissection

┌─────────────────────────┐
│ **Be careful!** Dissection │
│ tools are SHARP!   │
└─────────────────────────┘

Supplies:

NOTE: The course website discussed in the "Student Notes" section at the beginning of this book has several pictures of a perch dissection.

- Dissecting tools and tray that came with your dissection kit
- Perch specimen
- Magnifying glass
- Laboratory notebook
- Water
- Small bowl

Object: To become more familiar with the anatomy of the perch through dissection

Procedure:

1. Examine your specimen. Identify all of the structures indicated in Figure 13.9. List in your laboratory notebook any structures in that figure which you could not identify.
2. Open the specimen's mouth and examine the teeth. Write a description of them in your laboratory notebook.
3. Examine the tongue. Where is it attached to the mouth? Record this in your laboratory notebook.
4. Examine the anterior dorsal fin. Fins can be supported by one of two structures: rays or spines. The way to tell the difference is to put your finger on the tip of the supporting structure and push lightly. If the structure bends, it is a ray. If it does not bend, it is a spine. Raise the anterior dorsal fin so that it looks like the anterior dorsal fin in Figure 13.9. Now put your finger on the tip of one of the supporting structures and push gently. Don't push too hard, or you will get stuck! Is the anterior dorsal fin supported by rays or spines?
5. Repeat the previous step for all other fins and record in your notebook which are supported by spines and which are supported by rays.
6. Pull a scale off the specimen and observe it under the magnifying glass. Draw what you see. Can you make out the rings? These are growth rings, which show you how the scales grew as the fish grew.

7. Raise the right operculum and use your probe to count the gills. Now do the same for the left operculum. Record the number of gills on each side in your laboratory notebook.

8. Using your scissors, cut the left operculum away and remove one set of gills. Place the gills in the bowl and cover them with water. Note the structure. You should be able to see a strong **arch** in each gill. The arch should have a comb-like structure on one side and feathery extensions on the other. The "teeth" of that comb-like structure are called the **rakers**, while the feathery extensions are called the **filaments**.

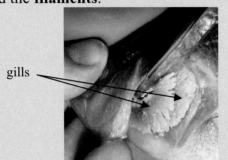

gills

rakers

gill arch

filaments

9. Draw a gill in your laboratory notebook, labeling the arch, rakers, and filaments.

10. Now you are ready to look at the internal structures of the specimen. To do so, you will want to make a "window cut," which is shown in the photo below step #13. Hold the specimen ventral side up with the head pointing away from you. Use your scalpel to make a cut from just anterior of the anus all the way to the operculum.

11. Turn the specimen so that its left side is facing you and so that the head is pointing left and the tail is pointing right. Now make a new cut from the point at which you left off in the previous step up towards the dorsal fins. Continue until your cut is just above the level of the fish's eye.

12. Make a similar cut from the anus straight towards the dorsal fins until that cut reaches essentially the same level as the cut in the previous step.

13. Lift up the body wall and completely remove the flap of body covering by making a final cut with your scalpel that runs from the cut you made in step #11 to the cut you made in step #12. Your fish should look something like the photo below.

liver

gonad
(This is an ovary.)

As shown in the photograph above, you should be able to see the following organs:

> **Liver** - It should be large and lie in the anterior region of the fish.
> **Gonad** - It may not be nearly as large as the one shown above. This female was almost ready to lay eggs, so her ovary is large. If your specimen is a female not ready to lay eggs, her gonad will be smaller. If it is a male, the gonad will be even smaller. Record the sex of the specimen.

14. If you gently raise the lobes of the liver, you should see the **gall bladder**, which looks like a deflated balloon, and the **stomach** (see the photos on the next page).

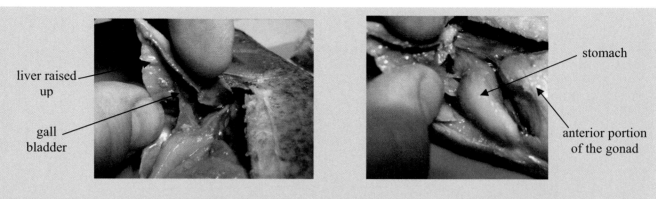

15. Try to follow the **esophagus**, which runs from the stomach to the gills.
16. Remove the liver, gonad, and stomach by cutting them out. If you want, cut open the stomach once you have removed it and see if you recognize the perch's last meal! You should now be able to find the **intestine** and **air bladder** (see photo below). The air bladder might be popped.
17. You might be able to see strands just ventral to the vertebral column. These make up the **kidney**.
18. Now find the **heart**. Notice that it has two chambers. The upper one is the atrium and the lower one is the ventricle.

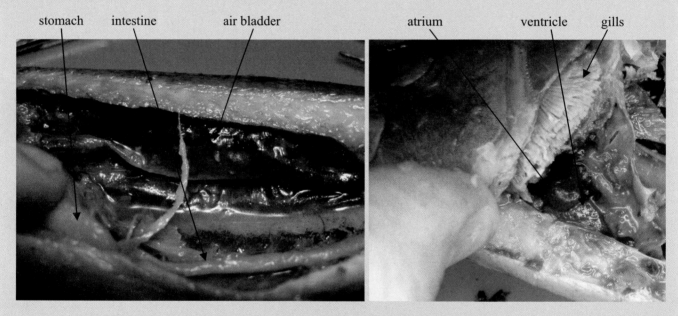

19. It is now time to find the fish's brain. Hold the fish with its dorsal side up and position its head so that it points away from you. Using your scalpel, cut the skin away from the skull.
20. Once you expose the skull, begin scraping it with your scissors to wear away the bone.
21. As the bone gets thinner, start picking it away with your forceps instead of scraping it. If you are careful, you can expose the brain in this manner.

22. Once the brain has been exposed, look at Figure 13.5 and the photo on the previous page to help you identify the following regions of the brain:

Olfactory lobes - Two small bulges in the front of the brain
Cerebrum - The two lobes behind the olfactory lobes
Optic lobes - The largest lobes just behind the cerebrum
Cerebellum - A single lobe behind the optic lobes
Medulla oblongata - Just underneath the cerebellum

23. Draw the brain as you see it in your specimen, labeling all parts that you can identify. List those parts you could not identify.
24. Clean up everything. Wash and dry your tools and tray. Put everything away.

Class Amphibia

In Greek, the term "amphi" means "both sides" while "bio" means life. Thus, "amphibia" (am fib' ee uh) means "dual life." This is an excellent description of the organisms in class **Amphibia**. Most amphibians begin their life cycle in the water. The young hatch from eggs that are allowed to develop outside of the mother. They typically begin life in the larval stage. These larvae have gills and breathe in the water. When metamorphosis occurs, the gills degenerate and an air-breathing respiration system develops. Fins disappear and legs appear. The resulting adult is an air-breathing, land-dwelling creature that looks nothing like the larva. The adult female lays eggs which are externally fertilized by the male in the water, and the whole process begins again.

Although this metamorphosis life cycle is common to all amphibians, it takes on a variety of different forms in this class. The larva of a salamander actually looks a lot like the adult, but it has gills for respiration, whereas the adult does not. The larva of a frog, on the other hand, looks nothing like the adult. This larva, often called the **tadpole**, looks like some sort of bulbous eel. It has an oval body that tapers down to a tail. When it matures into an adult, the tail disappears, as do the gills. Limbs are formed, and lungs take the place of the gills.

Amphibians have several characteristics that separate them from the rest of subphylum Vertebrata:

➤ Endoskeleton made mostly of bone
➤ Smooth skin with many capillaries and pigments (no scales)
➤ Two pairs of limbs with webbed feet (usually)
➤ As many as four organs for respiration
➤ Three-chambered heart
➤ Oviparous with external fertilization

These characteristics make amphibians a very interesting class of animals to study!

Perhaps the most interesting aspects of amphibians are their respiratory and circulatory systems. As was mentioned above, amphibians have as many as four respiratory organs. First, all amphibians have gills during their larval stage. When the gills degenerate, they are often replaced by

lungs, which allow the exchange of gases between the blood and the atmosphere to occur internally. Lungs, however, are generally used only as a backup. Like the earthworm, the main organ of respiration for amphibians is their skin, which is rich in capillaries. For some amphibians, more than 90% of their respiration is accomplished through their skin. Finally, some amphibians (like the frog) use the lining in their mouths as a respiratory organ. Like the skin, the lining of a frog's mouth is rich with capillaries. A large amount of gas exchange can take place there.

Because amphibians have so many means of respiration at their disposal in their adult form, they usually are comfortable in or out of water. Most amphibians can go for very long periods of time without filling their lungs. As a result, they can breathe through their skin while they are under water. This allows them to stay submerged for long periods of time. Thus, even in their adult form, amphibians can spend a lot of time in the water, even though they are technically land dwellers.

Like fish, amphibians are ectothermic. Thus, the temperature of their surroundings affects them greatly. When the temperature drops, as it does in the winter, a typical amphibian will burrow into the mud. As the mud gets colder, so do the internal organs and blood in the amphibian. This slows down all of the chemical reactions in the amphibian's body, putting it in a deep state of low activity which we call **hibernation**.

Hibernation – A state of extremely low metabolism and respiration, accompanied by lower-than-normal body temperatures

Since the amphibian's life processes are dramatically slowed during hibernation, it does not need to eat. It continues to breathe through its skin, and can simply "wait out" the cold weather. When warm weather returns, the chemical reactions in the body speed up, and the amphibian resumes its normal life.

The circulatory system of a frog is similar to that of a fish, but there are some major differences. Predominant among these differences is the amphibian's heart, which is composed of three chambers, not two, as shown in the figure below.

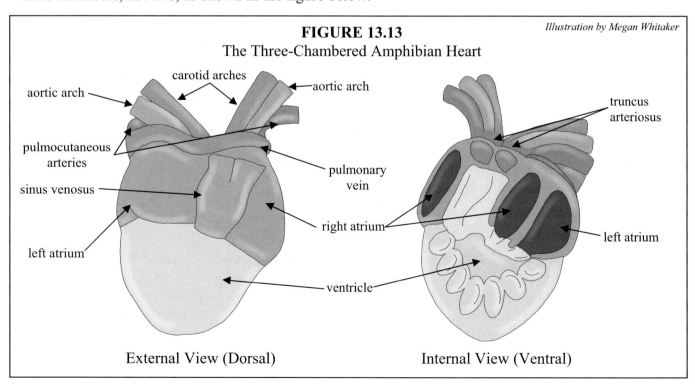

FIGURE 13.13
The Three-Chambered Amphibian Heart

Illustration by Megan Whitaker

carotid arches

aortic arch

aortic arch

pulmocutaneous arteries

truncus arteriosus

sinus venosus

pulmonary vein

left atrium

left atrium

right atrium

ventricle

External View (Dorsal)

Internal View (Ventral)

A three-chambered heart has a single ventricle and two separate atria (plural of atrium), the **left atrium** and the **right atrium**. The left atrium of the heart is filled with oxygen-rich blood that has come from the lungs, mouth, or skin. This oxygen-rich blood is dumped into the left atrium by a large vein called the **pulmonary vein**. Three large veins fill the **sinus venosus** (veh no' sus) with oxygen-poor blood that has come from the various internal organs of the amphibian. The sinus venosus then empties the oxygen-poor blood into the right atrium.

Once filled, both atria dump their contents into the ventricle. This mixes the oxygen-rich blood with the oxygen-poor blood. The ventricle then pumps this mixture out into a single artery, the **conus** (cah' nus) **arteriosus** (are teer' ee oh sis), which then splits into two smaller arteries called the left and right **truncus** (trunk' us) **arteriosus**. Each of these arteries splits into three smaller arteries. The first, the **carotid** (kuh raht' id) **arch**, sends blood to the head. Since there are two truncus arteriosus arteries, there are two carotid arches. Each of them provides blood to its side of the head. The second pair of arteries is composed of the **aortic arches**, which fuse into a single dorsal aorta, like that of a fish. Finally, the last pair of arteries is composed of the **pulmocutaneous** (pull moh kyoo tay' nee us) **arteries**. These arteries take blood to the lungs, skin, and mouth to get reoxygenated.

Specific Creatures in Class Amphibia

The majority of amphibians can be lumped into two orders: **Caudata** (caw dah' tuh) and **Anura** (uh nur' uh). Members of each of these orders are shown in the figure below.

a. Photo © Karel Igm
b. Photo © IT Stock

FIGURE 13.14
Three Amphibians

c. Photo © Matthias Bandemer

a. Salamander (Caudata)   b. Frog (Anura)   c. Toad (Anura)

Salamanders, like the one pictured in Figure 13.14a, are often mistaken for lizards. Unlike lizards, however, salamanders have the webbed feet of an amphibian and do most of their respiration through their skin. As a result, they cannot be grouped with the reptiles. Instead, they are in order Caudata of class Amphibia.

Most salamanders are small and have little color. There are, of course, exceptions to this general rule. The giant salamander of Japan can reach lengths of up to five feet, and the tiger salamander gets its name from the bright markings on its skin. Most salamanders have no lungs. More than 90% of their respiration occurs through their skin. As a result, salamanders can spend a great deal of time under water.

Frogs, like the one pictured in Figure 13.14b, and toads (Figure 13.14c) belong to order Anura. What's the difference between frogs and toads? Well, frogs have smooth, shiny skin that dries easily. Since frogs must keep their skin wet, they spend a great deal of time in the water. Toads, on the other hand, have dry, bumpy skin. They return to the water only to reproduce. Despite many ideas to the contrary, toads *do not* cause warts. They got that reputation simply because of the appearance of their skin.

ON YOUR OWN

13.22 An amphibian is breathing with gills. Is it in its larval or adult stage?

13.23 A frog that has overexerted itself leaves its mouth wide open. Why?

13.24 You find a creature from order Anura. If you find it far from any body of water, what (most likely) is it?

EXPERIMENT 13.2
Frog Dissection

Be careful! Dissection tools are SHARP!

If you purchased the dissection kit, you will want to dissect the frog at this time. In order to give you some experience in doing things more like a real scientist would, we will not give you any directions in how to perform this dissection. After all, a real scientist does not have directions at her disposal! In order to perform experiments, real scientists do literature research to find experimental protocols that they can use. That's what we ask you to do. Go to your local library or look on the Internet for a protocol that will tell you how to dissect the frog. Use a protocol you find and try to learn as much as you can about the frog's anatomy. Compare that to how much you learned in the other dissections you did. On the course website we discussed in the "Student Notes" section at the beginning of this book, you will find pictures of (but not instructions for) a frog dissection. They should help you a bit.

One of the most important things that experimental biologists can do is make **field studies**. In these studies, biologists examine a part of nature, looking at the organisms that inhabit that region and the relationships those organisms have with one another. You have already made one field study. In Module #2, you went to a pond and observed the life that existed near the edge of that pond. We now want you to perform your second field study. If you did the dissection experiments in this module, you can skip this experiment.

ALTERNATE EXPERIMENT FOR MODULE #13
Field Study II

Supplies:

♦ Lab notebook and pencil
♦ Colored pencils
♦ Magnifying glass (if available)
♦ Field guide (This will help you identify the organisms that you see. Most libraries have field guides. Try to find one on plants and one on animals.)

Object: To observe how God's wonderful systems work together to sustain life in the world around us

Procedure:

1. Locate an area that is known for its wildlife like a creek, a pond, a lake area, or some woods. Suggested field trip areas would be a state or city park or wildlife preserve near your home. (Take the whole family and make it a day outing.)
2. Upon arriving, find a relatively isolated area and sit down with your notebook and pencils. Remain in the area one to two hours.
3. Note every living thing, plant and animal, you see as you sit quietly.
4. Try to identify each specimen with your field guides.
5. Make up a chart in your notebook including the following:
    i. Date
    ii. Time
    iii. Weather Conditions
    iv. Plants (Identify if possible and note the number seen.)
        Trees
        Shrubs
        Other Plants
    v. Animals (Identify if possible and note the number seen.)
        Insects
        Birds
        Fish
        Amphibians
        Other animals
6. Try to draw a few of the organisms that you see. The drawings need not be artistic. When biologists draw organisms, it forces them to look at the organism more analytically than usual. This allows them to notice details that they otherwise would not notice.
7. Write a summary approximately one page long and tell what you observed and how different organisms interacted. This will serve as the summary that you normally write at the end of an experiment.

Summing Up

Even though we have not finished phylum Chordata (we want to discuss three more classes), we will actually end our discussion of the vertebrates for now. Never fear, however. We will pick up our discussion again in Module #16 when we study reptiles, birds, and mammals. For this particular module, you should be most familiar with the bony fish. You need to have the organs and systems illustrated in Figures 13.10 and 13.11 committed to memory, as well as a strong knowledge of how these systems work. For the rest of the classes discussed here, you need to have a more general knowledge of the facts presented. As always, use the study guide and "On Your Own" problems to indicate what you need to concentrate on for the test.

ANSWERS TO THE "ON YOUR OWN" PROBLEMS

13.1 <u>The sea squirt is not classified with sponges because, other than its feeding habits and the fact that it stays immobile as an adult, it has nothing else in common with them.</u> For example, sponges never have a notochord, do not support themselves with leathery coverings, do not have one specific entrance and another specific exit for water, and do not have the internal organs that a sea squirt has.

13.2 <u>The tunic, or leathery covering, in the adult supports the organism.</u> That's what the notochord does in the larva.

13.3 <u>The oral cirri were not functioning properly.</u> Usually, large debris does not clog up the slits of the pharynx because the oral cirri filter it out of the water that the lancelet is taking in.

13.4 <u>Blood cells are most affected by this procedure.</u> After all, the bone marrow makes blood cells. If you replace a person's bone marrow, new blood cells will be created.

13.5 <u>A shark's skeleton is made of cartilage.</u> Since the shark is in subphylum Vertebrata, its endoskeleton must be made of either bone or cartilage. Since the human skeleton is made of bone, and since cartilage is more flexible than bone, the shark's skeleton must be made of cartilage.

13.6 <u>The blood most likely came from an artery.</u> Arteries carry blood away from the heart and, in most arteries, the blood is oxygen-rich and is therefore bright red. Veins carry blood back to the heart, and in most veins, the blood is oxygen-poor and is dark red. Please note that there are exceptions to this general rule, however. In people, for example, the pulmonary arteries carry blood away from the heart, but that blood is oxygen-poor because it is going from the heart to the lungs. The pulmonary veins carry oxygen-rich blood back to the heart from the lungs.

13.7 <u>It is in a vein.</u> Arteries carry blood away from the heart, while veins bring it back.

13.8 <u>The optic lobes are most likely not working.</u> In order to see, the eyes must detect light and send signals to the brain, which it interprets as an image. If the optic lobes are not working, the second half of the process fails.

13.9 <u>The brain sends signals down the spinal cord to nerves that branch out to the limbs. That's how a person moves his limbs. If the spinal cord is cut, the signals cannot travel any farther than the cut. Thus, any limbs below that cut will never receive the signals sent to them.</u>

13.10 <u>The placenta is not working.</u> In viviparous animals, the young develop in the mother and are nourished through the placenta. If the placenta does not work, the young will starve.

13.11 This reproduction involves <u>external fertilization and oviparous development.</u>

13.12 <u>No.</u> Remember, lampreys either live their entire lives in fresh water or migrate back to fresh water to reproduce. Thus, no reproduction occurs in salt water.

13.13 <u>The phrase should really be "adult lampreys are parasites."</u> In the larval stage, the lamprey feeds on plankton and other small organisms.

13.14 <u>The olfactory lobes will most likely be larger</u>. Vertebrates tend to have large brain lobes that control the most important processes. Since olfactory lobes control the sense of smell, they will be proportionally larger in any creature that uses the sense of smell as a primary means of survival.

13.15 <u>The shark will not be able to stay upright in the water</u>. The dorsal fins allow the shark to balance itself in the water. Without them, the shark would tumble around, not being able to stay upright.

13.16 The hardest parts of a shark are its teeth, because they are calcified. Thus, <u>its teeth are most likely to be fossilized</u>. Shark teeth are, in fact, the most abundant shark fossil that can be found.

13.17 If the battery is not quite dead, but is instead very weak, it still emits an electrical signal. If the electrical signal just happens to be like that of the prey a shark is hunting, <u>the shark can mistake the electrical signal from the battery for the electrical signal of its intended prey</u>.

13.18 <u>It is a ray</u>. Skates have thick, fleshy tails.

13.19 <u>The student is not necessarily correct. Just because a part of the skeleton is cartilage, it does not preclude membership in class Osteichthyes</u>. Many bony fish have ribs that are partly made of cartilage, so the student needs to examine more of the skeleton before classifying the fish.

13.20 <u>The liver is not functioning properly</u>. The liver produces bile, which aids in the digestion of fats. Without bile, the fish cannot digest fats very well. You could also say gall bladder here, since it concentrates the bile, making the bile more effective.

13.21 The ventral aorta is an artery, because it carries blood away from the heart. However, the blood has not reached the gills yet, so it is still oxygen poor. Also, afferent brachial arteries hold oxygen-poor blood until they reach the gills. Thus, <u>either the ventral aorta or the afferent brachial arteries</u> are acceptable answers.

13.22 <u>The amphibian must be in the larval stage</u>. Amphibians use gills only in their larval stage.

13.23 <u>The frog can breathe through the lining of its mouth</u>. By keeping the mouth open, the frog is breathing extra hard.

13.24 <u>It is most likely a toad</u>. Only frogs and toads are in order Anura. Frogs, however, stay very close to water to keep their skin moist.

STUDY GUIDE FOR MODULE #13

1. Define the following terms:

a. Vertebrae
b. Notochord
c. Endoskeleton
d. Bone marrow
e. Axial skeleton
f. Appendicular skeleton
g. Closed circulatory system
h. Arteries
i. Capillaries
j. Veins
k. Olfactory lobes
l. Cerebrum
m. Optic lobes

n. Cerebellum
o. Medulla oblongata
p. Internal fertilization
q. External fertilization
r. Oviparous development
s. Ovoviviparous development
t. Viviparous development
u. Anadromous
v. Bile
w. Atrium
x. Ventricle
y. Ectothermic
z. Hibernation

2. Assign the following creatures to one of these classifications: subphylum Urochordata, subphylum Cephalochordata, class Agnatha, class Chondrichthyes, class Osteichthyes, class Amphibia

a. Frog   b. Shark   c. Lancelet   d. Carp   e. Sea squirt   f. Lamprey eel

3. What do sea squirts, lampreys, and amphibians have in common?

4. What is the difference between cartilage and bone?

5. You see a blood vessel from a creature. You have no idea what creature and you have no idea where it came from. You do notice, however, that the blood vessel wall is very thin. What kind of blood vessel is this?

6. What do red blood cells do?

7. What protein gives red blood cells their color?

8. Frogs and toads are quite uncoordinated. They move their muscles in a very jerky manner. Which brain lobe is small in amphibians?

9. An owl has very sensitive vision. Which brain lobes are larger in the owl compared to the "average" vertebrate?

10. A creature reproduces when the female receives sperm from the male and then lays an egg which hatches. Is fertilization internal or external? What kind of development is this?

11. Which has the most inflexible skeleton: a ray, a lamprey, or a salmon?

12. What do Atlantic salmon and many lampreys have in common?

13. What is the shark's most sensitive means of finding prey?

14. What function does the lateral line perform in sharks and bony fish?

15. What function do the dorsal fins perform in both sharks and bony fish? What function does the anterior dorsal fin play only in bony fish?

16. What is the major difference between the tail of a ray and the tail of a skate?

17. Identify the structures in this figure:

Illustration by Megan Whitaker

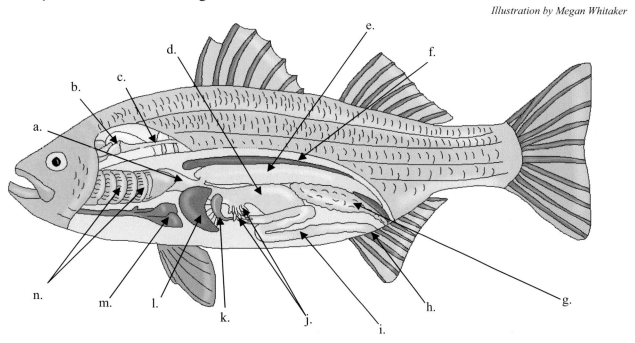

18. Describe the basic function of each organ in problem #17.

19. Identify the structures in this figure:

Illustration by Megan Whitaker

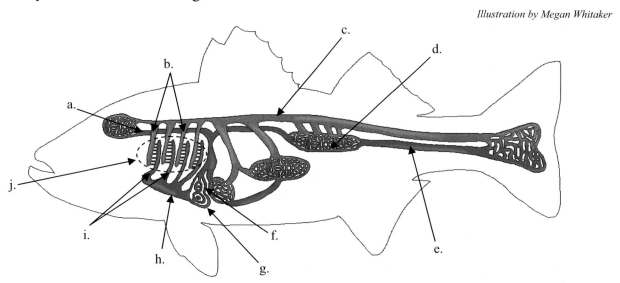

20. Of the structures listed in problem 19, which are veins, which are arteries, and which are neither?

21. List the six common characteristics of amphibians.

22. What is the difference between a toad and a frog?

23. For most amphibians, what is the major respiratory organ?

Cartoon by Speartoons

MODULE #14: Kingdom Plantae: Anatomy and Classification

Introduction

We are now going to take a break from studying animals and begin a study of plants. Kingdom **Plantae** (plan' tay) is large and diverse, so we will spend two modules studying it. In this module, we will look at basic plant anatomy and classification. In the next module, we will turn our eyes to the life processes of plants. After you get through these two modules, you should have a much deeper appreciation for the plants that God has placed in His creation.

The study of plants is usually referred to as **botany**, and biologists that study plants are called **botanists**.

<u>Botany</u> – The study of plants

Since most people know that plants perform photosynthesis, they assume that plants are our main source of oxygen. It turns out that this idea is a common misconception among science students and even some scientists. Although plants do contribute to our planet's oxygen supply, they are not the principal producers of oxygen. As we have mentioned before, the majority of oxygen comes from phytoplankton that live in the oceans, lakes, and streams of the planet.

Basic Plant Anatomy

There are many, many different ways to look at and analyze plants. One very useful way to examine plants is to determine whether or not they are **woody** or **herbaceous** (her bay' shus). If a plant has woody parts, such as trunks and/or woody stems, it typically grows year after year. We call these plants **perennials** (per en' ee uhls).

<u>Perennial plants</u> – Plants that grow year after year

Plants that do not have woody parts are called herbaceous, and they typically live for only one year. These plants are called **annuals**.

<u>Annual plants</u> – Plants that live for only one year

A few types of herbaceous plants are **biennial** (bye en' ee uhl). These plants live for two years. Typically, they store food during the first season of growth and then reproduce in the second season.

<u>Biennial plants</u> – Plants that live for two years

When you begin to study a given plant, you will notice that, much like an animal, it has certain organs and tissues. In general, plant organs can be categorized as either **vegetative organs** or **reproductive organs**.

<u>Vegetative organs</u> – The parts of a plant (such as stems, roots, and leaves) that are not involved in reproduction

<u>Reproductive plant organs</u> – The parts of a plant (such as flowers, fruits, and seeds) involved in reproduction

Have you ever wondered whether a food item (like a tomato) is a fruit or a vegetable? Well, from a biological point of view, the answer is rather simple. If a food item is a reproductive plant organ, it is a fruit. If it is a vegetative organ, it is a vegetable. Thus, a tomato (which contains seeds) is definitely a fruit, because it is a reproductive organ. We know it is a reproductive organ because it encases seeds. Something that comes from a root (like a carrot), however, is definitely a vegetable. Lettuce is composed of leaves, so it is a vegetable, and celery stalks are stems, which makes them vegetables as well. However, many of the foods that we call "vegetables" are, in fact, seeds or containers of seeds. Thus, they are technically reproductive organs and therefore fruits. Peas and corn, for example, are seeds, which makes them fruits, not vegetables.

Plants have four basic kinds of tissues: **meristematic** (mehr uh stem ah' tik) **tissue, ground tissue, dermal tissue**, and **vascular** (vas' kyoo luhr) **tissue**. Meristematic tissue contains cells that are undifferentiated.

<u>Undifferentiated cells</u> – Cells that have not specialized in any particular function

Since these cells are undifferentiated, they can develop into any tissue that the plant needs. As a plant grows, it produces new cells via mitosis. This mitosis takes place in the meristematic tissues. That way, the new cells produced can mature into any kind of cell that the plant needs.

Ground tissue is the most common tissue in a plant. The cells that make up this tissue have a wide variety of functions. Some of them provide storage for starches and oils that the plant needs. Others are involved in metabolism. The cells that do photosynthesis, for example, are a part of the ground tissue, as are the cells that are primarily responsible for making the proteins that the plant needs. Cells of the ground tissue also help to support the plant.

Dermal tissue is generally made out of a single layer of cells. It protects the plant by providing a shield between the environment and the plant's internal tissues. This shield can stop many pathogenic bacteria and fungi, and it can also prevent needed chemicals inside the plant from leaking out into the environment. The dermal tissue in the roots of a plant is also responsible for absorbing water and minerals that the plant needs.

Vascular tissue is not present in all plants. However, the majority of plants have it, and in those plants, it is used to carry water and dissolved material throughout the plant. In a way, these tissues are much like the blood vessels in animals. Whereas blood vessels transport blood that carries nutrients throughout the animal, vascular tissues transport water and nutrients throughout the plant. Vascular tissue is generally either **xylem** (zy' luhm) or **phloem** (floh' ehm).

<u>Xylem</u> – Nonliving vascular tissue that carries water and dissolved minerals from the roots of a plant to its leaves

<u>Phloem</u> – Living vascular tissue that carries sugar and organic substances throughout a plant

The cells that make up the xylem have thicker walls than those that make up the phloem. In addition, as the definition for xylem implies, xylem cells die when they mature. Phloem cells continue to live after they mature.

ON YOUR OWN

14.1 A gardener plants a group of flowers that grow beautifully over the course of a year. In order to have the same flowers each year, however, he must replant. Are these most likely woody plants or herbaceous plants?

14.2 A carrot is the root of a carrot plant. What kind of organ is it?

14.3 A section of plant no longer has any mitosis going on. What kind of tissue should be absent in that section?

The Macroscopic Structure of a Leaf

In the next module, we will concentrate on the reproductive organs that are present in most plants. In this module, we want to spend some time on the vegetative organs in a plant. We begin with the **leaf**. Leaves come in all shapes and sizes. The major parts of a leaf are shown in the figure below.

FIGURE 14.1
The Major Parts of a Leaf

Photo by Kathleen J. Wile

petiole

blade

stipules

apex

The primary portion of the leaf is called the **blade**, and the very tip of the blade is called the **apex**. The blade is attached to the stem with a small stalk called the **petiole** (peh' tee ohl). At the base of the petiole, most plants have **stipules** (stihp' yoolz). These small stalklike or leaflike growths are usually the structure that covered the leaf when it first began to grow.

There are two basic kinds of leaves: **simple leaves** and **compound leaves**. As shown in the figure below, a simple leaf is one leaf attached to the stem of the plant by a single petiole. A compound leaf has several leaflets attached to a single petiole.

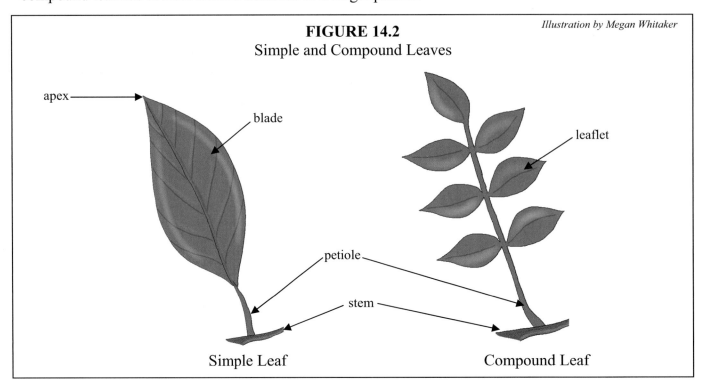

FIGURE 14.2
Simple and Compound Leaves

Illustration by Megan Whitaker

apex

blade

leaflet

petiole

stem

Simple Leaf   Compound Leaf

The most important job that the leaf performs is photosynthesis, so that the plant can obtain the food it needs. As a result, leaves are usually arranged in such a way as to expose the largest amount of surface area to the sun. The arrangement of leaves on the stem of a plant is called the **leaf mosaic**.

Leaf mosaic – The arrangement of leaves on the stem of a plant

There are many different leaf mosaics in creation, but we will concentrate on the three main types, which are illustrated in the figure below.

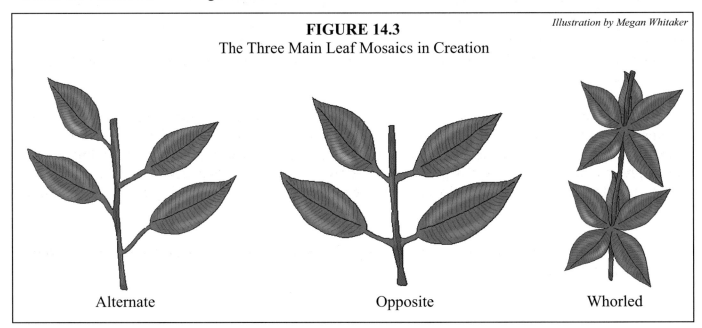

FIGURE 14.3
The Three Main Leaf Mosaics in Creation

Illustration by Megan Whitaker

Alternate   Opposite   Whorled

Whereas the alternate leaf mosaic is probably the most common among plants, tropical plants tend to favor the whorled mosaic.

Looking at the leaf in detail, there are three characteristics that botanists can use to classify the plant from which it comes: **shape, margin**, and **venation** (ven ay' shun). Let's start with the shape of a leaf. Although leaves come in many shapes, we can show you the common ones in creation.

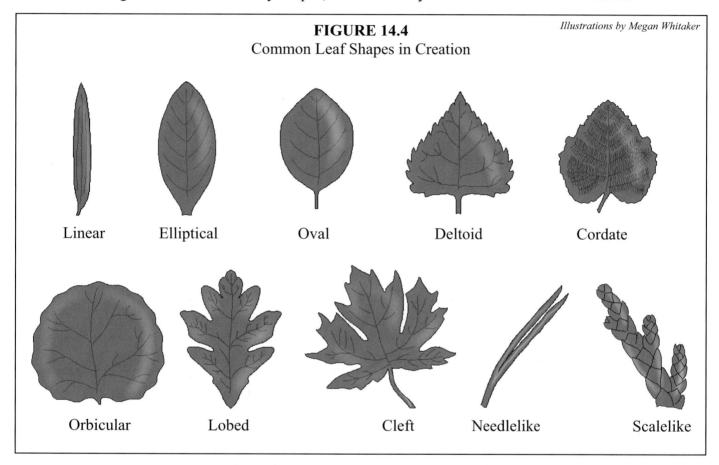

FIGURE 14.4
Common Leaf Shapes in Creation

Illustrations by Megan Whitaker

Linear   Elliptical   Oval   Deltoid   Cordate

Orbicular   Lobed   Cleft   Needlelike   Scalelike

Sometimes leaves are **linear**, which means that they are about the same width from the top of the leaf to the bottom. If the leaf tapers at both ends, but is still longer than it is wide, it has an **elliptical** (ee lip' tik uhl) shape. If a leaf is egg-shaped, we say that it has an **oval** shape. A leaf that is triangular in shape is called a **deltoid** (del' toyd) leaf. If it looks like an upside-down heart, it is a **cordate** (kor' dayt) leaf. If the leaf is nearly circular, we call its shape **orbicular** (or bik' yoo ler), which means circular.

There are some leaf shapes that are more irregular than the ones we have discussed so far. For example, there are **lobed** leaves that have deep indentations in the leaf. If the indentations are very deep and tend to be sharp, the leaf has a **cleft** shape. The needles on pine trees are actually leaves, and although they are linear, we usually give them their own name, calling them **needlelike** leaves. Some plants have leaves that look almost like the scales on a fish. Not surprisingly, they are called **scalelike** leaves.

Although the shape of a leaf is rather straightforward, the term margin is probably new to you, at least in its relationship to botany.

<u>Leaf margin</u> – The characteristics of the leaf edge

The figure below illustrates the common leaf margins in creation.

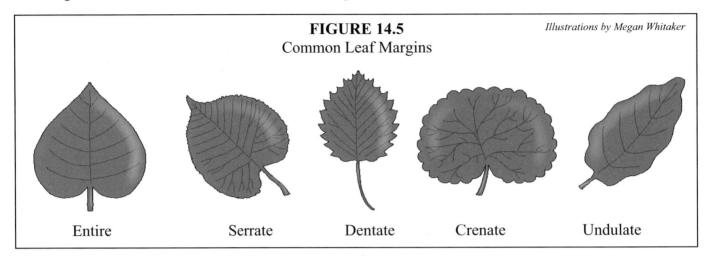

FIGURE 14.5
Common Leaf Margins

Illustrations by Megan Whitaker

| Entire | Serrate | Dentate | Crenate | Undulate |

If the outer edge of a leaf is smooth with no indentations or teeth, it has an **entire** margin. If, on the other hand, the leaf has tiny, sharp teeth along its outer edge, it has a **serrate** (seh' rayt) margin. With serrate margins, the teeth usually point upwards towards the apex of the leaf. If a leaf's outer edge has more pronounced teeth that also point outward rather than just towards the apex, it has a **dentate** margin. If the teeth are rounded rather than pointed, the margin is called **crenate** (kree' nayt). Finally, if the leaf's edge doesn't have teeth but tends to be wavy, we say that it has an **undulate** (un' joo layt) margin.

The other characteristic used to classify leaves is leaf venation. If you look closely at a leaf, you will find that there are veins that run through it. These veins (composed of xylem and phloem) form different patterns in different leaves. Generally, a leaf has a large, central vein that extends from the petiole. It is called the **midrib**. When a leaf's veins run up and down the leaf parallel to the midrib, we say that it has **parallel venation**. When a leaf's veins all branch out from the midrib, we say that the leaf has **pinnate venation**. Finally, when a leaf's veins not only branch out from the midrib, but those branches also have branches on them, the leaf has **palmate venation**. These three types of venation are illustrated below.

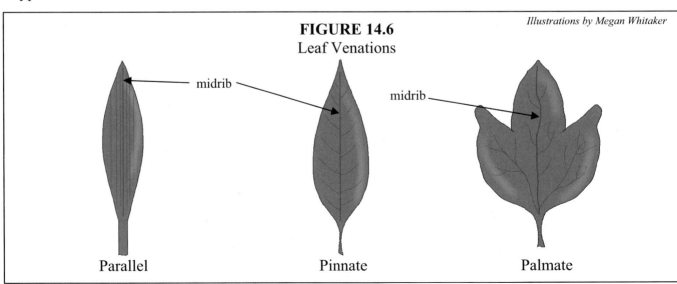

FIGURE 14.6
Leaf Venations

Illustrations by Megan Whitaker

midrib

midrib

| Parallel | Pinnate | Palmate |

It turns out that the venation of a leaf can tell us something about the taxonomy of the plant from which it came. In Module #1, you were presented with a biological key that allowed you to determine what class a plant belonged to based on the venation of its leaf. If the venation is parallel, the plant belongs to class Monocotyledonae and is typically called a **monocot**. On the other hand, if the venation is netted, the plant belongs to class Dicotyledonae and is typically called a **dicot**. We will learn later that the distinction between monocots and dicots actually depends on the structure of seed that is produced. However, since monocots have parallel venation and dicots have netted venation, the kind of venation you see on a leaf can tell you what kind of seed the plant produces.

Since the venation (and other physical characteristics) of the leaf can tell you so much about a plant, it is important for you to be able to properly identify the shape, margin, and venation of a leaf. Thus, try your hand at using these terms in classifying the leaves shown below.

ON YOUR OWN   *Illustrations from www.clipart.com*

14.4 Determine the shape, margin, and venation of the following leaves.

a.   b.   c.   d.

The leaves that a tree produces can generally be used to classify the tree. To get more experience identifying leaves and to see how this is done, perform the following experiment. Depending on your school schedule and the climate in your area, you may have to put this one off for a while. You should perform this experiment in the spring, once new leaves have grown on the trees.

EXPERIMENT 14.1
Leaf Collection and Identification

<u>Supplies:</u>

♦ Leaf press or old newspapers or old telephone books
♦ Laboratory notebook
♦ Tree identification book (from the library)

Object: To become familiar with the various trees in your area

<u>Procedure:</u>

1. Begin collecting leaf specimens from trees in your area. You should have around 20 different types of leaves. Make good notes (see step #6) about each tree from which you collect a leaf.
2. Allow sufficient time for the leaves to dry. Leaves that are still damp will mold later.
3. Press your leaves in a leaf press or between the pages of old magazines.

4. If you are using magazines, make sure you add weight to the stack of magazines to help press the leaves.

5. For those in a hurry, leaves can be pressed between two layers of waxed paper with a low-temperature iron. Put a paper towel between the ironing board and the waxed paper as well as between the iron and the waxed paper to avoid getting wax on your iron or ironing board. Trim the waxed paper a short distance from the leaf before mounting.

6. Attach the leaves carefully to the pages of your laboratory notebook. Each page should contain one leaf and the following information:

> ⇒ **Leaf mosaic**: alternate, opposite, or whorled
> ⇒ **Shape of the leaf**: deltoid, linear, elliptical, oval, cordate, lobed, cleft, orbicular, needlelike, or scalelike
> ⇒ **Venation of the leaf**: parallel, pinnate, palmate
> ⇒ **Leaf margins**: entire, undulate, serrate, crenate, or dentate
> ⇒ **Bark**:
>   • color
>   • rough or somewhat smooth
>   • other outstanding characteristics
> ⇒ **Any evidence of fruits** (including nuts)
> ⇒ **Identification of the tree** (if possible)

7. Be sure to clean up any mess you have made.

The Microscopic Structure of a Leaf

If we take a closer look at a leaf with a microscope, we can actually see that it is made up of a rather complex structure, as shown in the figure below.

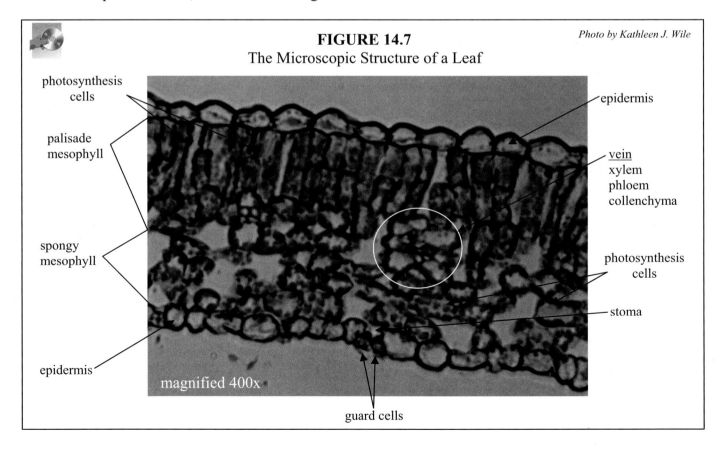

FIGURE 14.7
The Microscopic Structure of a Leaf

Photo by Kathleen J. Wile

photosynthesis cells

epidermis

palisade mesophyll

vein
xylem
phloem
collenchyma

spongy mesophyll

photosynthesis cells

stoma

epidermis

magnified 400x

guard cells

The top and bottom of a leaf are covered with a single layer of cells called the **epidermis**, which protects the inner parts of the leaf. Sometimes, the epidermis secretes a waxy substance called a **cuticle** (kyoo' tuh cuhl). Leaves that have cuticles are often shiny in appearance. If the epidermis of a leaf does not secrete a cuticle, it often grows hairs. These hairs give the leaf a velvety appearance. The African violet, for example, has velvety-looking leaves that get their appearance from the epidermal hairs. Many plants use these hairs as a defense mechanism. Stinging nettles, for example, have hairs that stick into your skin and inject an irritant. This is an effective means of keeping people and animals from walking through them.

On the underside of most leaves, there are tiny holes called **stomata** (stoh mah' tah). These stomata allow for the exchange of gases with the atmosphere, which is absolutely necessary for the survival of the plant. Each stoma (singular of stomata) is flanked by two cells called **guard cells**. These amazing little cells open and close the stoma. They contain chlorophyll and do photosynthesis to produce specialized sugars. These sugars cause the water pressure inside the cell to change, swelling or shrinking the cell. When the cells swell, they bend out, opening the stoma. When they shrink, the cells collapse, closing the stoma.

The opening and closing of the guard cells is primarily controlled by the level of light that the plant receives. When there is a lot of light, photosynthesis can take place, so the guard cells open the stomata. This allows the leaf to take in carbon dioxide for photosynthesis and release oxygen. When light levels are low, however, the leaf cannot afford to have the stomata open, because water evaporates from the leaf through the stomata. Remember from Module #10 that this is called transpiration. Since photosynthesis cannot take place when light levels are low, there is no reason for the stomata to be open, so the guard cells close the stomata. This keeps the leaf from losing too much water through transpiration. When the light level increases again, the guard cells open up, allowing photosynthesis to begin. There are other factors that play a role in the opening and closing of guard cells, but the amount of light the leaf receives is the most important one.

Under the epidermis on both sides of the leaf are **parenchyma** (pair en' kye muh) **tissues**, which are composed of cells that do the photosynthesis. The parenchyma tissue is composed of two layers, the **palisade** (pal ih sayd') **mesophyll** (mez' uh fil) and the **spongy mesophyll**. These tissues are made of oval-shaped cells that are rich in chloroplasts. In the palisade mesophyll, the cells are packed tightly, so as to maximize the number of cells in the tissue. At first, you might think that this reduces the amount of sunlight that the cells can absorb, because they all block each other's light. The Designer of these cells, however, has worked out an ingenious way around this problem.

The chloroplasts in the cells are constantly moving due to cytoplasmic streaming. This cytoplasmic streaming directs the chloroplasts to the top of the cell. Once at the top of the cell, a chloroplast absorbs all of the sunlight that it can handle and then is moved to make room for the next. In the end, then, the chloroplasts absorb the energy they need when they reach the top of the cell. That way, the cells can be packed tightly, to have as many food-producing cells as possible in the leaf.

Now wait a minute, you might be thinking. In order to perform photosynthesis, the food-producing cells need to absorb carbon dioxide. If the cells are packed together very tightly, there will not be room for the carbon dioxide to get into the cells. Once again, the leaf's Designer has worked a way around this problem. On the bottom of the leaf, the parenchyma is made up of the spongy mesophyll. This tissue has plenty of room for air, because the food-producing cells are packed very loosely. Thus, the stomata open into the spongy mesophyll so that air can come in and fill the space there. The cells in the spongy mesophyll are arranged so that the air that occupies the space provided for it touches *every food-producing cell in the leaf*! That way, each cell can absorb carbon dioxide for photosynthesis and can release the oxygen that the photosynthesis produces.

The veins that you see running through a leaf are made up of three tissues. These tissues come in two types: the vascular tissue, comprised of xylem and phloem, and another tissue called the **collenchyma** (kuh leng' kye muh). The collenchyma is made up of thick-walled cells that support the vein. Towards the end of the leaf, the veins get so small that there is no collenchyma any more.

Think about what you have just read. The leaf is designed in an incredibly efficient way so as to maximize its food-producing capabilities. It packs cells so that the total number of food-producing cells is as large as it can possibly be without hampering the availability of the chemicals and sunlight necessary for photosynthesis. At the same time, the leaf has stomata, which open when the leaf can use the gases in the air and close when the leaf cannot use them! That way, the leaf is exposed to the outside only when it benefits the plant. If exposure to the air does not benefit the plant, the plant shuts down! This is an amazingly efficient, well-designed food-producing machine! How in the world can anyone think that it developed as a result of random processes? Its design tells you that it had an incredible Designer.

ON YOUR OWN

14.5 A leaf cannot get the carbon dioxide that it needs for photosynthesis. What, most likely, is wrong?

14.6 Why can't the parenchyma be made of two layers of palisade mesophyll?

14.7 In the first section of this module, you learned about the four basic tissues in a plant: dermal, ground, meristematic, and vascular. In this section, you learned about the epidermis, the parenchyma, and the collenchyma. Classify each of these tissues within one of the four basic tissue types.

Leaf Color

Most leaves are green. Why? Well, the chlorophyll in their chloroplasts gives them their color. The more chlorophyll-containing cells there are in a section of leaf, the darker green that section is. Many leaves, for example, have a much deeper green color on the top of the leaf than on the underside. This is because the top of the leaf has the palisade mesophyll, where the chlorophyll-containing cells are packed together tightly. The underside of the leaf, however, contains the spongy mesophyll, where the chlorophyll-containing cells are not packed as tightly. As a result, the underside of the leaf is usually a lighter shade of green than is the top side of the leaf.

Of course, not all leaves are green, are they? Many plants have leaves that are not green at all or are only partially green. In addition, many leaves lose their green color in the fall. Consider, for example, the leaves shown in Figure 14.8 on the next page.

a. Photo from www.clipart.com
b. Photo © Jean-Yves Benedeyt

FIGURE 14.8
Leaves That Are Not Green or Are Only Partially Green

c. Photo © Corbis, Inc.

a. *Caladium* Leaf   b. European Beech   c. Autumn Leaves

In Figure 14.8a, we show the leaf of a caladium plant. Although there is clearly a lot of green on the leaf, there is also some red. In Figure 14.8b, we show a specific species of European beech. While many European beech trees have green leaves, some species (like the one in the photograph) have dark red or purple leaves. Finally, in Figure 14.8c, we show trees in autumn. Many of the leaves are no longer green at all, but are a mixture of red and yellow.

Where do all of these colors come from? Well, as we said before, the green is due to chlorophyll. Remember, in Module #6, we called chlorophyll a "pigment." It gives the leaf color, although its real job in the leaf is to absorb sunlight for photosynthesis. Plants contain pigments other than chlorophyll, however. In Module #6, we learned that cells contain **plastids**. These plastids hold pigments, starches, and oils. The plastids of some plants contain a group of pigments called **carotenoids** (kuh rot' en oydz). These carotenoids typically have yellow or orange hues. The orange color of pumpkins and carrots, for example, comes from carotenoids. The yellow in yellow peas or corn comes from other carotenoids. It turns out that most leaves contain carotenoids as well. In most leaves, the green color of the chlorophyll overwhelms the oranges and yellows of the carotenoids. In other plants, however, the amount of carotenoids is greater, and the leaf picks up some color from them.

In certain leaf tissues, there is another set of pigments called **anthocyanins** (an tho sye' uh ninz). These interesting pigments have different colors, depending on the pH of the leaf tissue. To give you an idea of what this means, perform Experiment 14.2.

EXPERIMENT 14.2
How Anthocyanins and pH Help Determine Leaf Color

<u>Supplies</u>
♦ Red (some people call it purple) cabbage (just a few leaves)
♦ Stove

- ♦ Stirring spoon
- ♦ Pot
- ♦ White vinegar (It must be clear. Apple cider vinegar will not work for this experiment.)
- ♦ Clear ammonia solution (This is sold in grocery stores with the cleaning supplies.)
- ♦ Water
- ♦ 2 eyedroppers
- ♦ 2 small cups or glasses
- ♦ 1 small glass (It must be see-through!)
- ♦ A sheet of white paper (preferably without lines)
- ♦ Measuring cups (1 cup and ¼ cup)
- ♦ Tablespoon

Object: To see how anthocyanins change color with differing pH levels

Procedure:

1. Put one cup of water and a few leaves of red cabbage into the pot.
2. Turn on the heat and bring the water to a boil. Allow the water to boil for five minutes, stirring continuously. If you start running out of water, just add a little more.
3. Allow the water to cool for a few minutes.
4. While the water is cooling, add two tablespoons of ammonia to one of the small glasses. This does not need to be a see-through glass.
5. Use the measuring cups to add 1¼ cups of water to the two tablespoons of ammonia. Swirl the solution to mix it.
6. Place an eyedropper in the small glass, and use it to stir the solution.
7. Pour some white vinegar into another small glass. It need not be a see-through glass.
8. Place an eyedropper in the glass of vinegar.
9. Once the boiled water has cooled, pour it into the measuring cup until you have ¼ cup of the boiled water solution. Be sure to remove any leaves that got into the cup.
10. Place the see-through glass on the white sheet of paper and pour the ¼ cup of the boiled water solution into it. Observe the color. It should be some shade of blue or green, but depending on the nature of your tap water, it could be purple or pink. Record that color in your notebook.
11. Add five drops of the ammonia solution you made earlier to the boiled water solution and swirl so that the ammonia solution mixes in. Is your solution green yet? If it is, stop adding the ammonia solution. If not, add the ammonia solution five drops at a time, swirling each time after adding the drops, until you get a green color. Even if your water was green as soon as you poured it into the glass, you still need to add those first five drops of ammonia solution. Record how many drops of ammonia solution you added and that the color is now green.
12. Add two drops of vinegar. Swirl the glass so that the vinegar mixes in. Note the color. Record the fact that you added two drops of vinegar and record the color.
13. Continue to add vinegar two drops at a time, swirling each time. Record the total number of drops of vinegar added and the color after adding those drops. Continue to do this until the solution has reached a pink hue and has not noticeably changed from that shade of pink for three additions of vinegar.
14. How many different colors did you see? You should have seen four distinct colors: green, blue, purple, and pink. If you want, you can reverse what you did by adding more of the ammonia

solution. With enough ammonia solution, you should be able to get back to the green color with which you started.
15. Clean up your mess.

What happened in this experiment? Well, red cabbage contains anthocyanins. When you boiled the red cabbage leaves in water, you extracted the anthocyanins from the leaves and dissolved them in the water. Ammonia is a base, so the ammonia solution you made was a base. As we learned in Module #5, bases have a pH greater than seven. In fact, the pH of the ammonia solution was probably about ten. When you added ammonia to the boiled water solution (which contained anthocyanins), you gave it a pH of more than nine. At that pH, the anthocyanins are green; thus, the solution was green. Vinegar is an acid. As you began adding acid, you started lowering the pH. When the pH dropped below nine, the anthocyanins turned blue. As you continued to add more vinegar, the pH continued to drop. When it dropped below seven, the anthocyanins turned purple. Pure vinegar has a pH of about three or four. As you added more and more vinegar, you got the pH closer and closer to four. Once it got near four, the anthocyanins turned pink. We see from this experiment, then, that a single set of pigments (anthocyanins) can give a leaf different colors depending on the pH of the leaf.

So now you know why some leaves are never completely green. Depending on the amount of carotenoids in the leaf, it might have an orange or yellow hue. If this orange or yellow hue mixes with the green of chlorophyll, another color might develop. Finally, depending on the pH of the leaf, anthocyanins will provide a blue, purple, or pink color to the leaf.

Of course, we still haven't explained why leaves that are green in the spring and summer turn different colors in the fall. The explanation for that is partly rooted in what we have discussed, but there is another part to the explanation as well. Before we launch into that explanation, however, we need to make one thing clear. As you are well aware, not all leaves change colors in the autumn. The needles on evergreen trees, for example, never change colors in the autumn. Also, the leaves on some tropical trees do not change color in the autumn. Only certain trees have leaves that change color in the autumn. They are the trees that lose all of their leaves for winter. We call those trees **deciduous** (duh sid' yoo us) trees.

Deciduous plant – A plant that loses its leaves for winter

Deciduous plants lose their leaves in order to conserve water throughout the winter.

At the base of each petiole in a deciduous tree, there is a thin layer of tissue called the **abscission** (ab cih' shun) **layer**. These cells perform a very specialized task. When the days start getting shorter, the cells in the abscission layer begin to block the xylem and phloem running through the petiole of the leaf. Once they have succeeded in blocking the xylem and phloem, they begin to die. As the cells in the abscission layer die, the layer itself begins cracking. The cracks eventually become so severe that the leaf falls off the tree under its own weight. Thus, the cells in the abscission layer have only one job: wait for the days to turn shorter, block the leaf's supply of water and nutrients, and then die. If you have a deciduous plant indoors, you might notice that it loses its leaves in the autumn as well. This is because the abscission layer is not affected by temperature. It begins doing its job when the time it is exposed to light begins decreasing.

Once the abscission layer has blocked the water and nutrient supply from the leaf, the leaf can no longer produce chlorophyll. As a result, the green color slowly fades away. This reveals whatever pigments were covered by the vibrant green of the chlorophyll. At the same time, when the leaf stops doing photosynthesis because it has no chlorophyll, other chemical reactions begin to occur. Under the right conditions (sunny and cool weather), these chemical reactions will produce anthocyanins and change the pH in the leaves. This, in turn, changes the color of the anthocyanins in the leaf. In areas of the world in which there are sunny, cool autumns, you will see brilliant reds, yellows, and purples in the leaves. This is mostly due to the anthocyanins in the leaf and the pH that the leaf has as a result of the chemical reactions that are occurring in it. In parts of the world where the autumn is cloudy and warm, the leaves do not have these brilliant colors because anthocyanins have not been produced.

Some leaves do not produce pigments other than chlorophyll. As a result, when they die, they just turn brown. This brown color is the result of a chemical called **tannic** (tan' ik) **acid**. Tannic acid is one product of the breakdown of plant cell contents. With no pigments to mask the color of tannic acid, the leaf is brown. If the leaf contains a high concentration of tannic acid, the leaf can be boiled, and the tannic acid will dissolve into the water, making tea. Since all dead leaves have tannic acid in them, you could make tea out of any leaf. However, when you boil leaves, chemicals other than tannic acid dissolve in the water as well. If your taste buds respond pleasantly to those chemicals, a pleasant-tasting tea is made. If not, the tea can be quite revolting. For some leaves, it can even be toxic. The next time you drink tea, just remember that you are actually drinking the remains of dead leaves!

ON YOUR OWN

14.8 There are a few leaves, such as the floating leaves of a water lily plant, in which the stomata and the spongy mesophyll are on the top side of the leaf while the palisade mesophyll is on the bottom. In these leaves, which side will have the darker green color?

14.9 If a leaf isn't green, does that mean there is no chlorophyll in it?

14.10 If a green leaf has no abscission layer, what color will the leaf be in the winter?

Roots

The roots of a plant perform three very important functions. First, they absorb water and nutrients from the plant's surroundings and transport them to where they are needed. Second, they anchor the plant. Finally, they are often used as a place for food storage. Many times, we exploit that last function for our own good. When you eat a carrot, you are actually eating the root of the carrot plant. The substances in that carrot that provide you with nutrition are actually food the plant had stored for later use.

Most roots are below ground, in the soil. This is not true for all plants, however. The epiphytic (ep uh fih' tik) orchid, for example, winds around the branches of tropical trees. It has roots that grow underground, but it also has aerial roots that allow it to cling to the branches around which it winds. These aerial roots absorb the materials that collect in the cracks of the branches. In addition, parasitic plants, such as the mistletoe, sink their roots into a host. These roots steal the nutrients that the host has absorbed. Other plants, such as ivy, have roots that hold them to rough surfaces like brick walls or rough tree bark. Even though we will discuss roots as if they all exist under the soil, it is important to note that there are exceptions like the ones we just discussed.

There are basically two kinds of root systems in plants: **fibrous root systems** and **taproot systems**. When a seed begins to sprout, the first root that comes out is called the "primary root." If the primary root continues to grow and stays the main root, the plant has a taproot system. Carrot plants, for example, have a taproot system. The carrot is the primary root, and it continues to grow as the plant's main root. Aside from a few tiny branches, the carrot is the predominant root in the plant. If, on the other hand, the primary root begins branching and branching until the root system looks like an underground "bush," the plant has a fibrous root system.

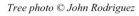 

FIGURE 14.9
Root Systems

This tree has a fibrous root system.   Each of these carrots is the taproot
                                              system of a carrot plant.

Most plants need significantly more surface area in their root system than in their leaves. Thus, the root systems of most plants are significantly larger than the part of the plant that exists above the soil. Corn, for example, usually reaches a height of eight to ten feet. The fibrous root system, however, has so many roots that, if you attached them end to end, they would stretch more than 150 feet! This fact is especially true of trees. The roots of a tree, if attached end to end, will generally be five to ten times longer than the tree is tall.

Now that you know a bit about the macroscopic features of roots, we want to discuss some microscopic features. A magnified image of a **longitudinal cross section** of a young primary root is shown in Figure 14.10 on the next page. In a longitudinal cross section, we take a slice of the root along its length. Then, we look at it as if we are looking at the side of the root.

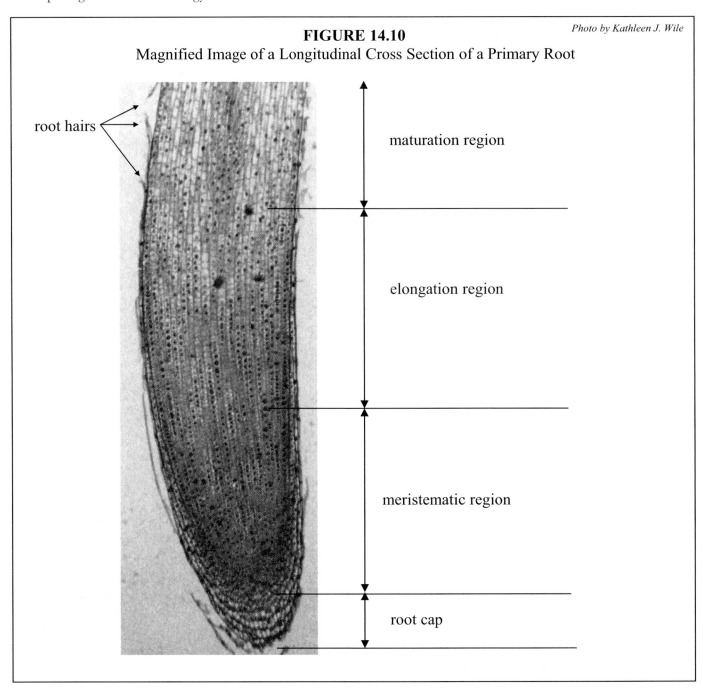

FIGURE 14.10

Magnified Image of a Longitudinal Cross Section of a Primary Root

Photo by Kathleen J. Wile

root hairs

maturation region

elongation region

meristematic region

root cap

Notice that the root is split into four basic regions: the **root cap**, the **meristematic region,** the **elongation region**, and the **maturation region**. The root cap is composed of dead, thick-walled cells. These cells protect the root as it shoves its way down into the soil. Just above the root cap is the meristematic region. In this part of the root, undifferentiated cells carry on mitosis. This is where most of the growth of the root takes place. After all, in order to grow, a root needs more cells. It makes sense, then, that the majority of growth will take place where mitosis occurs. In the elongation region of the root, cells are beginning to differentiate into specific kinds of cells. They stretch out, filling their central vacuoles with water. Since the cells are stretching out, some growth takes place in this portion of the root as well. Finally, in the maturation region, cells are becoming fully differentiated. Often, **root hairs** are produced. These hairs increase the surface area of the root, allowing it to absorb more water and nutrients from the soil.

Although a longitudinal cross section of the root, as shown in Figure 14.10, is instructive, another interesting way to look at a root is by taking a **lateral cross section** of the maturation region. In a lateral cross section, we slice a thin layer of the root from side to side so that the center of the root is the center of the slice. We then look at it as if we are looking straight down the root. The figure below shows such a cross section.

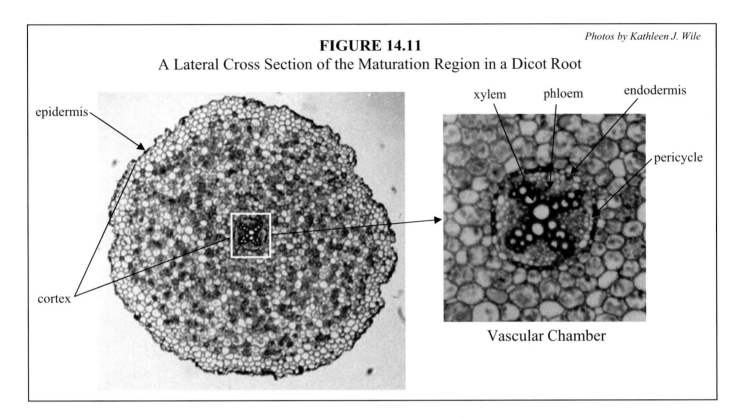

FIGURE 14.11

A Lateral Cross Section of the Maturation Region in a Dicot Root

Photos by Kathleen J. Wile

Vascular Chamber

As you can see from the figure, the root is protected by an epidermis. The cells inside the epidermis are called the **cortex**. This is where substances are stored for later use. Towards the center of the root, there is another one-cell-thick layer called the **endodermis**. These cells surround the **vascular chamber**, so that any substances which go to the xylem and phloem must first pass through them. This is the way the root controls what substances enter the plant. If the endodermis does not let a substance through, there is no way it can reach any other part of the plant. Thus, the endodermis guards the xylem and phloem, keeping out unwanted substances.

Inside the vascular chamber are the xylem and phloem, which transport nutrients and water throughout the plant. In the root of a dicot (the plants that have leaves with netted venation), the xylem usually form an X-shape. The phloem then fit between the arms of the "X." Between the xylem and the phloem lies the **vascular cambium** (cam' bee uhm). This tissue can become either xylem or phloem, depending on what the root needs. Just inside the endodermis, you can find the **pericycle**. These cells are undifferentiated and are technically a part of the vascular cambium. Rather than forming new xylem or phloem, however, the cells of the pericycle can form new branches for the root. If the root sends out a branch, the branch will form in this tissue and then "break out" of the endodermis, eventually traveling through the cortex and breaking out of the epidermis, thus forming a new root branch.

ON YOUR OWN

14.11 A 12-foot high plant has a root system that travels to a soil depth of only three feet. Does this plant have a taproot system or a fibrous root system?

14.12 Classify each tissue labeled in Figure 14.11 as ground tissue, dermal tissue, meristematic tissue, or vascular tissue.

14.13 If a root contains little cortex tissue compared to similar roots, what function can you conclude the root does **not** perform?

Stems

Plant stems have a variety of forms. They can be erect (as they are in trees and flowers), climbing (as they are in creeping vines), prostrate (as they are in watermelons and cucumbers), or even subterranean (as they are in the potato). They can be either woody or herbaceous. Regardless of the variety, however, stems perform three basic functions in a plant. First, they support and manufacture the plant's leaves. Second, they conduct water and nutrients to and from the leaves. Finally, they carry on photosynthesis. In some plants, stems only carry on photosynthesis when they are young. In other plants, like the cactus, the stems are actually the primary organs of photosynthesis.

Herbaceous Stems

The herbaceous stems of dicots and monocots are slightly different. In the figure below, lateral cross sections of a herbaceous monocot stem and a herbaceous dicot stem are shown.

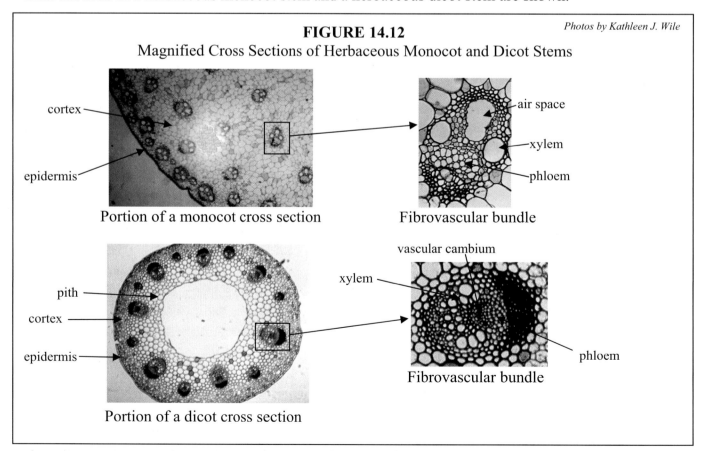

FIGURE 14.12

Photos by Kathleen J. Wile

Magnified Cross Sections of Herbaceous Monocot and Dicot Stems

Portion of a monocot cross section

Fibrovascular bundle

Portion of a dicot cross section

Fibrovascular bundle

Although there are clearly differences between the two types of stems, there are also some similarities. Both are covered with an epidermis, which protects the stem. They also both have **fibrovascular** (fye broh vas' kyoo ler) **bundles**. These bundles contain the xylem and phloem that transport substances throughout the plant. Each stem has a cortex, which contains photosynthetic cells and also stores starch. In the dicot stem, the cortex is found between the epidermis and the fibrovascular bundles, while in monocot stems, it is found throughout the stem.

The main differences between monocot and dicot stems are the fibrovascular bundles and the ways in which they are arranged. In monocot stems, the bundles are located throughout the stem. In dicot stems, however, they tend to form a ring near the outer part of the stem. Another difference between the two stems is the presence of **pith** tissue. In dicots, the pith tissue is the ground tissue inside the ring of fibrovascular bundles. A portion of the pith often breaks down, forming a hollow stem, as is shown in the figure.

Notice that the xylem and phloem within the fibrovascular bundles have different shapes in the two stems. In a monocot stem, the xylem and air spaces are arranged to form what almost looks like a monkey's face. The phloem form the forehead of the face. In the dicot stem, however, there is no illusion of a monkey face. Instead, xylem are clumped in one region of the fibrovascular bundle, and phloem are clumped in another region.

One other difference between monocot and dicot stems is the presence of a **vascular cambium**. In the fibrovascular bundles of a dicot stem, the vascular cambium can form new xylem or phloem if the stem needs the ability to transport more materials. The amount of new xylem and phloem that can be produced is limited, however. The epidermis of a herbaceous dicot stem cannot grow once it is mature. Thus, if too many new xylem or phloem are created, the stem will crack through the epidermis, exposing its inner tissues to the environment. This will usually kill the stem.

Woody Stems

Woody stems are different from herbaceous stems, as can be seen in the microscopic image below.

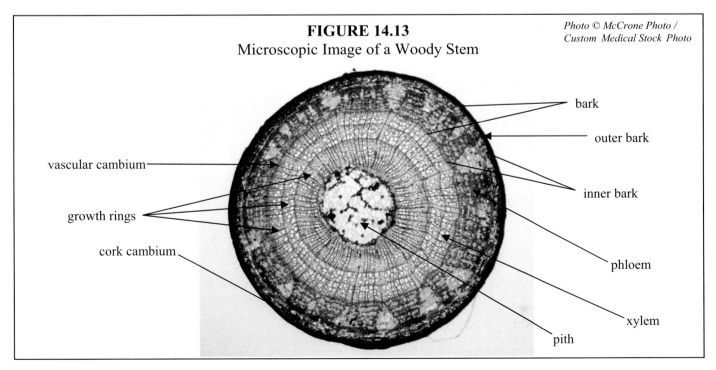

FIGURE 14.13
Microscopic Image of a Woody Stem

Photo © McCrone Photo / Custom Medical Stock Photo

bark
outer bark
vascular cambium
inner bark
growth rings
cork cambium
phloem
xylem
pith

One of the major differences between herbaceous stems and woody stems is the **bark**. The bark of a woody stem is actually composed of two layers: the **inner bark** and the **outer bark**. The inner bark is composed of phloem and cortex tissue. Between the inner bark and the outer bark is a layer of tissue called the **cork cambium**. This layer continually produces **cork cells**. These cork cells die quickly and are impenetrable to water, gases, and most parasites. They form the **outer bark**.

The formation of bark allows a woody stem to continue to grow, unlike most herbaceous stems. Remember, if a herbaceous stem grows too much, the epidermis cracks, exposing the inner parts of the stem to the environment. In a woody stem, however, the growth causes the outer bark to crack and break, but that's okay, because the cork cambium simply produces new outer bark to protect the stem. Now as far as we are concerned, the trunk of a tree is really just the main stem of the tree. So now you know why the bark on the outside of a tree is cracked and rough. It is the result of the tree "outgrowing" its bark shield and breaking through it.

Not all trees have rough bark, however. The white birch tree, commonly found in Canada but also found in the United States, has thin, white bark that peels off. When this bark peels off, the cork cambium just produces new bark to take its place. This gives the stems of the white birch a very smooth feel. Because the bark is so thin, it is an excellent fire starter. Boy Scouts and Girl Scouts are taught to find a white birch and peel off its bark in order to get easy-to-light kindling for a fire.

Because there is really no limit to the growth of a woody stem, new xylem and phloem must always be produced inside the stem. This is done by the **vascular cambium**, which produces phloem on its outer side and xylem on its inner side. During the spring when there is plenty of water, the xylem produced are quite large. As the summer goes on and water becomes scarce, the xylem produced become smaller. When the xylem cells die, they form what we call the wood of the tree. Since the xylem produced in the spring are much bigger than those produced in the later part of the growing season, the wood of the tree is lighter when it is produced in the spring and darker when it is produced in the late summer. These alternating areas of light and dark form what we call **annual growth rings**. If you look at a lateral cross section of a woody stem, you can tell how old it is by counting these rings. Also, the thickness of the rings can tell you a lot about the weather conditions during that year of formation.

Notice the presence of pith in the woody stem. This should tell you something. The stem in the figure is a dicot, because only dicots have pith. It turns out that monocots do not produce woody stems, because of the arrangement of their vascular bundles. After all, a dicot's vascular bundles are all on the outer edge of the stem. That way, the vascular cambium can produce xylem on one side and phloem on the other. The phloem becomes the inner bark, while the xylem becomes the wood, increasing the diameter of the stem. Since a monocot's vascular bundles are scattered throughout the stem, the formation of bark in this manner is just not possible. Thus, monocot stems are not woody.

In case this little fact has slipped past you, we need to emphasize it. In a woody stem, the phloem are always near the outside of the stem, while the xylem are always in the inner part of the stem. This leads us to a discussion of **girdling**.

<u>Girdling</u> – The process of cutting away a ring of inner and outer bark all the way around a tree trunk

When a tree is girdled, at first it appears that the tree is okay. For quite a while, the leaves of the tree will stay green and the tree will grow.

Remember, the xylem (which carry water and minerals up the stem) are in the inner part of a stem (or tree trunk), while the phloem (which carry food and organic materials) are a part of the inner bark. If you cut away a ring of the inner bark, the water and minerals can still move up the tree (in the xylem), but the food and organic materials cannot move down the tree past the ring, because all of the phloem have been cut. As a result, the roots can still send water and nutrients up to the leaves, but the leaves cannot send food back down to the roots. Eventually, the roots (and all cells below the ring) will starve to death, and that will kill the rest of the tree. This process can take several years, however, so the damage is not readily apparent.

Specialized Stems

As we mentioned before, some stems actually exist underground. They are sometimes mistaken for roots, but they are, nevertheless, stems. The onion plant, for example, produces underground **bulbs**. These bulbs are not roots. They are, instead, a collection of underground leaves that sprout from an underground stem. These leaves do not perform photosynthesis, of course, because there is no light underground. Instead, the leaves are used for storage. Another stem often mistaken for a root is the **tuber**. Potato plants produce tubers, which are underground stems that are used to store excess food. We harvest these tubers, of course, and we call them potatoes.

ON YOUR OWN

14.14 If a stem has no cork, is it woody or herbaceous?

14.15 If a stem has no limits to its growth, is it woody or herbaceous?

14.16 A stem has xylem and phloem packed together in fibrovascular bundles scattered throughout the stem. Is it woody or herbaceous?

To better familiarize yourself with the roots, stems, and leaves of plants, perform the following microscope experiment. If you do not have a microscope, read through the experiment to learn what you can.

EXPERIMENT 14.3
Cross Sections of Roots, Stems, and a Leaf

Supplies:

♦ Prepared slide: *Zea mays* (corn) cross section of stem
♦ Prepared slide: *Zea mays* (corn) cross section of root
♦ Prepared slide: *Ranunculus* (buttercup) cross section of stem
♦ Prepared slide: *Ranunculus* (buttercup) cross section of root
♦ Prepared slide: Leaf cross section with vein
♦ Microscope
♦ Lab notebook
♦ Colored pencils

Object: To observe the microscopic structure of a leaf and to compare the microscopic structures of monocot and dicot stems and roots

Procedure:

A. Observation of a leaf cross section

1. Set up your microscope as you have for other experiments.
2. Observe the prepared slide of a leaf cross section with low magnification at first.
3. As you get familiar with the slide, you can increase the magnification to find the optimum magnification with which to view the slide (100x is usually best).
4. Although your leaf cross section might not look exactly like the one shown in Figure 14.7, you should be able to use the figure as a guide to find the following structures:

   a. Upper epidermal cells
   b. Palisade mesophyll
   c. Veins
   d. Spongy mesophyll
   e. Lower epidermal cells
   f. Guard cells
   g. Stomata

5. Draw the cross section you have been studying, labeling all of the structures that you can find.

B. Observation of a lateral cross section of a *Ranunculus* root

1. Observe the *Ranunculus* (a dicot) root cross section. Once again, start on low magnification.
2. Draw what you see. Using Figure 14.11 as a guide, identify the following structures:

   a. Epidermis
   b. Cortex

3. Center on the vascular chamber and increase to high magnification. Draw what you see. Using Figure 14.11 as a guide, identify the following structures:

   a. Xylem
   b. Phloem
   c. Pericycle
   d. Endodermis

C. Observation of a lateral cross section of a *Zea mays* root

1. *Zea mays* is the binomial name of corn, which is a monocot. Observe the slide on low magnification. Draw what you see. We did not show you a figure for the root of a monocot. However, you should still be able to identify these structures:

   a. Epidermis
   b. Cortex

c. Endodermis

2. Center on the vascular tissue, which is just on the inside of the endodermis. Increase magnification and see if you can distinguish the xylem from the phloem.
3. A monocot root has a section of pith. Can you identify the pith in your specimen?
4. If you need help identifying the structures, go to the course website that we discussed in the "Student Notes" section of the book. There are some links to pictures that will help you.
5. Now that you have studied both a monocot root and a dicot root, note the differences between the two.

D. Observation of a lateral cross section of a *Zea mays* stem

1. Observe the *Zea mays* stem on low magnification.
2. Draw what you see. Using Figure 14.12 as a guide, label the following structures:

   a. Epidermis
   b. Cortex
   c. Fibrovascular bundle

3. Center on one of the fibrovascular bundles and increase to high magnification.
4. Draw what you see. Using Figure 14.12 as a guide, label the following structures:

   a. Xylem
   b. Phloem
   c. Air space

E. Observation of a lateral cross section of a *Ranunculus* stem

1. Observe the *Ranunculus* stem on low magnification.
2. Draw what you see. Using Figure 14.12 as a guide, label the following structures:

   a. Pith
   b. Epidermis
   c. Cortex
   d. Fibrovascular bundle

3. Center on one of the fibrovascular bundles and increase to high magnification.
4. Draw what you see. Using Figure 14.12 as a guide, label the following structures:

   a. Xylem
   b. Phloem
   c. Vascular cambium

5. Now that you have studied both a monocot and a dicot stem, note the differences between the two.
6. Clean up your mess and put everything back in its proper place.

Classification of Plants

There is some controversy about how to classify plants. Thus, the classification scheme that you learn here might be different from other ones you might study later on. In fact, some classification schemes that you learn might even put algae in kingdom Plantae. After all, since algae are photosynthetic, they do have something in common with the plants. In this book, however, we consider plants to be multicellular creatures. Thus, most algae are not really plants. Even the algae that are multicellular do not have the specialized tissues that plants do, so we do not consider them plants. Nevertheless, some biologists do. Depending on the next biology book you read, then, there may be differences between the classification scheme that we give you and the one that it gives you.

Plants can be split into two basic groups: plants with vascular tissue and plants without vascular tissue. Plants without vascular tissue are typically called either **bryophytes** (bry' oh fytes) or **nonvascular plants**. Plants with vascular tissue are often called **tracheophytes** (tray' key oh fytes), because trachea are tubes inside of animals, and the xylem and phloem can be thought of as tubes inside of plants. Although many biologists use the term tracheophytes, others simply call these plants **vascular plants**. We will begin our discussion of plant classification with the bryophytes.

The Bryophytes

Because bryophytes do not have vascular tissue, they cannot grow to be very tall. After all, the taller the plant, the farther nutrients must be transported. Without vascular tissue, it is difficult to transport nutrients very far, so the size of bryophytes is limited.

Three phyla of bryophytes exist, but the most representative phylum is phylum **Bryophyta** (bry oh fye' tuh), which contains the mosses. Although it is tempting to call any small, green clump a moss, to a biologist, the term **moss** means something quite particular. A moss is composed of many tightly packed individual plants. These plants are composed of **leafy shoots** and **rhizoids**. The leafy shoots are tiny stems with even tinier, leaflike structures. These leaflike structures are but one cell thick, and they directly absorb the nutrients that they need from the environment while they perform photosynthesis. Nutrients can travel through the leafy shoot by falling through the spaces in between the cells of the stem, much like water can travel through a paper towel.

The rhizoids of a moss plant look like roots, but they are not. Roots can absorb nutrients and then send them to the other parts of the plant. Since mosses have no vascular tissues, the rhizoids cannot send nutrients anywhere. Instead, the rhizoids are simply strands of tissue used to anchor the moss. Since the rhizoids cannot absorb water from the soil and send it anywhere, mosses are dependent on absorbing the water that collects on their leafy shoots. This means that in order to survive, mosses need very moist environments.

Mosses have an interesting life cycle which biologists often describe as **alternation of generations**.

<u>Alternation of generations</u> – A life cycle in which there is both a muticellular diploid form and a multicellular haploid form

This interesting life cycle is illustrated in the figure below.

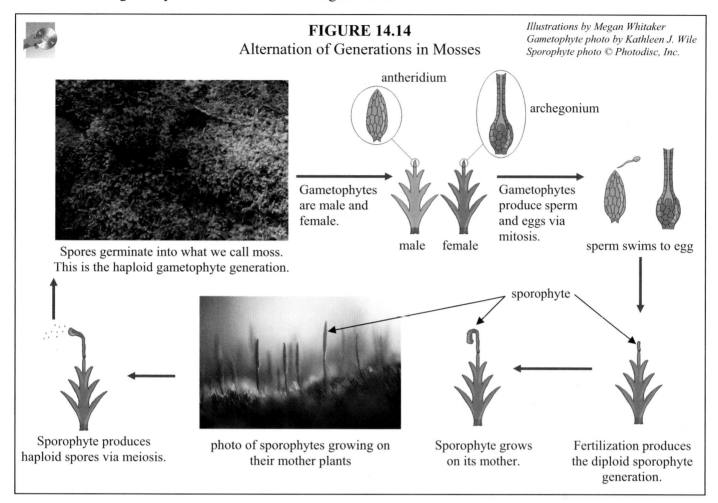

FIGURE 14.14
Alternation of Generations in Mosses

Illustrations by Megan Whitaker
Gametophyte photo by Kathleen J. Wile
Sporophyte photo © Photodisc, Inc.

antheridium

archegonium

Gametophytes are male and female.

Gametophytes produce sperm and eggs via mitosis.

male   female

sperm swims to egg

Spores germinate into what we call moss. This is the haploid gametophyte generation.

sporophyte

Sporophyte produces haploid spores via meiosis.

photo of sporophytes growing on their mother plants

Sporophyte grows on its mother.

Fertilization produces the diploid sporophyte generation.

What you and I call moss is actually only one part of the moss life cycle: the **gametophyte** (gam ee' toh fyte) **generation**. This generation gets its name from the fact that it has individual males and females that produce gametes (sperm and eggs). However, unlike what you have studied so far, those sperm and eggs are made through *mitosis*, not meiosis! How in the world can gametes be made from mitosis? Well, the gametophyte generation of moss is actually *haploid*! This means that every cell in a moss plant during this generation has only one chromosome from each homologous pair. Since mitosis makes copies of cells, and since all of the cells of a moss plant in this generation are haploid, the gametes that this generation produces are made through mitosis.

The sperm made by the male plants in the gametophyte generation are made in structures called **antheridia** (an thuh rid' ee uh – singular is antheridium). The female plants produce eggs in structures called **archegonia** (ar kee goh' nee uh – singular is archegonium). When the antheridia release their sperm, the sperm travel through the water that has collected on the moss until they can reach an archegoinum that contains an egg. When fertilization takes place, the result is a diploid zygote, which develops into the **sporophyte** (spoor' oh fyte) **generation** of the moss life cycle.

The sporophyte actually grows right on top of the mother plant that held the egg that was fertilized. When it is ready to reproduce, it forms spores, which is why this is called the sporophyte generation. Now remember, the sporophyte is made up of diploid cells, and the spores that it makes will give rise to the haploid gametophyte generation. Thus, the sporophyte makes its spores via meiosis, which turns diploid cells into haploid cells. The sporophyte then releases the spores, and if they land in a spot with the appropriate conditions, they will germinate into the haploid gametophyte generation, which starts the life cycle all over again.

Do you see what's going on here? The leafy shoots that make up moss are composed of haploid cells. They reproduce sexually, forming haploid gametes via mitosis. As a result, the leafy shoots are a part of the gametophyte generation. The offspring from this sexual reproduction are diploid, because the two haploid gametes join their DNA during fertilization. This diploid generation produces haploid spores via meiosis. Thus, this is called the sporophyte generation. When the spores begin to develop, they form leafy shoots again, which are part of the gametophyte generation.

Do you see why we call this "alternation of generations?" In the first generation, the moss is a haploid organism that produces gametes. In the second generation, it is diploid organism that produces spores. In the end, then, the generations alternate between haploid and diploid! Since the leafy shoot is what we typically see when we examine mosses, we say that the gametophyte is the **dominant generation**.

Dominant generation – In alternation of generations, the generation that occupies the largest portion of the life cycle

Alternation of generations, although seemingly exotic, happens in phylum Pterophyta (a phylum we will study in a moment) as well. In fact, some biologists apply the alternation of generations life cycle to all plants. While we think that this is overstating the importance of the alternation of generations life cycle, you will see what we mean as we study the other plants in our classification system.

Although mosses are rather common, they have few uses for people. About the only kind of moss that is of economic importance to us is peat moss. This moss grows floating at the top of a pond. When the moss dies, it sinks to the bottom. Eventually, the dead moss fills up the pond, turning it into a bog. Eventually, plants begin to grow in the peat moss of the bog and, after a while, there is no evidence that a pond was there at all. Farmers use peat moss to make their soil more fertile. It can also be used to pack plants for shipment. Finally, dried peat moss can be burned as fuel.

Mosses are important to creation as a whole in many ways. They provide a place to live for many tiny creatures, and they are useful to birds for building nests. Most animals don't usually eat moss, because it does not have a lot of nutritional value. However, some animals such as bears, deer, and turtles will eat it if they can't find any other kind of food. There is one animal, however, that eats it regularly: the reindeer. Reindeer live in cold climates, and moss has a special chemical that helps keep the fluids inside the reindeer from freezing, even on the coldest of days. Moss, then, is a kind of reindeer antifreeze.

ON YOUR OWN

14.17 You see a plant that is two feet tall. Can it be a bryophyte?

14.18 You study a moss that reproduces by making spores. Is it composed of diploid or haploid cells? Does it make the spores using mitosis or meiosis? If you study the offspring of this reproduction, will its cells be haploid or diploid? What kind of cells will it make in order to reproduce? Will it use mitosis or meiosis to make these cells?

Seedless Vascular Plants

The vast majority of plants are vascular plants, or tracheophytes. Because they have vascular tissues, these plants can grow to be quite large. Sequoia trees in California for example, can grow more than 250 feet tall and have trunks that are as large as 85 feet in circumference! There are many phyla of vascular plants, but we want to concentrate only on the three that contain the vast majority of living species in creation. In this section, we will discuss a phylum that contains vascular plants that do not produce seeds. Not surprisingly, they are often called the **seedless vascular plants**.

The members of phylum **Pterophyta** (ter uh fye' tuh) are commonly called "ferns." Ferns can be found either on the forest floor, growing on other trees, or with slender trunks that can reach heights of up to 60 feet. Ferns are separated from other vascular plants because they do not produce seeds. Like mosses, they have an alternation of generations life cycle. Some examples of ferns are shown in the figure below.

FIGURE 14.15
Various Types of Ferns

© Painet Photographic Arts and Illustration NETwork

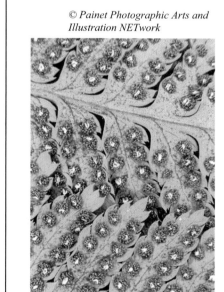

a. Underside of a fern leaf covered with sorri

b. Ferns growing on the forest floor
© Jim Lundgren

c. Tall, treelike fern
© Creatas, Inc.

Ferns usually grow from a stem that is either underground or attached to a tree. If the stem is underground, it produces roots. If the stem is attached to a tree, the fern does not act as a parasite. Instead, the stem produces root hairs that absorb the water and nutrients that have collected in the cracks of the tree's outer bark. There are even a few species of fern that look like a tree. They have slender trunks that support the stems and leaves. Unlike tree trunks, fern trunks are not a solid woody mass. Instead, they are made up of a network of hard stems. Ferns are typically delicate and often die as a result of even slight environmental changes. Although ferns are popular houseplants, many people find it hard to keep them alive, because of their delicate nature.

Ferns are also characterized by their alternation of generations life cycle. In the fern, however, the sporophyte generation is the dominant generation, as illustrated below.

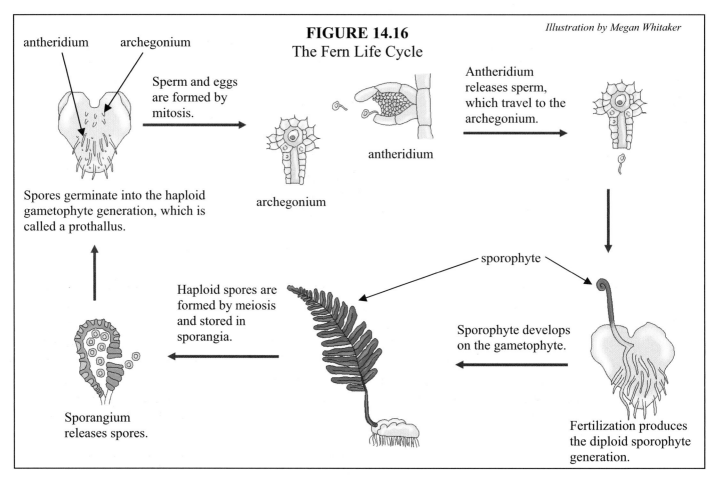

FIGURE 14.16
The Fern Life Cycle

Illustration by Megan Whitaker

antheridium   archegonium

Sperm and eggs
are formed by
mitosis.

Antheridium
releases sperm,
which travel to the
archegonium.

antheridium

archegonium

Spores germinate into the haploid
gametophyte generation, which is
called a prothallus.

sporophyte

Haploid spores are
formed by meiosis
and stored in
sporangia.

Sporophyte develops
on the gametophyte.

Sporangium
releases spores.

Fertilization produces
the diploid sporophyte
generation.

This should look familiar. In the fern life cycle, the haploid gametophyte generation is called a **prothallus** (pro thal' us). It has both male (antheridia) and female (archegonia) parts, and the gametes are produced by mitosis. The antheridia release their sperm, and fertilization occurs when a sperm successfully fuses with an egg in an archegonium. This produces a diploid sporophyte, which develops right on the prothallus. This is what we call a fern. The fern eventually produces spores by meiosis, and those spores are stored in vessels called **sporangia** (spoor an' gee uh). The sporangia are arranged in large clumps called **sorri** (sor' eye) on the underside of the leaves. If you look at Figure 14.15a, you will see a fern leaf with sorri on it. The spores, when released and conditions are right, will mature into a prothallus, which is the gametophyte generation of the fern. Notice, then, that the alternation of generations life cycle of ferns is very similar to that of mosses, except that the dominant generation in the case of ferns is the sporophyte generation, while in mosses, the dominant generation is the gametophyte generation.

ON YOUR OWN

14.19 A biologist studies a fern leaf with sorri. Is the fern leaf made of diploid or haploid cells?

14.20 A student sees a fern growing on the branch of another tree. The student says that the fern is obviously a parasite. Is the student correct? Why or why not?

Seed-Making Plants

So far, we have discussed nonvascular plants (bryophytes) and seedless vascular plants. However, the majority of plants make seeds in order to reproduce. There are several phyla of seed-making plants, but we will concentrate on the two most populated phyla: **Coniferophyta** (con uh fur' oh fye' tuh) and **Anthophyta** (an thoh fye' tuh). The evergreen tree is the general name for members of phylum Coniferophyta. These trees, typically called "conifers," reproduce by making cones.

A conifer actually produces two kinds of cones. The first, which everyone seems to notice, are called "pine cones." These are what biologists call **seed cones** (see Figure 14.17), and they are the female reproductive organs of the pine tree. The seed cones contain the egg cells. On the tips of most pine branches, little knobs develop. Those knobs are actually **pollen cones** (see Figure 14.17), which contain **pollen**.

Pollen – A fine dust that contains the sperm of seed-producing plants

When the pollen cones release their pollen, it is carried by the wind. Some of the pollen will land in seed cones. If the seed cone happens to be on the same tree as the pollen cone was, the result is called **self-fertilization**, which we talked about in Module #8. The seed cone then closes, and the sperm attempt to fertilize the egg. This may take several months. Once fertilization occurs, a seed begins to form. When the seed is ready and conditions are right, the seed cone opens again, releasing the seeds. You can actually call this an alternation of generations life cycle, since the structures in the seed and pollen cones are multicellular and haploid. If you look at the conifer life cycle in that way, then, the tree is the sporophyte generation, which is dominant, and the structures in the seed and pollen cones are the gametophyte generation.

Seed cone photo by Dr. Jay L. Wile   **FIGURE 14.17**   *Pollen cones photo © David Foltz*
The Reproductive Organs of a Conifer

Please note that these two pictures are not on the same scale. The pollen cones are much smaller than the seed cone.

Seed Cone

Pollen Cones

Although pine trees are the best-known conifers, there are others. The giant redwood trees and sequoia trees in California are conifers. The oldest trees in creation, the bristlecone pines, are also conifers. Remember that woody stems have annual growth rings. Because of this, you can look at a cross section of a tree trunk and count the rings to determine the tree's age. Using this method, the oldest known tree in creation is a bristlecone pine that is approximately 4,700 years old. This is particularly interesting, as biologists indicate that there is no theoretical age limit on these trees. As a result, they should be able to live much, much longer than 4,700 years. Why is the oldest known tree in creation only 4,700 years old? We cannot say for sure, but it is interesting to note that the worldwide flood as described in the Bible happened about 4,700 years ago. For those who do not believe in the worldwide flood, the age of the oldest known tree in creation is a real problem.

The last phylum we want to discuss is phylum Anthophyta. Members of this phylum are often called the "flowering plants." This is because anthophytes all have flowers in which their seeds are produced. Although you usually think of flowers as beautiful blossoms, it is important to note that not all flowers are brightly-colored and beautiful. If you pass a corn field towards the end of the corn season, you will see tassels on top of the corn. Those are actually the flowers of the corn plant.

The flowers on an anthophyte produce both sperm and eggs. The sperm are produced on structures called stamens, while the eggs are produced in a carpel and covered in a structure called the **ovary**. When the egg is fertilized, the ovary develops into the **fruit** of the plant, which encases the resulting seed. Like the conifers, many anthophytes can also self-fertilize. Once again, you can interpret this as an alternation of generations life cycle, with the plant being the sporophyte generation and the multicellular structures in the flower being the gametophyte generation.

Phylum Anthophyta is split into two classes: **Monocotyledonae** (mon uh kaht' uh lee' doh nay) and **Dicotyledonae** (dye kaht' uh lee' doh nay). These are the monocots and dicots that you read about earlier. Although there are many differences between monocots and dicots (stem structure, root structure, and leaf venation), the fundamental difference between them is the way that the seed is constructed. Seeds produced by anthophytes have either one or two **cotyledons** (cot uh lee' dunz).

Cotyledon – A "seed leaf" which develops as a part of the seed. It provides nutrients to the developing seedling and eventually becomes the first leaf of the plant.

Monocots (such as corn) have a single cotyledon, whereas dicots (such as bean plants) have two cotyledons.

This fundamental difference between monocots and dicots leads to a host of other differences. As you already know, the venation of a leaf in a monocot is parallel, whereas it is netted in a dicot. The structures of the stem are different in monocots and dicots. Typically, monocots have fibrous root systems, whereas dicots have taproot systems. Finally, monocots usually produce flowers with petals in groups of three or six, while the petals of dicots flowers are usually in groups of four or five.

Reproduction of monocots and dicots is a bit complicated, so we will study the process in detail in the next module. When we do that, you will learn more about the parts of a flower, how anthophytes make seeds, and what cotyledons are. Thus, if some of what we have covered in regard to the anthophytes is unclear, it will probably make a lot more sense after you read the next module.

ON YOUR OWN

14.21 Suppose a conifer self-fertilizes. From a genetic point of view, is this the same as asexual reproduction? Why or why not?

14.22 Construct a biological key for the classification of plants. Assume that the specimen you are examining is definitely a part of kingdom Plantae and is in one of the phyla we have discussed. For phyla other than Anthophyta, simply stop at the phylum level. If the plant is a part of phylum Anthophyta, classify it down to the class level, using the fundamental distinction between monocots and dicots.

ANSWERS TO THE "ON YOUR OWN" PROBLEMS

14.1 <u>They are most likely herbaceous plants</u>, because herbaceous plants are typically annuals.

14.2 <u>It is a vegetative organ.</u> All roots are vegetative organs.

14.3 <u>Meristematic tissue will not be present.</u> Tissue that is undergoing mitosis is the meristematic tissue. If a region of a plant is not undergoing mitosis, no meristematic tissue will be there.

14.4

Letter	Shape	Margin	Venation
a.	Deltoid	Serrate	Pinnate
b.	Lobed	Entire	Palmate
c.	Cleft	Dentate	Palmate
d.	Linear	Entire (It might look like undulate, but the leaf is just wrinkled around the edges.)	Parallel

14.5 <u>The stomata are not opening.</u> This is a problem with the guard cells.

14.6 <u>If both layers were palisade mesophyll, there would be no room for carbon dioxide to get into the leaf or oxygen to get out.</u>

14.7 <u>The epidermis is the dermal tissue.</u> Remember, dermal tissue is one cell thick and is on the outside of the plant. <u>The parenchyma and collenchyma are ground tissues.</u> Remember what ground tissues do. Among other things, they perform photosynthesis (which is what the parenchyma does), and they support other cells (which is what the collenchyma does). You might be tempted to say that the collenchyma is a part of the vascular tissue, because it can be found in the vein. However, remember that vascular tissue is composed only of xylem and phloem. Also, remember what the collenchyma does: it supports the vascular tissue. Support is a function of the ground tissue.

14.8 <u>The underside of the leaf will be darker</u>, because in these leaves, the chlorophyll-containing cells are packed more tightly on the underside of the leaf.

14.9 <u>No.</u> The color of the chlorophyll may be masked by other pigments.

14.10 Without an abscission layer, there is nothing to cut off the flow of nutrients. This means the leaf will not fall off the tree and will remain <u>green all winter</u>.

14.11 Since almost all plants have roots that, when stacked end to end, are much longer than the plant itself, this plant must have a <u>fibrous root system</u>. That's the only way a root system that goes only three feet deep will be longer than the plant.

14.12 <u>The epidermis is dermal tissue</u>, because it is on the outside of the plant. <u>The cortex is ground tissue</u>, as it stores nutrients. <u>The endodermis is ground tissue</u>, as it supports the xylem and phloem by regulating what can enter the vascular tissues. <u>The pericycle is meristematic tissue</u>, as it is undifferentiated. <u>The xylem and phloem are vascular tissues.</u>

14.13 With little cortex tissue, the root will <u>not store substances</u>.

14.14 Without cork, the stem is <u>herbaceous</u>. All woody stems have cork, which makes up the outer bark.

14.15 Herbaceous stems have limits to their growth. This must be a <u>woody stem</u>, because the construction of bark poses no limits to growth.

14.16 <u>The stem is herbaceous</u>, because monocots have their fibrovascular bundles distributed that way, and monocots have herbaceous stems.

14.17 <u>No</u>. Bryophytes must be tiny due to their lack of vascular tissue.

14.18 Mosses have an alternation of generations life cycle. If you are studying the moss that reproduces using spores, you are studying the sporophyte generation. This means that <u>the organism you are studying is composed of diploid cells and uses meiosis to make its spores</u>, since the spores are haploid. <u>The offspring of this reproduction will be haploid</u>, as it grows from haploid spores. <u>It will produce gametes</u> (sperm and egg), and since it is already haploid, <u>it produces its gametes through mitosis</u>.

14.19 <u>It is made of diploid cells</u>. If it has sorri, it is the sporophyte generation, which is diploid.

14.20 <u>The student is wrong. Ferns do grow on other trees, but they do not take nutrients from the trees. Instead, they take nutrients that gather in the cracks of the tree's bark.</u>

14.21 <u>This is quite different from asexual reproduction.</u> Remember, in asexual reproduction, the genetic code is exactly the same in parent and offspring. In Module #8, however, we saw that when a Tt plant self-fertilizes, it can make a TT plant, a Tt plant, or a tt plant (see Figure 8.4). Since this can happen for every allele in the genetic code, an offspring from self-fertilization will be genetically different from its parent.

14.22 Your key may look different. You need to ask the same types of questions, but yours can be in a different order than ours.

1. Vascular Tissue **2**
    No Vascular Tissue..*phylum Bryophyta*

2. Produces seeds **3**
    Does not produce seeds ....*phylum Pterophyta*

3. Seeds produced in flowers ....*phylum Anthophyta*............ **4**
    Seeds produced in cones ....*phylum Coniferophyta*

4. Single cotyledon in seed .... *class Monocotyledonae*
    Two cotyledons in seed .... *class Dicotyledonae*

NOTE: This is by no means a complete biological key for plants. There are phyla that we did not discuss, and, of course, we haven't gone anywhere close to classifying down to the species level!

STUDY GUIDE FOR MODULE #14

1. Define the following terms:

a. Botany
b. Perennial plants
c. Annual plants
d. Biennial plants
e. Vegetative organs
f. Reproductive plant organs

g. Undifferentiated cells
h. Xylem
i. Phloem
j. Leaf mosaic
k. Leaf margin
l. Deciduous plant

m. Girdling
n. Alternation of generations
o. Dominant generation
p. Pollen
q. Cotyledon

2. If a portion of a plant is producing new cells, what type of plant tissue will be in that region?

3. What do we call the structure that attaches the blade of the leaf to the stem?

4. Identify the leaf mosaics in the pictures below:

Illustrations by Megan Whitaker

a.

b.

c.

5. Determine the shape, margin, and venation of the following leaves:

Illustrations from www.clipart.com

a.

b.

c.

d.

e.

f.

6. In a leaf, what is the function of the following tissues?

 a. palisade mesophyll b. spongy mesophyll c. epidermis d. xylem e. phloem f. chollenchyma

7. What controls the opening and closing of the stomata on a leaf?

8. Why is the bottom of a leaf typically a lighter shade of green than the top of the leaf?

9. Name two types of pigments that cause leaves to be a color other than green.

10. If a tree has no abscission layer, will it be deciduous?

11. Where is the abscission layer?

12. Name the four regions of a root. Which region contains undifferentiated cells?

13. State which of the following stem cross sections came from a monocot and which came from a dicot:

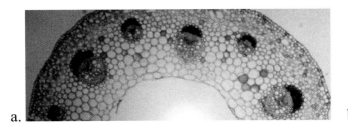
a.

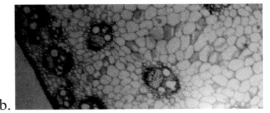
b.

Photos by Kathleen J. Wile

14. What allows woody stems to have no limits to their growth, unlike herbaceous stems?

15. What is the function of vascular cambium?

16. If a stem has cork cambium, is it woody or herbaceous?

17. What kind of vascular tissue makes up most of the wood in a woody stem? What kind of vascular tissue is found in the inner bark of a woody stem?

18. What is the dominant generation in the moss life cycle? Is it haploid or diploid?

19. A fern has antheridia and archegonia. Which part of the fern life cycle is it in? Is this the dominant generation?

20. Why are plants from phylum Bryophyta relatively small?

21. If a 15-foot tall plant has a root system that goes four feet deep, is it a fibrous or taproot system?

22. What are the male and female reproductive organs in a tree from phylum Coniferophyta?

23. What is the fundamental difference between monocots and dicots?

24. Name another difference between monocots and dicots.

25. A plant produces seed cones and pollen cones. Is it vascular? To what phylum (of the ones that we discussed) does it belong?

26. A plant produces flowers. To what phylum does it belong?

MODULE #15: Kingdom Plantae: Physiology and Reproduction

Introduction

In the previous module, we discussed kingdom Plantae from an anatomy and classification point of view. We will now conclude our discussion of plants by looking at their **physiology** (fiz ee ah' luh gee).

Physiology – The study of life processes in an organism

The definition of physiology tells us that we are going to spend time talking about how a plant actually functions from day to day. Since reproduction is a part of how a plant functions, it can actually be included in the term "physiology." However, in the title of this module, we have included reproduction separately, because a large fraction of this module will be spent discussing how reproduction occurs in phylum Anthophyta.

How a Plant Depends on Water

We already know some of the aspects of the life processes in a plant. After all, we know that a plant manufactures its own food through photosynthesis. We know that this process involves absorbing carbon dioxide from the atmosphere and releasing oxygen back into the atmosphere. We also know that a plant absorbs water and nutrients through its root system and sends them to the various parts of the plant. There is so much more to the life processes of a plant, however, that we must spend a little more time studying how a plant functions. We will start with something that you probably think you already understand: how a plant depends on water. You might be surprised at the complexity of this subject!

We know all plants need water. However, some plants need a lot more water than others. Some species of flowering plants need to be watered several times each week, whereas other plants (like household ferns) typically need to be watered only once per week or even less. In addition, plants like cacti (the plural of cactus) need to be watered once a month or even less. Each species of plant has been designed by God to fill a particular role in creation. As a result, the water needs of plants are different, depending on what ecosystem they have been designed to occupy.

Regardless of how much water a plant requires, water is used for essentially four processes: **photosynthesis, turgor pressure, hydrolysis,** and **transport**. Now, believe it or not, we have discussed all of these processes before. It has been a while, however, so we should spend some time reviewing them. In Module #5, we showed you that to make its own food (glucose – $C_6H_{12}O_6$), a plant uses carbon dioxide and water in a process called photosynthesis. In fact, for each molecule of glucose, the plant needs six molecules of water and six molecules of carbon dioxide. Obviously, then, without water, the plant would not be able to manufacture its own food and would end up starving to death.

Although we typically think of photosynthesis as the main function of the water in a plant, without the other three processes mentioned above, the plant would die. In Module #6, we talked about turgor pressure. In a plant cell, there is a large central vacuole that fills with water by osmosis. As more and more water fills the vacuole, the cell becomes pressurized. This pressure, called turgor pressure, keeps the plant (especially the stems and leaves) stiff. This is why plants wilt when deprived of water. Without turgor pressure, plants cannot stand up.

Turgor pressure is also responsible for some of the motion that we observe in plants. For example, some flowers open their petals during the day and close them at night. This is an example of **nastic** (nas' tik) **movement**, and it happens because cells near the base of the flower change their turgor pressure.

Nastic movement – A plant's response to a stimulus such that the direction of the response is preprogrammed and not dependent on the direction of the stimulus

The crocus flower, for example, opens at temperatures of about 60° F or higher, and it closes at temperatures lower than that. This is accomplished by a change in the turgor pressure of cells that control the petals of the flower. In the same way, the evening primrose flower opens at dusk due to changes in turgor pressure that are triggered by a change in the amount of light the plant receives.

One of the more interesting examples of nastic movement comes from the species *Mimosa pudica*. Often called "sensitive plants," these plants actually react to touch. If you touch them, they rapidly change the turgor pressure in certain cells to close their leaves, as shown in the figure below.

FIGURE 15.1
The Sensitive Plant

Photos by Kathleen J. Wile

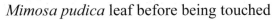
Mimosa pudica leaf before being touched   *Mimosa pudica* leaf after being touched

You might wonder what is meant by the part of the definition that says nastic movement is "independent of the direction of the stimulus." Well, it turns out that plants can move in another way as well. For example, plants tend to grow towards light sources. This kind of movement is typically called a **tropism** (trohp' izm) rather than a nastic movement, because the direction of the movement is in the direction of the stimulus (towards the light). We will discuss tropisms later. Often, the distinction between a tropism and a nastic movement is hard to make, but think about it this way: nastic movements occur in a preprogrammed direction, such as a flower opening or closing. The direction of a tropism, on the other hand, depends on the direction of the stimulus. If light comes from the east, the tropism will cause the plant to grow towards the east. However, if the light comes from the west, the tropism will cause the plant to grow towards the west.

In Module #6, we also talked about hydrolysis. When a cell takes in large molecules, it must first break them down into smaller molecules in order to use them. The cell accomplishes this by adding water to the large molecules. In the presence of specialized enzymes, the addition of water will break down the large molecule. This process is called hydrolysis.

What large molecules must a plant break down? Well, the glucose that it manufactures is a monosaccharide (we discussed this kind of molecule in Module #5). The plant can use that monosaccharide right away for energy, or, if there is plenty of food, it can store the glucose for later use. In order to be efficient, however, the plant will not store glucose as a monosaccharide. The glucose can be stored in a smaller volume if it is first converted into a polysaccharide (also discussed in Module #5) that we call **starch**. Thus, the stored food reserves of plants are large polysaccharide molecules. When a plant needs to use its stored food supply, it must first break the polysaccharides down into monosaccharides. That's one of the many things a plant does with hydrolysis.

Although a plant uses water for all three of the processes discussed so far, the vast majority of water that a plant takes in is used for transportation. You see, in order to live, plants need a lot more than just carbon dioxide and water. These two substances are all that the plant requires to make its food, but there are a host of other chemicals that a plant must manufacture in order to stay alive. After all, the plant cells must carry on all sorts of biosynthesis, which requires a lot of raw materials. These raw materials are absorbed from the environment and transported throughout the plant. This transportation is accomplished by dissolving the materials in water and allowing the water to carry them up the xylem. This actually requires transpiration! As a result, plants use transpiration quite heavily. In the next two sections, we will discuss how plants absorb water, and you will see how the motion of water through a plant requires transpiration.

ON YOUR OWN

15.1 If water is in short supply, a plant wilts but does not die unless the water stays in short supply for a long time. Based on this, which one of the four water-based processes that we discussed can be temporarily neglected?

<u>Water Absorption in Plants</u>

Water is absorbed by a vascular plant through its roots. Contrary to what many people think, a vascular plant cannot absorb water that falls on its leaves. It can absorb a small amount of *water vapor* through its leaves if the stomata are open, but the vast majority of the water absorbed by a vascular plant comes in through the roots. Although there are exceptions, most vascular plants have their root systems in the soil, so most roots absorb the water that is in the soil.

The composition of soil is rather complex, as there are both organic and inorganic components to it. The organic components of the soil include various living creatures such as molds, bacteria, protozoa, worms, and yeasts. In addition, there are organic components in the soil that come from the decaying remains of once-living organisms. The inorganic components of the soil are **gravel**, **sand**, **silt**, and **clay**. The big difference between these inorganic components is the *size of the particles* in that component. Gravel, for example, is composed of relatively large particles with diameters of one to two millimeters (0.04 to 0.08 inches). Sand is made up of smaller particles, while silt is made up of even smaller particles, and clay is made up of the smallest particles.

One important role that these components play is forming the **pore spaces** of the soil.

Pore spaces – Spaces in the soil that determine how much water and air the soil can hold

You see, when a soil is made up of large particles, such as gravel and sand, the soil has large pore spaces. These spaces allow the soil to absorb lots of water in a short time, but the water passes through the soil quickly, so that the soil dries out soon after it gets wet. When soil is made up of very small particles, however, the pore spaces are small. This means that the soil takes a long time to absorb water, but once it is wet, it holds the moisture for a long time.

Given these facts, the smaller the pore spaces, the more fertile the soil, right? Wrong! Even though the roots of a plant need to absorb water, they need to absorb other things as well. Remember, the roots of a plant are made up of living tissue. Thus, the cells in the roots need to absorb oxygen for respiration. They also need to release carbon dioxide into the environment. This is one reason that overwatering a plant can kill it. If water fills up the pore spaces in the soil, there will be no oxygen for the roots to absorb, and if that condition lasts for too long, they could die.

The larger the pore spaces, the more oxygen the soil can hold and the more carbon dioxide the roots can release. Thus, truly fertile soil must have pore spaces large enough to contain plenty of oxygen, but small enough so that the soil can hold moisture for a reasonably long time. The way this can happen is for the soil to be a good **loam**.

Loam – A mixture of gravel, sand, silt, clay, and organic matter

Soil scientists say that the most fertile soil would be a loam consisting of a tiny amount of gravel, 40% silt, 40% sand, and slightly less than 20% clay. The rest should be organic matter. This mixture gives the ideal pore spaces for roots to absorb plenty of water while still being able to absorb oxygen and release carbon dioxide.

ON YOUR OWN

15.2 Some people think that putting plant roots underwater will allow the plant to absorb all of the water it needs. If you do this to most plants, however, it will kill them. Why?

Water Transport in Plants

Now that we know what conditions are necessary for roots to absorb water, the next thing we need to know is how that water makes its way to the other parts of the plants. We know from the previous module that vascular plants have xylem which allow water to travel up the plant, but that doesn't tell us *how* the water actually moves. After all, in the animals we have studied (with the exception of the earthworm), the creatures' blood vessels did not move the blood through the organism. They provided a "path" for the blood flow, but the organism needed a heart to actually pump the blood through the vessels to the various parts of the organism's body. Well, in a plant, the xylem provide a path for the water to travel, but what actually makes water move? Plants don't have a "water pump" like a heart. What, then, causes the water to move?

Believe it or not, we aren't completely sure how to answer this question. Although we have a good theory as to why this happens, the theory has not been tested enough to become a scientific law. In fact, there is a small amount of data that contradict the theory. Thus, the theory provides us with an explanation that is consistent with most of what we know about vascular plants, but we cannot say for sure that it is the right explanation. This theory, called the **cohesion-tension theory**, says that transpiration causes the water to move up the xylem in a plant.

Remember that back in Module #10, we defined transpiration as the evaporation of water from the leaves of a plant. We didn't go into much detail then because it wasn't necessary at that time. Now it is. In transpiration, water actually evaporates out of the inside of a leaf. It does this because in order to have carbon dioxide for photosynthesis, the leaves must have their stomata open. When the stomata are open, the water in the leaves is exposed to the air and begins to evaporate. Thus, transpiration is really just evaporation, but it is evaporation that occurs from inside the leaf. This evaporation is a consequence of the fact that a leaf must have its stomata open in order to perform photosynthesis. This is, of course, why a plant's stomata close when it cannot perform photosynthesis. After all, every time the stomata are open, the plant loses water. Excessive water loss is a problem; therefore, the stomata will stay open only when the plant needs them to be open.

How in the world can transpiration cause water to move up through a plant? Well, water has a very strong tendency towards **cohesion** (coh he' shun).

Cohesion – The phenomenon that occurs when individual molecules are so strongly attracted to each other that they tend to stay together, even when exposed to tension

Now this definition might be a little confusing to you, so we want to explain it by means of an illustration. Have you ever seen a water strider like the one pictured below?

FIGURE 15.2
A Water Strider Can Walk on Water Because of Water's Cohesion

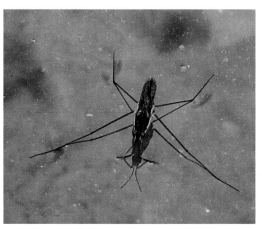

Why don't the strider's legs just sink into the water? Well, the strider's weight is pressing down on the surface of the water, evenly distributed by tiny hairs that extend from its legs. If, in response to this weight, the individual water molecules were to move away from each other, the

strider's legs would sink into the water. This doesn't happen, however, because of cohesion. The water molecules are attracted to one another and therefore tend to stay close together. The strider's weight, evenly distributed over the hairs jutting out of its legs, is not large enough to counteract this tendency to stay together. Thus, the water molecules do not move away from each other, and the strider's legs do not sink.

Now, of course, cohesion has its limits. When most animals step onto the surface of a body of water, they exert so much weight on the water that it counteracts the cohesion of the water molecules, and the animals sink. In order to make a difference, then, the strength of cohesion between the water molecules must be greater than whatever force is trying to pull them apart from one another.

How does this explain the transport of water in plants? When water evaporates from the leaves, there is a deficit of water in the leaves. This causes a **tension**, which pulls up on the water molecules just below where the deficit of water exists. These water molecules, in response to the tension, move up to replace the evaporated water molecules. The water molecules near the ones that move up, however, do not want to be separated from them. Because of cohesion, then, these water molecules move up to stay near the others that have moved up. This results in a "chain reaction," with all water molecules moving upwards due to their cohesion with the water molecules directly above them. Like a reverse domino effect, this motion goes all the way down to the roots, causing water in the roots to move upwards towards the other parts of the plant. You can visualize this by thinking of the water molecules as a string of beads that is being pulled out of the leaves and, as a result, up the plant.

Although we cannot yet be sure that this is the proper explanation for how water moves upwards in a plant, there is a lot of experimental evidence to back up this theory. As a result, the cohesion-tension theory has become a widely accepted explanation for water transport in plants.

Now, of course, a plant needs to transport things besides just water. Minerals and other raw materials for biosynthesis are absorbed by the roots. These must make it up to the remote parts of the plant as well. That's why we said that any materials the plant needs to transport upwards must be dissolved in water. Since the water flows upwards in accordance with the cohesion-tension theory, any substances dissolved in the water will move up as well.

Although the cohesion-tension theory explains how substances move upwards through the xylem in a plant, it does not explain how substances travel throughout the plant in the phloem. This process, called **translocation**, is quite different.

Translocation – The process by which organic substances move through the phloem of a plant

If you think about it, although water and minerals need to travel up the plant, the only substances that need to move down a plant are the products of photosynthesis and other types of biosynthesis. After all, the roots and stems are composed of living tissues that need food. The food is produced primarily in the leaves of the plant, so the food must move down from the leaves to the other parts of the plant, including the roots. As we have already learned, some plants store their excess food in their roots. Thus, the roots need food to stay alive and are often the repositories for excess food. As a result, the products of photosynthesis, which are organic chemicals, must move down the phloem in the plant when photosynthesis is occurring. When photosynthesis is not occurring, the stored food must move out of storage and into the rest of the plant via the phloem.

How is this accomplished? Do you remember one of the big differences we mentioned between xylem and phloem tissue? The cells that make up the xylem die when they are mature. This is because the xylem are simply tubes through which the water flows. Thus, the cells that make them up need not be alive, because they do nothing to promote the transport of water. Phloem cells, on the other hand, need to be alive, because they actively expend energy to guide the flow of the organic molecules throughout the plant. The details of this process are a bit beyond the scope of this course, so you just need to remember that this is the reason that phloem cells are alive, as opposed to xylem cells, which die when they are mature. Since the phloem cells actively participate in translocation, they must be alive to perform their duties.

ON YOUR OWN

15.3 A botanist has two samples of liquid. Sample A is composed primarily of organic materials, while sample B is composed mostly of water and minerals. Which liquid was extracted from the xylem of a plant and which came from the phloem?

15.4 Typically, a plant opens its stomata during the day and closes them at night. When would you expect transportation of water to occur, in the day or during the night?

Plant Growth

As a plant develops, it grows. As we mentioned before, much of this growth occurs in the meristematic tissue, where undifferentiated cells exist and mitosis takes place. Although you already know the ins and outs of mitosis (discussed in Module #8), you probably don't know what controls it. After all, if each cell in the meristematic tissue went through mitosis as often as it could, there would be *a lot* of growth in the plant. Most likely, the plant could not sustain so much growth. Thus, plants (and animals) must have some means of controlling the rate at which mitosis occurs.

In addition, once a new cell has formed, its development must be regulated so that it develops into the tissue that the plant needs. Therefore, plants also need chemicals that will regulate the development of new cells. Controlling mitosis and regulating plant cell development are accomplished with chemicals called **hormones**.

Hormones – Chemicals that circulate throughout multicellular organisms, regulating cellular processes by interacting with specifically targeted cells

In plants, there are at least five identifiable groups of plant hormones: **auxins** (awks' uhns), **gibberellins** (jib uh rehl' ins**), cytokinins** (sigh toh kye' nins), **abscisic** (ab sih' sik) **acid**, and **ethylene** (eth' uh leen). In addition, botanists suspect there is at least one more kind of plant hormone that, as of yet, we cannot identify with certainty. This mystery group of hormones is usually referred to as **florigen** (floor' uh jen).

Auxins were the first group of plant hormones discovered. These hormones regulate the development of cells, altering the amount that they elongate. This primarily affects the length of a plant's stems. The more the stem cells elongate, the longer the stem grows. Auxins are considered the driving force in **phototropism** (foh' toh trohp' iz uhm), **gravitropism** (grav' ih trohp' iz uhm), and **thigmotropism** (thig' muh trohp' iz uhm).

<u>Phototropism</u> – A growth response to light

<u>Gravitropism</u> – A growth response to gravity

<u>Thigmotropism</u> – A growth response to touch

If you have experimented with plants, you are probably familiar with these growth responses.

Do you remember the discussion we had about nastic movements? Well, these are the other kinds of movement that plants can have. The difference between tropisms and nastic movements is that while nastic movements occur in a preprogrammed direction, the direction of a tropism depends on the direction of the stimulus. Phototropism causes plants to grow *towards* the light. Gravitropism causes plants to grow *opposite* the force of gravity. Nastic movements, on the other hand, are the result of a general response that is independent of the direction from which the stimulus comes. If a flower closes at night, it does so regardless of the direction from which light is coming.

To give you a better idea of what a tropism is, consider placing a houseplant near a window that is the only source of light for the plant. Over time, the plant will grow so that its leaves are all pointed towards the window. If you then turn the plant around so that the leaves point away from the window, the plant will change its growth patterns so that, eventually, the leaves will face the window again. This is the phenomenon of phototropism. Most botanists believe that auxins are destroyed by light. Thus, they end up in highest concentration where the plant is exposed to the least light. This causes the plant tissue in the darker regions to grow quickly, while the plant tissue that is exposed to a lot of light doesn't grow much. The result of this growth imbalance is that the plant stems tend to bend toward the light.

Etiolation (ee tee uh lay' shun) might be considered an extreme form of phototropism. If a plant is placed in a rather dark area, none of its auxins are destroyed. As a result, the stems grow rapidly, becoming very long and thin. Since they are long and thin, the stems do not stand up well, but tend to crawl along the ground. Because there is little light, there is little need for chlorophyll, so the stems tend to have a very light color, and there tend to be few leaves. If the stems eventually grow into an area that does have light, the plant will produce leaves and chlorophyll there. If sufficient light is not found in a reasonable time frame, the plant dies.

The phenomenon of gravitropism is illustrated in the figure below.

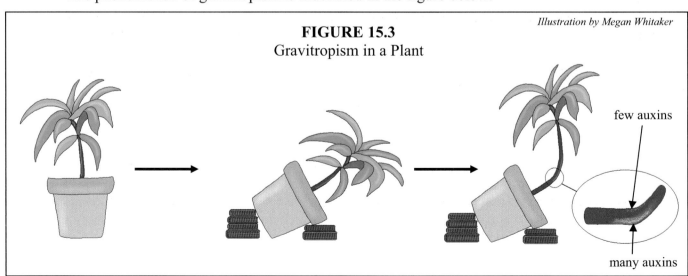

Illustration by Megan Whitaker

FIGURE 15.3
Gravitropism in a Plant

few auxins

many auxins

If you tilt a plant on its side and prop it so that it stands diagonally, the plant will bend in a few days so that the leaves are pointed upwards again. This is because the auxins in the plant are affected by gravity. They tend to collect wherever gravity causes them to fall. If a plant is tilted, the underside of the stem will have more auxins than the top side of the stem. This will cause a growth imbalance, with the tissues on the underside growing faster than those on the top side of the stem. This bends the stem so that the plant begins to grow upwards.

Thigmotropism is best illustrated by a vine. If you look at a vine, you will see that it tends to wrap around whatever it touches, and the leaves of the vine always point away from the structure upon which it is growing. This is due to thigmotropism. Auxins tend to move away from the surfaces that a plant touches. Thus, when a vine touches a structure, the auxins travel to the other side of the vine. That's why stems and leaves grow there. Also, the same growth imbalance that you see in phototropism and gravotropism occurs in this case. Thus, if the vine can, it tends to bend around the surface which it touches. Over time, this causes the vine to wrap around the surface to which it is attached.

The gibberellin family of hormones promotes elongation in stems, but it does other jobs as well. Gibberellins affect mitosis rates, and they can also induce seeds to germinate. Cytokinins affect mitosis rates and cellular differentiation, but in addition, they induce leaf cells to elongate, allowing the leaf to expand. They also affect the synthesis of chlorophyll. Abscisic acid's primary job is to inhibit the abscission layer so that it doesn't close off, but it also helps to control the leaf stomata. The last hormone, ethylene, promotes the ripening of fruits and causes the abscission layer to close off so that the leaves of a deciduous plant will fall. Since it promotes fruit ripening, it is often used by farmers to artificially ripen fruits that were picked before they had a chance to ripen on the plant.

As we mentioned above, botanists also believe in the existence of a heretofore undetected hormone called florigen. Why do botanists believe that this hormone exists even though it has not been discovered yet? Well, something controls the flowering of an anthophyte. Botanists believe that it must be some sort of hormone, but the search for this hormone has only recently produced results. In 2004, scientists from Cornell University published a paper indicating that "constans," a plant protein that they discovered, might, indeed, be the chemical that controls flowering. If this research pans out, botanists might be able to finally say what this mysterious "florigen" really is!

As we mentioned before, animals have hormones that promote the growth and development of certain tissues as well. For example, some athletes take specialized hormones called "anabolic steroids." These hormones promote cell growth and mitosis, resulting in the growth of muscle (and sometimes bone) tissue. This tends to "bulk" up the athlete, giving him strength that he would not normally have. This is a very dangerous practice, however, because each person's body chemistry is delicately balanced. By artificially adding hormones to their bodies, these athletes can become stronger, but they also throw off their body chemistry, resulting in liver problems, heart disease, kidney problems, and even reproductive problems.

ON YOUR OWN

15.5 Regardless of how you plant a seed, the seedling always sprouts up through the soil. What growth hormone is responsible for this amazing ability?

15.6 You see a bean plant that has grown abnormally long and is more yellow than green. What is the most likely reason for this bean plant's appearance?

Insectivorous Plants

The **insectivorous** (in sek tiv' or us) **plants**, sometimes called carnivorous plants, have a rather interesting physiology that is worth discussing. These plants have leaves that are designed to trap and digest insects. A common example of such a plant is shown in the figure below.

Photo on left © Jack Schiffer

FIGURE 15.4
Venus Flytrap

Photo on right © Painet Photographic Arts and Illustration NETwork

When an insect lands on the inside of a Venus flytrap's leaf, the leaf constricts, trapping the insect. That is about to happen to the fly on the right side of the figure. This is a nastic movement, as the leaf always constricts, regardless of the direction from which the insect lands. Digestive juices are then secreted from specialized cells in the leaf's tissue, decomposing the insect.

Contrary to popular belief, insectivorous plants *do not use the insects that they trap as food.* **Instead of using the insects for food, insectivorous plants have photosynthetic cells that produce the plant's food, just like other plants**. Why, then, do insectivorous plants trap and digest insects? Well, remember that a plant needs more than just food to survive. It needs certain raw materials for biosynthesis. Most plants get these raw materials from minerals and other substances in the soil. Insectivorous plants, however, have been designed by God to live in soils that have few or none of these raw materials. Thus, they must get them from another source. Insectivorous plants get their raw materials for biosynthesis from the chemicals that come from digesting the insects they catch.

One of the most important raw materials that a plant needs for biosynthesis is nitrogen. Now there are plenty of nitrogen atoms in the air around the plant. The atmosphere, after all, is 78% nitrogen gas. As you learned in Module #10, however, plants do not have the ability to use this nitrogen gas. As a result, plants must get their nitrogen from some other source. Most plants get their nitrogen from ammonia, which is produced in the nitrogen cycle (Figure 10.12). In soils that contain little ammonia, plants such as the Venus flytrap flourish. Interestingly enough, if you place a Venus flytrap in soil that is rich in ammonia and other minerals, it will not produce many insect-trapping leaves.

ON YOUR OWN

15.7 In order to be healthy, a Venus flytrap needs insects, water, and what else?

Reproduction in Plants

Many forms of reproduction can be found in kingdom Plantae. As you have already learned, bryophytes and seedless vascular plants produce spores in one generation and gametes in another generation. Conifers reproduce sexually by producing pollen in pollen cones and eggs in seed cones. Flowering plants use their flowers for sexual reproduction, as we will discuss in a little while. Most plants also have the ability to reproduce asexually. This is often accomplished through a process called **vegetative reproduction**. Let's cover vegetative reproduction first. After that, we will spend the rest of our time on the sexual reproduction and development of flowering plants.

Vegetative Reproduction

Vegetative reproduction in plants comes in many different forms. Some plants, like the piggyback plant, grow small "plantlets" right on their leaves. If these plantlets reach the soil, they will grow into new, separate plants. Other plants produce underground stems that originate in their roots. These stems eventually grow into a new plant. If, for example, a gardener plants a single mint plant in her garden, dozens of mint plants will appear around the original. This happens because as the roots of the mint plant grow away from the plant, they develop specialized stems that grow into a new plant. Grass also reproduces in this way, as do many weeds. The Irish potato produces tubers, another specialized stem that is formed in the roots of the plant. The tuber is what most people think of as the potato itself. These tubers have many buds on them, which are often called "eyes." These eyes can produce new plants. Often, gardeners propagate potatoes by simply cutting up their tubers into several pieces, each with an eye. They then plant those pieces, and the eyes mature into new potato plants.

Other plants can vegetatively reproduce using above-ground stems. Plants such as the strawberry plant can produce **runners**. These above-ground stems are long and spindly, and they produce a small plantlet at the end. When the plantlet grows too heavy for the runner, it touches the ground, and the plantlet begins to form roots. Once the roots take hold, the runner dies, and the result is a new strawberry plant. This type of vegetative reproduction is illustrated in Figure 15.5 on the next page.

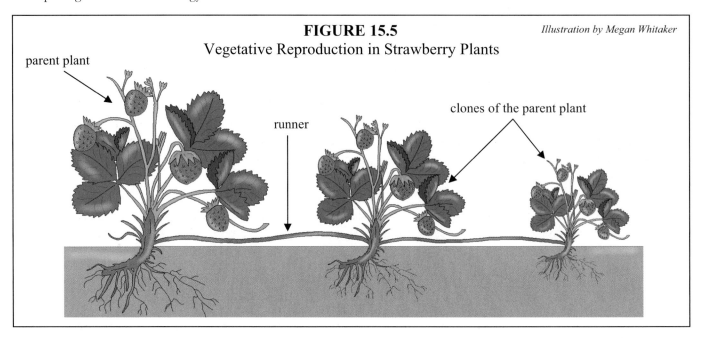

FIGURE 15.5

Illustration by Megan Whitaker

Vegetative Reproduction in Strawberry Plants

The regular stems of certain plants are also capable of asexual reproduction. In plants such as the blackberry, if a stem is broken or bent so that it touches the soil, it will often produce roots and begin to grow as a separate plant. If the original bend or break was severe enough to kill the stem, the stem will die, and the new plant will be separated from the original one. If the bend or break heals, however, the new plant will stay attached to the original one. Sometimes, gardeners and farmers use this method to propagate their plants. They will cut a stem from a plant, often referring to it as a "stem cutting," and then replant the stem. The stem will form roots, just as if a break had occurred, and a new plant will form.

Yet another form of vegetative reproduction comes from the leaves of certain plants. A leaf from an African violet, for example, can be planted in soil, and the meristematic tissue in the leaf can actually form roots and eventually mature into a new African violet plant. In principle, any leaf *should* be able to do this. Under normal conditions, however, most leaves die long before the meristematic tissue has time to form roots. If an expert manipulates the conditions under which a leaf is planted in the soil, however, many plants can be reproduced in this way.

Now remember, all of these forms of vegetative reproduction are asexual. This means that the offspring are genetic copies (clones) of the parent. Thus, if an African violet reproduces when a leaf is broken off of the plant and drops into the soil, the new plant will be genetically identical to the original plant. This kind of reproduction can be very useful to commercial gardeners. For example, suppose a gardener spends a lot of money on a rose plant that produces ideal-looking roses. If that plant were to sexually reproduce with itself or another rose plant, the resulting offspring would not be genetically identical to the original rose plant. As a result, the gardener will not be able to guarantee that the roses produced by the new plant will be as ideal. However, if the gardener vegetatively reproduces the rose plant, the roses produced by the offspring will most likely be the same ideal roses produced by the original plant. Remember, environmental conditions play a role as well, so you cannot *guarantee* that the roses produced by the offspring will be identical to those of the parent. Nevertheless, ensuring that the offspring has the same genes as the parent certainly makes the outcome more likely. Thus, vegetative reproduction is often used by commercial gardeners to ensure that their plants are as close to ideal as possible.

One other process that is sometimes considered vegetative reproduction is **grafting**. In grafting, a stem is cut from one plant and attached to another. The stem is called the **scion** (sye' un), and the plant to which it is attached is called the **stock**. Typically, the scion is attached to the stock so as to ensure that the vascular cambiums of the two are as close to one another as possible. This allows the xylem and phloem to merge, making a seamless interface between scion and stock. The point at which the graft takes place is usually wrapped in tape and wax to make an airtight, watertight seal. In order for grafting to work, the stock and scion generally have to be rather similar. Usually, this means that they must come from the same genus. Also, the grafting should take place when both plants are not growing.

The neat thing about grafting is the fact that no genetic material is exchanged in the process. Thus, the stock remains the same as before the graft, and the scion remains the same as well. Thus, if the stock has stems, those stems will produce leaves, flowers, and fruits in accordance to its species, and the scion will produce leaves, flowers, and fruits in accordance to its species! Grafting is often used in orchards. Gardeners take stems from a certain tree that produces nice fruit. They will then take those stems to a second tree that produces fruits that aren't nearly as nice. The stems of the second tree are cut away, and the scions from the first tree are then grafted on to the second tree. In a few years, the second tree will produce the same, nice fruits as the first tree does. This allows the caretaker of the orchard to ensure a quality crop of fruits from all of the trees.

The New Testament provides an excellent illustration of grafting. In Romans 11:16-24, we read that Gentiles are like "scions" that have been grafted onto the "tree" of Israel. The Jews who believe in Christ are like the stems from the original "stock," serving the Lord as members of the original chosen race. Gentiles who believe (the grafted stems) are also members of the chosen race. After all, even though they come from a different race, they have been grafted onto the "stock" of Israel. Thus, Jews who believe in Christ and Gentiles who believe in Christ are both members of the chosen people, even though they originally came from different stock!

ON YOUR OWN

15.8 A gardener is experimenting with one of her African violet plants. This plant produces large, deeply colored flowers. She shows you two offspring from the plant. The first produces small, lightly colored flowers, while the second produces large, deeply-colored flowers. Which offspring was produced with vegetative reproduction? Which one was produced with sexual reproduction?

15.9 A gardener wants to graft a limb from a McIntosh apple tree onto another tree so that he can have more McIntosh apples. Should the gardener graft this limb to his Red Delicious apple tree or to his wild cherry tree?

Sexual Reproduction in Phylum Anthophyta

In the previous module, we spent some time discussing the sexual reproduction that occurs in the alternate generations of phyla Bryophyta and Pterophyta. In addition, we discussed the sexual reproduction in phylum Coniferophyta that occurs by seed cones and pollen cones. When we got to phylum Anthophyta, however, we decided to put off a discussion of sexual reproduction because we said it was a little complicated. Well, we are going to have that discussion now.

Members of phylum Anthophyta are referred to as the "flowering plants." This is, of course, because all members of this phylum produce flowers at some point in their life cycle. These flowers contain the reproductive organs that allow the flowering plants to reproduce sexually. Although the color, size, odor, and shape of flowers differ greatly from species to species within phylum Anthophyta, there are some general characteristics that all flowers share. They are illustrated in the figure below.

FIGURE 15.6
The Structure of a Flower

Illustration by Megan Whitaker

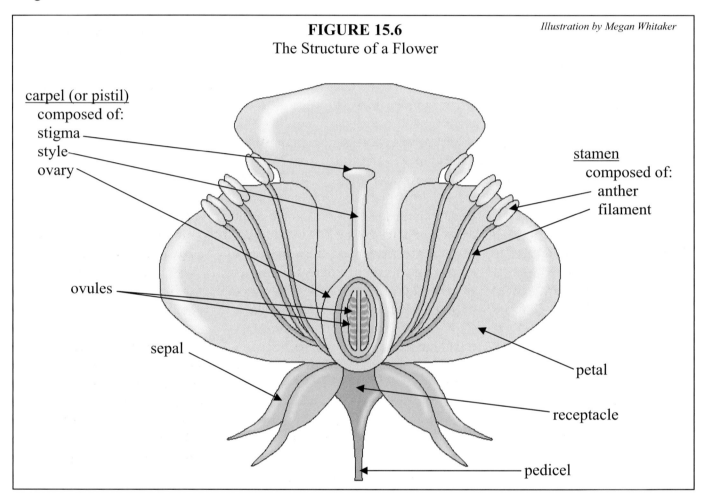

Notice from the figure that there are really only five basic parts to a flower. The **pedicel** (ped' uh sil) is the stem that holds the flower; the **sepals** (see' puls) are the green leaflike structures above the pedicel; the **petals** are the leaflike structures above the sepals; the **stamens** (stay' mens) are the male reproductive organs of the flower; and the **carpel** (car' pul) is the female reproductive organ. The carpel is referred to as **pistil** in many biology books, but newer books tend to use the term carpel. You should be familiar with both terms, however, since the next biology book you use might refer to the female reproductive organ in a flower as a pistil. Look at how these five structures are arranged in the flower. If you were to "fold up" the flower, the reproductive organs would be enclosed in the petals, which would be enclosed in the sepals. The flower is designed this way in order to protect its sexual organs before the flower opens.

If you look again at Figure 15.6, you will see that the carpel and stamens are actually composed of other structures. The stamens have long **filaments** that serve as "stalks" which support the **anther** (an' thur). The anther contains **pollen grains**, which we will discuss more in a moment. These pollen grains contain haploid cells that function as the sperm of the plant. The carpel is composed of the **stigma, style,** and **ovary**. The stigma is covered in a sticky substance designed to catch pollen grains; the style is an extension of the ovary designed to hold the stigma up where it can be exposed to pollen grains; and the ovary contains the **ovule** (ov' yool), where the **embryo sac** develops. The embryo sac contains the eggs of the plant, and it is where the seed develops. We will talk more about that in a moment.

It is important to note that while these five structures make up the "typical" flower, not all flowers have all of them. As you should be aware of by now, there are exceptions to nearly every rule in biology. The structure of a flower is no different. Some flowers do not have both stamens and carpels. Flowers that have both reproductive organs are called **perfect flowers**, while those that have either stamens *or* carpels are called **imperfect flowers**.

Perfect flowers – Flowers with both stamens and carpels

Imperfect flowers – Flowers with either stamens or carpels, but not both

Remember in Module #8 when you first learned that plants can sexually reproduce with themselves (self-fertilize)? In Module #14, it became apparent how conifers can do so, and now it should be clear how flowering plants can do so. After all, if the pollen grains (which contain sperm) produced by the stamen of a flower fertilize the embryo sac (which contains the egg) in the ovary of a flower on the same plant, that plant will have sexually reproduced with itself! Remember, however, that when a plant self-fertilizes, the offspring is not a genetic copy of the parent.

It is also important to note that many plants cannot self-fertilize. In some plants, the male and female flower parts mature at different times. That way, the carpel on a plant will not be ready to accept pollen when the stamens on the same plant release their pollen. Other plants have chemical mechanisms that reject their own pollen. The plant's carpel will happily accept pollen from another plant of the same type, but it rejects its own pollen. In other plants, the entire plant has imperfect flowers that are either all male or all female. As a result, a plant of this type that produces pollen cannot accept pollen, because it has no carpels. Nevertheless, many plants (including a lot of the plants that we use for food) do have the ability to self-fertilize.

While the function of the carpels and stamens is rather obvious, what are the functions of the other three parts of the flower? Well, the pedicel holds the flower up. It actually has a swelling on its tip called the **receptacle.** This receptacle holds the flower bud as it develops and supports the flower. The sepals enclose the bud of the flower, protecting it while it develops. When the flower is properly developed, the sepals peel back, revealing the petals. The petals also peel back, revealing the stamens and carpel.

What function do the petals perform? Are they just an extra layer of protection for the carpel and stamens as they form? Absolutely not. The petals contain carotenoids and other pigments that often result in beautiful colors. In addition, the epidermis of the petals often contains oils that let off a fragrant smell. Why the bright colors and fragrant smells? Well, in order for the pollen grains of one flower to reach the ovaries of another, the pollen grains must be carried from flower to flower. This is often accomplished by wind, but it is also common for insects and birds to carry pollen from flower to flower. The bright colors and fragrant smells of flowers are designed to attract these animals so that they can transport pollen. Thus, even though we enjoy them, the petals of flowers are not just pleasing to the eyes and nose; they are designed to be very functional!

Although Figure 15.6 shows you the basic anatomy of a flower, please realize that there are many variations on this basic structure. As we have already mentioned, imperfect flowers have only one of the two reproductive organs. In addition, some flowers have several carpels (instead of just one as is shown in the figure). A single carpel can also have just one ovule instead of several. Finally, some flowers are actually **composite flowers**, which are made up of several individual flowers. The sunflower, for example, looks like a single flower from far away. If you examine it closely, however, you will see that it is actually numerous tiny flowers that form in a single receptacle.

ON YOUR OWN

15.10 Is there any way a plant that produces imperfect flowers can sexually reproduce with itself? Why or why not?

15.11 Suppose a flower has a carpel with a very short style. Which would you consider to be the best means of transferring pollen grains to the carpel: the wind or an animal?

Before going on to see how flowers are used in the reproductive process, perform the following experiment, which will give you more experience with the anatomy of a flower. Even though this experiment calls for a microscope, the instructions tell you what to do if you do not have a microscope. As a result, everyone should do this experiment.

EXPERIMENT 15.1
Flower Anatomy

Supplies:

♦ Sharp scissors (If you have the dissection kit, use the scissors in it.)
♦ Sharp blade (If you have the dissection kit, use the scalpel in it.)
♦ Slides and coverslips
♦ Water
♦ Eyedropper
♦ Magnifying glass
♦ Microscope (optional)
♦ Lab notebook and colored pencils
♦ A variety of flowers (Most flower shops will save old flowers for you if you contact them ahead of time and tell them why you want them. They do not need to be fresh, but you should get a good variety. An example of a good variety would be: a rose, a carnation, a daisy, a lily and a tulip. At least one of them, preferably more, should have stamens and at least one carpel that are easy to see. In the list above, the lily and tulip will have easily visible stamens and a carpel. The rose and carnation will have them as well, but they will be harder to find. Look in the very center of the flower. The daisy is a composite flower, so its reproductive organs will be even harder to see.)

Object: To observe various types of flowers and compare their differences and similarities

Procedure:

1. Choose a flower that has easily visible reproductive organs.
2. Begin dissecting the flower by carefully pulling off the petals and sepals on one side, making sure you **do not disturb** the other parts.
3. Use your knife to cut the carpel vertically through the middle. This will expose the ovary and ovules if they are developed enough. Try to do this without breaking off the carpel, so that you see a good cross section of the flower intact.
4. Make a drawing of the flower and label the parts.
5. The following is a list of things to look for and label in your drawings:

   a. **Pedicel** – stalk that supports the flower
   b. **Receptacle** – "bulge" at the tip of the pedicel
   c. **Sepals** – leaflike structures at the base of the flower, collectively called the **calyx**
   d. **Petals** – collectively called the **corolla**
   e. **Stamens** - male reproductive parts
   f. **Filament** – stalk of a stamen
   g. **Anther** - forms and holds pollen
   h. **Pollen grains** – dustlike grains on (or in) the anther which contain the sperm nuclei
   i. **Carpel** - female reproductive part
   j. **Stigma** - top of carpel, receives pollen
   k. **Style** - supports stigma
   l. **Ovary** - contains ovules
   m. **Ovules** - holds egg, develops seed

6. Now choose a **composite flower**, like a daisy, aster, sunflower, or dandelion. Composite flowers are made up of hundreds of complete individual flowers. Use the magnifying glass to see if you can identify the carpels and stamens on your composite flower.
7. Repeat steps #1-6 for all of the flowers that you have.
8. Scrape some pollen from each flower.
9. If you do not have a microscope, do steps #10-12. If you have a microscope, skip to step #13.
10. Observe each pollen sample using the magnifying glass. See if you can note any differences between the pollen samples.
11. Use the magnifying glass to look at any other parts of the flowers that you wish to see better.
12. Skip to step #18.
13. Place a sample of pollen on a slide and add a drop of water.
14. Cover the wet sample with a coverslip.
15. Observe the pollen sample with the microscope and draw what you see.
16. Repeat this process for every sample of pollen, noting the varying shapes and colors of the pollen.
17. Investigate with the microscope any other part of the flower that you wish to see better.
18. Clean up and put away all of the equipment.

The Reproductive Process in Anthophytes, Part 1: Forming Pollen and Embryo Sacs

We already know that the stamens produce the pollen grains (which contain sperm) of a flowering plant, while the embryo sacs (which contain the eggs) are produced in the carpel. How does this happen? Well, we need to look at the formation of each of these gametes individually, because God has designed the process of pollen grain formation quite differently from the process of egg formation. We will start with pollen grain formation, which is illustrated in the figure below.

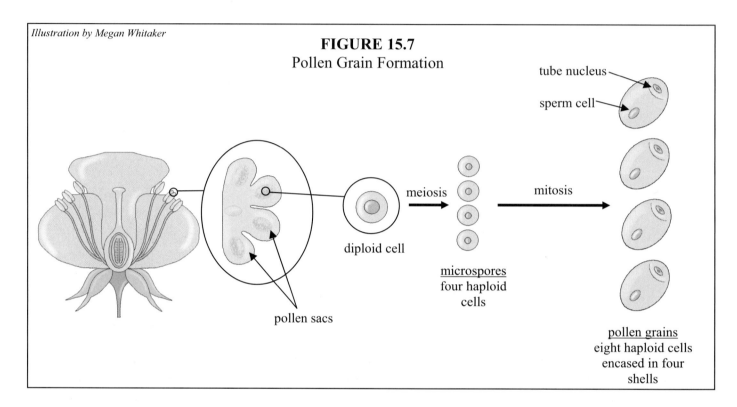

Illustration by Megan Whitaker

FIGURE 15.7
Pollen Grain Formation

tube nucleus
sperm cell
diploid cell
meiosis
mitosis
microspores four haploid cells
pollen grains eight haploid cells encased in four shells
pollen sacs

Each anther in a flower's stamen has regions called **pollen sacs**. In these sacs, diploid cells undergo meiosis in order to form haploid cells. Unlike the process of sperm formation in animals, however, this is just one step in the formation of pollen grains. Once meiosis has occurred, there are four haploid cells, which are often referred to as **microspores**. Each microspore becomes encased in a thick wall that is designed to protect it from unfavorable conditions. Once the microspore is encased, it undergoes mitosis. This results in two haploid cells inside the casing. In some plants, one of the cells inside the casing goes through mitosis again, making a total of three haploid cells inside the casing.

Once there are at least two cells in the casing, one of the cells differentiates and becomes a **tube nucleus**. The other cell differentiates into a **sperm cell**. If there is a third cell, it differentiates into a sperm cell so that there are two sperm cells and one tube nucleus. Remember, the tube nucleus and the sperm are still encased in a tough cell wall, so they are protected from unfavorable conditions. These cells, encased in their protective coating, make up a pollen grain. Each pollen sac in a mature anther contains hundreds or even thousands of pollen grains.

In the carpel, a completely different process, illustrated in Figure 15.8, is responsible for the formation of egg cells.

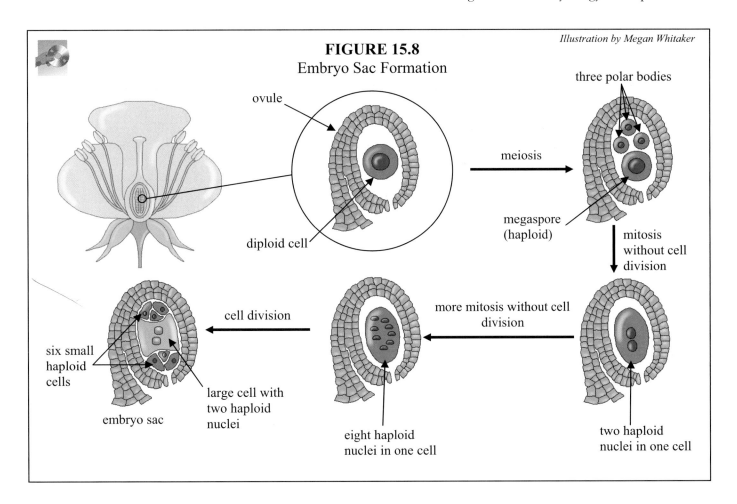

FIGURE 15.8
Embryo Sac Formation

Illustration by Megan Whitaker

On the inside wall of the ovary, a swelling begins to grow. This is the beginning of the ovule. As the ovule grows, a single diploid cell undergoes meiosis, resulting in four haploid cells. As is the case in animal female meiosis, three of these haploid cells (polar bodies) are useless and eventually disintegrate. The remaining haploid cell, called a **megaspore**, then undergoes a very interesting form of mitosis. In this form of mitosis, the DNA is copied and a new nucleus is made, but the cell never divides. In then end, then, this form of mitosis produces a single cell with two haploid nuclei. These nuclei continue mitosis, making a total of eight haploid nuclei in a single cell.

Once the megaspore has eight haploid nuclei, cell division takes place. Cell walls and membranes form around six of the individual nuclei, and a cell wall and membrane encases the remaining two nuclei. This results in a large, central cell with two nuclei and six smaller cells, three of which are on one side of the large cell and three of which are on the other. This arrangement of seven cells is called an **embryo sac**.

Do you remember from Module #14 that some biologists consider all plants as having the alternation of generations life cycle? Well, since the pollen grains contain at least two cells and since the embryo sac contains seven cells, both of them are multicellular structures. They each produce gametes, so you *can* interpret this as an alternation of generations life cycle. Looking at the anthophyte life cycle in this way, the plant itself is the sporophyte which, through meiosis, produces the pollen grain and embryo sacs, which form the gametophyte generation. Once the two gametes participate in fertilization, a diploid zygote results, which gives rise to the sporophyte generation.

ON YOUR OWN

15.12 If a pollen sac contains 1,000 diploid cells before the formation of pollen grains, how many pollen grains could there be in the pollen sac once pollen formation is complete?

15.13 After the original cell of an ovule undergoes meiosis and forms one viable megaspore, how many times does mitosis occur in order to get a total of eight nuclei in the megaspore?

The Reproductive Process in Anthophytes, Part 2: Pollination

By early spring, most flowers have completed forming pollen grains and embryo sacs and are ready to begin the process of **pollination**.

Pollination – The transfer of pollen grains from the anther to the carpel in flowering plants

If you have hay fever, you are already acutely aware of the process of pollination. When the anthers in a flower release their pollen grains, the pollen grains often travel on the wind in order to reach the carpel of another flower. Many people are allergic to these pollen grains. Thus, when they breathe the pollen-laden air in the spring, they have an allergic reaction to the pollen grains, causing stuffed-up noses, coughing, congestion, and respiratory difficulties. Although many pollen grains travel on the wind in order to reach the carpel of another flower, God has designed several other means of transport in order to ensure that the pollen grains of one plant make it to the carpel of another.

The plants that produce flowers send some of the glucose produced by photosynthesis up to the flowers. The flowers store this glucose in a sticky sweet solution called **nectar**. This nectar attracts birds, bees, moths, and butterflies that eat the nectar for food. As these creatures eat the nectar in the flower, the pollen grains in the anther get on them. As they travel from flower to flower, then, they will brush up against the carpels of the flowers, and the pollen on them will brush off onto the sticky surface of the carpel. This completes the pollination process. Each flower that these creatures visit receives pollen from the other flowers that they have visited and donates pollen for the next flowers that will be visited. This results in an incredibly efficient means of making sure that the pollen of a flower reaches the carpels of many other flowers.

It turns out that the process of pollination is even more efficient than we have described up to this point. As we mentioned previously, the bright colors and fragrant smells of flowers have been designed to attract creatures for help in pollination. What we didn't mention before, however, was that *specific flowers are designed to attract specific creatures!* For example, birds are typically attracted to certain types of red flowers which store nectar in long tubes. Birds with long, thin beaks (like hummingbirds) are designed to fly to the flower and suck up the nectar. Insects are not drawn to these flowers, however, because there is so much nectar in the flower that most insects can easily drown in the substance that they would normally use for food. Thus, God has designed insects to ignore such flowers, but He has designed birds to be attracted by them. Interestingly enough, birds have a poor sense of smell. As a result, flowers that have been designed to be pollinated by birds don't always have a fragrant smell, because the smell would serve no purpose. Now, of course, not all red flowers have no smell. Some red flowers (roses, for example) do not depend on birds for pollination; thus, they have a fragrance. In general, however, red flowers are less likely to be fragrant than flowers that are not red, because they are more likely to depend on birds for pollination.

Flowers that attract bees for pollination, on the other hand, usually produce fragrant, sweet odors. These odors attract the bees to the flower. As the bee feeds on the nectar of the flowers, the anthers release pollen onto its body. Before the bee leaves the flower, it grooms itself, placing the pollen that has been scattered over its body into specialized structures (also called pollen sacs) formed by the hairs on its legs. They take much of that pollen back to their hives, and they use it to make a substance called "bee bread," a mixture of pollen and nectar that is used to feed the bee larvae in the hive.

Beetles are also attracted by smell, but the smell is different from that which attracts bees. Beetles feed on dung and decaying matter. Thus, the flowers that God has designed to be pollinated by beetles actually have the foul odor of organic wastes and decaying organisms! These flowers are generally produced by short plants or plants that produce runners which run along the surface of the soil. As beetles crawl around them, they smell the odor of the flower and think it is food. Thus, they crawl all over the flower looking for dung or decaying matter, and all they get is a body covered by pollen. When they head to the next flower, some of that pollen ends up sticking to the carpel, completing pollination.

Moths and butterflies also aid in flower pollination. Butterflies typically forage for food in the daytime. They are attracted to flowers that have fragrant smells but also a wide, relatively horizontal surface upon which they can land, like daisies. Moths, on the other hand, tend to forage at night. They tend to be attracted to white or otherwise pale colors, because those colors show up better at night. Fragrant odors also attract moths. Once again, when these insects land on a flower, pollen is scattered on their bodies. As they travel from flower to flower, they transfer this pollen to the carpels of the flowers upon which they land.

Why has God gone to all of the trouble to design flowers so that birds, bees, beetles, butterflies, and moths can transport pollen from one plant to another? If many plants can self-fertilize, why did God bother to design flowers to ensure that pollen from one plant can get to the carpel of another plant? Why didn't He just design all flowers to self-fertilize? Well, remember what we learned in Module #9. God has designed His creation to be able to change and adapt in order to survive in a changing environment. One way He has seen fit to do this is to design a genetic code that is incredibly flexible. In order to retain this flexibility, though, the genetic codes must be continually mixed among individuals. If a plant simply reproduced with itself, only a certain number of phenotypes would be possible in the offspring. If a radical change in environment took place, the offspring of such plants might not be able to adapt to it. If, however, several different individual plants reproduce with each other, the genetic codes of many individuals are involved. This means there are a lot more possibilities of phenotypes, which ensures that the plants can adapt to changes in the environment.

ON YOUR OWN

15.14 Suppose you were blindfolded and asked to smell several flowers. If you encountered a flower with little or no smell, what would be a good guess as to its predominant color?

15.15 Suppose you were on an island with plenty of plant life but essentially no insects or birds. Would you expect to see many flowering plants? Why or why not?

The Reproductive Process in Anthophytes, Part 3: Fertilization

When pollination occurs by any one of the several means God has designed, the pollen grains stick to the sticky substance at the top of the stigma. Although the "sperm" of an anther has been transferred to the carpel of the flower, fertilization has not yet taken place. After all, the sperm cells have to make it to the egg cell in order to really have fertilization. Well, in order to get to the egg cells, these sperm cells are going to have to travel down the carpel to the ovule. How does this happen? The process is illustrated in the figure below.

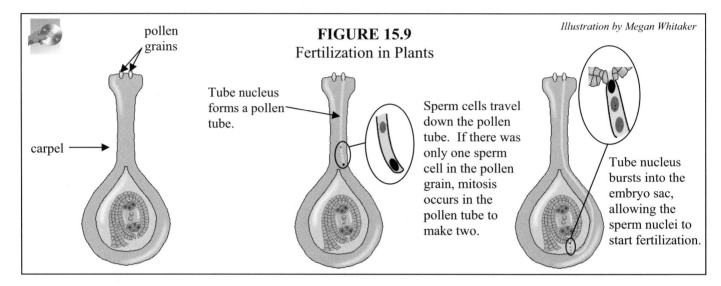

FIGURE 15.9
Fertilization in Plants

Illustration by Megan Whitaker

pollen grains

carpel

Tube nucleus forms a pollen tube.

Sperm cells travel down the pollen tube. If there was only one sperm cell in the pollen grain, mitosis occurs in the pollen tube to make two.

Tube nucleus bursts into the embryo sac, allowing the sperm nuclei to start fertilization.

Remember the tube nucleus that formed in the pollen grain? Its job is quite different from the sperm cell. When pollen lands on the stigma, the tube nucleus begins to undergo mitosis and differentiation. This results in a tube, which is called the **pollen tube**. The pollen tube grows through the tissue of the ovary, eventually finding its way to the embryo sac. This is the path that the sperm cells take to get to the ovary. If the pollen grain contains only one sperm cell (such as those in the Figure 15.7), mitosis occurs in the pollen tube so that there are now two sperm cells. If there were already two sperm cells in the pollen grain, they both travel down the tube nucleus. Either way, two sperm cells make their way to the ovule via the pollen tube.

Once the two sperm cells have reached the embryo sac, a rather unusual thing happens. One of the sperm cells fuses with one of the six small cells (the egg) in the embryo sac. When this fertilization takes place, a zygote is formed, and the other five small cells disintegrate. What happens to the other sperm cell? It travels to the large, double-nucleus cell in the center of the ovule and fuses with its two nuclei. Remember, since each of the two nuclei in the large cell is haploid, and since the sperm cell that fuses with them is also haploid, the cell that results, called the **endosperm**, contains extra genetic information. This extra genetic information causes the endosperm to develop differently from a normal zygote. Whereas the zygote begins to develop into the seedling of the plant, the endosperm begins to develop into a food source for that zygote.

Since two sperm nuclei fuse with two different cells in the ovule, we say that plants go through **double fertilization**, which is unique to kingdom Plantae.

Double fertilization – A fertilization process that requires two sperm to fuse with two other cells

ON YOUR OWN

15.16 In Module #7, we used the letter "n" to represent haploid cells, and "2n" to represent diploid cells. Using this notation, how would you represent the zygote formed during double fertilization? How would you represent the endosperm?

15.17 The process of a seedling sprouting from a seed is often called "germination." One of the steps in the fertilization process described above is often called germination as well. What part of the fertilization process discussed above could be called germination?

Seeds and Fruits

Once fertilization takes place, the endosperm and the zygote begin to develop by undergoing mitosis. The endosperm develops into the food source for the developing zygote by absorbing food that the plant continues to send and holding it until the embryo absorbs it. Early on, the zygote forms a stalk of cells that anchor to the endosperm and begin to absorb nutrients from it. This powers the mitosis that begins to form the embryo of the plant. The next structure that forms in the plant is the **cotyledon**. We mentioned this structure in the previous module. It is the fundamental characteristic that separates monocots from dicots. Monocots produce one cotyledon in the seed, while dicots produce two. The cotyledons either absorb the remaining endosperm and then provide nutrition to the embryo, or they produce enzymes that help the embryo to absorb more nutrients from the endosperm.

Once the embryo has fully formed its cotyledon or cotyledons, it is considered mature. Examples of mature monocot and dicot seeds are shown below.

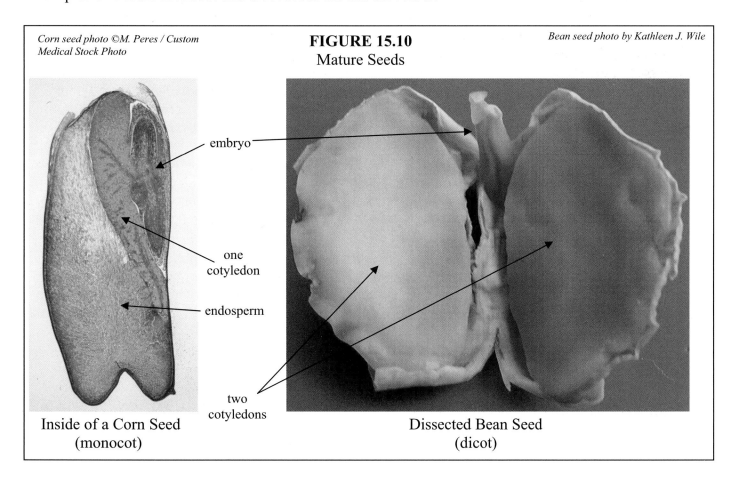

Corn seed photo ©M. Peres / Custom Medical Stock Photo

FIGURE 15.10
Mature Seeds

Bean seed photo by Kathleen J. Wile

embryo

one cotyledon

endosperm

two cotyledons

Inside of a Corn Seed
(monocot)

Dissected Bean Seed
(dicot)

Notice the differences and similarities between these two seeds. They each contain an embryo, of course, which will develop into the new plant. The corn seed has only one cotyledon, because it is a monocot, while the bean seed has two, because it is a dicot.

The other big difference between these two seeds is the function of the cotyledons. In the mature corn seed, the endosperm is still present. For this seed, then, the cotyledon simply aids in the transfer of food from the endosperm to the embryo. In the mature bean seed, however, there is no endosperm. That's because the two cotyledons have already absorbed it. In this seed, then, the cotyledons are the food source for the embryo. They do not transfer food from the endosperm to the embryo, rather, they have absorbed the endosperm and become the embryo's food.

Once a seed is mature, the embryo needs no more food, because it stops its development. Thus, the cotyledon no longer gives food to the embryo. Instead, the endosperm (or the cotyledons if they have absorbed the endosperm) is a food reserve ready for the embryo when it bursts forth from the seed and grows. The plant then severs its connection with the ovule, and the ovule begins to develop a protective coating, called a **testa**. Once the testa has been formed, the ovule is called a **seed**.

Although the seed has no connection to the plant, it is still inside the ovary. The ovary begins to swell and mature, becoming the **fruit** of the plant. Before we discuss the different kinds of fruits that result from the sexual reproduction of flowering plants, it is important to make sure you understand the difference between a fruit and a seed.

Seed – An ovule with a protective coating, encasing a mature plant embryo and a nutrient source

Fruit – A mature ovary that contains a seed or seeds

So you see that a fruit contains a seed, but the fruit does not grow into a plant. The seed that is contained in the fruit is what eventually grows into a plant.

Plants in phylum Anthophyta produce a wide variety of fruits. There are many ways that we can classify them. First of all, if a fruit forms from a single ovary, it is considered a **simple fruit**, whereas fruits that form from many ovaries are called **compound fruits**. Compound fruits can be further classified into either **aggregate fruits** or **multiple fruits**. If the fruit is formed by several ovaries from the same flower, it is an aggregate fruit.   If the collection of ovaries comes from different flowers, it is a multiple fruit. Raspberries and strawberries, for example, are aggregate fruits. The multiple ovaries are all a part of the same flower on a raspberry bush or strawberry plant. Pineapples, figs, and mulberries, however, form when ovaries from several different flowers fuse together to form a multiple fruit.

Simple fruits can be further classified into many, many different categories. To begin with, a simple fruit can be a **fleshy fruit** or a **dry fruit**. Fleshy fruits such as apples, tomatoes, cucumbers, and cantaloupes have fleshy tissue that forms between the seeds and the ovary covering. Dry fruits, on the other hand, have no fleshy tissue between the seeds and the ovary's covering. Nuts, grains, and pods are examples of dry fruits.

Fleshy fruits are further classified as **pomes** (pohms), **drupes** (droops), **berries**, or **modified berries**. Pomes are formed when the fleshy part of the fruit is not formed from the ovary, but instead comes from the tip of the pedicel. The ovary just forms a leathery covering for the fleshy tissue.

Apples and pears are examples of pomes. If the ovary itself develops the fleshy tissue of the fruit, there are two possible structures for the fruit. The ovary can form a hard covering around the seed and then enclose that covering in fleshy tissue. These fruits are classified as drupes. Olives and peaches are examples of drupes. Typically, the seed encased in the covering is called the "pit" of the drupe, and the fleshy tissue is called the "fruit." If the ovary doesn't form a hard covering around the seed but instead simply encloses the seed or seeds in fleshy tissue, it is some form of berry. If the ovary has a thin covering around this fleshy tissue, it is just called a berry, whereas if the ovary has a thick, tough covering, the fruit is called a modified berry. While tomatoes and grapes are examples of berries, cucumbers and oranges are modified berries.

Dry fruits can also be classified into several different groups. If the ovary has a single chamber containing many seeds, it is called a **pod**. Beans and peanuts are examples of pods. If the ovary has many chambers, each of which contains many seeds, we call the fruit a **capsule**. Poppy seeds form in capsules. If the fruit forms a long, thin wing from the ovary wall, it is called a **samara** (suh mar' uh). Maple tree fruits are examples of samaras. If the ovary forms a hard, woody covering around a single seed, it is called a **nut**. Walnuts and acorns are examples of nuts. It is important to note that although a peanut contains "nut" in its name, it is not a nut. It is a pod, because its single chamber contains two seeds. If the ovary wall is actually connected to the seed, the fruit is called a **grain**, while an ovary that is separated from the seed forms an **achene** (uh keen'). Corn is a grain, while the fruit of a sunflower is an achene.

To give you some more experience with the classification of fruits, perform the following experiment.

EXPERIMENT 15.2
Fruit Classification

Supplies:

♦ Sharp blade (If you have the dissection kit, use the scalpel in it.)
♦ Lab notebook
♦ A variety of different fruits (suggested fruits: apple, plum, orange, tomato, walnut, sunflower seed, maple seed, pea in pod, strawberry, and raspberry)

Object: To observe the various types of fruits and compare their many differences

Procedure:

1. Use the knife to dissect and investigate each fruit.
2. Examine each fruit and place it in a class, recording the reasoning you used to come up with the class that you assigned. If you have trouble, check out the course website we discussed in the "Student Notes" section at the beginning of this book. There are several links to websites with more details and examples of the various types of fruits.   Here is an abbreviated list of each fruit class and its description:

   a. **Aggregate fruit**: several separate ovaries of a single flower which ripen individually
   b. **Multiple fruit**: several ovaries from separate flowers which ripen fused together
   c. **Pod**: single-chambered ovary with more than one seed, opens along its side when ripe

      d. **Capsule**: multiple-chambered ovary, each with many seeds
      e. **Samara**: fruit with thin wings
      f. **Nut:** thick, hard, woody ovary wall enclosing a single seed
      g. **Grain**: ovary wall fastened to a single seed
      h. **Pome**: fleshy portion develops from receptacle, ovary forms leather core with seeds inside
      i. **Drupe**: ovary forms hard covering around seed and encloses that covering in fleshy tissue
      j. **Berry**: thin-skinned fruit with ovary divided in sections that contain the seeds
      k. **Modified berry**: like a berry, but with a thick, tough skin

3. Clean up your mess.

Fruits have been designed to move the seeds of a plant away from the parent. After all, if a seed ends up on the ground near the parent plant, the offspring will compete with the parent for nutrients and resources. If, instead, the seed can end up on the ground some distance away from the parent, the developing plant will not compete with its parent. Fruits disperse the seeds far from the parent plant in many ways. Samara, for example, will float on the wind because of their wings. This allows them to drop on the ground far from the parent plant. Nuts and pods are often carried away from the plant by animals. Animals then bury the nuts and pods as food reserves. They tend to bury more nuts and pods than they use, so some of them grow into new plants.

Compound fruits, as well as many fleshy fruits, are often eaten by animals. The animals usually carry the fruits away from the plant before eating them. When they eat the fruit and discard the seeds, they have succeeded in moving the seed away from the plant. Other fruits are eaten, seeds and all, by animals. The seeds have hard enough coverings to prevent them from being digested. Thus, when the animal excretes its undigested food, the seeds come out as well, ready to begin development.

Now remember, the growth of the embryo stops when the seed is mature. When the seed is dispersed, however, the embryo can "turn on" its development again. Typically, this happens when the seed embryo begins to absorb a significant amount of water. There are other factors, however. Often, seeds need the right temperature or amount of oxygen in the soil before the embryo will restart its development. Once these factors are in place, and the embryo starts growing again, it breaks through the seed wall. Gravitropism allows the plant to grow its roots down into the soil and allows the seedling to grow up, breaking through the soil surface. Typically, this process is referred to as **germination**, and we will discuss it in the final section of this module.

Since the embryo does not grow or consume food after the seed is mature, it is in a kind of suspended animation. It can stay in that state for quite some time and still successfully germinate. For example, bean seeds were found in King Tutankhamen's tomb in Egypt. Since he was buried more than 3,000 years ago, those seeds were more than 3,000 years old. Nevertheless, when they were planted, some of them germinated into bean plants. Thus, a seed need not germinate right after it matures. It can wait a long time until conditions are right and then germinate into a new plant.

ON YOUR OWN

15.18 The mature bean seed shown in Figure15.10 has no endosperm. Did it ever have one? If so, what happened to it?

Germination and Early Growth

When reasonably warm water surrounds the seed, the outer covering of the seed (the testa) loosens. This allows water to get inside the seed, which triggers the growth processes in the embryo. This starts the germination process.

When the germination process begins, the lower portion of the embryo, called the **radicle** (rad' ih kul), sprouts out of the seed. The radicle will develop into the roots of the plant. After that, the middle portion of the embryo, called the **hypocotyl** (hi' puh kot' uhl), begins to develop into the stem of the plant. During this time, the embryo is still underground. As a result, no photosynthesis is taking place. All of the food that the embryo needs is being supplied to it by the cotyledon or cotyledons, depending on whether the plant is a monocot or a dicot.

When the plant breaks through the soil, it is typically called a seedling. At this point, the cotyledon or cotyledons begin to open, forming leaflike structures. As a result, cotyledons are often called **seed leaves**. In many plants, the cotyledons will perform photosynthesis to continue feeding the plant. In other plants, they simply wither and fall off as they run out of nutrients. The true leaves of the plant begin to form at this time as well. They form from the top portion of the embryo, which is called the **epicotyl** (ep' uh kot' uhl). In some plants, the embryo is so "eager" to begin photosynthesis that it forms the beginnings of its first true leaves when it is still in the seed. That way, the plant's true leaves will develop quickly, allowing it to grow and mature rapidly. If an embryo does form the beginnings of its true leaves in the seed, that structure is usually called the **plumule** (ploom' yool). If you look back to Figure 15.10, the tiny flap at the top of the embryo is the plumule. The germination process for a dicot is shown in the figure below.

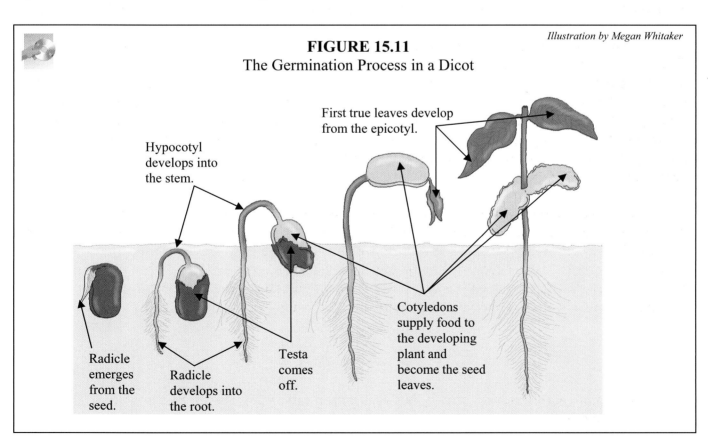

FIGURE 15.11
The Germination Process in a Dicot

Illustration by Megan Whitaker

Of course, this is a generalized discussion of seed germination, so the germination of an individual plant might occur differently from the way we described. For example, in some plants, the cotyledons (or cotyledon) stay below ground and never break the surface. These seed leaves, then, never do any photosynthesis. They just supply the seedling with food from the seed until the true leaves can begin photosynthesis.

Other plants have seeds with very thick testas. Because of this, it takes more than just some warm water to break the testa and begin the germination process. Some seed testas are so thick that they must be gnawed by animals before the germination process can begin. Others must freeze and thaw several times so that the testa breaks due to the repeated expansion and contraction that occurs.

Other plants have tiny seeds that do not have a lot of food for the embryo. Some of these seeds have to be planted very close to (if not on) the surface of the soil so that photosynthesis can occur very soon so that the developing embryo gets what it needs. Some small seeds have been designed to interact with other organisms to get what they need. Most orchid seeds, for example, do not contain much food for the developing embryo, and the initial stage of orchid development cannot do photosynthesis. In order for the developing embryo to get the food it needs, orchid seeds participate in a symbiotic relationship with certain fungi. The hyphae of the fungi penetrate the seed, but then the seed acts as a parasite, extracting the sugars it needs from the fungus. If you try to plant orchids in soil without the necessary fungi, the orchids will not grow unless you continue to add to the soil the sugars that the orchid plant needs.

ON YOUR OWN

15.19 If you plant a seed too deep, the seedling will break through the seed and begin to grow. It will die before it reaches the surface, however. Why does the plant die?

"I can't marry you, Henry. You're too attached to your mother."

Cartoon by Speartoons

ANSWERS TO THE "ON YOUR OWN" PROBLEMS

15.1 <u>Turgor pressure can be temporarily neglected.</u> Since turgor pressure keeps a plant from wilting, as soon as the plant starts wilting, you know that it is not maintaining turgor pressure. Since a plant can wilt without dying, we know that a lack of turgor pressure is not necessarily fatal.

15.2 Plant roots also need oxygen for respiration. Underwater, the plants cannot absorb oxygen very well. Thus, <u>putting plant roots underwater deprives them of oxygen</u>. Some plants (like rice plants) actually have air spaces in their roots that allow them to transport air from the surface down to roots that are underwater. As a result, these roots can live underwater. However, most plants do not have these air spaces, so their roots will drown if covered in water for too long.

15.3 <u>Sample B came from the xylem and sample A came from phloem.</u> Xylem transfer water and minerals up the plant. Phloem transport organic materials (such as the products of photosynthesis) throughout the plant.

15.4 Water transport takes place due to evaporation of water through the leaves. When the stomata are closed, evaporation does not take place. Thus, <u>water transport takes place during the day</u>.

15.5 Auxins are responsible for gravitropism. Gravitropism allows a plant to "know" which way is up and which way is down. Thus, <u>auxins are responsible for a seedling "knowing" which way to grow</u>.

15.6 Most likely, <u>this plant did not get enough light</u>. Its abnormal length is a sign of etiolation, which generally results from a lack of light. The yellow color also indicates low light levels, since it tells you that not much chlorophyll was made.

15.7 <u>It also needs light and carbon dioxide</u>. Remember, the Venus flytrap still gets its food from photosynthesis. Thus, it needs water, light, and carbon dioxide for that. The insects will give it the substances it needs for biosynthesis. They will not give it food.

15.8 <u>The second was the result of vegetative reproduction, while the first was the result of sexual reproduction</u>. The offspring of vegetative reproduction are genetically identical to the parent plant. Thus, the flowers of the offspring should look like those of the parent plant.

15.9 Grafts work when the scion and stock are close to the same species. Thus, <u>the gardener should graft to the Red Delicious apple tree</u>.

15.10 <u>Yes, if one of its flowers can produce a stamen and another can produce a carpel</u>. If pollen goes from the stamens of one flower to the carpel of another on the same plant, the plant can sexually reproduce with itself. This is different from plants that form imperfect flowers of only one sex. If a plant produces only flowers with stamen, then it clearly cannot self-fertilize.

15.11 If the style is short, the carpel will not be exposed to the wind very well. Thus, <u>animals will be the best means of pollen transfer</u> for this plant.

15.12 When a diploid cell undergoes meiosis, four haploid cells are formed. In a plant, each of those haploid cells forms a pollen grain, making four pollen grains for each diploid cell. Thus, there could be <u>4,000 pollen grains</u> in the pollen sac.

15.13 When the first mitosis occurs, there will be two nuclei. When each of those undergo mitosis, there will be four. When each of those undergoes mitosis, there will be eight. Thus, <u>the megaspore undergoes mitosis three times</u>.

15.14 Red flowers tend to attract birds. Since birds have a poor sense of smell, they often do not have a fragrance. Thus, <u>if a flower has no smell, it is more likely to be red than any other color</u>.

15.15 <u>You would expect to see few flowering plants</u>. Birds and insects are vital to the pollination of flowering plants. Without them, the reproductive process of flowering plants is so inhibited that flowering plants will not flourish.

15.16 In this notation, "n" represents half of the genetic code. Thus, a haploid cell is "n" and a diploid cell is "2n." Since the zygote is the product of fusion between two haploid cells, <u>the zygote would be labeled as "2n."</u> The endosperm is the product of the fusion of three haploid nuclei, so <u>the endosperm would be labeled as "3n."</u>

15.17 When the pollen tube begins to form, it grows, much like a plant grows. Thus, <u>the formation of the pollen tube is often called germination</u>.

15.18 <u>The seed did have an endosperm; however, it was absorbed by the cotyledons while the seed was maturing</u>. Remember, cotyledons either form a link between the endosperm and embryo, or they absorb the endosperm and become food for the embryo. In a corn seed, the former happens, and in a bean seed, the latter happens.

15.19 <u>The plant dies of starvation</u>. Remember, the endosperm has some nutrients to allow the embryo to start growing. If the cotyledon or cotyledons of the embryo do not get to the surface and begin photosynthesis before the endosperm runs out of nutrients, the plant will have no more food and will die.

STUDY GUIDE FOR MODULE #15

1. Define the following terms:

a. Physiology
b. Nastic movement
c. Pore spaces
d. Loam
e. Cohesion
f. Translocation
g. Hormones
h. Phototropism

i. Gravitropism
j. Thigmotropism
k. Perfect flowers
l. Imperfect flowers
m. Pollination
n. Double fertilization
o. Seed
p. Fruit

2. Name the four processes for which plants require water. Which of these processes can be neglected for a short amount of time?

3. A biologist studies two plants. The flowers of the first plant open each morning and close each night. The second plant's flowers stay open all of the time. However, if the plant is placed so that one of its sides is in the shade and the other is in the sunlight, the plant will eventually grow so that all of its leaves point towards the sunlight. Which plant is using nastic movement and which is using phototropism?

4. Briefly describe the cohesion-tension theory of water transport in plants.

5. Do xylem cells need to be alive in order for xylem to do their job? Why or why not?

6. Do phloem cells need to be alive in order for phloem to do their job? Why or why not?

7. What substances do xylem contain? What substances do phloem contain?

8. Do insectivorous plants really eat insects? Why or why not?

9. From a genetic point of view, what is the difference between vegetative reproduction and sexual reproduction in plants?

10. A gardener says that one limb of his crabapple tree now produces normal-sized apples. What must the gardener have done to make this happen?

11. What is the male reproductive organ of a flower? What is the female reproductive organ?

12. Why are the pollen grains and embryo sacs of flowers sometimes considered the gametophyte generation in an alternation of generations life cycle?

13. What two types of cells are found in a pollen grain?

14. Typically, how many cells are in an embryo sac? How many of them get fertilized?

15. Identify the structures in the figure below: *Illustration by Megan Whitaker*

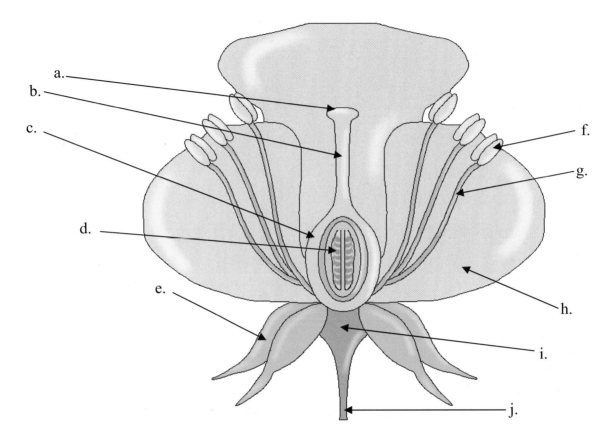

16. What structure is composed of parts a, b, and c from the drawing above?

17. What structure is composed of parts f and g from the drawing above?

18. What is the difference between pollination and fertilization?

19. How many sperm cells are used in plant fertilization?

20. Where does the endosperm come from? What is its purpose?

21. The cotyledon or cotyledons help provide food for the plant before and often after germination. How do cotyledons accomplish each task?

22. Name the three basic parts of the plant embryo and what each gives rise to in germination.

23. What is the purpose of a fruit?

24. Name at least three ways in which pollen is transferred from the stamens of one flower to the carpel of another.

25. Why are cotyledons sometimes called "seed leaves?"

MODULE #16: Reptiles, Birds, and Mammals

Introduction

In the previous two modules, we took a break from discussing the animal kingdom and concentrated instead on the plant kingdom. Well, in this, the final (yeah!) module of the course, we are going to revisit the animal kingdom, concentrating on reptiles, birds, and mammals. Those who believe in macroevolution call these the "higher" animals because they are under the mistaken impression that reptiles, birds, and mammals evolved from the other animals that we have already studied. Of course, you know that the whole idea of macroevolution is not scientifically sound, but nevertheless, you will probably hear the term "higher animal" in the future, so it is important for you to know where it comes from.

Class Reptilia

Class **Reptilia** (rep til' ee uh) contains the turtles, snakes, lizards, alligators, and crocodiles of creation. In addition, biologists put dinosaurs in this class. The members of class Reptilia have several characteristics in common:

➢ **Covered with tough, dry scales**
➢ **Ectothermic**
➢ **Breathe with lungs throughout their lives**
➢ **Three-chambered heart with a ventricle that is partially divided**
➢ **Produce amniotic (am nee ah' tik) eggs covered with a leathery shell, most oviparous, some ovoviviparous**

These characteristics serve to distinguish reptiles from other creatures in creation, so they are worth discussing.

First, reptiles are covered with tough, dry scales. Many people call snakes "slimy." However, since they belong to class Reptilia, they are not slimy at all. If you touch any reptile (including a snake), its skin generally feels cool, dry, and leathery. This is due to its scales. Unlike fish, the scales on a reptile are not covered in mucus. Instead, God has designed reptile scales to prevent water loss. Many reptiles live in regions where water is scarce. Thus, it is important that they do not lose water through evaporation. The scales make sure that this does not happen.

Scales serve another purpose as well. Reptiles are ectothermic, which means they have no internal mechanisms to keep themselves warm. This is why reptiles like to lie on warm rocks and sun themselves. This allows them to warm up their bodies. Once they are warm, their scales help insulate them, keeping the warmth in. Although reptiles love to sun themselves, it is possible for a reptile to get too much sun. If this happens, the reptile will actually get too hot and die. Thus, during the hottest part of the day, you rarely see reptiles sunning themselves. During that time, they are actually hiding from the sun in a cool, shady spot.

The scales on a reptile are not living tissue. As a result, they cannot grow with the creature. Much like the arthropods, reptiles must periodically shed their old skin (molt) in order to continue growing. In order to accomplish this, a new, larger set of scales begins growing before the old scales are shed. That way, the reptile is always covered with scales.

The next characteristic of reptiles is that they breathe with lungs. Remember, most adult amphibians have lungs, but their lungs are only one of three possible means of respiration. For reptiles, lungs are the only means of respiration. As a result, a reptile's lungs are more efficient and have more capacity than the lungs of a similarly sized amphibian. Surprisingly enough, even reptiles that live in the water (like water snakes and sea turtles) still breathe through lungs. Thus, it is possible for such reptiles to drown, despite the fact that they make the water their home!

Fish have two-chambered hearts, amphibians have three-chambered hearts, and reptiles have three-chambered hearts with ventricles that are partially divided. If you remember the amphibian heart you studied in Module #13, you will recall that it was composed of two atria and one ventricle. This resulted in three chambers. Well, the reptilian heart has a ventricle that is partially divided. As a result, we can think of reptiles as having an almost four-chambered heart.

The next characteristic of reptiles may use a term with which you are not familiar. Amniotic eggs are different from the eggs of fish and amphibians.

Amniotic egg – A shelled, water-retaining egg that allows reptile, bird, and certain mammal embryos
                    to develop on land

Remember, fish and amphibians lay their eggs underwater. These eggs are jellylike, with only a thin membrane covering them. That is okay, because they are always bathed in water. If reptile eggs were like that, they would dry out right away when they are laid on the ground. Thus, reptiles lay amniotic eggs. The figure below shows an illustration of a typical amniotic egg.

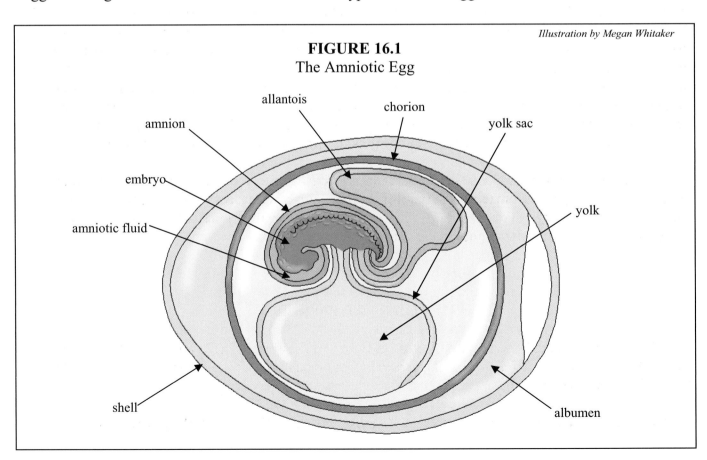

Illustration by Megan Whitaker

FIGURE 16.1
The Amniotic Egg

In reptiles, fertilization occurs inside the female. Once this happens, the developing zygote is encased in a protective **shell**. The shell is a remarkably engineered structure which is just porous enough to exchange gases with the environment but not porous enough for water to evaporate out of the egg! As the egg develops, four membranes form inside. The **amnion** (am nee' uhn) grows around the embryo, forming a fluid-filled sac in which the embryo floats. The next membrane is called the **yolk sac**. An amniotic egg has a clump of nutrients that is used to sustain the embryo during development. This clump of nutrients is called the **yolk**, and the yolk sac encloses it. As the yolk is used up by the developing embryo, it slowly disappears, and the yolk sac is drawn into the embryo until it disappears as well. The next membrane, the **allantois** (uh lan' toh is), is a sac of blood vessels that allows for the respiration and excretion of the embryo. Finally, the **chorion** (kor' ee ahn), is a membrane that envelopes all of the structures that we just described.

As we already mentioned, the shell of an egg is a marvelously engineered device, but that is only the beginning of the careful design apparent in an egg. Think about it. The egg is a self-contained unit. It provides a home, protection, and nourishment for the developing embryo. Since the egg is porous, the embryo can breathe. Of course, if it were too porous, water would escape, but the egg is not that porous. However, whenever a surface is porous enough to let in air, it is also porous enough to let in bacteria and other pathogens. This presents a problem, doesn't it? The embryo must breathe. At the same time, however, if oxygen can get into the egg, so can pathogens. Why don't bacteria and other pathogens infect eggs? Well, the bacteria and pathogens do get *into* the egg, but they don't usually make it to the tissues of the embryo or its food. Why?

When you break open an egg, you see a yellow part and a white part. The yellow part is the yolk. Most of the eggs with which you are familiar do not contain embryos, because the chickens that lay them are not allowed to mate. If the egg had an embryo in it, however, the embryo would feed off the yolk. The white part of the egg, technically called the **albumen** (al byoo' men), is what God put in the egg to protect the embryo. When bacteria and other pathogens enter the shell, they immediately encounter the egg white, which is full of chemicals that impede the growth of pathogens! Thus, the albumen is a defense mechanism for an egg, overcoming the problem of letting in pathogens with the air that is essential to the embryo's development. Isn't that amazing? If the egg never hatches for some reason, bacteria will eventually infect the yolk, as the defensive chemicals in the albumen can only do so much. Once this happens, the egg begins to smell, and we say that the egg is "rotten."

The albumen has other functions as well. It is a storehouse for water, keeping the egg from drying out. It also provides mechanical support for the chorion and its contents, even acting as a shock-absorber if the egg tumbles or falls. The albumen also stores proteins, which can be broken down into amino acids that the embryo can use in biosynthesis.

ON YOUR OWN

16.1 You are blindfolded and asked to feel the skin of two creatures. One is a fish and the other is a snake. If the first creature feels dry and leathery and the second feels slimy and rough, which is the fish and which is the snake?

16.2 If an amniotic egg develops without an allantois, will the embryo live? Why or why not?

498   Exploring Creation With Biology

16.3 Using X-ray technology, you observe the inside of an egg as the embryo develops. At first, the yolk sac and allantois are the same size, so you cannot tell them apart. After a few days, the one on the left is smaller than it was, and the one on the right is the same size as before. Which one is the allantois?

16.4 You have a tadpole and a baby water snake. Can either of them drown? If so, which one?

Classification of Reptiles

As is the case with all classes of vertebrates, there is plenty of diversity in class Reptilia. Figure 16.2 shows a few of the creatures contained in this class.

a-e Photos © IT Stock
f. Photo © Photodisc, Inc.

FIGURE 16.2
Some Members of Class Reptilia

g. Photo © Corel, Inc.
h. Photo © Tim Bushell
i. Photo © Emil Przygoda

a. Iguana (Squamata)　　b. Geckos (Squamata)　　c. Chameleon (Squamata)

d. Python (Squamata)　　e. Scarlet Snake (Squamata)　　f. Coral Snake (Squamata)

g. Hawksbill Turtle (Testudines)　　h. Giant Tortoise (Testudines)　　i. Crocodile (Crocodilia)

There are many orders within class Reptilia, but some of them contain only extinct reptiles. The reptiles living today belong to one of four orders: **Testudines** (test uh deen' eez), **Squamata** (sqwah mah' tuh), **Crocodilia** (crok uh dil' ee uh), and **Rhynchocephalia** (ring' koh suh fail' ee uh). Order Testudines is home to the turtles in creation, while snakes and lizards occupy order Squamata. Alligators and crocodiles make up order Crocodilia. The majority of the other orders in class Reptilia are used to classify dinosaurs, which are assumed to be extinct today. There is one order, however, that contains a single genus that is still living today: Rhynchocephalia. We will talk about this order first.

Order Rhynchocephalia

Order Rhynchocephalia contains many now extinct reptiles (including some dinosaurs) but it also contains one living reptile, called the **tuatara** (too uh tar' uh). Its name means "spine bearer," which points out one of its distinguishing characteristics: a tuatara has spines along its back. Tuataras typically grow to a maximum length of 30 inches and usually eat insects. They are largely nocturnal (active at night), but they spend a good part of the day basking in the sun near the entrance to their home, a burrow in the sand. Tuataras have a very slow reproductive process. They have a life span of 60 years but do not start reproducing until they are 20 years old. Once a tuatara lays an egg, the embryo must develop unattended for more than a year before hatching.

An interesting feature of the tuatara is a third eye, which is on the top of its head. Although this eye, called a **parietal** (puh rye' uh tul) **eye**, does have a small lens and light-sensitive cells that make up a retina, it is covered with scales, so it cannot send images to the brain. The eye is not useless, however, because it is connected to the brain with a nerve. Many biologists think this eye functions as a light sensor that helps the tuatara determine how long to bask in the sun before it retreats into its burrow.

Tuataras live on a few small islands near New Zealand. Their population is dwindling, however, because when the Europeans landed on these islands, they brought along rats in their ships. These rats found their way off the ships and onto the islands, throwing the ecosystem out of balance. The rats began eating tuatara eggs, which previously had no real predators. Since the eggs lie unattended for more than a year, they are relatively easy pickings for the rats. As a result, not nearly as many tuataras are born now as previously. To correct for this, environmental organizations are moving tuataras to islands that are rat-free, hoping that they will thrive there.

ON YOUR OWN

16.5 We assume that the tuatara uses its parietal eye to limit the time that it basks in the sun. Why must it limit its time in the sun?

Order Squamata

Snakes and lizards make up order Squamata. Although they belong to the same order, there are several differences between them:

➤ **Lizards have two pairs of limbs, while snakes have none.**
➤ **Lizards have ears and can hear, while snakes are deaf.**
➤ **Lizards have the same type of scales all over their bodies, while snakes have specialized scales on their bellies for locomotion.**
➤ **Most lizards have eyelids and can therefore close their eyes. Snakes' eyes are always open because they have no eyelids.**

These differences allow us to distinguish between lizards and snakes.

Lizards

Lizards have four legs, can close their eyes, can hear, and have scales on their bellies that are the same as the scales on the rest of their bodies. This separates them from snakes. The **iguana** (ih gwah' nuh) (Figure 16.2a) is one of the larger reptiles, often reaching a length of four to six feet. There are many different species of iguana, inhabiting many different ecosystems. Some, like the green iguana pictured in the figure, live high in the trees of a rainforest. Others, like the desert iguana, live in the desert. There is even one type of iguana, the marine iguana, that lives on the Galapagos Islands and actually swims out into the ocean to feast on seaweed! Although iguanas are large, they are not the largest of the lizards. The world's largest lizard is the **Komodo dragon**. Males of this species can reach lengths of up to ten feet.

Unlike Komodo dragons and iguanas, most lizards are small. The **gecko**, for example, is a small climbing lizard that tends to live in trees. It rarely reaches lengths greater than nine inches. This interesting lizard is an excellent climber, however. It is so good at sticking to surfaces that it can run upside down on polished glass! How does it do this? Biologists have only recently found the explanation. The gecko has tiny hairs on its feet. These hairs are about one hundredth of a centimeter long, and they are packed so that *five million of them would fit in one square inch*! The end of *each hair* has several hundred branches that end in extremely tiny spatula-like structures. These structures are so tiny that they can exploit the attractive forces (called "Van der Waals forces") that exist between all molecules.

Once biologists learned of how the gecko is able to stick to surfaces, researchers have been trying to duplicate the structures to make "gecko tape." The "gecko tape" that has been made so far, while strong, is not nearly as durable as the gecko's feet. While the gecko can stick and unstick its feet to a surface over and over again, the human-made "gecko tape" cannot be stuck and unstuck more than a few times before losing its sticking ability. The incredible design of the gecko's feet is just another in a long list of features that tell us that our world and the life in it have been designed by an amazing Designer!

Like the gecko, the **chameleon** (kuh meel' yuhn) (Figure 16.2c) is a tree-dwelling lizard that can change color in response to several different stimuli. Its colors provide camouflage so that it can blend in with its surroundings, and it can use its ability to change colors to communicate with other chameleons. This fascinating lizard has a long tongue that it shoots out to catch insects. The tongue of a chameleon can be longer than the chameleon itself! When it shoots its tongue out of its mouth in order to catch its prey, the tongue accelerates at an astonishing *fifty times* the acceleration due to gravity. To give you some idea of just how astonishing this is, fighter pilots typically pass out when they accelerate at ten times the acceleration due to gravity! The design that the chameleon needs to make its incredible tongue work is yet another testament to the awesomeness of its Designer.

The **horned toad** is a common lizard in the western United States. It has spines on its head and body. When it is afraid, this lizard swells up, making its spines very obvious. In addition, it actually *squirts blood* through its eyes at its attacker! These actions are designed to frighten its would-be attacker, because as fearsome as the horned toad looks, it does not fight well at all. Thus, its best bet is to frighten away its attacker. In fact, many people keep horned toads as pets, an indication of how harmless the lizard really is.

Some lizards are not small or harmless, however. The **Gila** (he' luh) **monster** is the largest lizard in the United States. It grows to a length of almost two feet. Although this is not nearly as large as some of the other reptiles, it is large for a lizard. In addition, the Gila monster is poisonous. It is the only poisonous lizard in the United States. Interestingly enough, a hormone in the venom of the Gila monster is currently being investigated as a possible treatment for diabetes.

Snakes

Most people are afraid of snakes on one level or another. This is probably due to the fact that they have the reputation of biting and poisoning people. Although this is the popular view of snakes, most of them are quite harmless and many are beneficial. After all, snakes feed on rodents and insects that plague humans. Without these useful reptiles, the populations of these pests would quickly rise out of control!

As we pointed out previously, snakes do not have legs. Instead, they have specialized scales on their bellies. These scales, called **scutes** (scoots), have the ability to grasp the ground upon which the snake is crawling. This enables the snake to get traction on the ground, allowing it to push itself in the direction it wants to go. This isn't the only type of locomotion that the snake can use, however. The snake can also wind itself into a tight "S"-shape and then extend itself forward. Also, if there are plenty of twigs, stones, and the like, the snake can simply wind its way around them, pushing against the twigs and stones to move itself forward. Many snakes also live in the water and can swim.

Snakes are carnivorous, feeding on live prey such as rodents, insects, lizards, eggs, and even other snakes. Although their eyes are always open (as we pointed out previously), they are not especially sharp-sighted. In addition, snakes are also deaf. They have no ears, so they simply cannot hear. They can, however, sense vibrations that travel through the ground. Snakes are drawn to regular ground vibrations, as they represent possible sources of prey.

Since snakes cannot hear and have poor eyesight, they use other senses to hunt. Snakes have a keen sense of smell and typically rely heavily on it to find prey. A snake has two nostrils on the front of its head that lead to the olfactory lobes in its brain. To augment this sense of smell, however, most snakes have sensory pits called **Jacobson's organs** in their mouths. When a snake sticks out its tongue, it collects chemicals suspended in the air. It then pulls its tongue back into its mouth, transferring the collected chemicals to the Jacobson's organs, which send nerve signals to the brain. The brain correlates these signals with those coming from the nostrils, and the result is one of the keenest senses of smell in creation. Between the nostrils and the Jacobson's organs, snakes can pick up even the faintest trace of an odor! Interestingly enough, a snake also uses its tongue and Jacobson's organs to sample the ground in order to sense the trail of prey it is tracking. This might be what God was talking about in Genesis 3:14 when He told the snake, "…and dust you will eat all the days of your life."

Once prey has been detected, the snake captures the unfortunate creature. The simplest means by which this happens is for the snake to simply grab onto the creature with its mouth and begin swallowing it. Some snakes wind themselves around their prey and then squeeze them to death. This is called **constriction**. Other snakes use their bite to inject a poison into their prey. This poison, manufactured in poison glands connected to the snake's fangs, can be powerful enough to cause death in humans.

If a snake uses the constriction method of capturing prey, it is often called a "constrictor." Two of the more famous constrictors are the **anaconda**, which is found in South America, and the **python** (Figure 16.2d), which lives in tropical and subtropical climates. Some python species can grow very long (up to 40 feet) and have been known on rare occasions to constrict and eat humans. Other species can be much smaller. The **ball python**, for example, is commonly sold as a pet, because it generally does not grow longer than five feet. Another popular pet constrictor, the **boa constrictor**, usually reaches lengths of only ten feet or so.

Although poisonous snakes are probably the reason that many people are so frightened of this order of reptiles, they actually make up a small percentage of the total number of snakes in creation. Typically, poisonous snakes produce either **neurotoxins** (noor oh tahk' sinz) or **hemotoxins** (hee muh tahk' sinz).

Neurotoxin – A poison that attacks the nervous system, causing blindness, paralysis, or suffocation

Hemotoxin – A poison that attacks the red blood cells and blood vessels, destroying circulation

Neurotoxins are fast-acting, while hemotoxins take longer to do damage. Hemotoxins are by far the deadlier of the two, however.

There are basically two types of poisonous snakes. The first kind has short, fixed fangs that deliver the poison while the snake chews on its prey. Those in this category predominately produce neurotoxins as their poison. The second type of poisonous snake has long fangs that fold away into pockets of the mouth when they are not in use. These snakes typically produce hemotoxins as their poison, and they usually inject a large amount of the poison through their fangs in a single strike.

Examples of the short-fanged poisonous snakes are the **sea snake, coral snake** (Figure 16.2f), and **cobra**. As we mentioned before, even though sea snakes live in the sea, they must breathe air through lungs. Thus, sea snakes must raise their heads out of the water periodically to get gulps of air. Some species of sea snake have such large lungs that they can stay underwater for more than an hour with just one gulp of air. Thus, to the short-term observer, sea snakes may look like they do not need to breathe air. If you watch them long enough, however, you will see that they do. Although generally not aggressive, the sea snake has perhaps the most powerful poison of any of the short-fanged snakes.

The coral snake lives in many subtropical areas, including the southern part of the United States. It is easy to distinguish, being marked with red, yellow, and black bands. Like most snakes, it is not really aggressive. In fact, if you are outfitted properly, the coral snake isn't even dangerous. You see, the fangs of a coral snake are so short that they cannot penetrate thick clothing. Thus, people who hike in the southern parts of the United States need only to wear shoes with thick socks and thick pants to protect themselves from these snakes.

Interestingly enough, many nonpoisonous snakes mimic poisonous snakes so as to scare predators. Notice the scarlet snake in Figure 16.2e. It looks a lot like a coral snake, since it has bands of the same colors (red, yellow, and black). The arrangement of these bands is slightly different, however, which allows people to tell the poisonous snakes from the nonpoisonous ones. All you have to do is remember the simple rhyme, "Red on black, poison lack; red on yellow kills a fellow." Notice that the scarlet snake has its red bands touching its black bands. That tells you it is not poisonous. The coral snake has its red bands touching its yellow bands. That tells you it is poisonous.

Cobras typically live in parts of Asia. These snakes are best known for their "hoods," which are expandable regions of skin behind their heads. When a cobra is ready to strike, it expands this hood as a warning. Some cobras, like the **black-necked cobra** of Africa, can spit their venom. They aim for their opponent's eyes, hoping to blind the creature.

Long-fanged poisonous snakes are commonly called **vipers**, and they are common in Africa, Europe, and Asia. In the Americas, a special kind of viper called the **pit viper** exists. These vipers have a special means of detecting prey. In addition to their nostrils, pit vipers have two **heat-sensing pits**. These pits, located between the nostrils and the eyes, allow the snake to sense warm-blooded creatures, even if there is no light with which to see them and no way to detect their smell. Most pit vipers like to hunt in total darkness, because their heat-sensing pits give them an advantage in the dark. Common examples of pit vipers in the United States are **water moccasins** (which live in southern swamps and lakes), **copperheads** (which live in the woods on the East Coast), **rattlesnakes** (which are common throughout the United States), and the **western diamondback** (which is found from Texas to eastern California).

Regardless of how the prey is captured, snakes usually swallow their prey whole, even if the creature is significantly larger than the snake. This can happen because the jaw of the snake is not attached to its skull. Instead, it is attached to the **quadrate bone**, which acts as a hinge. This arrangement, along with its elastic skin, allows the snake to open its mouth very wide so that it can swallow creatures that are much, much larger than it is. An African python, for example, was discovered with a 130-pound impala (an African antelope) inside it!

Order Testudines

Turtles and tortoises are found in order Testudines. The term "turtle" refers to the members of this order which live in water, while the term "tortoise" is used to refer to those that live on land. Members of this order are distinguished by the large shells that they carry on their backs. Most members of this order can pull all of their appendages (legs, neck, and head) into the shell and close off the openings with little "trap doors." This allows them to become veritable fortresses, almost impervious to attack.

Most members of this order have short, thick legs that end in five-clawed toes, although some turtles have flippers instead. These animals do not have teeth. Instead, their jaws form a tough beak that can be used to snap their food into bits. The nostrils of a turtle are placed very high on the snout, allowing it to breathe air while keeping the vast majority of its body submerged underwater. Because turtles live predominantly in the water but can come onto the shore periodically, they are often confused with amphibians. Turtles are not amphibians, however. They have all of the characteristics of reptiles.

Most turtles and tortoises are rather small. However, the marine-dwelling **hawksbill turtle** (Figure 16.2g) typically reaches a weight of 270 pounds. As you can see in the figure, the hawksbill turtle has flippers rather than feet, and it is an excellent swimmer. The **giant tortoise** (Figure16.2h) is larger than the hawksbill turtle and can weigh more than 500 pounds. The largest member of order Testudines is the **leatherback sea turtle**, which can weigh as much as 1,500 pounds!

Order Crocodilia

Alligators and crocodiles make up order Crocodilia. What's the difference between an alligator and a crocodile? Well, alligators have thick, blunt snouts. When they close their mouths, their teeth all fit inside. Crocodiles, on the other hand, have thinner, more pointed snouts. In addition, not all of their teeth fit into their mouths. Notice the crocodile in Figure 16.2i. You can tell it is a crocodile because despite the fact that its mouth is closed, you can still see its teeth.

Members of this order are considered to be the largest reptiles in creation. The crocodiles that swim in the salty waters of Australia, for example, can reach lengths of up to 30 feet and weights of up to 3,000 pounds.

Crocodiles and alligators are stealthy predators. Their eyes and nostrils are placed on the top of their heads, so they can glide through water very quietly with only the topmost portion of their heads breaking the surface of the water. One of their favorite means of catching prey is to float in the water with just their eyes and nose above the surface. They float so still that most creatures mistake them for logs. When a creature gets close enough, however, it soon finds out its mistake! By then, it is too late.

Like most reptiles, members of this order are oviparous. They tend to lay their eggs and cover them with a mound of rotting vegetation. The heat generated by the decomposers eating the rotting vegetation warms the eggs, allowing them to be unattended. Although the mother does not attend the eggs, she does visit them from time to time, because after about two weeks the eggs hatch but the young are buried in vegetation. Once hatched, they begin making peeping noises that tell the mother to dig them out and carry them to the water.

ON YOUR OWN

16.6 A reptile has no eyelids and is deaf. Is it a lizard or a snake?

16.7 Why do biologists say that human beings are responsible for the tuatara's dwindling population, even though humans do not hunt or kill tuataras?

16.8 If a snake generally hunts at night, what is it most likely using to detect its prey?

16.9 Why do some biologists say that snakes smell with their tongues?

16.10 You see a member of order Testudines that has flippers. Is it a turtle or a tortoise?

16.11 You see a member of order Crocodilia with its mouth closed. You cannot see any teeth. Is this an alligator or crocodile?

Dinosaurs

We could hardly leave a discussion of reptiles without mentioning dinosaurs. The term "dinosaur" actually means "terrible lizard," so we can conclude that dinosaurs are considered part of class Reptilia. Indeed, many of the orders in class Reptilia exist simply to classify dinosaurs. Unfortunately, discussing dinosaurs is a very difficult task because we have never seen them. The only data that we have regarding the dinosaurs are their fossilized remains. This is a serious problem. You see, the most likely part of a vertebrate to fossilize is the skeleton. Fleshy tissue, organs, scales, hair, etc. rarely, if ever, fossilize. Thus, what we know of the dinosaurs is rather limited.

There is also another problem we face when trying to understand dinosaurs. Of the fossilized remains that we do have, there is precious little. Remember from Module #9 that the vast majority of all fossils that we recover are of mollusks. The vast majority of the remaining fossils are of marine vertebrates and insects. Only a tiny, tiny percentage of the fossils that we have discovered are of reptiles, plants, and mammals. Thus, dinosaurs fossils are a tiny, tiny, tiny fraction of the fossils that we have.

Despite the fact that there are precious little data, paleontologists have been able to learn something about some of the dinosaurs that once existed on earth. Figure 16.3 shows artists' renderings of what we think some dinosaurs might have looked like. Remember, since all we have is a few of the creatures' bones, these drawings are, at best, educated guesses.

FIGURE 16.3
A Few Types of Dinosaurs

Illustrations from www.clipart.com

Apatosaurus: This dinosaur belongs to a group known as the **sauropods** (sor' uh pods). These were probably the largest of the dinosaurs. Their bodies were more than 60 feet long, and they could have weighed more than 30 tons. Biologists think that these dinosaurs were swamp-dwelling herbivores.

Stegosaurus: This dinosaur belongs to a group known as the **thyreophorans** (thy ree' uh for ans), which ranged in size from small to fairly large. Stegosaurus probably reached a length of 30 feet or so and weighed a few tons. It had a tiny brain, armored plates on its back, and a spiked tail for defense. Biologists think that thyreophorans were herbivores.

Triceratops: This dinosaur belongs to a group known as the **marginocephalia** (mar' gin uh suh fail' ee uh), which means "fringe heads," because they have a shelf or frill at the back of the skull. Triceratops probably reached a length of 30 feet or so and weighed from five to ten tons. It had three horns for defense. Biologists think that the marginocephalia were herbivores.

Tyrannosaurus: This dinosaur is one of the **theropods** (ther' uh pods), which ranged from very small to very large. Tyrannosaurus, often called the "king of the dinosaurs," was probably the biggest theropod, reaching heights of up to 40 feet and weights of seven tons. With six-inch teeth and eight-inch hind claws, biologists think that Tyrannosaurus was a fierce predator.

Plesiosaurus: Although technically not a dinosaur, this creature is a part of an extinct group of marine reptiles called the **sauropterygia** (sar op' ter uh jee' uh), which means "lizard flippers." Since they were large (spanning 10-60 feet), extinct reptiles, however, we put them with the dinosaurs. Biologists assume that these creatures were fierce marine predators.

Pteranodon: A flying reptile that had a wingspan of up to 30 feet. Once again, this is technically not a dinosaur, but an extinct flying reptile. The fossil remains indicate that it had no teeth, but it probably hunted by soaring over the water and catching fish, much like pelicans do today. Although very large, it probably weighed less than 50 pounds because its bones were quite delicate.

Although these drawings might make it look like we know a lot about dinosaurs, we really know very little. Were they ectothermic like other reptiles? We assume so, but we really have no evidence. Some of the latest theories in paleontology are that at least some dinosaurs were warm-blooded, like mammals and birds. Was the tyrannosaurus really a ferocious predator? We assume he was because of his huge teeth and claws, but many paleontologists have noted that while his teeth are really quite sharp, they have very shallow roots and could be dislodged from the skull rather easily. In addition, his tiny arms would make it very difficult for him to grab onto prey. These features might indicate that he was more of scavenger. The point is, we really know very little about these creatures.

Despite the fact the we have very little data regarding dinosaurs, paleontologists have jumped to some rather wild conclusions about them. For example, most paleontologists believe that dinosaurs lived long before human beings. Why? Well, there are really two reasons. First, most paleontologists are committed to macroevolution. Thus, they have to believe that the earth is ancient and that dinosaurs were a step in macroevolution that happened millions and millions of years ago. Second, they think that since we have never found dinosaur fossils and human fossils together, dinosaurs and humans must have existed at different times. Of course, based on the fact that there are so *few* fossils of dinosaurs and even *fewer* fossils of human beings, there is no reason to expect that you would find them together. It is so rare to find them by themselves that it is virtually impossible to find them together. Some claim to have found fossils of dinosaurs and humans together, but so far, none of the claims can be reliably verified.

There are two big problems with the idea that dinosaurs lived long before human beings. First, there is no reason to believe in the idea that the earth is really ancient. If you are committed to the hypothesis of macroevolution, then you must *assume* it is. However, we have already learned that macroevolution is not supported by scientific data. Thus, why should we believe in it? Second, archaeologists have found examples of ancient artwork that contain incredibly accurate drawings of dinosaurs. The Natural Bridges National Monument in the United States, for example, contains ancient artwork on rocks, all of which were drawn prior to the year A.D. 500. One of those pieces of artwork is a very good representation of apatosaurus. In the San Rafael Swell, there are ancient rock drawings from well over 1,000 years ago. One drawing accurately depicts a pteranodon. Arizona's Havasupai Canyon contains an ancient drawing of a creature that has the unmistakable proportions, stance, and tail of a theropod dinosaur. This is unlike any other kind of creature known to man. How in the world could people who lived long before paleontology accurately portray these creatures unless they saw them or were drawing them based on descriptions from those who saw them?

In the end, then, there is really no reason to think that dinosaurs and man did not live together at one time. In fact, the Bible supports the idea that they did. In Job 40:15-24, God describes to Job a creature (behemoth) that sounds like apatosaurus. Also, in Job 41:1-8, He describes a creature (leviathan) that sounds like plesiosaurus. In each case, God seems to imply that Job should *recognize* these creatures. If this is true, Job must have lived with them. Most creation scientists believe that dinosaurs and humans did live together on earth before Noah's flood. They were taken on the ark (there was plenty of room for young dinosaurs, which were considerably smaller than adult dinosaurs), but they probably did not adapt well to the changes in climate that took place after the worldwide flood. Thus, they slowly died out and are, most likely, extinct today.

Class Aves

Birds have always been a fascinating subject for scientists to study. Their ability to fly inspired scientists and engineers to build airplanes so that human beings could fly as well. As a result, the members of class **Aves** (aye' veez) have been studied closely for centuries. To be a member of this well-studied class, a creature must possess the following characteristics:

➢ **Endothermic**
➢ **Heart with four chambers**
➢ **Toothless bill**
➢ **Oviparous, laying an amniotic egg that is covered in a lime-containing shell**
➢ **Covered with feathers**
➢ **Skeleton composed of porous, lightweight bones (not a characteristic for all birds)**

Notice that the ability to fly is *not* on this list. That's because several birds, such as penguins and ostriches, do not fly. They are, nevertheless, members of class Aves.

The first characteristic of birds refers to their body's internal temperature. Birds are endothermic.

<u>Endotherm</u> – An organism that is internally warmed by a heat-generating metabolic process

Unlike reptiles and fish, then, birds have internal mechanisms that keep their bodies warm. They need not lie out in the sun or use some other means to regulate their bodies' temperature. Instead, their bodies have mechanisms by which they convert food into energy and use that energy to keep their internal organs and tissues warm. When a creature has such mechanisms, we say that the creature is "warm-blooded."

The second characteristic of birds refers to the circulatory system. Birds have four-chambered hearts. Remember, amphibians have three-chambered hearts, with a single ventricle and two atria. In reptiles, the ventricle has a partial separation; thus, the reptile heart is almost four-chambered. Well, birds have a true four-chambered heart, with two atria *and* two ventricles. This allows a complete separation between oxygen-rich and oxygen-poor blood. In a bird's circulatory system, the right atrium receives oxygen-poor blood and dumps it into the right ventricle. The right ventricle then pumps the oxygen-poor blood into the lungs. Once the oxygen level in the blood is enriched by the lungs, it travels from the lungs to the left atrium. The left atrium then dumps the oxygenated blood into the left ventricle, which pumps the blood to the various tissues of the body.

The next characteristic of birds relates to the mouth. A bird's mouth ends in a toothless bill. The structure of a bird's bill varies depending on the way in which it gathers its food. For example, predatory birds such as hawks and eagles have sharp, hooked bills that are ideally designed for tearing meat. Flamingos, on the other hand, have long bills with sieves. They feed by sucking water into the bill and then forcing it out through the sieves. Tiny shrimp and other small creatures get trapped in the sieves so the flamingo can eat them. Finally, birds such as hummingbirds that feed off the nectar of flowers have long, slender bills ideal for seeking the sticky sweet substance inside a flower.

Another characteristic of birds involves their method of reproduction. Like reptiles, fertilization is internal in birds, and birds are oviparous, laying amniotic eggs. Unlike reptiles, however, the covering of a bird's egg is not leathery. Instead, it is much harder, because it contains lime, a calcium-containing compound. Remember when we discussed bones, we made the point that bone tissue is hard because it contains calcium. Much like bone tissue, then, the covering of an egg is hard because it is made up of a calcium-containing compound. To get an idea of how a bird embryo develops in its egg, perform the following experiment.

EXPERIMENT 16.1
Bird Embryology

Supplies:

♦ Micro slide: The Chick Embryo
♦ Magnifying glass
♦ Microscope (optional)
♦ Desk lamp
♦ Lab notebook
♦ Colored pencils

Object: To observe bird embryology by studying the chick embryo

Procedure:

1. Pull the micro slide of the chick embryo out of its cardboard booklet. Hold the slide in your right hand so that the copyright section is closest to your hand and you can read the words in the copyright section.

2. Hold the micro slide up so that the light from the desk lamp shines through the images on the slide. With this illumination, you should be able to see the image in this section fairly well with your naked eye. To get a closer look, use your magnifying glass.

3. There are eight frames on the slide. Observe each section and read the information about that section in the cardboard booklet which came with the micro slide. Observe with both your naked eye and the magnifying glass. Be careful not to stare at the image for too long, as your eyes will tire from the light of the lamp.

4. Draw and label the image in each frame. Remember, each frame represents a certain stage in the embryo's development. The total time it takes for all of this development is only 96 hours!

5. Important things to note:

    a. The brain is forming in the 18th hour.
    b. By the 21st hour, the mouth and digestive system have started forming.
    c. In the 28th hour the circulatory system is beginning.
    d. By the 38th hour, the heart is forming and beating. The brain now has five regions.
    e. By the 56th hour, the four chambers of the heart are forming, along with the ears and eyes. Additionally, many organs like the lungs and kidneys are developing.
    f. By the 96th hour, the chick has every part, including the beginning of the wings.

6. (Optional) Observe each frame under the lowest power of your microscope. This slide was not really meant for a microscope, so you will have to hold it in your hands and move it up and down to find a good focus. Even though this method is a bit annoying, it will allow you to see some really nice detail.

7. Clean up your mess and put all of the equipment away, leaving it in good condition.

ON YOUR OWN

16.12 Fresh samples of blood are taken from a bird and a reptile. If the samples were taken during a cool night, which creature's blood will be the warmest?

16.13 A biologist shows you a sample of blood taken from a creature's ventricle. If this blood is a mixture of oxygen-rich and oxygen-poor blood, did it come from an amphibian or a bird?

16.14 You find an egg that has a very hard shell. Is it a bird egg or a reptile egg?

A Bird's Ability to Fly

As we mentioned before, people have always been fascinated by a bird's ability to fly. Thus, we have studied birds a great deal in order to unlock the secrets of flight. As a result, we have some pretty good ideas about what characteristics of birds make it possible for them to fly. The three most important of these are feathers, wings, and skeletal structure. We will first look at feathers, then we will discuss wings, and then we will look at skeletal structure.

Feathers

Regardless of whether or not a bird can fly, it must have feathers. Indeed, the presence of feathers is one of the unique characteristics that we listed for class Aves. The feathers of a bird, which are both very strong and very light, grow from small structures called **follicles**, which reside on the skin. The basic structure of a feather is shown in the figure below.

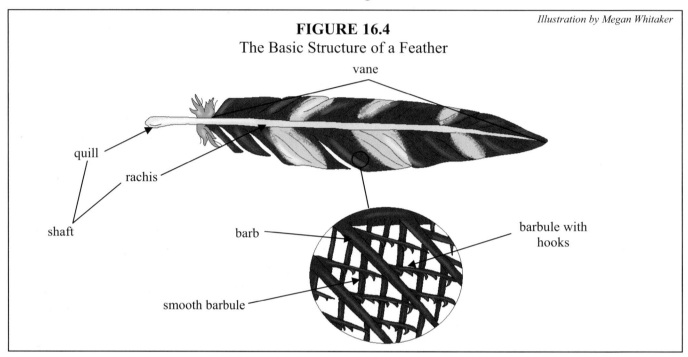

FIGURE 16.4
The Basic Structure of a Feather

Illustration by Megan Whitaker

vane

quill

rachis

shaft

barb

barbule with hooks

smooth barbule

Notice from the figure that a feather has two basic parts: a **shaft** and a **vane**. The shaft is generally divided into two sections. The bare portion of the shaft that connects to the follicle is called the **quill**, while the portion of the shaft that holds the vane is called the **rachis** (ray' kis). The vane is composed of parallel **barbs** that originate on the rachis and extend outward. These barbs have two types of **barbules** (barb' yoolz) that extend away from the barb. One type of barbule is smooth, and the other type has hooks. These barbules are arranged so that the hooks grab onto and slide along the smooth barbules.

Think about this design for a moment. Since the hooked barbules attach to the smooth barbules, the feather is supported by a network of interlocking strands. This gives the feather a lot of strength. At the same time, however, the feather is flexible. After all, the hooked barbules can slide up and down the smooth barbules, allowing the feather to flex a great deal. If the feather flexes so much that the hooked barbules actually slip off the smooth barbules, that's no problem. The bird can reattach the hooks to the smooth barbules by stroking the edge of the vane with its beak. In the end, then, the feather possesses an *ideal* compromise of weight, strength, and flexibility.

In order to retain ultimate flexibility in the wing, the hooked barbules must slide along the smooth barbules very easily. Thus, there needs to be some sort of lubricant covering the barbules. Well, it turns out that a bird produces oil in a gland near the base of the tail. When a bird wants to "oil" its feathers, it dips its bill into that gland and then runs the bill over its feathers. This is called **preening**, and by doing this, the bird can spread oil all over its wings. This allows the hooked barbules to slide freely along the smooth barbules and, as a side benefit, the oil that the bird spreads on its feathers is usually waterproof. This keeps the bird's wings from getting wet, which in turn keeps the bird warm. It also keeps the bird from getting so waterlogged that it becomes too heavy to fly.

Now it turns out that there are two basic types of feathers: **down feathers** and **contour feathers**.

Down feathers – Feathers with smooth barbules but no hooked barbules

Contour feathers – Feathers with hooked and smooth barbules, allowing the barbules to interlock

Obviously, then, the feather pictured in Figure 16.4 is a contour feather. Since the barbules in a down feather cannot hook together as they do in a contour feather, down feathers are fluffy and soft rather than strong and flexible. They are found underneath the layer of contour feathers on a bird, and they mostly serve as insulation for the bird. People use down feathers as pillow or cushion stuffing because they are very soft. Contour feathers are the feathers you are used to seeing on a bird. The contour feathers on the bird's body give it shape and color, while the contour feathers on its wings are essential for flight.

Like scales, feathers are made of nonliving tissue. Thus, they cannot grow or repair damage. As a result, birds must molt their feathers like reptiles shed their skin. Most birds do not molt their feathers all at once, however, because they would lose the ability to fly if they did so. In fact, flight is such a difficult skill that a bird cannot lose more feathers on one side of its body than on the other. If that were to happen, the bird would become unbalanced and could not stay in the air for any length of time! To avoid such problems, then, the molting of a bird is a very precise, delicately balanced process. Basically, a bird molts its feathers in pairs. Each feather on one side of the bird's body will molt *at exactly the same time as the corresponding feather on the other side of its body*! This process occurs in small groups, so that the bird is never really missing more than a few pairs of feathers at any given time. Although this is true for most birds, many waterfowl molt all at once. As a result, they become flightless for weeks at a time.

Wings

Feathers are, indeed, marvelously engineered structures. They provide strength and flexibility at a fraction of the weight of even the most advanced structures made by human science. Nevertheless, feathers do not allow for flight until they are attached to wings. Scientists recognized rather early that a bird's wings were the key to its flight ability, but they had the mistaken notion that wings were the sole key to a bird's ability to fly. As a result, they made all manner of contraptions with wings attached, hoping that these wings would give their contraptions the ability to fly as well. Of course, it took centuries of painstaking study to determine *how* a bird's wings give it the ability to fly. After we learned this secret from studying birds, we were able copy the design and apply it to airplanes that actually could fly.

Now it turns out that the principle behind how wings allow birds to fly is just a bit too complicated to cover in this course. If you take physics in college, you will learn about something called Bernoulli's principle. It describes how a bird's wing or an airplane's wing makes flight possible. Even though a discussion of this principle is beyond the scope of this course, there is one very important aspect of a bird's wing that we want to discuss here.

As we said before, scientists and engineers studied bird wings and bird flight in order to determine how to make airplanes fly. Through the genius of the Wright brothers at Kitty Hawk, man was finally able to leave the confines of the ground and fly through the air, albeit relatively slowly and only for a few minutes. Eventually, however, airplane technology began to improve, and airplanes began flying faster and for longer periods of time. As planes got to traveling faster and faster, however, engineers suddenly discovered a problem.

You see, airplanes experience **turbulence**, which is essentially "rough air." This turbulence shakes the plane as the plane passes through it. If you have ever ridden on a plane, you have probably experienced episodes of turbulence. Well, it turns out that turbulence is particularly bad when a plane lands. This was a big problem when people were designing early airplanes. The turbulence during landing was so severe that the planes' wings would shake violently and even fall off. Clearly, this was a problem! Engineers fretted and worried over this problem and could not figure out how to fix it, until someone had the bright idea to go back and study birds again. Clearly, birds didn't have the problem that man-made airplanes had. How do birds get around the problem? The figure below shows you how.

FIGURE 16.5
A Bearded Vulture Landing

Photo © Photodisc, Inc.

alula   alula

A bird's wing is equipped with a set of feathers called the **alula** (al' yuh luh). When the bird lands, the alula is lifted so that a gap exists between the wing and the alula. In flight, the alula lies flush against the wing. Engineers had *no idea* how this got rid of the turbulence problem, but since the action of the feather was related to landing, they decided that it *must* do something! Thus, they created a similar mechanism on an airplane's wing, called a "wing slot." Guess what? This solved the turbulence problem! Now remember, even though the mechanism that engineers copied from the bird's wing had fixed the problem, they still *had no idea why it worked.* They only knew that it did work. It took the power of a supercomputer for physicists to finally explain *how* this mechanism reduces air turbulence. The explanation is far too complex to discuss here. The point should be clear, however. The bird possesses mechanisms for flight that we could not understand without years of effort and mammoth computing power!

Skeletal Structure

Most birds have a skeleton made up of porous, light bones. Penguins don't have such bones, but all birds that fly do. You see, one of the really tricky parts about flight is the fact that any flying object must be very lightweight but at the same time very strong. We've already seen that feathers meet that condition very nicely, but if a creature's skeleton is too heavy, all of the feathers in the world will not make a difference. It turns out that a skeleton made of vertebrate bones like that which we studied in Module #13 would simply be too heavy for flight. Thus, the bones of birds must be unique in subphylum Vertebrata.

Unlike other vertebrates, birds have bones that are full of air-filled cavities. These cavities make the bones very light so that the skeleton does not hinder a bird's ability to fly! Now you might wonder if a bone that is full of air-filled cavities can be strong. Well, the answer is no, it cannot. However, a bird's bones are full of thin strands of bone tissue that traverse the air cavities. These strands serve as struts that strengthen the bones. It turns out that in order to build viable airplanes, this is another design feature that engineers had to copy from birds. Engineers realized early on that they needed to make airplanes out of lightweight materials, so they used hollow tubes, etc. to build the frames of airplanes. As airplanes got bigger and faster, however, these tubes would collapse, because they simply were not strong enough. When engineers studied bird bones, however, they discovered "struts" in the air-filled cavities, and they constructed similar struts, which they called "Warren's trusses," in their hollow tubes. These trusses kept the tubes from collapsing during flight.

Now think about all of these facts for a moment. The bird has elegant feathers that have a better combination of strength, flexibility, and weight than anything that modern human science can build. In addition, human scientists and engineers could not design a working plane until they copied the wing design, the turbulence-reducer design, and the skeletal design from birds. Even with all of that help, human science still cannot create a plane that is anywhere near as efficient as a bird! Shouldn't that tell you something about the Creator of the bird? In the end, all that human scientists did was copy the design of the Creator in order to make airplanes. Unfortunately, human science cannot copy the design as well as God can create, so the most sophisticated product of human science still cannot compare to what God's creation has always had!

ON YOUR OWN

16.15 A bird's feathers begin to lose their flexibility. What should the bird do in order to fix this problem?

16.16 You see a picture of a bird in a book. The alula on each wing is pushed against the wing so that there is no gap in between the alula and the wing. Is this bird most likely in the middle of its flight or in the process of landing?

16.17 Two bones look identical. The person showing you the bones says that one comes from a reptile and the other from a bird. How could you tell which came from the bird without damaging the bones in any way?

Classification in Class Aves

There are many, many orders of birds. Table 16.1 lists the major orders and examples of the birds which are found in them.

TABLE 16.1
The Major Orders of Birds

Order	Examples of Birds in the Order
Anseriformes (an suh ree' for meez)	swans, geese, ducks
Apodiformes (uh pod' uh for meez)	hummingbirds, swifts
Charadriiformes (chuh rad' rih for meez)	plovers, gulls, terns
Ciconiiformes (sih' kon ih for meez)	herons
Columbiformes (koh luhm' buh for meez)	pigeons, doves
Falconiformes (fal' kuhn uh for meez)	eagles, hawks, falcons
Galliformes (gal' uh for meez)	turkeys, quails, pheasants
Passeriformes (pass' er uh for meez)	robins, finches, sparrows, nightingales
Pelecaniformes (pel ih kahn' uh for meez)	pelicans
Phoenicopteriformes (foh nee' she opt' uh for meez)	flamingos
Procellariiformes (pro sel' uh rih for meez)	albatrosses, petrels
Psittaciformes (sit' uh suh for meez)	parrots, parakeets, lovebirds
Sphenisciformes (fen uh suh' for meez)	penguins
Strigiformes (strih' guh for meez)	owls
Struthioniformes (struh thee on' uh for meez)	ostriches

Now don't get wrapped up in trying to memorize these orders and their members. We present them just so you get an idea of bird classification.   Examples of birds in some of these orders are shown in the figure below.

a. Photo by Dawn Strunc
b. Photo © Phil Sigin-Lavdanski
c-e, g, i. Photos © Photodisc, Inc.

FIGURE 16.6
Members of Class Aves

f. Photo © Painet Photographic Arts and Illustration NETwork
h. Photo from www.clipart.com

a. Ostrich (Struthioniformes)   b. Penguin (Sphenisciformes)   c. Bald Eagle (Falconiformes)

d. Owl (Strigiformes)   e. Flamingo (Phoenicopteriformes)   f. Hummingbird (Apodiformes)

g. Quail (Galliformes)   h. Pelican (Pelecaniformes)   i. Heron (Ciconiiformes)

The majority of flightless birds are found in two orders: Struthioniformes and Sphenisciformes. These birds, such as **ostriches** (Figure 16.6a) and **penguins** (Figure 16.6b), have most of the characteristics of birds but simply do not fly. Instead, penguins are graceful and powerful swimmers, and ostriches are powerful runners. As we mentioned previously, penguins do not have the porous bones characteristic of class Aves, and only some of the ostrich's bones are porous. Note that they both have wings and feathers, however. Penguins use their wings as flippers to propel themselves in the water, while ostriches use their wings to make quick turns while running. They can also use them as "sails" when running with the wind in order to increase their speed. In both photos of flightless birds, an adult is pictured with one of its offspring. Note the difference in feathers between the young penguin and the adult penguin. The parent penguins will care for their offspring until its feathers change into those of an adult. At that point, the young penguin is ready to live independently.

The members of orders Falconiformes (such as the **bald eagle** in Figure 16.6c) and Strigiformes (such as the **owl** pictured in Figure 16.6d) are often called **birds of prey**. They have keen eyesight used to spot their prey from a tall perch or while in flight. When they have their prey in sight, they swoop down on it in a high-speed (up to 70 mph) dive and grasp the unfortunate creature in their sharp talons. Usually, the fast impact of the birds' talons on the prey instantly stuns or kills the victim, and the birds then take their feast somewhere that they can eat in peace.

Although we generally think of birds of prey capturing mice and other land-dwelling creatures, a large number of these birds actually hunt fish. They fly over bodies of water looking for a fish swimming near the surface of the water. They then swoop down and grab the fish before it can swim away. The bald eagle, symbol of the United States of America, typically feeds on fish. The name "bald eagle" is a bit misleading. The bald eagle is not bald. Instead it has distinctive white feathers on both its head and tail. The term "bald" comes from an obsolete use of the word "piebald," which can mean "blotched with white." It turns out that the population of bald eagles is dwindling, mostly because of overhunting. Although it is illegal to kill a bald eagle, hunters typically shoot them by accident, because a bald eagle does not develop its distinctive white head and tail feathers until it is three or four years old. As a result, many young bald eagles are shot down by hunters who honestly do not recognize them as bald eagles.

Members of order Phoenicopteriformes are commonly known as **flamingos** (Figure 16.6e). We already told you that flamingos eat by taking water into their bills and then straining out the tiny creatures in the water for food. Although you mostly see them standing in water, they do fly. Since they are very social birds, they tend to fly in groups, usually forming a "V" pattern in the air. They also tend to forage in groups. Interestingly enough, the pink color you associate with flamingos comes from their diet of shrimp or cyanobacteria, which is usually rich in carotenoids. In zoos, flamingos rarely eat such food, so the zoos usually add carrots, beets, or a supplement containing carotenoids to their diet so that they will still develop their pink color.

Order Apodiformes contains **hummingbirds** (Figure 16.6f). The term "hummingbird" is actually a generic name for hundreds of species of small (as little as three inches long) birds that feed on the nectar found in certain flowers. These birds are known for the rapid rate at which they beat their wings. This fast motion causes a humming sound from which they derive their common name. They feed by hovering near a flower and probing it for nectar and small insects. It turns out that members of this order are the only birds in creation that have the ability to fly backwards. This is a necessary skill for hummingbirds, as they must back away from a flower after feeding on it.

Members of orders Galliformes (like the **quail** in Figure 16.6g) are often called **game birds**. They have tender white meat that is, in fact, composed of their flying muscles. Found predominantly in the breast, the flight muscles in these birds receive less oxygen than the flight muscles in other birds, which means they have less myoglobin, a compound used to store oxygen in muscles. It is the lack of myoglobin that causes white meat to be lighter than dark meat. As a result of having low oxygen storage capabilities, the flight muscles on game birds tire quickly, and they are only capable of quick bursts of flight, after which they must rest their muscles. This makes the meat derived from these tissues more tender than typical bird meat. Since their meat is so tender, game birds are the most common birds to find on a dinner table.

Many birds inhabit water in one way or another. Members of order Anseriformes, for example, are typically called **swimming birds**, because they are at home swimming in the water. Examples would include ducks, swans, and geese. The oil they produce for preening is especially waterproof, so that they dry off quickly after leaving the water. Their feet are webbed for efficient swimming. The order Pelecaniformes contains **pelicans** (Figure 16.6h), while Procellariiformes contains albatrosses and petrels. These are often called **diving birds**, because they are specially designed to dive into the water and catch fish. They do not grab onto their prey with talons; instead, they dive headlong into the water and catch their prey with their bills. Order Ciconiiformes contains birds such as the **heron** (Figure 16.6i). These birds, called **wading birds**, wade in the water, looking for small fish and other aquatic organisms to eat.

Order Passeriformes contains what many people call the **song birds**. All birds have the ability to make a chirping sound because of a special organ called the **syrinx** (sihr' inks). Members of this order, such as sparrows and nightingales, tend to have well-developed syringes (sihr' in jeez – plural of syrinx) which are capable of making beautiful sounds.

To get an idea of what birds inhabit your region of the world, perform the following experiment.

EXPERIMENT 16.2
Bird Identification

Supplies:

♦ Bird field guides (available at your local library)
♦ Binoculars (if available)
♦ Bird seed
♦ Lab notebook

Object: To become familiar with birds common to your locality

Procedure:

1. This is a two-fold exercise. The first part should be done at your own home.
2. Buy a small bag of bird seed.
3. Spread some of the seed on a large, cleared area on the ground. Place some in a higher place like a bird feeder or a window sill.
4. Observe for several days. Note the birds that come to feed on the ground and the ones at the higher place. You should observe that some birds tend to be ground feeders, and others will eat at the feeder or window sill.
5. Use the field guides to identify these birds. Note the birds that you see and their classification in your lab notebook.
6. Plan a trip to a state or city park that has an outside bird sanctuary. Plan to spend at least an hour observing the various types of birds and their habits. Note the birds that you see and their classification in your lab notebook.

ON YOUR OWN

16.18 The roseate spoonbill feeds by standing in water and sweeping its open bill back and forth, catching small fish, shrimp, snails, and insects. In which order does it belong?

16.19 The red jungle fowl has tender white breast meat when cooked. It flies in short bursts, spending the rest of its time on the ground. In which order does it belong?

Class Mammalia

We conclude our biology course with a discussion of the class that contains people, class **Mammalia**. It turns out that class Mammalia (muh mail' ee uh) is one of the smaller classes in subphylum Vertebrata, but it nevertheless is probably the class that most people associate with the word "animal." This is probably because mammals occupy virtually every ecosystem with which people are familiar. As a result, "animal" has come to mean "mammal" in many people's minds. It is important that you do not make such a mistake, however. The term "animal" also includes such creatures as sponges, hydras, fishes, amphibians, reptiles, and birds.   The following characteristics separate mammals from the other vertebrates.

➢ **Hair covering the skin**
➢ **Reproduce with internal fertilization and usually viviparous**
➢ **Nourish their young with milk secreted from specialized glands**
➢ **Four-chambered heart**
➢ **Endothermic**

Notice that the last two characteristics of mammals are the same as those of birds. The first three characteristics, however, set mammals apart from any other creatures in phylum Chordata.

The one characteristic of mammals that most students learn in grade school is that mammals have **hair**. In most mammals, the hair is really obvious. In some mammals, however, it is much harder to see. Elephants have hair, but unless you examine them closely, you might miss the ragged tufts that make a patchwork of hair across the skin. Whales are mammals, and it is even harder to find their hair. It is there, nevertheless. Some have whiskers on their snouts, others have hair on their chin, and others have hair behind their blowholes.

Hair, like the scales on reptiles and the feathers on birds, is made of nonliving cells. These cells are produced by **hair follicles**, which are tiny structures in the skin. Most mammals have two types of hair, **underhair** and **guard hair**. As its name implies, underhair is found under the guard hair. It is a soft, insulating layer of fur next to the animal's skin. The hair that we see when we look at mammals is typically the guard hair. This hair is coarser than the underhair and is typically longer as well. This is the hair that gives most mammals their colors and distinctive markings.

Hair has many functions. Underhair insulates the mammal and keeps it warm. In a beaver or sea otter, the underhair is so dense and well-insulating that these mammals can swim in ice-cold water without decreasing their body temperature significantly! Hair can also provide camouflage. Young deer, for example, tend to blend in with their surroundings because of the coloring and markings of their hair. Hair can also be used as a defense mechanism, as is the case with the porcupine, whose guard hair is stiff and pointed. Hair can also aid the senses. The whiskers on a cat are very sensitive to touch. They are used to detect objects in the dark or when the objects are outside the mammal's field of vision. A cat will never go through a passage (such as a hole in a fence) unless it can move through it without its whiskers being touched. If the cat's whiskers can pass through the hole, the cat knows that its entire body can pass through the hole. When dogs and cats are threatened, hairs on their neck or tail (or both) will stand up. When this happens, we usually say that the hair "bristles." Bristled hair is incredibly sensitive, and if that hair gets touched, the mammal's immediate instinct is to jump away then turn and attack in the direction of the touch.

Mammals reproduce with internal fertilization. In the vast majority of mammals, the embryo's development is viviparous. There are exceptions (of course) to the viviparous rule. One order of mammals (which we will discuss in a moment) is oviparous, while another (which we will discuss in a moment) has a pouch in which the embryo develops. These two orders of mammals are called **nonplacental** (non pluh sent' uhl) **mammals**. The other orders contain **placental mammals**, which have young that develop viviparously.

In placental mammals, a **placenta** forms so that the embryo can develop inside the mother.

Placenta – A structure that allows an embryo to be nourished with the mother's blood supply

The placenta is rich in blood vessels from both the developing embryo and the mother. These blood vessels get very close to one another, close enough to exchange gases and nutrients, but the blood of the mother and the blood of the developing offspring never actually mix. The mother's blood brings oxygen and nutrients to the placenta, which allows the embryo's blood vessels to absorb them. At the same time, the embryo's blood vessels give waste products to the placenta, which transfers them to the mother's blood vessels so that they can exit via the mother's excretory system. The blood of the infant travels back up from the placenta through an **umbilical** (um bil' ih kul) **cord** that connects the developing offspring to the placenta. Your "belly button" shows you where your umbilical cord attached to you while you were an embryo.

As the offspring develops inside the mother, we say that it is in **gestation** (jes tay' shun).

Gestation – The period of time during which an embryo develops before being born

The gestation time for mammals varies greatly. Mice have a gestation period of about three weeks, while horses have a gestation period of almost one full year. The practical result of long gestation periods is that the offspring is much more developed when it is born. Mice, which have a short gestation period, are born hairless, blind, and weak. They grow hair, develop sight, and gain strength after birth. Horses, however, are born with a full coat of hair, and within hours of their birth, they can stand, walk, and even run.

Once the offspring is born, all mammals (placental or not) care for their young. The female provides milk to nourish the young. This milk is produced in **mammary** (mam' uh ree) **glands** that are specially designed to produce milk.

Mammary glands – Specialized organs in mammals that produce milk to nourish the young

The mammary glands of female mammals are always present, but they tend to be active only shortly before and for some length of time after birth. When a female mammal's mammary glands are active, we say that the female is **lactating** (lak' tayt ing). Typically, the young get their mother's milk by sucking on nipples which attach to the mother's mammary glands. Most mammals have some sort of "family" structure in which the parents teach the young survival skills before the young leave the protection of the parents.

As is the case with birds, mammals are endothermic and have four-chambered hearts. Thus, they are able to regulate their own internal temperature and keep it constant. Typically, a mammal's underhair aids in the process, insulating the mammal so that only a small amount of the internally-generated warmth escapes in cold climates. Because mammals have a four-chambered heart, oxygen-poor blood never mixes with oxygen-rich blood in the blood vessels.

ON YOUR OWN

16.20 Which mammal would have thicker underhair, a polar bear (which lives in a very cold climate) or a prairie dog (which lives in a warmer climate)?

16.21 A cat with its whiskers cut off tends to run into objects a lot more frequently than the same breed of cat whose whiskers are not cut off. Why?

16.22 Two different species of mammals have different gestation periods. The first has a long gestation period and the second has a short gestation period. Which mammal will have offspring that are more developed at birth?

Classification in Class Mammalia

The members of class Mammalia are divided into many orders. The major ones are summarized in the table below.

TABLE 16.2
The Major Orders of Mammals

Order	Examples of Mammals in this Order
Artiodactyla (art' ee oh dak' til uh)	sheep, deer, bison, goats, pigs, cattle
Carnivora (kar nih' vor uh)	cats, dogs, bears
Cetacea (seh tay see' uh)	whales, dolphins
Chiroptera (kye rop' tur uh)	bats
Dermoptera (derm ahpt' uh ruh)	flying lemurs
Edentata (ee den' tah tuh)	anteaters, tree sloths, armadillos
Hyracoidea (hi ruh koy' dee uh)	hyraxes
Insectivora (in sek' tiv or uh)	shrews, moles, hedgehogs
Lagomorpha (lag' uh mor fuh)	rabbits, hares, pikas
Marsupialia (mar soop ee ay' lee uh)	kangaroos, wallabies, koalas, opossums
Monotremata (mon' uh trem' ah ta)	duck-billed platypuses, spiny anteaters
Perissodactyla (puh' ris' uh dak' til uh)	horses, zebras, rhinos, tapirs
Pholidota (foh lih doh' tuh)	pangolins
Primates (pri' mates)	monkeys, lemurs, humans
Proboscidea (pro boh seed' ee uh)	elephants, mammoths (extinct)
Rodentia (roh den' chee uh)	squirrels, rats, mice, porcupines, prairie dogs
Sirenia (sir ee' nee uh)	manatees and other sea cows
Tubulidentata (tuh boo' luh den tah' tuh)	aardvarks

Once again, do not worry about memorizing these orders; they are shown for the sake of completeness. Examples of creatures from some of these orders are shown in the figure below.

FIGURE 16.7
Members of Class Mammalia

a, c, d, f, g-h. Photos © Photodisc, Inc.
b, i. Photos © Comstock, Inc.
e. Photo by Dawn Strunc

a. Koala (Marsupialia)   b. Lion (Carnivora)   c. Dolphin (Cetacea)

d. Elephant (Proboscidea)   e. Zebra (Perissodactyla)   f. Mountain Goat (Artiodactyla)

g. Prairie Dog (Rodentia)   h. Snowshoe Hare (Lagomorpha)   i. Baboon (Primates)

Let's start our discussion of mammals with the two orders of "odd" mammals. We call them "odd" mammals because they are not viviparous. Members of order Monotremata, for example, are often called the **egg-laying mammals**. These mammals lay eggs and attend them like birds do. Once the offspring hatch, they are nourished by their mother's milk. Interestingly enough, however, members of this order do not have nipples. Instead, their mammary glands are evenly distributed along the female's underside. These glands dump milk directly onto the skin, and the young lap the milk from the mother's fur.

The best known member of order Monotremata is the **duck-billed platypus**. When Europeans originally made it to Australia, where the duck-billed platypus lives, they reported seeing an odd creature that is covered with fur yet has a bill like a duck, lays eggs, and has webbed feet. Scientists in Europe did not believe the travelers, because they could not imagine a mammal (an animal with hair) like that. Even when a dead one was brought to Europe, scientists thought it was a fake. They would not believe that such a strange creature existed until live specimens were brought back to Europe. The duck-billed platypus uses its bill to scoop invertebrates out of the mud to eat. Interestingly enough, its bill is sensitive to the electrical signals of other creatures, and that's the sense it uses to find its prey. Sought for its novel fur, it was nearly hunted to extinction. Now that the Australian government protects the duck-billed platypus, their numbers are on the rise.

Order Marsupialia contains the other group of nonplacental mammals. In these creatures, such as **kangaroos, koalas** (Figure 16.7a), and **opossums**, the fertilized egg develops within the mother for only a few days. Because this phase of development is short, no placenta is formed. Instead, in just a few days, the tiny, immature offspring crawls out of the mother and makes its way to a pouch, which contains nipples. The offspring must attach its mouth to a nipple and then continue its development in the pouch, drawing nourishment from the mammary gland to which the nipple is attached.

Most members of this order live in or near Australia. The United States has a marsupial, however. It is the opossum. This mammal, which resembles a large rat, is best known for its tendency to play dead. When it senses a potential predator nearby, an opossum will flop over on its side, appearing dead. Its eyes take on a glassy look, and its face contorts into a grimace. This tends to make the predator think that the animal has been dead for some time, and most predators instinctively avoid such easy prey, because long-dead animals are often carriers of disease. Although opossums play dead, they are also fierce fighters if forced into a fight. They typically prey on small mammals, birds, or insects.

It is important to note that many biology books do not use order Marsupialia anymore. Those that do not use this order typically split the marsupials into seven different orders, and they call all seven of those orders part of subclass Metatheria (met uh thur' ee uh). We do not see the need to do this, because there are less than 300 known species of marsupials, so it seems extreme to split them up into seven different orders. Nevertheless, the next biology book you read may, indeed, do this.

Members of order Chiroptera are often called the **flying mammals**, and the best-known example is the **bat**. Bats look like mice with wings. At one time, they were actually classified in the same order as mice, but today, they are considered unique enough to deserve their own order. Bats tend to live in dark environments like caves, barns, or dark forests. They live in these environments because of their unique ability to "see" in the dark, which gives them a distinct advantage over other animals.

Bats "see" in the dark with an elegantly designed sonar system. They emit sound waves from their mouths. If the sound waves encounter an obstruction or another creature, these sound waves bounce back to the bat. The bat receives these reflected sound waves and interprets them in order to determine the size and shape of what they bounced off. Somehow, the bat knows which sounds are its own and which ones come from other sources, even other bats. Laboratory experiments indicate that even when their reflected sound waves are 2,000 times weaker than other sounds reaching the bat's ears, the bat can still recognize them and use them to "see." This sonar system is so efficient that a bat can pinpoint the precise location of a fruit fly (one of the tiniest flies in creation) from as far as 100 feet away! Bats can use this sonar so effectively that they can eat up to five fruit flies each minute! In the words of one author, "Ounce for ounce, watt for watt, it is millions of times more efficient and more sensitive than the radars and sonars contrived by man" (Michael Pitman, *Adam and Evolution*, [London: Rider & Company, 1984], 220). The bat stands as another testimony that life was *designed*.

Members of order Carnivora include **cats**, **dogs**, and other mammals that eat meat. These mammals have distinctive teeth. They have **canine fangs** that are designed to tear meat and sharp molars that are designed to chew flesh. In addition, their feet typically have sharp claws used to capture prey. Cats such as **lions** (Figure 16.7b), **tigers**, and even domesticated cats use their claws to pounce on prey and hold them. **Bears**, on the other hand, often use their claws to catch fish right out of streams.

Each family in order Carnivora has its own distinctive behaviors that allow its members to catch prey. Cats, for example, rely on stealth and cunning to sneak up on their prey. Dogs, on the other hand, require teamwork. Dogs like **wolves**, for example, hunt in packs and share the food that the pack catches. Bears are actually omnivorous. They eat tree bark, honey, plants, and animals (especially fish). Thus, members of order Carnivora are not necessarily carnivores. Some are omnivores. They all eat at least some meat, however.

Aquatic mammals belong to orders Sirenia and Cetacea. All of these mammals breathe with lungs; thus, they must come to the surface in order to breathe. **Whales** and **dolphins** (Figure 16.7c), for example, have blow holes on the top of their heads which they use to breathe air. Order Cetacea contains the largest known animal in creation, the **blue whale**, which reaches weights of 170 tons and lengths of 110 feet. Even though the blue whale is large, it is a gentle creature that feeds on plankton in the water.

Dolphins and **porpoises** are interesting aquatic mammals. Porpoises look like dolphins, but there are distinguishing characteristics that separate the two. Porpoises are usually smaller than dolphins, and they have rounded heads. Dolphins, on the other hand, have pointier heads that end in a pointed mouth resembling a beak. Also, porpoises have triangular dorsal fins, whereas dolphin dorsal fins are hooked. These mammals have two very interesting abilities.

First, like the bat, dolphins and porpoises navigate with sonar. They must do this because they swim very quickly, much too quickly to navigate by sight. This is because light travels poorly in water, so the range of sight is short under water. Sound, however, travels quickly and efficiently in water; therefore, dolphins and porpoises can use their sonar as effectively as the bat does.

Second, dolphins and porpoises are incredibly fast swimmers because of a special design feature that God gave them. You see, it is rather difficult to reach high speed under water because water tends to drag against any body moving in it. As a result, it slows the motion of the body down.

Now it turns out that the faster a body moves in water, the harder water drags against it. In the end, there comes a point at which no matter how much more energy is expended, speed cannot be increased significantly because almost all of the extra energy goes into overcoming the drag of the water. Thus, for most water creatures, there is an inherent limit to how fast they can swim. Porpoises and dolphins get around this problem because a spongy material within their loose, finely laced, layered skin beats rhythmically with the motion of the water to *strongly reduce water drag*. This allows the porpoise to travel incredibly quickly. As with most of the elegantly designed structures in creation, human science cannot develop anything similar to this material. If the navy could manufacture something like this material, think about what could be done with it!

Order Proboscidea contains the **elephants** (Figure 16.7d). These creatures are distinguished by their long trunks. This boneless, muscular feature of elephants is actually a greatly elongated upper lip and nose. It is used to pick up grasses, leaves, and water and bring these substances to the elephant's mouth. An extremely versatile organ, the trunk is also used to trumpet calls, pull down trees, rip off foliage, and draw up water for bathing. It is also a highly sensitive organ, which the animals raise into the air to detect wind-borne scents. By means of fingerlike lobes on the end of the trunk and by the sucking action of the two nostrils, elephants can pick up and examine small objects. Very hairy versions of these creatures, the mammoths and mastodons, are extinct today.

Orders Perissodactyla and Artiodactyla contain the **hoofed mammals**. The two orders are distinguished from one another by the type of hooves that their members possess. The mammals in order Perissodactyla have either a single hoof that encompasses the entire foot, or three enlarged toes that make up the hoof. As a result, these mammals, such as the **horse**, **zebra** (Figure 16.7e), and **rhino**, are sometimes called the **odd-toed hoofed mammals**. Members of order Artiodactyla, such as **sheep**, **deer**, **bison**, **goats** (Figure 16.7f), **pigs**, and **cattle**, have two or four toes that make up their hooves. Thus, they are often called the **even-toed hoofed mammals**.

Order Rodentia, which contains **squirrels**, **rats**, **mice**, **porcupines**, and **prairie dogs** (Figure 16.7g), has the largest number of species. These creatures are distinguished by large, sharp incisor teeth in both the upper and lower jaws. These teeth grow continually, and members of order Rodentia use them to gnaw and bite. In fact, a rodent must continually gnaw with its incisors, or those incisors will grow so long that they will interfere with the rodent's ability to close its own mouth! Thus, rodents have an instinct to gnaw and gnaw and gnaw. As a result, members of this order are often called the **gnawing mammals**.

Most rodents have essentially no defense mechanisms against predators. Consequently, they are constantly being eaten by carnivorous reptiles, birds, and mammals. How, then, do they keep from dying out? They reproduce in huge numbers! **Field mice**, for example, reproduce several times throughout the course of their life. The young are completely independent from their mother in as little as four weeks after birth. At the end of six weeks of life, most field mice have produced their first litter of offspring! Some species of mice that have been housed in laboratory conditions have produced as many as *80 offspring per year!*

Rabbits and **hares** (Figure 16.7h), which belong to order Lagomorpha, are often called the **rodentlike mammals**. They look and behave much like the mammals of order Rodentia, but they are not similar enough to belong to the same order. For example, while both lagomorphs and rodents have pairs of long incisors on their upper and lower jaws, lagomorphs have a second pair of peglike incisors on their upper jaw that rodents don't have. What's the difference between a rabbit and a hare?

Actually, the distinguishing characteristic is related to how they are born. The offspring of rabbits are born with no hair and with their eyes still closed. Hares, on the other hand, are born covered with fur (many believe that's where the name originated) and with their eyes open. In adults, hares can usually be distinguished from rabbits because hares have longer ears and are usually larger than the average rabbit. Like rodents, they survive simply because they reproduce prolifically. The typical rabbit can begin producing litters about six months after birth. They produce three to eight offspring in each litter, with a gestation period of a little more than a month. Since the average rabbit or hare lives about ten years, you can see how many offspring just one individual can produce!

Order Edentata, often called order Xenarthra, contains such creatures as the **anteater**. This interesting mammal has a long head with a long, tubular mouth and long tongue, but no teeth. It tends to live in forests and swampy areas and avoids highly populated regions. The animal actually walks on its knuckles, using its claws only for defense or for tearing apart anthills or termite mounds. Once it has torn the insects' home apart, its long tongue flicks rapidly in and out of its small mouth opening, scooping up termites or other insects on its sticky surface.

Members of order Tubulidentata, the **aardvarks**, are often mistakenly called anteaters. This is because they tend to eat ants and termites like the anteater and have a long mouth like that of an anteater. An aardvark's mouth, however, is not nearly as long as an anteater's. One of the main characteristics that separate aardvarks from anteaters, however, is the fact that aardvarks have teeth, while anteaters do not.

Members of order Insectivora include **shrews**, **moles**, and **hedgehogs**. These creatures have pointed snouts, designed for burrowing into the earth to find insects. Shrews and moles actually spend a great deal of their lives underground, burrowing tunnels and the like. Hedgehogs rarely burrow tunnels. As their name implies, they tend to live in hedges, hiding from predators during the day and emerging at night to hunt for food. While the defense mechanisms of moles and shrews are basically related to running underground to avoid predators, the hedgehog uses its guard hair which, like the porcupine's, is sharp and strong. When the hedgehog is threatened, it will actually roll itself up in a ball, and its spiny hairs will stick out in all directions, making it very difficult for a predator to attack without getting pricked.

The last order of mammals we want to discuss is order Primates. This order includes **monkeys** (like the baboon in Figure 16.7i), **apes**, **lemurs**, and **human beings**. These mammals are often called the **erect mammals**, because they all have the ability to walk on two legs. Humans are the only mammals whose natural position is to walk on two legs. The other primates spend the majority of their time walking on all fours, but they have the ability to stand erect if they want to.

Although you might object to being classified with monkeys, apes, and the like, you actually have several characteristics in common with these mammals. All primates have good depth perception. In addition, the size of a primate's brain is large compared to its body size. Most primates are omnivorous, although the vast majority of primates concentrate on eating vegetation and only occasionally eat meat. All primates have nails on their fingers and toes, and all but one type (spider monkeys) have five fingers on each hand and five toes on each foot. Most have opposable thumbs. Primates have long gestation periods for animals of their size, and they are highly social.

Now, even though we have many characteristics in common with apes and monkeys, there is no reason to think that we are *related* to them. That is the mistake that evolutionists make. They see similarities between animals and immediately think that these similarities come from common genes in a common ancestor. As we learned in Module #9, however, all genetic information that we have been able to acquire indicates that this simply isn't the case. The similarities among animals are the result of a common Designer, not a common ancestor.

Some people not only object to the fact that humans are classified in order Primates, but they also object that we are in kingdom Animalia. Instead, these people want a sixth kingdom solely for human beings. However, this just does not make sense. Humans share all of the characteristics of mammals. Why, then, would you not put them in class Mammalia in kingdom Animalia? Some would say that this implies that human beings are "animals" and that this demeans the status of human beings, who are made in the image of God. However, that's only true if you are referring to the *common* usage of the term "animal." The fact is, scientific terminology is much more precise than common terminology, and when a scientist says "animal," it means something quite different from when a nonscientist uses the term. By putting human beings in kingdom Animalia, phylum Chordata, class Mammalia, and order Primates, we are simply stating the obvious fact that human beings have features in common with other organisms classified in the same way. This does not imply anything about how human beings were created. Human beings were created in the image of God, and that makes us unique. Of course, this is why we have our own genus and species, *Homo sapiens*.

ON YOUR OWN

16.23 A mammal has a long snout, eats ants and termites, and has teeth. To which order does it belong?

16.24 A mammal lays eggs. To which order does it belong?

16.25 A mammal has a hoof with three toes. To which order does it belong?

Summing It All Up

Well, believe it or not, you have reached the end of your first high school biology course. We hope that you have learned a great deal about the world around you, but most of all, we hope that you have developed a deep appreciation for the creation that God has given you. If you think back for a moment on the things that you have learned, you will hopefully get a glimpse of how much creative power and energy the Almighty used to make this awesome world. From the inner workings of the smallest microorganism to the life processes of the mammals, you should see the work of a magnificent Designer. Indeed, if biology teaches us anything, it is that the world around us is simply too complex to have come about by chance. Consider the words of Dr. Robert Gange, a physicist who has won several awards for his research:

Everything we know tells us that machines are structures intelligence designs, and that accidents destroy. Therefore, accidents do not design machines. Intellect does. And the myriad of biological wonders that sprinkle our world testify to the design ingenuity of a Supreme Intellect. (Robert Gange, *Origins and Destiny*, [Dallas: Word Publishing, 1986], 100)

Indeed, creation cries out the wonders of God, and it is up to the scientist to recognize these cries and seek out the creator!

ANSWERS TO THE "ON YOUR OWN" PROBLEMS

16.1 <u>The first is a snake and the second is a fish.</u> Reptiles have dry, leathery scales, while fish scales are slimy due to mucus that covers them.

16.2 <u>The embryo will not live. It will die of suffocation.</u> The allantois allows for the respiration of the embryo.

16.3 <u>The allantois is the one on the right.</u> As time goes on, the yolk sac grows smaller because the yolk is being used up. The allantois does not get used up, so it either stays the same size or grows with the embryo.

16.4 <u>The baby water snake can drown.</u> Tadpoles are the larval stage of an amphibian. Thus, they have gills and can breathe under water. Reptiles, on the other hand, must breathe air using their lungs. Thus, the snake can drown.

16.5 <u>If the tuatara laid out in the sun too long, it could die of overheating.</u> Remember, reptiles are ectothermic. They must warm themselves in the sun. However, if they lay out in the sun too much, they will warm themselves too much.

16.6 <u>The reptile is a snake.</u> Lizards can hear and have eyelids.

16.7 <u>Although humans do not hunt tuataras, they brought rats to the tuatara's habitat, where rats did not previously exist. These rats are destroying the tuatara population because they eat tuatara eggs.</u>

16.8 <u>It is probably using heat-sensing pits.</u> Pit vipers general hunt at night because their heat-sensing pits give them an advantage.

16.9 <u>They use their tongues and Jacobson's organs to enhance their sense of smell, so, in a way, they are smelling with their tongues.</u>

16.10 <u>It is a turtle.</u> If a member of order Testudines has flippers, it must live in the water. All members of Testudines that live in the water are called turtles. Please note that many turtles do not have flippers; some have feet. However, *if* a member of order Testudines has flippers, it is definitely a turtle.

16.11 <u>It is an alligator.</u> Alligator teeth fit in an alligator's closed mouth; crocodile teeth do not fit into a crocodile's closed mouth.

16.12 <u>The bird's blood will be warmer.</u> Since birds are endothermic, their blood is warmer than the surroundings. Reptiles are ectothermic, so their blood temperature varies with the outside temperature. The cooler the outside temperature, the cooler you expect the blood of a reptile to be.

16.13 <u>It came from an amphibian.</u> A bird has a four-chambered heart, and the oxygen-rich blood does not mix with the oxygen-poor blood.

16.14 <u>It is a bird egg.</u> Bird eggs have hard, lime-containing shells, while reptile eggs have a more flexible, leathery shell.

16.15 <u>The bird should start preening itself</u>. In order to make sure the hooked barbules slide smoothly along the smooth barbules, the feathers must be oiled. Birds do this by preening.

16.16 <u>It is most likely in the middle of its flight</u>. The alula is usually used in landing to reduce turbulence.

16.17 <u>Weigh the bones. The lighter one comes from the bird</u>. Birds have air-filled cavities in their bones that make them lighter than similar bones in other vertebrates.

16.18 <u>It would go into order Ciconiiformes</u>, because it feeds like a wading bird.

16.19 <u>It should go into order Galliformes</u>, since white meat and spending a lot of time on the ground are both characteristic of this order.

16.20 <u>The polar bear would have thicker underhair</u>. The primary function of underhair is insulation. A polar bear needs that more than a prairie dog.

16.21 <u>A cat uses its whiskers to sense things outside its field of vision. Without the ability to do that, it will run into things that are outside its normal vision range</u>.

16.22 <u>The first will have offspring that are more developed at birth</u>. The longer the gestation time, the more developed the offspring is at birth.

16.23 <u>It belongs to order Tubulidentata</u>. Since it behaves and looks like an anteater, you might think that it belongs with the anteater. However, it has teeth, so it belongs with the aardvark.

16.24 All egg-laying mammals are in order <u>Monotremata</u>.

16.25 The odd-toed hoofed mammals are in order <u>Perissodactyla</u>.

STUDY GUIDE FOR MODULE #16

1. Define the following terms:

a. Amniotic egg   f. Contour feathers
b. Neurotoxin   g. Placenta
c. Hemotoxin   h. Gestation
d. Endothermic   i. Mammary glands
e. Down feathers

2. State the five characteristics that set reptiles apart from other vertebrates.

3. In this module, we studied reptiles, birds, and mammals. For each class, indicate whether they are ectothermic or endothermic.

4. Identify the parts of the amniotic egg: *Illustration by Megan Whitaker*

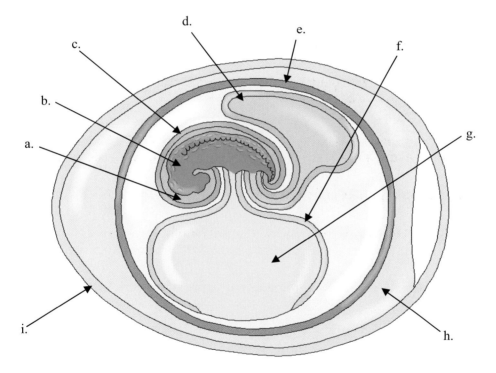

5. State the functions of the yolk, the allantois, and the albumen.

6. Reptiles have a growth-related characteristic in common with the arthropods. What is it?

7. What are the two most important functions of reptile scales?

8. These are the reptile orders that contain currently living reptiles:

Rhynchocephalia, Squamata, Crocodilia, Testudines

Place the following types of reptiles into their appropriate order:

a. snakes   b. tuataras   c. lizards   d. tortoises   e. alligators   f. turtles

9. State the six characteristics that set birds apart from other vertebrates.

10. Do all birds fly?

11. A blood sample comes from the ventricle of an animal that is either an amphibian or a bird. How can you tell which?

12. Which has a harder shell: the egg of a reptile or the egg of a bird?

13. You see some barbs from a feather. You have no idea whether they came from a down feather or a contour feather. Looking at the barbs under the microscope, however, you see that there are no hooked barbules. What kind of feather is it?

14. What type of feather (down or contour) is used for flight? What kind is used for insulation?

15. What is a bird actually doing when it is preening?

16. What is unique about a bird's method of molting?

17. What three things (at least) did flight engineers have to learn from birds to make practical flight possible?

18. Which is heavier, a bird's bone or the same size bone from an amphibian?

19. State the five characteristics that set mammals apart from other vertebrates.

20. What is the principal function of underhair?

21. What do we usually see when we look at a mammal, underhair or guard hair?

22. Name a nonplacental mammal.

23. What is the main difference between offspring born after a long gestation period and offspring born after a short gestation period?

GLOSSARY

The numbers in parentheses indicate the page number where the term was first discussed.

Abdomen – The body region posterior to the thorax (362)

Abiogenesis – The idea that long ago, very simple life forms spontaneously appeared through chemical reactions (15)

Absorption –The transport of dissolved substances into cells (161)

Activation energy – Energy necessary to get a chemical reaction going (182)

Active transport – Movement of molecules through the plasma membrane (typically opposite the dictates of osmosis or diffusion) aided by a process that requires energy (179)

Aerial hypha – A hypha that is not imbedded in the material upon which the fungus grows (100)

Aerobic organism – An organism that requires oxygen (42)

Allele – One of a pair of genes that occupies the same position on homologous chromosomes (233)

Alternation of generations – A life cycle in which there is both a muticellular diploid form and a multicellular haploid form (452)

Amniotic egg – A shelled, water-retaining egg that allows reptile, bird, and certain mammal embryos to develop on land (496)

Amoebocytes – Cells that move using pseudopods and perform different functions in different animals (333)

Anabolism – The sum total of all processes in an organism which use energy and simple chemical building blocks to produce large chemicals and structures necessary for life (2)

Anadromous – A life cycle in which creatures are hatched in fresh water, migrate to salt water as adults, and then go back to fresh water in order to reproduce (403)

Anaerobic organism – An organism that does not require oxygen (43)

Annual plants – Plants that live for only one year (429)

Anterior end – The end of an animal that contains its head (343)

Antibiotic – A chemical secreted by a living organism that kills or reduces the reproduction rate of other organisms (116)

Antibodies – Specialized proteins that aid in destroying infectious agents (220)

Anticodon – A three-nucleotide base sequence on tRNA (201)

Antigen – A protein that, when introduced in the blood, triggers the production of an antibody (251)

Appendicular skeleton – The portion of the skeleton that attaches to the axial skeleton and has the limbs attached to it (398)

Arteries – Blood vessels that carry blood away from the heart (399)

Asexual reproduction – Reproduction accomplished by a single organism (7)

Atrium – A heart chamber that receives blood (413)

Autosomal inheritance – Inheritance of a genetic trait not on a sex chromosome (253)

Autosomes – Chromosomes that do not determine the sex of an individual (247)

Autotrophs – Organisms that are able to make their own food (6)

Axial skeleton – The portion of the skeleton that supports and protects the head, neck, and trunk (398)

Biennial plants – Plants that live for two years (429)

Bilateral symmetry – An organism possesses bilateral symmetry if it can only be cut into two identical halves by a single longitudinal cut along its center which divides it into right and left halves. (330)

Bile – A mixture of salts and phospholipids that aids in the breakdown of fat (411)

Binomial nomenclature – Naming an organism with its genus and species name (27)

Biomass – A measure of the total dry mass of organisms within a particular region (304)

Biome – A group of ecosystems classified by climate and plant life (299)

Biosynthesis – The process by which living organisms produce larger molecules from smaller ones (142)

Bivalve – An organism with two shells (355)

Bone marrow – A soft tissue inside the bone that produces blood cells (397)

Botany – The study of plants (429)

Capillaries – Tiny, thin-walled blood vessels that allow the exchange of gases and nutrients between the blood and the cells of the body (399)

Carnivores – Organisms that eat only organisms other than plants (3)

Catabolism – The sum total of all processes in an organism which break down chemicals to produce energy and simple chemical building blocks (2)

Catalyst – A substance that alters the speed of a chemical reaction but is not used up in the process (141)

Cell wall – A rigid structure on the outside of certain cells, usually plant and bacteria cells (165)

Cellulose – A substance (made of sugars) that is common in the cell walls of many organisms (85)

Central vacuole – A large vacuole that rests at the center of most plant cells and is filled with a solution that contains a high concentration of solutes (169)

Centromere – The region that joins two sister chromatids (207)

Cephalothorax – A body region composed of the head and thorax fused together (362)

Cerebellum – The lobe that controls involuntary actions and refines muscle movement (400)

Cerebrum – The lobes of the brain that integrate sensory information and coordinate the creature's response to that information (400)

Change in chromosome number – A situation in which abnormal cellular events in meiosis lead to either none of a particular chromosome in the gamete or more than one chromosome in the gamete (255)

Change in chromosome structure – A situation in which a chromosome loses or gains genes during meiosis (254)

Chemical change – A change that alters the makeup of the elements or molecules of a substance (132)

Chitin – A chemical that provides both toughness and flexibility (101)

Chlorophyll – A pigment necessary for photosynthesis (75)

Chloroplast – An organelle containing chlorophyll for photosynthesis (75)

Chromatin – Clusters of DNA, RNA, and proteins in the nucleus of a cell (173)

Chromoplasts – Organelles that contain pigments used in photosynthesis (169)

Chromosome – DNA coiled around and supported by proteins, found in the nucleus of the cell (205)

Cilia – Hairlike projections that extend from the plasma membrane and are used for locomotion (78)

Circulatory system – A system designed to transport food and other necessary substances throughout a creature's body (344)

Closed circulatory system – A circulatory system in which the oxygen-carrying blood cells never leave the blood vessels (399)

Codon – A sequence of three nucleotide bases on mRNA that refers to a specific amino acid (201)

Cohesion – The phenomenon that occurs when individual molecules are so strongly attracted to each other that they tend to stay together, even when exposed to tension (467)

Collar cells – Flagellated cells that push water through a sponge (333)

Commensalism – A relationship between two organisms of different species where one benefits and the other is neither harmed nor benefited (77)

Community – A group of populations living and interacting in the same area (299)

Complete metamorphosis - Insect development consisting of four stages: egg, larva, pupa, and adult (384)

Compound eye – An eye made of many lenses, each with a very limited scope (363)

Concentration – A measurement of how much solute exists within a certain volume of solvent (135)

Conjugation – A temporary union of two organisms for the purpose of DNA transfer (48)

Consumers – Organisms that eat living producers and/or other consumers for food (4)

Contour feathers – Feathers with hooked and smooth barbules, allowing the barbules to interlock (511)

Cotyledon – A "seed leaf" which develops as a part of the seed. It provides nutrients to the developing seedling and eventually becomes the first leaf of the plant. (458)

Cytology – The study of cells (163)

Cytolysis – The rupturing of a cell due to excess internal pressure (180)

Cytoplasm – A jellylike fluid inside the cell in which the organelles are suspended (166)

Cytoplasmic streaming – The motion of cytoplasm in a cell that results in a coordinated movement of the cell's contents (166)

Cytoskeleton – A network of fibers that holds the cell together, helps the cell to keep its shape, and aids in movement (173)

Deciduous plant – A plant that loses its leaves for winter (441)

Decomposers – Organisms that break down the dead remains of other organisms (5)

Dehydration reaction – A chemical reaction in which molecules combine by removing water (145)

Diffusion – The random motion of molecules from an area of high concentration to an area of low concentration (135)

Digestion – The breakdown of absorbed substances (161)

Dihybrid cross – A cross between two individuals, concentrating on two definable traits (242)

Diploid cell – A cell with chromosomes that come in homologous pairs (212)

Diploid number (2n) – The total number of chromosomes in a diploid cell (212)

Disaccharides – Carbohydrates that are made up of two monosaccharides (145)

Dominant allele – An allele that will determine phenotype if just one is present in the genotype (234)

Dominant generation – In alternation of generations, the generation that occupies the largest portion of the life cycle (454)

Double fertilization – A fertilization process that requires two sperm to fuse with two other cells (484)

Down feathers – Feathers with smooth barbules but no hooked barbules (511)

Ecological pyramid – A diagram that shows the biomass of organisms at each trophic level (304)

Ecology – The study of the interactions between living and nonliving things (299)

Ecosystem – An association of living organisms and their physical environment (299)

Ectoplasm – The thin, watery cytoplasm near the plasma membrane of some cells (72)

Ectothermic – Lacking an internal mechanism for regulating body heat (413)

Egestion – The removal of nonsoluble waste materials (162)

Element – A collection of atoms that all have the same number of protons (128)

Endoplasm – The dense cytoplasm found in the interior of many cells (72)

Endoplasmic reticulum – An organelle composed of an extensive network of folded membranes that performs several tasks within a cell (168)

Endoskeleton – A skeleton on the inside of a creature's body, typically composed of bone or cartilage (396)

Endospore – The DNA and other essential parts of a bacterium coated with several hard layers (50)

Endotherm – An organism that is internally warmed by a heat-generating metabolic process (507)

Environmental factors – Those "nonbiological" factors that are involved in a person's surroundings such as the nature of the person's parents, the person's friends, and the person's behavioral choices (197)

Epidermis – An outer layer of cells designed to provide protection (333)

Epithelium – Animal tissue consisting of one or more layers of cells that have only one free surface, because the other surface adheres to a membrane or other substance (336)

Eukaryotic cell – A cell with distinct, membrane-bounded organelles (18)

Excretion – The removal of soluble waste materials (162)

Exoskeleton – A body covering, typically made of chitin, that provides support and protection (361)

Exponential growth – Population growth that is unhindered because of the abundance of resources for an ever-increasing population (47)

External fertilization – The process by which the female lays eggs and the male fertilizes them once they are outside of the female (401)

Extracellular digestion – Digestion that takes place outside of the cell (98)

Eyespot – A light-sensitive region in certain protozoa (75)

Fermentation – The anaerobic breakdown of sugars into smaller molecules (110)

Flagellate – A protozoan that propels itself with a flagellum (74)

Foot – A muscular organ that is used for locomotion and takes a variety of forms depending on the animal (355)

Fossils – Preserved remains of once-living organisms (270)

Fruit – A mature ovary that contains a seed or seeds (486)

Gametes – Haploid cells (n) produced by diploid cells (2n) for the purpose of sexual reproduction (213)

Ganglia (singular: ganglion) –Masses of nerve cell bodies (345)

Gemmule – A cluster of cells encased in a hard, spicule-reinforced shell (334)

Gene – A section of DNA that codes for the production of a protein or a portion of protein, thereby causing a trait (197)

Genetic disease carrier – A person who is heterozygous in a recessive genetic disorder (253)

Genetic factors – The general guideline of traits determined by a person's DNA (197)

Genetics – The science that studies how characteristics get passed from parent to offspring (195)

Genotype – Two-letter set that represents the alleles an organism possesses for a certain trait (234)

Gestation – The period of time during which an embryo develops before being born (519)

Girdling – The process of cutting away a ring of inner and outer bark all the way around a tree trunk (448)

Golgi bodies – The organelles where proteins and lipids are stored and then modified to suit the needs of the cell (171)

Gonad – A general term for the organ that produces gametes (371)

Gravitropism – A growth response to gravity (470)

Greenhouse effect – The process by which certain gases (principally water vapor, carbon dioxide, and methane) trap heat that would otherwise escape the earth and radiate into space (317)

Haploid cell – A cell that has only one representative of each chromosome pair (212)

Haploid number (n) – The number of homologous pairs in a diploid cell (213)

Haustorium – A hypha of a parasitic fungus that enters the host's cells, absorbing nutrition directly from the cytoplasm (100)

Hemotoxin – A poison that attacks the red blood cells and blood vessels, destroying circulation (502)

Herbivores – Organisms that eat only plants (3)

Hermaphroditic – Possessing both the male and the female reproductive organs (345)

Heterotrophs – Organisms that depend on other organisms for their food (6)

Heterozygous genotype – A genotype with two different alleles (234)

Hibernation – A state of extremely low metabolism and respiration, accompanied by lower-than-normal body temperatures (420)

Holdfast – A special structure used by an organism to anchor itself (88)

Homeostasis – Maintaining the status quo (162)

Homozygous genotype – A genotype in which both alleles are identical (234)

Hormones – Chemicals that circulate throughout multicellular organisms, regulating cellular processes by interacting with specifically targeted cells (469)

Hydrogen bond – A strong attraction between hydrogen atoms and certain other atoms (usually oxygen or nitrogen) in specific molecules (155)

Hydrolysis – Breaking down complex molecules by the chemical addition of water (146)

Hydrophobic – Lacking any affinity to water (148)

Hypertonic solution – A solution in which the concentration of solutes is greater than that of the cell that resides in the solution (179)

Hypha – A filament of fungal cells (98)

Hypothesis – An educated guess that attempts to explain an observation or answer a question (9)

Hypotonic solution – A solution in which the concentration of solutes is less than that of the cell that resides in the solution (180)

Imperfect flowers – Flowers with either stamens or carpels, but not both (477)

Incomplete metamorphosis - Insect development consisting of three stages: egg, nymph, and adult (384)

Inheritance – The process by which physical and biological characteristics are transmitted from the parent (or parents) to the offspring (7)

Intermediate filaments – Threadlike proteins in the cell's cytoskeleton that are roughly twice as thick as microfilaments (173)

Internal fertilization – The process by which the male places sperm inside the female's body, where the eggs are fertilized (401)

Interphase – The time interval between cellular reproduction (206)

Invertebrates – Animals that lack a backbone (329)

Ions – Substances in which at least one atom has an imbalance of protons and electrons (166)

Isomers – Two different molecules that have the same chemical formula (144)

Isotonic solution – A solution in which the concentration of solutes is essentially equal to that of the cell which resides in the solution (179)

Karyotype – The figure produced when the chromosomes of a species during metaphase are arranged according to their homologous pairs (212)

Leaf margin – The characteristics of the leaf edge (434)

Leaf mosaic – The arrangement of leaves on the stem of a plant (432)

Leucoplasts – Organelles that store starches or oils (168)

Loam – A mixture of gravel, sand, silt, clay, and organic matter (466)

Logistic growth – Population growth that is controlled by limited resources (47)

Lysosome - The organelle in animal cells responsible for hydrolysis reactions that break down proteins, polysaccharides, disaccharides, and some lipids (167)

Macroevolution – The hypothesis that processes similar to those at work in microevolution can, over eons of time, transform an organism into a completely different kind of organism (268)

Mammary glands – Specialized organs in mammals that produce milk to nourish the young (519)

Mantle – A sheath of tissue that encloses the vital organs of a mollusk, makes the mollusk's shell, and performs respiration (354)

Matter – Anything that has mass and takes up space (125)

Medulla oblongata – The lobes that coordinate vital functions, such as those of the circulatory and respiratory systems, and transport signals from the brain to the spinal cord (400)

Medusa – A free-swimming cnidarian with a bell-shaped body and tentacles (335)

Meiosis – The process by which a diploid (2n) cell forms gametes (n) (213)

Membrane – A thin covering of tissue (104)

Mesenchyme – The jellylike substance that separates the epidermis from the inner cells in a sponge (333)

Mesoglea – The jelly-like substance that separates the epithelial cells in a cnidarian (336)

Messenger RNA – The RNA that performs transcription (201)

Metabolism – The sum total of all processes in an organism which convert energy and matter from outside sources and use that energy and matter to sustain the organism's life functions (2)

Microevolution – The theory that natural selection can, over time, take an organism and transform it into a more specialized species of that organism (268)

Microfilaments – Fine, threadlike proteins found in the cell's cytoskeleton (173)

Microorganisms – Living creatures that are too small to see with the naked eye (13)

Microtubules – Spiral strands of protein molecules that form a tubelike structure (172)

Middle lamella – The thin film between the cell walls of adjacent plant cells (165)

Mitochondria – The organelles in which nutrients are converted to energy (167)

Mitosis – A process of asexual reproduction in eukaryotic cells (206)

Model – An explanation or representation of something that cannot be seen (126)

Molecules – Chemicals that result from atoms linking together (130)

Molt – To shed an old outer covering so that it can be replaced with a new one (362)

Monohybrid cross – A cross between two individuals, concentrating on only one definable trait (242)

Monosaccharides – Simple carbohydrates that contain 3 to 10 carbon atoms (145)

Mother cell – A cell ready to begin reproduction, containing duplicated DNA and centrioles (206)

Mutation – A radical chemical change in one or more alleles (254)

Mutation – An abrupt and marked change in the DNA of an organism compared to that of its parents (8)

Mutualism – A relationship between two or more organisms of different species where all benefit from the association (76)

Mycelium – The part of the fungus responsible for extracellular digestion and absorption of the digested food (98)

Nastic movement – A plant's response to a stimulus such that the direction of the response is preprogrammed and not dependent on the direction of the stimulus (464)

Nematocysts – Small capsules that contain a toxin which is injected into prey or predators (337)

Nervous system – A system of sensitive cells that respond to stimuli such as sound, touch, and taste (345)

Neurotoxin – A poison that attacks the nervous system, causing blindness, paralysis, or suffocation (502)

Notochord – A rod of tough, flexible material that runs the length of a creature's body, providing the majority of its support (393)

Nuclear membrane – A highly-porous membrane that separates the nucleus from the cytoplasm (172)

Nucleus – The region of a eukaryotic cell that contains the cell's main DNA (71)

Olfactory lobes – The lobes of the brain that receive signals from the receptors in the nose (400)

Omnivores – Organisms that eat both plants and other organisms (3)

Open circulatory system – A circulatory system that allows the blood to flow out of the blood vessels and into various body cavities so that the cells are in direct contact with the blood (364)

Optic lobes – The lobes of the brain that receive signals from the receptors in the eyes (400)

Organic molecule – A molecule that contains only carbon and any of the following: hydrogen, oxygen, nitrogen, sulfur, and/or phosphorous (142)

Osmosis – The tendency of a solvent to travel across a semipermeable membrane into areas of higher solute concentration (136)

Ovaries – Organs that produce eggs (339)

Oviparous development – Development that occurs in an egg that is hatched outside the female's body (402)

Ovoviviparous development – Development that occurs in an egg that is hatched inside the female's body (402)

Paleontology – The study of fossils (273)

Parasite – An organism that feeds on a living host (42)

Parasitism – A relationship between two organisms of different species where one benefits and the other is harmed (77)

Passive transport – Movement of molecules through the plasma membrane according to the dictates of osmosis or diffusion (179)

Pathogen – An organism that causes disease (37)

Pedigree – A diagram that follows a particular phenotype through several generations (238)

Pellicle – A firm, flexible coating outside the plasma membrane (75)

Peptide bond – A bond that links amino acids together in a protein (150)

Perennial plants – Plants that grow year after year (429)

Perfect flowers – Flowers with both stamens and carpels (477)

Phagocytic vacuole – A vacuole that holds the matter which a cell engulfs (169)

Phagocytosis – The process by which a cell engulfs foreign substances or other cells (169)

Phase – One of three forms - solid, liquid, or gas - which every substance is capable of attaining (133)

Phenotype – The observable expression of an organism's genes (234)

Phloem – Living vascular tissue that carries sugar and organic substances throughout a plant (430)

Phospholipid – A lipid in which one of the fatty acid molecules has been replaced by a molecule that contains a phosphate group (176)

Photosynthesis – The process by which green plants and some other organisms use the energy of sunlight and simple chemicals to produce their own food (3)

Phototropism – A growth response to light (470)

Physical change – A change that affects the appearance but not the chemical makeup of a substance (132)

Physiology – The study of life processes in an organism (463)

Phytoplankton – Tiny floating photosynthetic organisms, primarily algae (84)

Pinocytic vesicle – Vesicle formed at the plasma membrane to allow the absorption of large molecules (170)

Placenta – A structure that allows an embryo to be nourished with the mother's blood supply (519)

Plankton – Tiny organisms that float in the water (84)

Plasma membrane – The semipermeable membrane between the cell contents and either the cell wall or the cell's surroundings (166)

Plasmid – A small, circular section of extra DNA that confers one or more traits to a bacterium and can be reproduced separately from the main bacterial genetic code (48)

Plasmolysis – Collapse of a walled cell's cytoplasm due to a lack of water (179)

Pollen – A fine dust that contains the sperm of seed-producing plants (457)

Pollination – The transfer of pollen grains from the anther to the carpel in flowering plants (482)

Polyp – The sessile, tubular form of a cnidarian with a mouth and tentacles at one end and a basal disk at the other (335)

Polysaccharides – Carbohydrates that are made up of more than two monosaccharides (145)

Population – A group of interbreeding organisms coexisting together (299)

Pore spaces – Spaces in the soil that determine how much water and air the soil can hold (466)

Posterior end – The end of an animal that contains its tail (343)

Primary consumer – An organism that eats producers (301)

Producers – Organisms that produce their own food (4)

Prokaryotic cell – A cell that has no distinct, membrane-bounded organelles (18)

Pseudopod – A temporary, foot-like extension of a cell, used for locomotion or engulfing food (71)

Radial symmetry – An organism possesses radial symmetry if it can be cut into two identical halves by any longitudinal cut through its center. (330)

Radula – An organ covered with teeth that mollusks use to scrape food into their mouths (355)

Receptors – Special structures that allow living organisms to sense the conditions of their internal or external environment (7)

Recessive allele – An allele that will not determine the phenotype unless the genotype is homozygous in that allele (234)

Regeneration – The ability to regrow a missing part of the body (351)

Reproduction – Producing more cells (162)

Reproductive plant organs – The parts of a plant (such as flowers, fruits, and seeds) involved in reproduction (430)

Respiration – The breakdown of food molecules with a release of energy (161)

Rhizoid hypha – A hypha that is imbedded in the material on which the fungus grows (99)

Ribosomes – Non-membrane-bounded organelles responsible for protein synthesis (168)

Rough ER – ER that is dotted with ribosomes (168)

Saprophyte – An organism that feeds on dead matter (41)

Saturated fat – A lipid made from fatty acids that have no double bonds between carbon atoms (149)

Scientific law – A theory that has been tested by and is consistent with generations of data (10)

Secondary consumer – An organism that eats primary consumers (301)

Secretion – The release of biosynthesized substances (162)

Secretion vesicle – Vesicle that holds secretion products so that they can be transported to the plasma membrane and released (170)

Seed – An ovule with a protective coating, encasing a mature plant embryo and a nutrient source (486)

Semipermeable membrane – A membrane that allows some molecules to pass through but does not allow other molecules to pass through (136)

Sessile colony – A colony that uses holdfasts to anchor itself to an object (88)

Sex chromosomes – Chromosomes that determine the sex of an individual (247)

Sex-linked inheritance – Inheritance of a genetic trait located on the sex chromosomes (254)

Sexual reproduction – Reproduction that requires two organisms (7)

Shell – A tough, multilayered structure secreted by the mantle, generally used for protection, but sometimes for body support (354)

Simple eye – An eye with only one lens (363)

Smooth ER – ER that has no ribosomes (168)

Species – A unit of one or more populations of individuals that can reproduce under normal conditions, produce fertile offspring, and are reproductively isolated from other such units (21)

Spherical symmetry – An organism possesses spherical symmetry if it can be cut into two identical halves by any cut that runs through the organism's center. (330)

Spiritual factors – The factors in a person's life that are determined by the quality of his or her relationship with God (197)

Spore – A reproductive cell with a hard, protective coating (80)

Sporophore – Specialized aerial hypha that produces spores (100)

Statocyst – The organ of balance in a crustacean (370)

Steady state – A state in which members of a population die as quickly as new members are born (46)

Stolon – An aerial hypha that asexually reproduces to make more filaments (100)

Strains – Organisms from the same species that have markedly different traits (58)

Strata – Distinct layers of rock (270)

Structural homology – The study of similar structures in different species (282)

Symbiosis – A close relationship between two or more species where at least one benefits (76)

Taxonomy – The science of classifying organisms (27)

Tertiary consumer – An organism that eats secondary consumers (301)

Testes – Organs that produce sperm (339)

Thallus – The body of a plant-like organism that is not divided into leaves, roots, or stems (85)

The immutability of species – The idea that each individual species on the planet was specially created by God and could never fundamentally change (267)

Theory – A hypothesis that has been tested with a significant amount of data (10)

Thigmotropism – A growth response to touch (470)

Thorax – The body region between the head and the abdomen (362)

Transduction – The process in which infection by a virus results in DNA being transferred from one bacterium to another (50)

Transformation – The transfer of a DNA segment from a nonfunctional donor cell to that of a functional recipient cell (49)

Translocation – The process by which organic substances move through the phloem of a plant (468)

Transpiration – Evaporation of water from the leaves of a plant (311)

True breeding – If an organism has a certain characteristic that is always passed on to its offspring, we say that this organism bred true with respect to that characteristic. (228)

Undifferentiated cells – Cells that have not specialized in any particular function (430)

Univalve – An organism with a single shell (355)

Unsaturated fat – A lipid made from fatty acids that have at least one double bond between carbon atoms (149)

Vaccine – A weakened or inactive version of a pathogen that stimulates the body's production of antibodies which can aid in destroying the pathogen (220)

Vacuole – A membrane-bounded "sac" within a cell (72)

Vegetative organs – The parts of a plant (such as stems, roots, and leaves) that are not involved in reproduction (429)

Veins – Blood vessels that carry blood back to the heart (399)

Ventricle – A heart chamber from which blood is pumped out (413)

Vertebrae – Segments of bone or some other hard substance that are arranged into a backbone (393)

Vertebrates – Animals that possess a backbone (329)

Virus – A non-cellular infectious agent that has two characteristics:

>　　　(1) It has genetic material (RNA or DNA) inside a protective protein coat.

>　　　(2) It cannot reproduce on its own. (218)

Visceral hump – A hump that contains a mollusk's heart, digestive, and excretory organs (354)

Viviparous development – Development that occurs inside the female, allowing the offspring to gain nutrients and vital substances from the mother through a placenta (402)

Waste vacuoles – Vacuoles that contain the waste products of digestion (169)

Watershed – An ecosystem where all water runoff drains into a single body of water (312)

Xylem – Nonliving vascular tissue that carries water and dissolved minerals from the roots of a plant to its leaves (430)

Zooplankton – Tiny floating organisms that are either small animals or protozoa (84)

Zygospore – A zygote surrounded by a hard, protective covering (112)

Zygote – The result of sexual reproduction when each parent contributes half of the DNA necessary for the offspring (112)

APPENDIX A

A Simple Biological Key

1. Microscopic...2
    Macroscopic (visible with the naked eye)....................................3
2. Eukaryotic cell.. *kingdom Protista*
    Prokaryotic cell... *kingdom Monera*
3. Autotrophic..*kingdom Plantae*...................4
    Heterotrophic...5
4. Leaves with parallel veins ....*phylum Anthophyta*.................... *class Monocotyledoneae*
    Leaves with netted veins ....*phylum Anthophyta*.................... *class Dicotyledoneae*
5. Decomposer...*kingdom Fungi*
    Consumer...................................*kingdom Animalia*..................6
6. No backbone..7
    Backbone...*phylum Chordata*...................22
7. Organism can be externally divided into equal halves (like a pie), but it
    has no distinguishable right and left sides....................................8
    Organism either can be divided into right and left sides that are
    mirror images or cannot be divided into two equal halves..........................9
8. Soft, transparent body with tentacles .... *phylum Cnidaria*
    Firm body with internal support; covered with scales or spiny plates;
    tiny, hollow tube feet used for movement.......................... *phylum Echinodermata*
9. External plates that support and protect.......*phylum Arthropoda* 14
    External shell or soft, shell-less body.......................................10
10. External shell...............................*phylum Mollusca*....................11
    No external shell..12
11. Coiled shell... *class Gastropoda*
    Shell made of two similar parts... *class Bivalvia*
12. Wormlike body without tentacled receptors on head.............................. *phylum Annelida*
    Non-wormlike body or tentacled receptors on head....*phylum Mollusca*.... 13
13. Wormlike body with tentacled receptors on head..................................... *class Gastropoda*
    Non-wormlike body with 8 or more tentacles used for grasping................ *class Cephalopoda*
14. More than 3 pairs of legs..15
    3 pairs of walking legs.................................*class Insecta*....................16
15. 4 pairs of walking legs, body in two divisions................................. *class Arachnida*
    More than 4 pairs of walking legs... *class Malacostraca*
16 Wings...17
    No wings..21
17. All wings transparent..18
    Nontransparent wings...19
18. Capable of stinging from back of body.. *order Hymenoptera*
    Cannot sting (may be able to bite).. *order Diptera*
19. Large, sometimes colorful wings.. *order Lepidoptera*
    Thick, hard, leathery wings...20
20. Pair of hard wings covering a pair of folded, transparent wings................. *order Coleoptera*
    Pair of leathery wings covering a pair of transparent wings....................... *order Orthoptera*

21. Piercing, sucking mouthparts for obtaining blood..*order Siphonaptera*
    Mouthparts for chewing..*order Hymenoptera*
22. Jaws or beak..**23**
    No jaw or beak..*class Agnatha*
23. Skin covered with scales..**24**
    No scales on skin...**26**
24. Fins and gills...**25**
    No fins; breathes with lungs..*class Reptilia*
25. Mouth on lower part of body..*class Chondrichthyes*
    Mouth on front part of body...*class Osteichthyes*
26. No scales, no hair, no feathers; skin is slimy.......*class Amphibia*................**27**
    Feathers or hair..**28**
27. Tail..*order Caudata*
    No tail...*order Anura*
28. Feathers on body...*class Aves*
    Hair on body...*class Mammalia*.................**29**
29. Hooves..**30**
    No hooves...**31**
30. Odd number of toes...*order Perissodactyla*
    Even number of toes..*order Artiodactyla*
31. Carnivore...**32**
    Herbivore...**33**
32. Teeth..*order Carnivora*
    No teeth, eats insects...*order Insectivora*
33. Enlarged front teeth for gnawing..**34**
    No enlarged front teeth for gnawing...**35**
34. Legs for crawling..*order Rodentia*
    Hind legs for jumping..*order Lagomorpha*
35. Enlarged trunk, used for breathing and grasping..*order Proboscidea*
    Tendency to stand erect on two hind limbs...*order Primates*

APPENDIX B
MODULE SUMMARIES

Summary of Module #1
Review the vocabulary words listed in Question #1 of the study guide

<u>Fill in the blanks. Many blanks contain more than one word.</u>
Please note: We suggest that you actually write these paragraphs out rather than just filling in the blanks in the book. The act of writing these things out is a form of studying.

1. Four characteristics of life:

   a. All life forms contain _____, which is called _____.
   b. All life forms have a method by which they _____ from the surroundings and convert it into _____.
   c. All life forms can _____ in their surroundings and _____.
   d. All life forms _____.

2. DNA provides the _____ necessary to take a bunch of lifeless chemicals and turn them into _____.

3. _____ can be split into two categories: (1) _____, which involves using energy and simple chemical building blocks to produce large chemicals and structures and (2) catabolism, which involves _____.

4. The vast majority of energy that sustains life comes from _____. _____ use that energy to make food for themselves via a process called _____. Consumers get energy from the producers by _____. Consumers can be split into three categories: _____ (which eat only plants), _____ (which eat only nonplants), and _____ (which eat plants and nonplants). The energy of dead producers and consumers is recycled back into creation by the _____.

5. Producers are often called _____, the Greek roots of which literally mean "self-feeder." Consumers and decomposers are often called _____, which literally means "_____."

6. Living organisms are equipped with structures called _____, which receive information about their surroundings. God's creation is always _____, which is why these structures are necessary for survival.

7. In asexual reproduction, the characteristics and traits inherited by the offspring are, under normal circumstances, _____ to the parent. In sexual reproduction, under normal circumstances, the offspring's traits and characteristics are _____. When _____ occur, the offspring can possess traits that are incredibly different from those of the parent or parents.

8. In the scientific method, the scientist starts by _____ the world around him. He then forms a _____ to explain some aspect of how the world functions. He then _____ in an attempt to test his _____. If a large amount of _____ confirms the _____, it

becomes a _____, which is tested with even more _____. If it continues to be confirmed over several generations, it might become a _____.

9. Scientists once believed that life could spring from non-living things. This was called _____ _____, and it was refuted in the mid 1800s by a scientist named _____. The story of how the scientific community believed in it for so long demonstrates that science has _____.

10. The newest version of spontaneous generation is called _____, and it claims that long ago, _____.

11. The groups used in our classification scheme, from largest to smallest are: _____, _____, _____, _____, _____, _____, and _____.

12. The five kingdoms we use in this course are: _____, _____, _____, _____, and _____.

13. A cell with no membrane-bounded organelles is _____, while one with membrane-bounded organelles is a _____. Members of kingdom Monera are composed of _____.

14. A unit of one or more populations of individuals that can reproduce under normal conditions, produce fertile offspring, and are reproductively isolated from other such units is called a _____.

15. A series of questions that is designed to classify organisms is called a biological _____.

16. When we call wolves "*Canis lupus,*" we are using _____.

17. In the _____-_____ system of classification, the three basic groups are _____, _____ and _____. Members of kingdom Monera are placed in either _____ or _____, and all of the other kingdoms are placed in _____.

18. A creationist taxonomy scheme that attempts to classify organisms based on the kind of organisms that God made during creation is called _____.

19. Multicellular autotrophs are typically placed in kingdom _____.

20. Single-celled creatures made of eukaryotic cells are placed in kingdom _____.

21. Multicellular consumers are typically placed in kingdom _____.

22. Decomposers made of eukaryotic cells are mostly found in kingdom _____.

23. Organisms made of prokaryotic cells are found in kingdom _____.

Summary of Module #2
Review the vocabulary words listed in Question #1 of the study guide

<u>Fill in the blanks. Many blanks contain more than one word.</u>
Please note: We suggest that you actually write these paragraphs out rather than just filling in the blanks in the book. The act of writing these things out is a form of studying.

1. The term _____ is often used as a general term that applies to all members of kingdom Monera. Some are beneficial to humans, but some are _____, which means they cause disease.

2. Some bacteria have a _____ that surrounds the cell wall. It is composed of an organized layer of sticky sugars that _____. It is also a protective layer that _____.

3. Most bacteria have a _____ _____ that holds the contents of the bacterium together, regulates the amount of water that a bacterium can absorb, and holds the cell into one of three shapes: _____, _____, or _____. The absence or presence of a _____ _____ and its composition (if it exists) are used to _____ bacteria.

4. Underneath the cell wall (if it exists), there is a _____ _____, which regulates what the bacterium takes in from the outside world.

5. _____ exists throughout the interior of a cell, supporting the DNA and the ribosomes.

6. Many bacteria have fibrous bristles called _____, which are used for grasping. Locomotion is accomplished with a _____.

7. Ribosomes make special chemicals known as _____.

8. In terms of what they eat, most bacteria are d_____. As a result, they are called s_____.

9. Some bacteria are _____, which means they feed on a living host.

10. Autotrophic bacteria manufacture their own food by either _____ (using the energy of the sun to make food) or _____ (promoting chemical reactions that release energy).

11. Some bacteria are _____, which means they need oxygen in order to survive. Some are _____, which means they do not need oxygen. The latter bacteria either _____ dead organisms or _____ into chemicals that can be used by other life forms.

12. Asexual reproduction in bacteria is often called _____.

13. Typically, a population of bacteria starts off with _____ growth until it reaches a _____ _____ in which bacteria die as quickly as new ones are made. When population growth is controlled by resources, we call it _____. If resources begin to run out, the population will _____.

14. When bacteria exchange genetic information, we call it _____ _____, and it can occur in one of three ways: _____, _____, or _____.

15. In _____, bacteria link together to exchange circular strands of DNA called _____.
In _____, a bacterium can absorb a segment of _____ from a non-functional _____. In
_____, DNA can be transferred from one bacterium to another by a virus.

16. Bacteria can survive harsh conditions by forming _____.

17. A _____ is really just a simple association of individual bacteria. Bacteria in a
streptococcus colony have a _____ shape, while bacteria in a diplobacillus colony
have a _____ shape.

18. After a _____ _____, certain bacteria look blue when viewed under a microscope whereas
others looked _____.

19. _____ bacteria belong in phylum Gracilicutes, while _____ bacteria belong
in phylum Firmicutes. Bacteria with cell walls significantly different from those in these two phyla
belong in phylum _____, while bacteria with no cell walls belong in phylum _____.

20. Phylum Gracilicutes has three classes: _____ (non-photosynthetic bacteria),
_____ (photosynthetic bacteria that do not produce oxygen), and _____
(photosynthetic bacteria that produce oxygen). Phylum Firmicutes has two classes: _____
(cocci and bacilli bacteria) and _____ (bacteria of any other shape). Phylum
Tenericutes has only one class: _____. Phylum Mendosicutes has only one class:
_____. Many places that are uninhabitable to other organisms will be populated with
members of class _____.

21. Photosynthetic organisms called blue-green algae are more properly called _____.

22. *Clostridium botulinum* can cause _____. Undercooked eggs and poultry can give you
_____ poisoning. *Escherichia coli* bacteria live in your gut. There are pathogenic
_____ and non-pathogenic _____ of this bacterium.

23. For optimum growth, most bacteria need _____, _____, _____,
_____, and _____.

<u>Label all of the indicated structures on the bacterium below.</u>

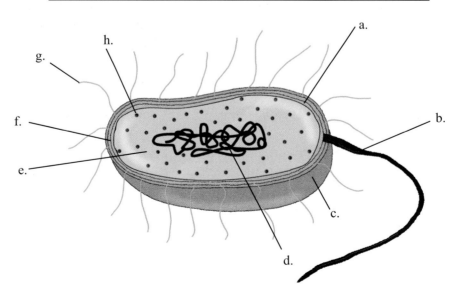

Summary of Module #3
Review the vocabulary words listed in Question #1 of the study guide

<u>Fill in the blanks. Many blanks contain more than one word.</u>
Please note: We suggest that you actually write these paragraphs out rather than just filling in the blanks in the book. The act of writing these things out is a form of studying.

1. Kingdom Protista is divided into two main groups: _____ (mostly individual, single-celled creatures with a form of locomotion) and _____ (mostly colonies of eukaryotic cells that have no form of locomotion).

2. Protozoa are split into four major phyla based on their locomotion: Mastigophora contains those that use _____, Sarcodina contains those that use _____, Ciliophora contains those that use _____, and Sporozoa contains those that have _____.

3. Algae are split into five major phyla based on habitat, organization, and cell wall. Chlorophyta contains those that live in _____, are composed of _____, and have cell walls made of _____. Chrysophyta contains those that live in _____, are composed of _____, and have cell walls made of _____. Pyrrophyta contains those that live in _____, are composed of _____, and have cell walls made of _____. Phaeophyta contains those that live in _____, are composed of _____, and have cell walls made of _____. Rhodophyta contains those that live in _____, are composed of _____, and have cell walls made of _____.

4. The main portion of a cell's DNA is stored in its _____. Membrane-bounded "sacs" in a cell are called _____. Two main types of vacuoles are _____, which store food, and _____, which regulate the amount of water in the cell.

5. The cytoplasm in a cell can be split into _____, which is thin and watery, and _____, which is more dense.

6. *Amoeba proteus* is a typical member of phylum _____, and it can form _____ to survive extreme conditions.

7. Genus *Euglena* contains organisms from phylum _____. When it comes to food, these creatures are both _____ and _____. They have firm but flexible shape-sustaining _____ and a light-sensitive region known as an _____.

8. Photosynthesis requires a pigment called _____, which cells store in _____.

9. Round, green colonies found in phylum Mastigophora are found in genus _____.

10. When organisms form a relationship in which at least one of them benefits, it is called _____. If all organisms involved benefit, it is specifically known as _____. If one benefits and the other neither benefits nor is harmed, it is specifically known as _____. If one benefits and the other is harmed, it is specifically known as _____.

11. Genus *Paramecium* contains organisms from phylum _____. Organisms in this genus have two _____. The _____ is the larger of the two, and it controls metabolism, while the _____ is the smaller of the two, and it controls reproduction.

12. Paramecia can exchange DNA through _____, but unlike this process in bacteria, the DNA exchange is _____.

13. Genus *Plasmodium* contains organisms from phylum _____ that cause _____. The organisms are transferred between people by the action of _____.

14. Members of phylum Sporozoa form _____ as a part of their normal lifecycle.

15. Tiny organisms that float in the water are called _____. Small animals and protozoa are called _____, while photosynthetic organisms (typically algae) are called _____

16. When conditions are ideal, algae will reproduce so rapidly that they essentially "take over" their habitat. This is referred to as an _____.

17. Members of phylum Chlorophyta have the pigment _____ and are often called _____.

18. _____ is a compound made of certain types of sugars that is common in many cell walls.

19. The members of phylum Chrysophyta are often called _____ and are responsible for a large amount of the photosynthesis that occurs in creation. When the cell wall remains of many of these organisms clump together, they form a crumbly, abrasive substance called _____.

20. A _____ is a colony that does not move and anchors itself to an object with a _____.

21. Members of phylum Pyrrophyta are often referred to as _____. They have two _____. One species in this phylum, *Gymnodinium brevis*, have blooms that are called _____.

22. Members of phylum Phaeophyta are often referred to as _____. Their cell walls contain _____ that is used as a thickening agent.

23. Members of genus _____ in phylum Phaeophyta are often called _____ or _____. They form _____ that allow them to anchor themselves to rocks which sit at the bottom of the ocean. Some can grow as long as 100 meters.

24. Members of phylum Rhodophyta are often called _____.

Summary of Module #4
Review the vocabulary words listed in Question #1 of the study guide

Fill in the blanks. Many blanks contain more than one word.
Please note: We suggest that you actually write these paragraphs out rather than just filling in the blanks in the book. The act of writing these things out is a form of studying.

1. The five features common to most fungi are _____, _____, _____, _____, and _____.

2. The cells in some hyphae are separated by cell walls. Thus are called _____. Other hyphae, called _____, have no separations between the cells, and the nuclei are spread throughout the hypha. Even septate hyphae have _____ through which _____ is exchanged.\

3. There are many forms of specialized hyphae. _____ are imbedded in the material on which the fungus grows. _____ are not imbedded in the material on which the fungus grows. If such a hypha produces spores, it is a _____, but if it asexually produces more filaments, it is a _____. In the case of a fungus that feeds on a living organism, a hypha that enters the cells of the living organism and draws nutrients directly from the cytoplasm of those cells is called a _____.

4. All fungi are assumed (but not all are confirmed) to have some _____ mode of spore formation.

5. Sexual reproduction in fungi usually involves forming specialized spore-forming structures called _____ _____, which are the result of sexual reproduction between compatible _____.

6. One mode of asexual reproduction in fungi involves the lengthening of a _____. After it reaches a certain length, it will begin to reproduce into hyphae that will form the _____ of a new fungus.

7. There are six major phyla in kingdom Fungi. Members of phylum Basidiomycota form sexual spores on clublike _____. Members of phylum Ascomycota form sexual spores in saclike ____. Members of phylum Zygomycota form sexual spores _____. Members of phylum Chytridiomycota form spores with _____. Members of phylum Deuteromycota have no _____ _____ and are sometimes called _____ _____. Members of phylum Myxomycota are placed in kingdom _____ by some biologists, because they resemble _____ for much of their lives.

8. Mushrooms make up most of the organisms in phylum _____. In the mushroom life cycle, mycelia grow from _____. Two mycelia will _____ for sexual reproduction. The resulting mycelium will grow, crowding out the _____. Eventually, many hyphae will enclose themselves in a membrane, forming the _____ stage of the mushroom. A _____ _____ will eventually emerge from the membrane, and it will be composed of three basic components: ____, __, and ____. The spores are formed on basidia found in the ____. Those basidia release spores, from which new mycelia will grow.

9. The mycelium of a mushroom tends to grow outward in _____ patches. Because it can run out of food in the center, the fruiting bodies of the fungus can sometimes form a _____ _____.

10. _____ are part of phylum Basidiomycota and grow spores inside a membrane. When disturbed, the spores are pushed out a _____ near the top of the membrane. _____ _____ are also in this phylum and grow shelflike structures on both dead and living wood.

11. Rusts and smuts are examples of _____ fungi in phylum Basidiomycota. Such a fungus tends to live most of its life cycle on one host, which is called its _____ _____. However, it must also spend a certain part of its life cycle on an _____ _____.

12. _____ are single-celled members of phylum Ascomycota that asexually reproduce by _____. They are used in _____, which anaerobically breaks down sugars into smaller molecules, such as _____ _____ and _____. The former can make bread dough _____, and the latter can be used to make _____ beverages.

13. Other members of phylum Ascomycota include _____ (that have fruiting bodies which look like sponges), ____ _____ (that have fruiting bodies which look like cups), _____ (that feeds on rye grain and is deadly to people), and tree parasites that cause _____ ____ _____ and _____ _____.

14. The spores formed by fungi in phylum Zygomycota are called _____, which are composed of a hard, protective coating around a _____.

15. Bread _____ is in phylum Zygomycota. It forms _____ spores in sporangiophores.

16. Fungi in genus _____ are in phylum Deuteromycota. They produce the first _____ ever discovered, which revolutionized the treatment of infections. Some bacteria have adapted to be _____ to these infection-fighting compounds, which means new ones must continually be found.

17. Members of phylum Myxomycota are often called _____ _____. They tend to resemble _____ during their feeding stage, and during this time, the mass of living tissue is called a _____. During their reproductive stage, these organisms resemble _____.

18. There is no such thing as an _____ scientist.

19. A _____ is a _____ relationship between a fungus and an alga. The alga produces _____ for itself and the fungus by means of _____, while the fungus gives _____ and _____ to the alga. Most lichens reproduce by releasing a dust-like substance called a _____, which contains spores of *both* the _____ and the _____ in a protective case.

20. The _____ is a mutualistic relationship between a fungus and a plant root. The fungus absorbs _____ from the roots and gives the plant _____ in return.

Summary of Module #5
Review the vocabulary words listed in Question #1 of the study guide

<u>Fill in the blanks. Many blanks contain more than one word.</u>
Please note: We suggest that you actually write these paragraphs out rather than just filling in the blanks in the book. The act of writing these things out is a form of studying.

1. _____ are the basic building blocks of matter. They are composed of _____ and _____ that form the _____ at the center of the atom as well as _____ that orbit the nucleus. Atoms have an equal number of _____ and _____, and the majority of an atom's properties are determined by the number of _____ it has.

2. An _____ is a collection of atoms that all have the same number of protons.

3. A carbon atom has six protons and eight neutrons. The complete name of this atom is _____, and it has _____ electrons.

4. The more important biological elements and their abbreviations (in parentheses) are: carbon (__), _____ (H), oxygen (__), _____ (N), phosphorus (__), and _____ (S).

5. "Sulfur-32" is the name of a specific _____, while "sulfur" is the name of an _____.

6. When atoms link together, they form _____. A molecule of ethyl alcohol, C_2H_6O, has _____ carbon atoms, _____ hydrogen atoms, and _____ oxygen atom.

7. Even though they both contain _____ and _____ atoms, CO and CO_2 are _____ molecules.

8. When sucrose is dissolved in water, a _____ change has taken place. On the other hand, when sucrose reacts with water with the help of an enzyme to make glucose and fructose, a _____ change has occurred. In general, _____ _____ are reversible, while _____ _____ are not.

9. All matter can exist in one of three phases: _____, _____, and _____. Adding energy turns _____ into _____ and _____ into _____, while taking away energy tends to reverse the process.

10. When salt is dissolved in water, __ is the solute, _____ is the solvent, and the solution is _____.

11. When a solute travels across a membrane in order to even out concentration, _____ has occurred. When the solvent travels across a membrane in order to even out concentration, _____ has occurred. _____ happens when a semipermeable membrane separates two solutions.

12. A cell sits in a solution that has a higher concentration of solutes than that found in the cell. Water will tend to travel _____.

13. In the balanced chemical equation:

$$C_{18}H_{32}O_{16} + 2H_2O \rightarrow 3C_6H_{12}O_6$$

____ molecule of $C_{18}H_{32}O_{16}$ reacts with _____ molecules of H_2O to make _____ molecules of $C_6H_{12}O_6$.

14. Photosynthesis requires _____ _____, , _____, , _____ , and _____
(which acts as a catalyst). It produces _____ and _____ via the chemical equation:

_____

15. Of the following molecules: $NaNO_3$, CH_2O, $C_6H_{15}N$, and $KSCN$, _____ and _____ are organic.

16. Glucose and fructose both have the same chemical formula, _____, which means they are
_____. They have different _____ formulas. A molecule can have more than one _____ _
formula. Glucose and fructose, for example, have both a _____ structure and a _____ structure.

17. A simple sugar is called a _____. Two such simple sugars can join to make a
_____. If three or more join, they form a _____. Simple sugars join together
through _____ reactions.

18. People and animals store excess sugars as a _____ known as _____. When they need
the simple sugars again, they break down this molecule into _____ via _____ reactions.

19. The pH scale runs from _____ to _____. A pH of _____ is neutral. A pH lower than
_____ indicates an ____ solution, while a pH greater than _____ indicates an _____ solution.

20. Lipids are formed in _____ reactions where three _____ are joined to one
_____ molecule. Lipids are _____, meaning they are not attracted to water. If the ____
_____ that make up the lipid have no double bonds between the carbon atoms, it is a
_____ ____ and is generally _____ at room temperature. If there are double bonds between the
carbon atoms, it is an _____ ____ and is generally _____ at room temperature.

21. Proteins are formed in _____ reactions where _____ are joined together. The bond
that forms between them is called a _____ _____. _____ make up a special class of proteins
that serve as _____ for many biologically-important chemical reactions, and they typically work
according to the _____, in which an active site complements the shape of a reactant.
Many of these molecules are quite _____, breaking down soon after they are formed.

22. _____ is a double chain of chemical units known as _____ that twist around one another in
a double helix. The units that make up these chains are composed of three basic constituents:
_____, a _____ _____, and a _____ _____. The double helix is held together
by _____ _____ that link certain _____ _____ together. In DNA, _____ can link only
to _____ (and vice-versa), while _____ can link only to _____ (and vice-versa).

Identify the following as a monosaccharide, fat, or acid.

a.

b.

c.

Summary of Module #6
Review the vocabulary words listed in Question #1 of the study guide

Matching: Match the organelle on the left to its description on the right.

1. Cell wall

2. Plasma membrane

3. Cytoplasm

4. Mitochondrion

5. Lysosome

6. Ribosome

7. Rough ER

8. Smooth ER

9. Leucoplast

10. Chromoplast

11. Central vacuole

12. Food vacuole

13. Waste vacuole

14. Contractile vacuole

15. Pinocytic vesicle

16. Secretion vesicle

17. Golgi bodies

18. Centrioles

19. Nucleus

20. Cytoskeleton

a. The organelle in which nutrients are converted to energy

b. Extensive network of folded membranes in which substances like lipids and hormones are made.

c. The "control center" of the cell, holding the cell's main DNA

d. A large vacuole at the center of plant cells that is responsible for making turgor pressure

e. A network of fibers that holds the cell together, helps the cell to keep its shape, and aids in movement

f. Small, membrane-bounded "sac" that holds secretion products so that they can be transported to the plasma membrane and released

g. Organelle that stores starches or oils

h. Membrane-bounded "sac" that controls the amount of water in a cell

i. Membrane-bounded "sac" that holds the waste products of digestion

j. Usually found in plant and bacteria cells, it is a rigid structure on the outside of the cell that provides protection and support to the cell.

k. A jelly-like substance in the cell in which organelles are suspended

l. Small, membrane-bounded "sac" formed at the plasma membrane to allow the absorption of large molecules

m. Organelles that produce microtubules which form cilia and flagella (in cells that have them) and also aid in the process of asexual reproduction in cells

n. Found in all cells (even prokaryotic cells), it is the organelle that synthesizes proteins.

o. Membrane-bounded "sac" that stores food

p. Organelles where proteins and lipids are stored and then modified to suit the needs of the cell

q. Made of phospholipids, cholesterol, proteins, and other chemicals, it regulates what can come into and what can leave the cell.

r. Organelles that contain pigments used in photosynthesis

s. Extensive network of folded membranes dotted with ribosomes that produces specialized proteins secreted by certain cells.

t. Found in animal cells, it is responsible for hydrolysis reactions that break down proteins, polysaccharides, disaccharides, and some lipids.

Fill in the blanks. Many blanks contain more than one word.
Please note: We suggest that you actually write these paragraphs out rather than just filling in the blanks in the book. The act of writing these things out is a form of studying.

21. Cells must perform at least eleven main functions in order to support and maintain life:
_____, _____, _____, _____, _____, _____,
_____, _____, _____, _____, and _____.

22. When a cell is placed in an isotonic solution, water _____.
When a cell is placed in a hypertonic solution, water _____,
which can result in _____. When a cell is placed in a hypotonic solution, water _____
_____, which can result in _____.

23. Cells store energy in little "packets" by converting _____ and _____ to _____. A gentle
release of energy occurs with the _____ is converted back to _____ and _____.

24. Aerobic cellular respiration occurs in four steps. The first is _____, in which _____ is
converted to _____ and _____ atoms. This takes _____ of energy,
but it produces _____ of energy, for a net gain of _____. The next step, the _____
_____, reacts _____ with _____ to make
_____, _____, and _____ atoms. It
produces _____ of energy. The third step, _____, reacts _____
_____ and _____ to make _____,
_____, and _____ atoms. It produces two ATPs of energy. The final
step, _____, takes the _____ atoms made in the previous steps and
reacts them with _____ to make _____. It produces _____ of energy.

25. If an animal cell performs respiration in aerobic conditions, it can make _____ for each
glucose molecule. Under anaerobic conditions, it can make only _____ for each glucose molecule.

Identify the structures pointed out below.

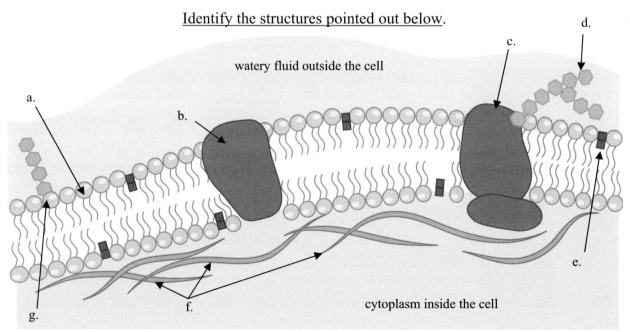

Illustration by Speartoons

Summary of Module #7
Review the vocabulary words listed in Question #1 of the study guide

<u>Fill in the blanks. Many blanks contain more than one word.</u>
Please note: We suggest that you actually write these paragraphs out rather than just filling in the blanks in the book. The act of writing these things out is a form of studying.

1. In determining your traits, your _____ sets a range of possible characteristics, called your _____ _____, and your activities determine what portion of that range is manifested in your body.

2. The three factors that determine the characteristics of a person are _____, _____, and _____.

3. Of the three factors listed above, _____ are laid down first. The information in an organism's DNA is split up into little groups known as _____.

4. By and large, the tasks that a cell can complete are dependent on the _____ that it produces.

5. Like DNA, the nucleotides of _____ join together in long strands. Unlike DNA, however, they do not form a _____ _____. Like DNA, _____ has four nucleotide bases: _____, _____, _____,and _____. Of these four nucleotide bases, _____links to _____ (and vice-versa), and _____links to _____ (and vice-versa). Unlike DNA, the sugar that makes up the foundation of the nucleotides in _____ is _____.

6. One of RNA's jobs is to take a "negative image" of the cell's DNA out of the _____ and to the _____. The RNA that does this job is called _____.

7. Protein synthesis in the cell can be split into two basic steps: _____ and _____.

8. In _____, _____ make a "negative image" of the cell's DNA. It does this by building a molecule that has _____anywhere the DNA has a thymine, _____ anywhere DNA has an adenine, _____anywhere that DNA has a guanine, and _____anywhere that DNA has a cytosine. The newly-built _____ molecule then leaves the _____ and heads to the _____.

9. The information in the _____ that leaves the nucleus is contained in a three-nucleotide-base sequence called a _____. Each _____specifies an _____ in the protein that is to be made.

10. Translation occurs at the _____, which is surrounded by a lot of RNA molecules that have an _____ attached. This type of RNA is called _____. The tRNA has a three-nucleotide-base sequence called an _____. If one of these _____can link to a _____ on the mRNA, it will do so, pulling the _____ along with it. This causes _____ to line up in the particular sequence specified by the mRNA, and the amino acids are then linked to form a _____.

11. A _____ that has the sequence adenine, uracil, guanine will attract a tRNA molecule with an _____ that has the sequence _____, _____, _____.

12. Eukaryotic genes contain sections called _____ (instructions for making a protein) and _____ (best described as "spacers"). Between _____ and _____, the mRNA must be processed. This processing removes the _____ and splices the _____ together.

13. The _____ of a eukaryotic cell is tightly bound together with a network of proteins. Certain proteins, called _____, act as spools, which wind up small stretches of _____. The DNA wrapped around these _____ form what could be described as "beads on a string," which we call _____. Other proteins stabilize and support these spools, making a complex network of DNA coils and proteins. This network is called a _____.

14. Asexual reproduction in eukaryotic cells is called _____. The stages of this process, in order, are _____, _____, _____, and _____. When the cell is not reproducing, it is said to be in _____. This process takes one _____ cell and produces two _____ cells.

15. In a _____ cell, chromosomes come in _____ pairs. A _____ cell has only one member from each pair.

16. The familiar "X" shape a chromosome takes occurs because _____ _____. Nevertheless, this still counts as only _____ chromosome.

17. Genetically, sex is determined by _____, which typically come in two forms: __ and __.

18. _____ is the process by which _____ are produced for the purpose of sexual reproduction. It starts with one _____ cell and produces _____. In males, _____ of the products are viable and are called _____. In females, _____ of the products is viable, and it is called an ___.

19. _____ occurs in two broad steps, called _____ and _____. In the first step, the chromosomes are all duplicated, and then the _____ are separated, resulting in _____ with _____ chromosomes. In the second step, the _____ are separated from the _____, resulting in a total of _____ with _____ chromosomes.

20. An organism has diploid cells that contain 20 chromosomes. If one of these cells undergoes mitosis, ____ cells with ___ chromosomes each will be produced. If it undergoes meiosis, _____ cells with ___ chromosomes each will be produced.

21. When two _____ cells fuse in fertilization, the result is a _____ cell, which is called a _____.

22. A _____ is a chemical entity that is not truly alive, but can infect cells via the _____. Although it is not alive, it has either _____ or _____ as its genetic material, and that material is housed in a protective _____. These chemical entities cause a wide range of _____ in people and animals. One particular type of _____, the _____, even infects bacteria.

23. Your body has several means by which it can protect itself from them, included _____ that engulf pathogens and _____, which are specialized proteins that help to ward off pathogens. Once your body produces _____against a particular pathogen, it can remember how to make them in case you are infected again. This is the principle behind a _____, which is one of the most effective means of protecting yourself against certain viruses.

Summary of Module #8
Review the vocabulary words listed in Question #1 of the study guide

Fill in the blanks. Many blanks contain more than one word.
Please note: We suggest that you actually write these paragraphs out rather than just filling in the blanks in the book. The act of writing these things out is a form of studying.

1. Gregor Mendel's life story shows what can happen when a person has a true desire to _____. His sacrifices for his _____ allowed him to unlock one of the deep _____ of God's creation. Mendel's story also shows that when you _____, you should not give up. Finally, his willingness to put all of that away in order to defend the _____ against an attack from the _____ shows that Mendel had the proper set of priorities.

2. Since animal cells have _____ pairs of _____, we know genes come in _____, with one ____ on each _____ chromosome. Each gene that makes up one of these pairs is called an ____.

3. The _____ of an organism is essentially a list of its alleles. The _____ of an organism is the observable expression of those alleles.

4. Mendel's principles in updated terminology:

1. _____
2. _____
3. _____
4. _____

5. A diagram that follows a particular phenotype through several generations is called a _____.

6. When you cross two individuals concentrating on only one trait, you are performing a _____ _____. A _____ still deals with two individuals, but it concentrates on two separate traits.

7. Some traits are sex-linked, which means the alleles that define those traits are found on the _____ _____ rather than the _____.

8. Many traits are caused by the interaction of *several* genes. This is called _____. Some traits are controlled by alleles that exhibit _____, which means the alleles tend to "mix" rather than one dominating the other. In some cases, one set of alleles might affect how another set of alleles is expressed. This is called _____. Sometimes, a single gene can affect multiple observable traits, which is called _____. Often, there are _____ for a gene, such as the human gene for blood type, which has A, B, and O alleles. In blood type, A and B are dominant over O but not dominant over each other. This is an example of _____.

9. When an antigen is introduced into the blood, the body's response is to produce an _____.

10. The "+" and "-" in blood type refers to the _____, which is controlled by a single gene with two alleles. The "__" allele is dominant, while the "__" allele is recessive.

11. There are at least five means by which genetic abnormalities occur. In _____, a genetic abnormality is passed through the autosomes. In _____, a genetic abnormality

is passed through the sex chromosomes. In _____, one of the alleles of a gene is chemically changed. In _____, a chromosome can gain or lose genes. In _____ _____, a cell can wind up with too many or too few of a specific chromosome.

<u>Answer the following questions.</u>

12. In Mendel's experiments, he determined that the allele for tallness (we'll call it "T") is dominant whereas the allele for shortness (we'll call it "t") is recessive. Let's suppose a homozygous tall pea plant was crossed with a heterozygous pea plant. What will be the genotypes and phenotypes of the offspring?

13. The allele for a pea plant to produce yellow peas (Y) is dominant over the allele for green peas (y). A heterozygous pea plant is crossed with a pea plant that produces green peas. What is the percent chance of each genotype and phenotype produced by this union?

14. In pea color and pea texture, yellow (Y) is dominant and green (y) is recessive. In addition, smooth peas (S) are dominant and wrinkled peas (s) are recessive. A pea plant that produces green peas and is heterozygous in pea texture is crossed with a plant that is heterozygous in pea color and produces wrinkled peas. What are the percentages of the phenotypes in the offspring?

15. Eye color in fruit flies is a sex-linked trait. Red (R) is dominant; white (r) is recessive. If a red-eyed male and a white-eyed female are crossed, what percentage of males and females are red-eyed?

16. Consider the following pedigree:

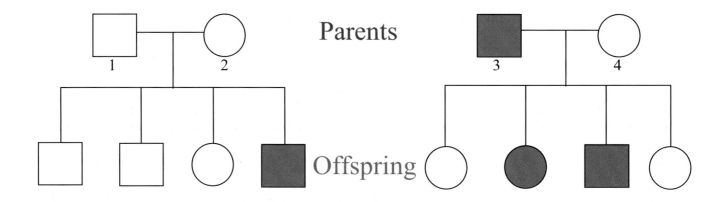

This pedigree is for people, concentrating on the ability to roll their tongues. Squares are males, circles are females. If the circle is filled, that represents someone who cannot roll his or her tongue. If the circle is empty, the person **can** roll his or her tongue. Which is recessive, which is dominant, and what are the genotype of the parents?

17. A man that has A+ blood and is heterozygous in each allele marries a woman whose blood is AB-. What are the possible blood types for their children, along with the percentage chance of each?

Summary of Module #9
Review the vocabulary words listed in Question #1 of the study guide

Fill in the blanks. Many blanks contain more than one word.
Please note: We suggest that you actually write these paragraphs out rather than just filling in the blanks in the book. The act of writing these things out is a form of studying.

1. In 1859, Charles R. _____ published a book entitled (in brief) _____. In his book, he proposed a theory that attempted to explain the diversity of life on earth with no reference to God.

2. At one part in his life, Darwin believed the literal words of the _____. However, he later compared it to the works of barbarians. The voyage of the _____ was a major turning point in his life.

3. Two scientists whose works had a profound effect on Darwin were _____, who studied populations, and _____, who studied geology. _____ inspired Darwin's idea that life is a constant _____, and _____ impressed upon him that _____.

4. Darwin _____ recant his theory and become a Christian on his deathbed.

5. Many science historians credit the _____ that live on the _____ as inspiring Darwin's theory of evolution through natural selection.

6. According to Darwin, if an organism is born with a trait that makes it _____, it will most likely live long enough to reproduce, and it will pass that trait on to its offspring. This is called _____, because nature is "selecting" certain traits in certain species. Over time, this allows populations of species to _____ to changes in their surroundings.

7. Darwin masterfully showed that organisms do _____ to changes in their environments, which can cause some species to _____ significantly over time. This destroyed the age-old idea of the _____, which essentially said that organisms never change significantly over time.

8. The idea that species can adapt to changes in the environment is referred to as _____, and it is a well-established scientific theory. The idea that one species can change into a radically different species over time is referred to as _____, and it is (at best) an unconfirmed hypothesis.

9. Fossils are generally found in _____ rock, which is usually laid down in layers called _____.

10. The _____ is a hypothetical construct of what some geologists think all of the fossil-bearing strata in the world would look like if they _____. If you believe that rocks form according to the speculations of _____, you find that the geological column is evidence ____ evolution. If you believe that most of the fossil-bearing rocks that we see today were formed in global catastrophes such as a _____, the geological column is evidence _____ evolution.

11. The vast majority of the fossil record is made up of _____. What we see in a typical geological column represents only about _____ of all fossils.

12. The myriad of transitional forms that the hypothesis of _____ predicts simply _____ _____ in the fossil record. The few that _____ try to pass off as examples of transitional forms are so close to _____ that their

status of a transitional form is _____. Two such examples are _____, which is supposed to link reptiles to birds, and _____, which is supposed to link apes to humans.

13. The fact that representatives from every major animal phylum can be found in Cambrian rock is called the _____. It is a real problem for macroevolution, because there is not nearly enough _____ in the Cambrian era to produce so much evolution, and there are no _____ linking one species to another.

14. The study of similar structures in different species is called _____, and it was once considered great evidence for evolution. After all, you could explain these similar structures by assuming that the species had a _____ which possessed the structure. However, it has been shown that similar structures are specified by _____ genes in different species. This tells us that they are not _____. Thus, the study of similar structures in different species is now evidence _____ macroevolution.

15. There are certain _____, such as hemoglobin and cytochrome C, that are found in most organisms. However, the sequence of _____ in these chemicals is slightly different from one species to the next.

16. In comparing the cytochrome C found in bacteria, yeasts, fish, and horses, a macroevolutionist would expect that the _____ and _____ would have the most similar cytochrome C, and the _____ and _____ would have the most differences. In fact, _____ and _____ have cytochrome Cs that are slightly *more similar* than those of _____ and _____. 99% of the data obtained by comparing similar proteins in difference species provide evidence _____ macroevolution.

17. Once the details of genetics were known, it was clear that the natural variations which occur between parent and offspring _____ produce macroevolution. Thus, a new version of macroevolution, _____ was formed. In this hypothesis, _____ caused radical changes between parent and offspring. Most such changes were _____ or _____, but a few were _____. The _____ are supposed to have produced macroevolution. This hypothesis _____ the number of transitional forms expected in the fossil record, and it provided a means by which _____ could be added to the genetic code of a species.

18. Although it is possible for a _____ to make an organism more likely to survive under certain conditions, it does not _____ to the genetic code. The mutations that have been observed to make an organism more likely to survive under certain conditions result in a _____ in information.

19. In the _____ hypothesis of macroevolution, mutations add genetic information to the genetic code, but they add it in steps that occur over _____ time intervals. In between these time intervals, _____ occurs. As a result, any transitional forms that result are _____ and unlikely to be _____. This allows macroevolutionists to "explain away" the problem that there are _____ in the fossil record.

Summary of Module #10
Review the vocabulary words listed in Question #1 of the study guide

Fill in the blanks. Many blanks contain more than one word.
Please note: We suggest that you actually write these paragraphs out rather than just filling in the blanks in the book. The act of writing these things out is a form of studying.

1. _____ is a study of the interactions between living and non-living things. In the taxonomy of ecological studies, a _____ is a group of interbreeding organisms coexisting together. A _____ is a group of populations living and interacting in the same area. An _____ is an association of living organisms and their physical environment. A _____ is a group of ecosystems classified by climate and plant life.

2. The continent of _____ was nearly overrun by _____ because a Rancher brought some from England to Australia for the purposes of hunting.

3. The four major trophic levels in a food web are _____ (which makes its own food), _____ _____ (which eats producers), _____ _____ (which eats primary consumers), and _____ _____ (which eats secondary consumers). In general, an ecosystem has the greatest amount of biomass in producers and the least amount of biomass in _____ _____. This is because energy is _____ as you travel up trophic levels in an ecosystem.

4. The _____ of an ecological pyramid is always the widest level.

5. We studied three _____ symbiotic relationships in this module. The _____ swims in the tentacles of the _____. The former _____ to the latter, and the latter _____ the former. There is also some evidence that the former _____ the latter by scaring away one of its major predators. The _____ lives in a hole that is dug by the _____. The former _____ while the latter cleans the hole. The _____ cleans the teeth of the _____, and the material on the latter's teeth is _____ for the former. This kind of mutualism is very _____ to explain in macroevolutionary terms.

6. If you were to try to explain the relationship between the blue-streak wrasse and the Oriental sweetlips using the hypothesis of macroevolution, you would be forced to believe that all of the _____ needed by the Oriental sweetlips and all of the _____ needed by the blue-streak wrasse evolved _____ in the two different animals.

7. The _____ of an ecosystem is made up of all the non-living things in the ecosystem.

8. A _____ is an ecosystem where all water runoff drains into a single body of water.

9. In the water cycle, water can enter the atmosphere by _____ and _____. It can leave the atmosphere by _____. In a watershed, the water can leave the ecosystem by _____ _____. Because of the water cycle, water is continually _____ between the atmosphere and various bodies of water.

10. In the oxygen cycle, oxygen can enter the atmosphere by _____ _____, and _____. It can leave the atmosphere through _____ _____,

and _____. The greatest contributor of oxygen to the atmosphere is _____, and the greatest removal of oxygen from the atmosphere comes as a result of _____.

11. In the carbon cycle, carbon dioxide enters the atmosphere by _____, _____, _____, and _____. It leaves the atmosphere by _____ and _____. _____ puts the most carbon dioxide into the atmosphere, and _____ takes the most out.

12. The _____ is the process by which energy that is being radiated by the earth is trapped in the atmosphere. This process is _____ for the existence of life on earth.

13. Because people burn a lot of fossil fuels, the amount of _____ in the atmosphere is rising. This has caused some to worry about _____.

14. Although _____ levels in the atmosphere have been rising steadily, the _____ _____ of the earth has not risen since 1925. A small increase (about 0.5 °C) occurred prior to 1925, but that was before _____ levels had risen significantly.

15. More than 200 scientific studies have shown that the average temperature of the earth was higher in the _____ than it is today.

16. Although parts of the earth are getting warmer, other parts of the earth are getting _____. As a result, the _____ has not changed significantly over the past 75 years.

17. The process by which nitrogen gas (N_2) is converted to a form that is more useful to most organisms on the planet is called _____. Although some of this is done by the _____ _____, most if it is done by _____.

18. The more biologically-useful forms of nitrogen are _____, _____, and _____. These are produced by _____, _____, _____, and the _____ of living organisms. Sometimes, these more biologically-useful forms of nitrogen are converted back into nitrogen gas. This process is called _____.

19. _____ absorb the biologically-useful forms of nitrogen for their biosynthesis. Consumers get the nitrogen they need by _____.

Summary of Module #11
Review the vocabulary words listed in Question #1 of the study guide

Fill in the blanks. Many blanks contain more than one word.
Please note: We suggest that you actually write these paragraphs out rather than just filling in the blanks in the book. The act of writing these things out is a form of studying.

1. Although not official taxonomy groups, biologists use the terms _____ and _____ to refer to animals with and without backbones, respectively.

2. The three basic forms of symmetry among organisms in creation are _____ (any cut through the center produces two equal halves), _____ (and longitudinal cut through the center produces two equal halves), and _____ (only one longitudinal cut through the center produces two equal halves). A round organism like *Volvox* has _____, a sea anemone has _____, and people have _____.

3. Phylum _____ contains the sponges, which have __ symmetry. These creatures pull _____ into their bodies, cleaning the water of _____ that they eat. They are composed of an outer layer of cells called the _____ and an inner layer of cells. These layers are separated by a jellylike substance called the _____. They support themselves with either _____ or _____, the latter of which results in _____ sponges. Water is pulled through their bodies by _____, and specialized cells called _____ digest food and transport it throughout the body. In addition, they take in _____ from the inner cells and travel to the epidermis to release them. These cells also exchange _____ with the surroundings and produce the lime or silica that makes up the _____. Sponges can produce a _____, which can survive through a long period of inclement weather.

4. Phylum _____ contains jellyfish, sea anemones, and hydra. These creatures have two basic forms: the _____ and the _____. Some, like the jellyfish, take on _____ forms during different parts of their life cycles. The bodies of these creatures have two layers of _____ separated by a jellylike layer called a _____. They have _____ symmetry, and their tentacles (and sometimes bodies) are covered with stinging _____.

5. Hydra have the _____ form. They asexually reproduced by _____, and they sexually reproduce using _____ that produce eggs and _____ that produce sperm. When hydras were first discovered, they were thought to be a transitional form between _____ and _____. We now know they are 100% _____.

6. While the _____ of the hydra are triggered by pressure, the _____ of the sea anemone are trigger by a complex _____.

7. Corals are also members of phylum _____ that use the _____ form.

8. In the life cycle of a jellyfish, sexual reproduction occurs in the _____ form, and asexual reproduction occurs in the _____ form.

9. Segmented worms are in phylum _____.

10. In order to move, the earthworm uses its posterior _____ to anchor its posterior end while it contracts its _____ muscles, stretching the earthworm out. This pushes the _____ end forward.

Once these muscles have contracted completely, the earthworm then uses it anterior _____ to anchor its anterior end releases its posterior _____. After that, the _____ muscles contract, causing the _____ end to move forward.

11. The earthworm ingests soil into its mouth with its _____. The soil is then passed into the ____, where it is stored for a while. Eventually, the soil makes it to the _____, where it is ground into small pieces. The worm _____ edible materials in the _____, and the inedible materials are pushed out through the _____. The digested food is absorbed by _____ that circulates through the walls of the intestine. The blood _____ the digested food to cells throughout the worm's body, pushed by contractions of the _____. Some wastes are also gathered in small organs called _____ which release the waste through tiny holes called _____.

12. Earthworms absorb oxygen and release carbon dioxide through a moist layer in the epidermis called a _____. Its nervous system is controlled by two _____ in the anterior end of the body. Signals come to and from the ganglia through a _____.

13. Earthworms are _____, meaning they have both male and the female reproductive organs. During mating, two earthworms deposit sperm (stored in their _____) into each other's _____. The sperm are later used to fertilize the eggs that are stored in the _____. The zygote produced develops in a _____.

14. Flatworms like the planarian belong in phylum _____. They reproduce asexually by ripping themselves in half and _____. They also _____ reproduce much like earthworms. Its complex nervous system allows the planarian to have senses of _____, _____, _____, and the ability to sense _____. It senses _____ using two _____ on its anterior end.

15. Phylum _____ contains roundworms. Their bodies are essentially composed of a _____ within a _____, and many members of this phylum are _____.

16. Phylum _____ contains clams, oysters, snails, and squid. Most members of this phylum have a _____, a _____, a _____, a _____, and a _____. A mollusk with one shell is called a _____, while one with two shells is called a _____.

Identify the structures in the diagram below.

Illustration by Megan Whitaker

Summary of Module #12
Review the vocabulary words listed in Question #1 of the study guide

<u>Fill in the blanks. Many blanks contain more than one word.</u>
Please note: We suggest that you actually write these paragraphs out rather than just filling in the blanks in the book. The act of writing these things out is a form of studying.

1. Arthropods have the following five characteristics in common: _____, _____
_____, _____, _____, and an _____.

2. Because they have an exoskeleton, arthropods must _____ in order to grow. An arthropod's body
might consist of three segments: _____, _____, and _____, or it might consist of two segments:
_____ and _____. The eyes might be _____ (consisting of one lens) or _____
(consisting of many lenses).

3. The crayfish belongs in class _____ of phylum Arthropoda. It gets its sense of balance
mostly from its _____ and its sensitive senses of taste and touch from its _____. It
uses its _____ (claws) to grab onto prey and for defense. Its _____ and _____ on its posterior
end are used for swimming, and its _____ are used for swimming and reproduction.

4. A crayfish gets oxygen that has been dissolved in the water by passing water over its _____. Blood
distributes the oxygen throughout the body. It collects in the _____ around the _____. It
enters the _____ through one of three openings which close when the _____ is ready to pump. The
_____ pumps blood through blood vessels that are open at the other end. These vessels dump the
blood _____. Gravity causes the blood to fall into the _____, where it is
collected by blood vessels that are open at one end. These vessels carry the blood back towards the
_____. On its way there, the blood is passed through the gills where it can release _____
_____ and absorb _____. The blood also passes through a _____, which cleans it.

5. In order to eat, the crayfish uses its _____ to break the food into small chunks. The food enters
a short _____ and goes into a _____ that has essentially two regions. The first region _____
_____. The second region sorts the particles. If they are small enough, they are sent
directly to _____ which secrete enzymes, completing the digestion process. If the particles
are too large to be digested immediately, they are sent to the _____. Anything that remains at the
other end of the intestine is considered indigestible and is expelled out the _____.

6. A crayfish's brain is comprised of two _____ that each has a _____. At the base of
each antennule is a _____, which gives the crayfish its sense of balance.

7. In male crayfish, _____ are formed in the _____ and transferred to the first and second pairs of
_____. Male crayfish deposit their _____ into special containers that the female has.
The female will then store the sperm until spring. In the spring, the female produces _____. They are
fertilized by the _____ and go to the _____. The fertilized eggs _____
_____ and develop for approximately six weeks, at which point the eggs hatch.

8. Spiders are in class _____. Members of this class have these five characteristics in common:
_____, _____, _____, _____,
and _____. They spin webs make of _____ that is produced in _____
_____ and spun with _____.

9. Centipedes, aggressive predators with _____ legs per segment, are found in class _____.
Millipedes, docile herbivores with _____ legs per segment, are found in class _____.

10. Members of class Insecta have four characteristics in common: _____,
_____, _____, _____. They do not
have lungs, but breathe through _____ that form an intricate network throughout the body. They all
go through some kind of _____ in their life cycle. _____ has four stages: ____,
_____, _____, and _____. _____ has three stages: ____, _____, and _____.

11. Butterflies and moths belong to order _____, while the ants, bees, and wasps belong in
order _____. These are also called _____, because they exist in a complex
society. Beetles belong to order _____. Flies, gnats, and mosquitoes belong in order
_____. Grasshoppers and crickets go in order _____.

<u>Identify the structures pointed out in the figures below.</u>

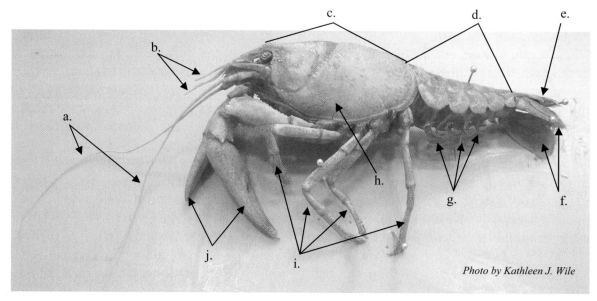

Photo by Kathleen J. Wile

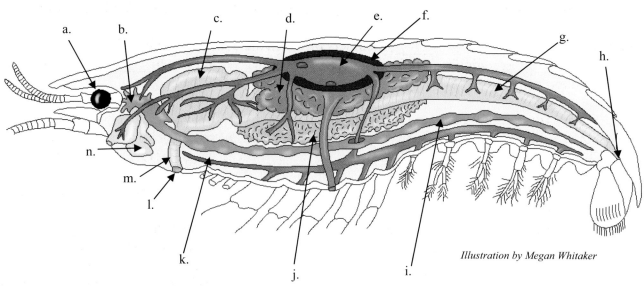

Illustration by Megan Whitaker

Summary of Module #13
Review the vocabulary words listed in Question #1 of the study guide

Fill in the blanks. Many blanks contain more than one word.
Please note: We suggest that you actually write these paragraphs out rather than just filling in the blanks in the book. The act of writing these things out is a form of studying.

1. Members of phylum Chordata have either a _____ or a _____. The sea squirt is a member of subphylum _____, and it has a _____ only in its larval stage. As an adult, it is supported by a leathery _____. The lancelet is a member of subphylum _____, and it has a _____ throughout its life. Animals with backbones are members of subphylum _____.

2. There are three types of bone cells: _____, _____, and _____. The bones they comprise are made of fibers of a protein called _____ that have been hardened by _____ _____. Bone tissue comes in two types: _____ (composed of tightly-packed fibers) and _____ (composed of loosely-packed fibers). These tissues are surrounded by a membrane called the _____ that contains blood vessels and nerves. In the very center of the bone is a cavity that holds _____, which produces blood cells.

3. A typical vertebrate endoskeleton is made up of an _____ (that supports and protects the head, neck, and trunk), and an _____ (has the limbs attached to it).

4. The vertebrate circulatory system is _____, and it is composed of _____ (that carry blood towards the heart), _____ (that carry blood away from the heart), and _____ (that allow gases to be exchanged with the tissues). Oxygen is carried by _____ that get their color from _____.

5. Most vertebrate brains have five sets of lobes: _____ (for smell), _____ (for integrating sensory information and making responses), _____ (for sight), _____ (for involuntary actions and refining muscle movement), and the _____ (for vital functions). Signals are sent down the medulla oblongata and into the _____.

6. Vertebrate reproduction can occur by _____ or _____ fertilization, and the embryo's development can be _____ (in an egg outside the female's body), _____ (in an egg inside the female's body), or _____ (inside the female, connected to her by a placenta).

7. Lamprey eels are in class _____, members of which are commonly called the _____. They are _____, which means they hatch in _____, migrate to _____ as adults, then go back to _____ in order to reproduce.

8. Sharks, rays, and skates belong to class _____. They have _____ endoskeletons, which are more flexible than human endoskeletons. Sharks hunt using many senses, including a _____ ___ that detects vibrations, and a keen _____ that detects minute amounts of electricity.

9. Members of class _____ are typically called "_____," since most or all of their skeletons are hardened with calcium. A member of this class gets its sense of taste from _____ on its tongue. Food passes through a _____ (throat) and into the _____, which leads to the _____. There, the food is broken down and stored. It is then sent to the _____, where it is digested. Any undigested remains leave the fish through the _____. The fish has a ___ that produces ___, which is concentrated in the _____ and helps digest fats. The _____ helps it control its depth in the water.

10. Bony fish have a _____ heart composed of an _____ and a _____. Blood from the tissues enters the _____ and is then transferred to the _____, which pushes it out. It then passes over the _____ to get _____ and release _____. It then goes to the tissues. While it passes through the body, it gets cleaned in organs called _____.

11. All of the organisms in this module are _____, which means they are "cold-blooded."

12. Amphibians have these six characteristics in common: _____, _____(no scales), _____, _____, _____, and _____.

13. In its larval stage, an amphibian breathes with _____. As an adult, it breathes with _____, _____, and _____. It deals with cold weather by _____.

Identify the structures in the figures below.

Illustrations by Megan Whitaker

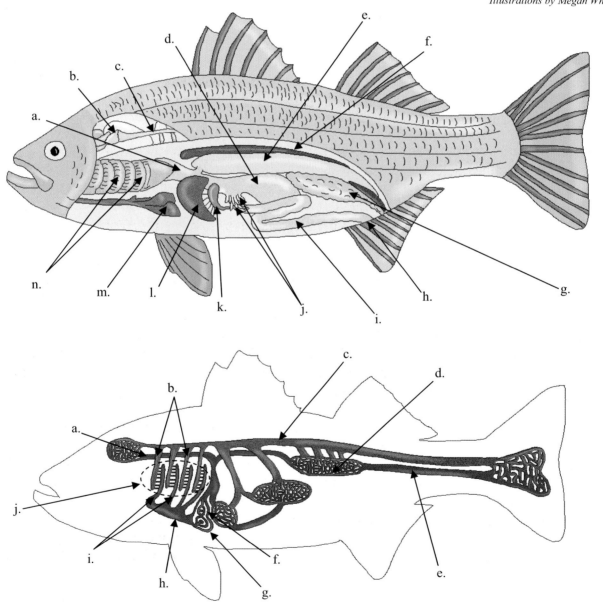

Summary of Module #14
Review the vocabulary words listed in Question #1 of the study guide

You must also be familiar with the mosaics, shapes, margins, and venations in Figures 14.3 – 14.6!

<u>Fill in the blanks. Many blanks contain more than one word</u>.
Please note: We suggest that you actually write these paragraphs out rather than just filling in the blanks in the book. The act of writing these things out is a form of studying.

1. The study of plants is called _____. Plants that grow year after year are _____, while plants that live for only one year are called _____. Plants that live for two years are _____.

2. Plants have two types of organs: _____ and _____. The former are considered vegetables, while the later are considered _____. These organs are made of up to four different types of tissues: _____ (containing undifferentiated cells), _____ (the most common), _____ (the outer layer), and _____ (which carry water and nutrients). If the vascular tissue carries water and minerals, it is called _____ and is made of dead cells. If it carries food and other organic chemicals, it is called _____ and is made of living cells.

3. The primary portion of a leaf is called the _____, and the very tip of the blade is called the _____. The blade is attached to the stem with a small stalk called the _____. At the base of the petiole, most plants have stipules. A simple leaf is one leaf attached to the stem of the plant by a single petiole. A _____ leaf has several leaflets attached to a single petiole.

4. Plants in class _____ (typically called _____) have leaves with _____ venation, seeds with one cotyledon, typically have _____, and produce flowers with petals in multiples of _____ or ____. On the other hand, if the venation is _____, the plants belong to class _____ and are typically called _____. These plants have seeds with ___ cotyledons, have _____, and produce flowers with petals that are usually in groups of _____ or _____.

5. The top and bottom of a leaf are covered with a single layer of cells called the _____, which protects the inner parts of the leaf. Sometimes, the epidermis secretes a waxy substance called a _____. Tiny holes called _____ are found on the _____ of most leaves. They allow for the exchange of _____ with the atmosphere. Each stoma is flanked by two cells called _____, which open and close the stoma. Under the epidermis on both sides of the leaf are _____, which are composed of cells that do the photosynthesis. These tissues are composed of two layers, the _____ (which has densely-packed cells) and the _____ (which has loosely-packed cells). The veins in a leaf are made up of three tissues: _____, _____, and _____.

6. Leaves have _____, which makes most of them green. Some leaves, however, have _____ as well, which have yellow or orange hues. In most leaves, the color of the _____ overwhelms the colors of the _____, but in some plants, the leaf picks up some color from them. In certain leaf tissues, there is another set of pigments called _____, which have different colors, depending on the pH of the leaf tissue. When a leaf falls off the tree and dies, the _____ decays, allowing its other pigments to show. Plants that lose their leaves for the winter are called _____. At the base of each petiole in a deciduous tree, there is a thin layer of tissue called the _____. As the days get shorter, this tissue blocks off the _____ and _____ running through the petiole.

7. Roots have three primary functions: _____, _____, and _____.

8. There are basically two kinds of root systems: _____ (which looks much like an underground bush) and _____ (in which the primary root continues to grow and stays the main root).

9. Longitudinally, a root is split into four regions: the _____ (made of dead, thick-walled cells), the _____ (where undifferentiated cells carry on mitosis), the elongation region (where cells differentiate and stretch), and the _____ (where the cells are becoming fully differentiated). _____, which increase the surface area of the root, are produced in the _____.

10. In a lateral cross section of a root, the _____ is the outer layer. The cells inside the epidermis are called the _____, where substances are stored for later use. Inside the root, there is another one-cell-thick layer called the _____. These cells surround the _____, which contains the _____ and _____.

11. Plant stems can be either woody or _____. They perform three basic functions: _____ _____, _____, _____.

12. In herbaceous stems, dicots and monocots can be easily distinguished by looking at their _____ bundles. The bundles of a monocot look like a _____, and the bundles are distributed _____. The bundles of a dicot are found in a ring near the edge of the stem. In dicots, a _____ can form new xylem or phloem, while a mature moncot has no such tissue.

13. Woody stems have an outer layer of _____ that is actually composed of two layers: the _____ _____ (composed of phloem and cortex tissue) and the _____ (composed of dead cork cells). Between these two layers is the _____, that continually produces _____. The formation of bark allows a woody stem to _____, unlike most herbaceous stems. New _____ and _____ must always be produced inside a woody stem. This is done by the _____, which produces phloem on its _____ side and xylem on its _____ side. This forms a pattern of alternating areas of light and dark wood, which forms what we call _____, which you can count to determine the age of the stem.

14. Some stems appear to be roots. _____ are underground stems that store food, and _____ are underground leaves that sprout from an underground stem.

15. Plants can be split into two basic groups: _____ (plants without vascular tissue) or _____ (plants with vascular tissue). _____ must be small.

16. Mosses are found in phylum _____ and are composed of _____ and _____. They have an _____ life cycle in which the plant we recognize as moss is the _____ generation and is composed of _____ cells. This generation produces _____ through _____. When fertilization occurs, the result is the _____ generation, which is made of _____ cells. This generation produces spores by _____, which germinate into the _____ generation. The _____ generation is dominant in mosses.

17. Ferns are members of phylum _____. They also have an _____ life cycle, but the dominant generation is the _____ generation, which is made of _____ cells.

18. Evergreens are members of phylum _____ and produce seeds that are the result of fertilization between _____ made in _____ and eggs made in _____. Phylum _____ contains the flower-making plants, which are either monocots or dicots.

Summary of Module #15
Review the vocabulary words listed in Question #1 of the study guide

Fill in the blanks. Many blanks contain more than one word.
Please note: We suggest that you actually write these paragraphs out rather than just filling in the blanks in the book. The act of writing these things out is a form of studying.

1. The study of life processes in an organism is called _____.

2. In plants, water is used for essentially four processes: _____, _____, _____, and _____. One of these processes (_____) is responsible for a particular kind of motion in plants. This motion, which is in response to a stimulus but independent of the direction of the stimulus, is called _____. The other type of motion, called a _____, is not controlled by water and is dependent on the direction of the stimulus.

3. Plants store excess food in the form of a polysaccharide known as _____.

4. A good soil for plant growth is a _____, which is a mixture of gravel, sand, silt, clay, and organic matter. It should have large enough _____ to allow for plenty of oxygen, but small enough _____ to allow for the retention of water.

5. The _____ describes how water moves up a plant through the xylem. In this theory, _____ causes a _____, pulling nearby water molecules up to fill the space left behind. The _____ of the water molecules further down the xylem causes them to be pulled up as well.

6. The transport of organic chemicals and food through the phloem is called _____, and it depends on the fact that the phloem are composed of _____ cells.

7. Controlling mitosis and regulating plant development is done by at least five groups of _____. _____ affect the way that cells elongate, which influences how the plant grows. _____ (Growing towards light), _____ (growing against gravity), and _____ (a growth response to touch) are all controlled by these hormones. _____ promote elongation in stems, affect mitosis rates, and induce seeds to germinate. _____ affect mitosis rates, influence cellular differentiation, induce leaf cells to elongate, and affect the synthesis of chlorophyll. _____ _____ inhibits the abscission layer so that it doesn't close off, and it also helps to control the stomata. _____ promotes the ripening of fruits and causes the abscission layer to close. In addition, botanists suspect there is at least one more _____, called _____, which controls flowering in an anthophyte.

8. _____ trap and digest insects, but they ____ ____ use them for food. Instead, they use them to get materials for _____ that cannot be found in the soil in which they grow.

9. Asexual reproduction in plants is typically called _____. Stems and roots can sometimes develop _____ and grow into new plants if put in soil. Some plants produce _____ that originate in their roots and become new plants. Other plants have specialized stems like _____ that can grow roots and become a new plant. Some plants produce _____ that grow along the ground and then sprout a new plant on the end. Also, the stem of one plant can be _____ onto a different plant. When this happens, the stem is called the _____ and the other plant is called the _____.

10. Phylum _____ contains the flowering plants, and the flowers are the _____.

11. The female part of the flower is called the _____, which is sometimes called the _____. It is made up of three parts: _____, _____, and _____. The male part of the flower is the _____, and it is made up of the _____ and _____. If a flower has both, it is called a _____. If it has one or the other, it is an _____.

12. The anther contains pollen sacs which produce _____. _____ cells in the pollen sacs go through _____ to produce four _____ cells. Those cells each go through _____ to produce a total of eight haploid cells. The cells are encased in four shells, and they differentiate into a _____ and a _____. In some plants, the sperm cell undergoes mitosis so that there are two sperm cells.

13. In the _____, which is inside the ovary, a _____ cell undergoes _____ to produce four _____ cells, three of which are _____ and die. The remaining cell, called the _____, undergoes _____ without cell division three times to make eight _____ nuclei. Cell division then occurs, producing six small _____ cells and one large cell with _____ nuclei.

14. To reproduce, a flower releases its _____. They travel to another flower by wind, insects, or birds and land on the _____ of the carpel. The _____ then digs a _____ through the style to the ovary. If there was only one sperm cell in the pollen grain, it undergoes _____ to make two. Both sperm cells enter the ovule. One fertilizes one of the _____ cells, and the other fertilizes the _____. The former produces the zygote, and the latter produces the _____, which becomes the nutrient source for the developing plant embryo. This process, called _____, is unique to kingdom Plantae.

15. When a seed is mature, it has either one or two _____, which makes the plant a monocot or dicot. In some seeds, the _____ consumes the _____ and becomes the food source for the embryo. In other seeds, it simply transfers nutrients from the _____ to the embryo. The ovary then ripens into a _____ that encases the seed.

16. When the seed germinates, the _____ is the first to emerge, and it becomes the roots of the plant. The _____ develops into the stem, and the first true leaves of the plant develop from the _____. In some plants, the first leaves are not true leaves, but are "seed leaves," which are the _____. Some _____ actually perform photosynthesis for a while until the true leaves develop.

<u>Identify the structures in the figure below.</u>

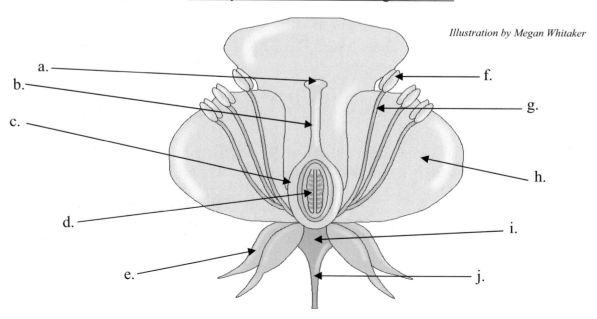

Illustration by Megan Whitaker

Summary of Module #16
Review the vocabulary words listed in Question #1 of the study guide

<u>Fill in the blanks. Many blanks contain more than one word.</u>
Please note: We suggest that you actually write these paragraphs out rather than just filling in the blanks in the book. The act of writing these things out is a form of studying.

1. Reptiles have the following six characteristics in common: _____,
_____, _____, _____
_____, _____, _____.

2. Scales prevent _____ and _____. Since scales are not made of living tissue,
reptiles must _____ in order to grow.

3. An amniotic egg is covered in a protective _____. The _____ grows around the
embryo, forming a fluid-filled sac in which the embryo floats. The _____ contains the _____,
which feeds the embryo. The _____ is a sac of blood vessels that allows for the respiration
and excretion of the embryo. The _____ is a membrane that envelopes these four structures.
The _____ protects the embryo from infection, is a storehouse for water, provides mechanical
support for the chorion, and stores proteins, which can be broken down into _____ that the
embryo can use in _____.

4. The _____ is in order Rhynchocephalia, which also contains many extinct reptiles.

5. Lizards and snakes are in order _____. Lizards have two pairs of limbs, while snakes have
_____. Lizards have ears and can hear, while snakes are _____. Lizards have the same type of
scales all over their bodies, while snakes have specialized scales on their _____ for locomotion.
Most lizards have eyelids and can therefore close their eyes, but snakes _____.

6. A snake has two nostrils for smelling, but it augments this sense with sensory pits called _____
_____. When a snake sticks out its tongue, it collects _____. It then
pulls its tongue back into its mouth, transferring them to the _____, which send signals
to the brain. Poisonous snakes generally produce _____ (which are fast-acting) or _____
(which take longer to do damage but are more deadly). _____ are poisonous snakes that have
two heat-sensing pits which allow them to sense _____, even if there is no light.

7. Turtles and tortoises are found in order _____. If it lives in the water, it is a _____,
but if it lives on land, it is a _____.

8. Alligators and crocodiles make up order _____. When an _____ mouth is closed,
you cannot see its teeth. _____ teeth can be seen even when the animal's mouth is closed.

9. Although a few reptiles that are classified as dinosaurs (like the tuatara) are still living, most
dinosaurs are _____. Although we know little about these creatures, their fossils do tell us some
things. Apatosaurus was one of the _____, which were probably the largest dinosaurs.
Stegosaurus was one of the _____, which were probably herbivores. Triceratops was one of
the _____, which means "fringe heads." Tyrannosaurus was one of the _____,
which stood on hind legs and ranged from very small to very large. Plesiosaurus was one of the

_____, which were technically not dinosaurs, but were marine reptiles with flippers. Pteranodon was also not technically a dinosaur. Instead, it was a large _____.

10. Birds (members of class Aves) have the following six characteristics in common: _____, _____, _____, _____ _____, _____, and _____.

11. The three most important features that allow birds to fly are _____, _____, and _____ _____. Feathers are composed of a _____ from which _____ extend. Contour feathers (used for flight) have interlocking _____ and _____, while down feathers (used for insulation) have only _____. Because feathers are not made of living tissue, they must be _____, but the process occurs very precisely so that the bird is never out of _____. The wings are equipped with turbulence-dampening devices called _____ that were copied so that airplane wings would not fall off. Birds that fly have _____ bones that are lightweight but strong, because they are reinforced with _____ that were also copied in aircraft design.

12. Mammals have the following five characteristics in common: _____, _____ _____, _____, _____, _____.

13. Mammal hair comes in two forms: _____ on top of _____. The latter serves as insulation.

14. Since most mammals are viviparous, the embryo is attached to the mother by a _____. The _____ mammals, however, are either _____, or their embryos develop in a _____. The length of the _____ determines how well developed the offspring is when it is born.

15. Female mammals secrete milk to nourish their young from special glands called _____.

Identify the structures in the following figure.

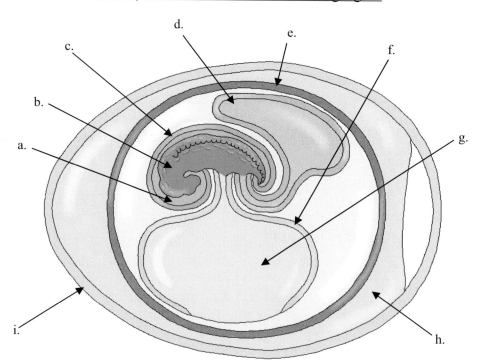

APPENDIX C
A COMPLETE LIST OF LAB SUPPLIES

Items in boldface, blue type are found in the laboratory equipment sets that are sold for the course. The other materials are available at supermarkets, hardware stores, or drug stores. If the bullets are black, the items are used in an experiment that employs only household items. Green bullets accompany items used in microscope experiments. Red bullets accompany items used in dissection experiments.

Module #1

- **Microscope**
- **Lens paper**
- **Slides**
- **Coverslips**
- **Eyedropper**
- **Methylene blue stain**
- Water
- Small pieces of bright thread
- Cotton swabs

Module #2

- **Microscope**
- **Slides**
- **Coverslips for the slides**
- **4 eyedroppers**
- Four jars with lids (You do not want a lot of light to get into the jars. Thus, jars with darkened glass or plastic work really well. If you can't find that kind of jar, cover your jars with paper or foil to keep the light out.)
- A small amount of chopped hay (Dried grass will work as a substitute.)
- Uncooked white rice (Brown rice will not work as well.)
- Egg yolk (uncooked)
- A small amount of rich soil
- A long-handled ladle (A good one can be made by attaching a kitchen ladle to a broom handle with duct tape.)
- The 4 culture jars made in Experiment 2.1
- A small amount of cotton (from a cotton ball, for example)
- A pond or small body of water (A still creek will do in a pinch, but it will not be ideal.)
- Something to rest your lab notebook on while you draw in it
- Colored pencils

Module #3

- **Microscope**
- **Slides**
- **Coverslips**

- **Prepared slide: amoeba**
- **Prepared slide: paramecium**
- **Prepared slide: euglena**
- **Prepared slide: volvox**
- **Prepared slide:** *Spirogyra*
- **Prepared slide: Diatoms**
- The 4 culture jars used in Experiment 2.2
- 4 eyedroppers (one for each jar)
- A small amount of cotton (from a cotton ball, for example)

Module #4

- **Microscope**
- **Slides**
- **Coverslips**
- **Eyedropper**
- **Methylene blue**
- Camembert cheese (available at large supermarkets)
- Roquefort cheese (available at large supermarkets)
- Water
- Magnifying glass
- Needle (or probe from your dissection kit)
- Mushrooms
- Puffballs
- Shelf fungi
- Gloves
- Packet of active dry yeast (can be purchased at a grocery store)
- Tablespoon
- Measuring cup
- Glass that holds at least 2 cups of water
- Sugar
- Bread, jelly, and/or fruit mold (Only one specimen is necessary, but if you observe more than one specimen, you will learn more!)
- Knife

Module #5

- Sugar
- Tablespoon
- Water
- A small glass
- A paper napkin
- Cellophane tape
- Plastic wrap
- Three coffee mugs
- One fresh, raw egg
- A measuring cup for liquids

- A tape measure
- White vinegar
- Clear sugar syrup (like Karo® syrup)
- Distilled water (You can purchase this at any large supermarket.)
- Part of a *fresh* pineapple (It cannot be canned. It must be fresh.)
- A blender or fine cheese grater
- Three small bowls
- A small box of Jell-O® gelatin mix - any flavor (Generic brands work just as well.)
- Pot
- Stove
- Refrigerator
- Two tablespoons

Module #6

- **Microscope**
- **Lens paper**
- **Slides**
- **Coverslips**
- **Eyedroppers**
- **Iodine**
- **Prepared slide: Hydra**
- **Prepared slide: Ranunculus root**
- **Prepared slide: Zea mays root**
- Water
- Onion
- Cork (Any cork item can be used.)
- Knife or scalpel
- Anacharis leaves (Anacharis is a generic term for aquatic plants that come from genus *Elodea* and genus *Egeria*. Sometimes called "water weeds," they can be purchased at aquarium stores. If you live in a state that has outlawed the sale of these plants due to their tendency to take over an ecosystem, ask the salesperson at the aquarium shop what they are selling in place of Anacharis. You could also use thin leaves from another plant like Impatiens. The main point is that the leaves need to be alive and very thin.)
- Banana
- Cotton swab
- Salt water (1 tablespoon of salt in about $\frac{1}{8}$ cup of water)

Module #7

- Blender
- Plastic bowl
- Toothpick
- Clear liquid hand soap or dish soap (The liquid hand soap tends to work just a bit better, and colorless will work a bit better than soap that is tinted with a color.)
- Salt
- Water

- Strainer
- Small glass
- Meat tenderizer (Make sure it has been bought within the last year or so.)
- Rubbing alcohol
- ½ cup of split peas
- Measuring cups and spoons
- Flashlight
- **Microscope**
- **Prepared slide of _Allium_ (onion) root tip**
- **Prepared slide of _Ascaris_ Mitosis**

Module #8

- Your parents, grandparents, aunts, uncles, and siblings (Even if your grandparents, aunts, and uncles are not living or are living far away, you might be able to find pictures of their ears, which is all that you need for the experiment. If you don't have many siblings or cannot determine the earlobe characteristics of your grandparents, aunts, and uncles, you might consider studying another family as well so that you can get even more information.)
- Mirror
- 60 radish seeds (purchase locally)
- 2 shallow pans or dishes
- Potting soil
- Clear plastic wrap
- Box to cover one dish
- Water
- Lab notebook
- Magnifying glass (if available)
- **Eyedropper**

Module #9

There are no experiments in Module #9

Module #10

- Thermometer (It must be able to read temperatures from room temperature to at least 100 degrees Fahrenheit. The smaller the thermometer, the better.)
- A large, clear Ziploc® freezer bag (It must be large enough for the thermometer to fit inside once it is zipped.)
- Sunny windowsill (If it's not sunny today, just wait until it is.)
- Plastic, two-liter soda pop bottle
- Vinegar
- Baking soda
- Teaspoon

Module #11

- **Microscope**
- **Prepared slide: sponge**
- **Prepared slide: *Hydra***
- **Prepared slide: planarian**
- Colored pencils
- Natural sponges (optional)
- **Dissecting tools and tray that came with your dissection kit**
- **Earthworm specimen**

Module #12

- **Dissecting tools and tray that came with your dissection kit**
- **Crayfish specimen**
- Magnifying glass
- Insect specimens (collected from outside or from someone's insect collection)

Module #13

- **Dissecting tools and tray that came with your dissection kit**
- **Perch specimen**
- **Frog specimen**
- Magnifying glass
- Water
- Small bowl
- Colored pencils
- Magnifying glass
- Field guide (Get one that covers both plants and animals in your area, or get one that covers plants and another that covers animals.)

Module #14

- Leaf press or old newspapers or old telephone books
- Tree identification book (from the library)
- Red (some people call it purple) cabbage (just a few leaves)
- Stove
- Stirring spoon
- Pot
- White vinegar (It must be clear. Apple cider vinegar will not work for this experiment.)
- Clear ammonia solution (This is sold in grocery stores with the cleaning supplies.)
- Water
- 2 eyedroppers
- 2 small cups or glasses
- 1 small glass (It must be see-through!)
- A white sheet of paper (preferably without lines)

- Measuring cups (1 cup and ¼ cup)
- Tablespoon
- **Microscope**
- **Prepared slide: Zea mays (corn) cross section of stem**
- **Prepared slide: Zea mays (corn) cross section of root**
- **Prepared slide: Ranunculus (buttercup) cross section of stem**
- **Prepared slide: Ranunculus (buttercup) cross section of root**
- **Prepared slide: Leaf cross section with vein**
- Colored pencils

Module #15

- Sharp scissors (If you have the dissection kit, use the scissors in it.)
- Sharp blade (If you have the dissection kit, use the scalpel in it.)
- Slides and coverslips
- Water
- Eyedropper
- Magnifying glass
- **Microscope** (optional)
- Colored pencils
- A variety of flowers (Most flower shops will save old flowers for you, if you contact them ahead of time and tell them why you want them. They do not need to be fresh, but you should get a good variety. An example of a good variety would be: a rose, a carnation, a daisy, a lily and a tulip. At least one of them, preferably more, should have stamens and at least one carpel that are easy to see. In the list above, the lily and tulip will have easily-visible stamens and a carpel. The rose and carnation will have them as well, but they will be harder to find. Look in the very center of the flower. The daisy is a composite flower, so its reproductive organs will be even harder to see.)
- A variety of different fruits (suggested fruits: apple, plum, orange, tomato, walnut, sunflower seed, maple seed, pea in pod, strawberry, and raspberry)

Module #16

- **Micro slide: The Chick Embryo**
- **Microscope (optional)**
- Magnifying glass
- Desk lamp
- Colored pencils
- Bird field guides (available at your local library)
- Binoculars (if available)
- Bird seed

INDEX

594 Exploring Creation With Biology

Venus flytrap, 472, 473
Venus flytrap (fig), 472
vertebrae, 393, 396
vertebral column, 398
Vertebrata, 393, 396, 419, 513
vertebrate, 329, 393, 396, 397
vertebrate nervous system (fig), 400
vertebrate circulatory system, 399
vertebrate reproduction, 401
vesicle, 170, 209
Victoria, 249
Vienna, 10
Vienna, University of, 227
vine, 471
vinegar, 131, 441
violet, 474
viper, 503
virus, 218-220, 300, 388
virus (fig), 218
visceral hump, 354
Vitamin K, 58
vitamins, 85
viviparous, 402
volvox, 62
Volvox, 75
vorticella, 62

-W-

wading birds, 517
Walcott, Charles, 280
walking legs, 366
wallabies, 520
walnuts, 487
Warren's trusses, 513
warts, 220
waste vacuole, 169
water, 133, 141, 188, 310, 311, 313, 315, 441, 463, 465, 468, 497
water cycle, 311, 313, 315
water cycle (fig), 311
water drag, 523
water drag, reduction, 523
water moccasin, 503
water strider, 381, 467
water transport, plants, 466
water vapor, 315, 317, 465
water, uses in plants, 463
watermelons, 446
watershed, 312, 313
wax, 386
web, 378
weeds, 473
west Antarctic ice sheet, 321
western diamondback, 503
Westminster Abbey, 263
whales, 520, 523
wheat, 107, 287, 288
wheat rust, 107
whey, 37

whiskers, 518
whiskey, 110
white birch tree, 448
white blood cell, 166, 170, 220
white oak, 27
whorled mosaic, 433
Wile, Jay L., 293
wind, 477
wine, 110
wing slot, 512
wings, 381, 385, 510, 511, 512
wings, ants, 387
wings, beetles, 387
wings, insect, 381
wolves, 523
woody, 429, 446
woody stem, 447, 448
woody stem (fig), 447
Word of God, 15
worm, Christmas tree, 347
worm, segmented, 342
wrasse, blue-streak, 307, 308
Wright brothers, 512

-X-

X chromosome, 212, 248, 249
Xenarthra, 525
xylem, 430, 438, 441, 445, 447, 448, 449, 465, 466, 468, 469, 475

-Y-

Y chromosome, 212, 248, 249
yeast, 97, 109, 110, 285, 286, 287, 288
yeast, baker's, 109
yolk, 497
yolk sac, 497

-Z-

zebra, 520, 524
zooplankton, 84
Zygomycota, 102, 103, 113
zygospore, 102, 112, 113
zygote, 112, 214, 217, 235, 334, 485